What the experts are saying about the

Solar Living
SOURCEBOOK

14th EDITION • REVISED & UPDATED

The *Solar Living Sourcebook* is a breath of fresh air for everyone interested in being part of the evolutionary revolution. John's book is the best single source I've ever found on the technologies, philosophies, and the lifestyle changes we must embrace in order for our species to survive and thrive into the future. Time is of the essence for all of us to start walking the talk and this book is the guide for taking those first steps

—**Woody Harrelson,** environmentalist, actor

Sourcebook is perfectly named, though it could easily be called the masterwork on re-inhabitation, an owner's manual for regenerating our earth, skies and water. Succinctly and clearly, *Sourcebook* delineates the pathways, tools, and knowledge we need to achieve the title of denizens, human beings who live in accordance with the laws of physics, botany, and biology, people who truly want to master the art of living as contrasted to a life of taking. The best manual you could possibly want.

—**Paul Hawken**, Project Drawdown, author *Blessed Unrest*

Sunlight is the source of all life, and if we don't harvest and manage it intelligently, life (for us humans) will become irrelevant. John Schaeffer's masterful and elegantly updated opus, the *Solar Living Sourcebook*, is the most comprehensive guide published for the practical management of sunlight. This book is our blueprint for both the present and future, if our species is to survive into the 22nd century.

—**Thom Hartmann**, NY Times Bestselling author and talk show host

John Schaeffer has been a pathfinder in spreading distributed solar energy and self-reliant life ways since the earliest days. In this astutely practical sourcebook, he provides an indispensable field manual at exactly the tipping point when these tools are finally becoming the norm. This masterful compendium pulls it all together and gives everyone what we need to get off the grid and democratize clean energy.

—**Kenny Ausubel**, CEO and Founder, Bioneers

I've been an early advocate for solar and renewable energy for over 40 years always doing my best to convert Hollywood to a lower carbon footprint. Through most of those 40 years I have always recommended the *Solar Living Sourcebook* as the #1 source, indeed the Bible, for all things solar and sustainable. Bravo to the 14th edition, it's the best one yet!

—**Ed Begley Jr.**, Actor/Actor Environment Activist

If you're interested in understanding alternative energy sources and sustainable lifestyles, Real Goods offers the most comprehensive, enlightening and entertaining resource in the world in its *Solar Living Sourcebook*. Building on many years of accumulated knowledge, Real Goods founder John Schaeffer has updated the *Sourcebook* in its freshest and most useful form ever. It's an indispensable resource on the bookshelf of anyone seriously interested in helping save our habitat.

—**Bryan Welch**, Publisher, Mother Earth News,
author of *Beautiful and Abundant: Building the World We Want*

The *Solar Living Sourcebook* is the must-have resource for consumers seeking the path to a sustainable lifestyle. This well organized and powerful book is still, after 30 years, one of the most useful resources on my bookshelf.

—**Rhone Resch**, President & CEO, Solar Energy Industries Association

Smart people who pay close attention to what's happening in the world around them fall into two groups: those who turn to whiskey and those who buy the *Solar Living Sourcebook*. Both have their merits. But if you have to choose, buy this book. It can change your life for the better. And the world, too.

—**Denis Hayes**, President, Bullitt Foundation

John is a sustainability and solar industry pioneer, who sold the first residential solar system and published *The Solar Living Sourcebook* more than three decades ago. His visionary actions and ideas then have now become a big, mainstream business—solar is the fastest growing industry in the U.S. and residential solar the fastest growing sector within it. That is good news for our environment, economy, and national security. Thirty years from now his recommendations for more sustainable living should improve our lives in the same ways.

—**Nat Kreamer**, Chairman of the Solar Energy Industries Association,
White House Champion of Change, and CEO of Clean Power Finance

Here in one book is everything you need—the information and the technology—to build a more sustainable lifestyle for yourself and your family. It's a treasure trove you'll find yourself returning to time and again.

—**Richard Heinberg**, Senior Fellow, Post Carbon Institute, Author, *The End of Growth*

Better than a bible! This book ranges from philosophy and history of why we need to go solar as a civilization to being a very hands-on and practical guide on how to do it, with great graphics and personal stories to boot. It will be looked back at as the key text of the greatest economic change the world will experience this century, creating jobs and clean energy for a prosperous and sustainable global community.

—**Danny Kennedy**, CoFounder, Sungevity and SfunCube,
and author, *Rooftop Revolution*

For a green economy and a just, sustainable world, it's crucial to go solar. And what's the best way to do that? Turn to the *Solar Living Sourcebook*!!

—**Alisa Gravitz**, President, Green America

This amazing book is just swimming in knowledge—and you can dive in again and again to get your daily fix of sustainable ideas. John is the solar sage and an inspiration to so many of us in the sustainability business.

—**Birchy**, co-founder and CEO, Sungevity

John Schaeffer's *Solar Living Sourcebook* is the authoritative guide to the solar home in the 21st century. REAL GOODS has been lighting the way to sustainability for nearly four decades. Read this book, install solar, eliminate your electric bill and declare independence from the grid forever!

—**David W. Orr**, author, *Hope is an Imperative*.

First there was the *Whole Earth Catalog*, and then there was the Real Goods *Solar Living Sourcebook*, and now…there is still the *Solar Living Sourcebook*. Now in its 14th edition, the *Sourcebook* continues to be jam-packed with exciting ideas and innovative products to make the good life a practical reality. Always a classic, but now even better.

—**Stephen Morris**, Publisher Green Living Journal and founder of The Public Press

The Solar Living Sourcebook, long a trusted source of accurately described energy-efficient products, has evolved into this remarkably comprehensive textbook for an education in Sustainability. You'll not only see what's available, but how to think about your choices as you learn the why and how to develop a lifestyle with minimal environmental footprint. Bucky Fuller wrote, "Philosophy, to be effective, must be mechanically applied." Here's the philosophy and most everything else you need to do so.

—**Jay Baldwin**, Senior Adjunct Professor of Industrial Design
and Sustainability, California College of the Arts (CCA) San Francisco

For all those dedicated to sane, reverent common sense—and to those not yet, here is a comprehensive, rousing guide to a renaissance of reverent ingenuity. Critique holds up a necessary mirror to our world—"look—ack!", with a wave of the hand we change the mirror into a window—"look how beautiful it could be!" then with another wave of the hand—the window becomes a door—"let's go and here's how!" John Schaeffer's *Solar Living Sourcebook* does all three. Let's spiral this forth to libraries, schools, neighborhoods, elected officials—at this time of dire beauty.

—**Caroline W. Casey**, "weaver of context" for The Visionary Activist Show radio show, founder of Coyote Network News, a mythological news service for the Trickster Redeemer within us all.

At a time when the Earth and its inhabitants are facing unprecedented challenges, John Schaeffer's *14th Edition Solar Living Sourcebook* offers an excellent comprehensive roadmap to sustainability. Reminiscent of The Whole Earth Catalog, which helped lead a generation to Earth friendly alternatives to conventional practices, *The Solar Living Sourcebook* is much more valuable and speaks to our vexing 21st Century issues. The complex problems we face can only be solved with a full menu of synergistic solutions. This book goes far and deep in laying the foundation for sustainable living. Globally we all live in the same neighborhood. By demonstrating what works, what is affordable, and what is sustainable, we can show, by example, and lead the way to a greener and brighter future for all of humankind. A greener future is ours alone to make. This book is an essential step in leading the way. Future generations are depending upon us now to make ecologically smart decisions and this book is an essential first step.

—**Paul Stamets**, D.Sc., Fungi Perfecti, and author, *Mycelium Running*

The solar photovoltaic cell—a unique California success story—was only 24 years old when John and his team at the Solar Living Institute first published this veritable bible of sustainable living. Today, hundreds of thousands of Californians have followed in John's footsteps, installing two nuclear power plants worth of solar power on rooftops from Humboldt to San Diego. Other states and countries now give California a run for its money in this race to the top where everyone wins. Whether you're an off-grid, hay-bale house, zero-carbon-footprint enthusiast or someone who's just interested in saving a little money with a home solar electric or hot water system, this is the resource book for you.

—**Bernadette Del Chiaro**, Executive Director, California Solar Energy Industries Association

The *Solar Living Sourcebook* continues to be the bible for practical uses of commercially-available solar, energy efficiency and other clean energy options. Now that more people than ever before are turning towards solar energy options, to me, this is the first resource to start with.

—**Scott Sklar**, President The Stella Group, Ltd. and Adjunct Professor, The George Washington University

Phenomenal. Once again, John and team have outdone themselves with this edition. The *Solar Living Sourcebook* has been, and always will be, the definitive guide for all things solar, off grid, and natural living and with the usual soulful philosophy thrown in for good measure. For earth! Congratulations and thank you!

—**Howard Wenger**, 30-year solar pioneer

John Schaeffer has been shining a light on the way forward for a long time—and his newest edition does not disappoint. If you think it's high time we started living as if we plan to be on this planet for a while, you'll find the *Solar Living Sourcebook* an essential reference.

—**Adam Browning**, Executive Director, Vote Solar

John has been working for decades to make solar accessible to everyone, and the Updated *Real Goods Solar Living Sourcebook* is another important contribution.

—**Billy Parish**, Founder & President, Mosaic

The 14th *Sourcebook* is written proof…the spell is broken! We can't continue to fight Mother Nature, and happily, our personal choices can make a greener world! This comprehensive encyclopedic guidebook inspires us all how-to move forward to the solar age.

—**Johnny Weiss**, Principal, Solar Consulting LLC, Co-Founder, Solar Energy International (SEI)

A rainbow of post-its flies out of the pages of our Ninth Edition *Solar Living Sourcebook*, which is a go-to guide here at our ecological architecture practice. We look forward to similarly flagging and referencing this new 14th edition, which is without doubt a far more up-to-date and complete version. You'd better get your own copy because we loan out some of the books at our office, but never this one!

—**David Arkin** AIA, LEED AP, Arkin Tilt Architects

This book was an inspiration to me in college and with solar scaling faster than cellphones today this latest edition is showing that we are the verge of the largest wealth creation opportunity of our generation—Climate Wealth.

—**Jigar Shah**, founder of Sun Edison

Since it first appeared in 1978, the *Real Goods Solar Living Sourcebook* has been one of the essential go-to volumes for anyone interested in solar energy and green alternatives of all kinds. This latest edition is even better than its predecessors, which is no small feat, and belongs on the bookshelf of anyone who recognizes that today's fossil-fueled lifestyles are past their pull date and is ready to embrace the challenges of the future.

—**John Michael Greer**, author of *Green Wizardry* and *After Progress*

John Schaeffer, the founder of Real Goods and an early pioneer in promoting resilient living strategies for self-reliant communities has given us in this *Sourcebook* the most comprehensive overview of how we can move forward in today's uncertain world of climate chaos towards a more ecological life style in tune with ourselves the living world.

—**Sim Van der Ryn**, Professor Emeritus UC Berkeley, Architecture & Ecological Design, and author, *Ecological Design* and *Design for an Empathic World*

The 14th edition of the *Solar Living Sourcebook* is a great resource for those that care about the health of natural systems. Both inspiring and practical, it offers resources and knowledge on how to live lighter on the land. Are you ready for the solar revolution ?

—**John W. Roulac**, Founder & CEO Nutiva, author, *Backyard Composting* and *Hemp Horizons*

EVERYTHING UNDER THE SUN

REAL GOODS

Solar Living
SOURCEBOOK

14TH EDITION • REVISED & UPDATED

Your Complete Guide to Living beyond the Grid
with Renewable Energy Technologies
and Sustainable Living

JOHN SCHAEFFER
FOREWORD BY BILL MCKIBBEN

new society
PUBLISHERS

Cataloging in Publication Data:

A catalog record for this publication is available from the National Library of Canada.

Cover design by Diane McIntosh.

Printed in Canada. First printing October 2014.

The Real Goods Solar Living Sourcebook is the fourteenth edition of the book
originally published as the *Alternative Energy Sourcebook*, with over 600,000 copies in print,
distributed in forty-four English-speaking countries.

The logo and name Real Goods™ are the trademark of Real Goods Solar Inc.

New Society Publishers acknowledges the financial support of
the Government of Canada through the Canada Book Fund (CBF) for our publishing activities.

Paperback ISBN: 978-0-86571-784-8   Ebook ISBN: 978-1-55092-579-1

Inquiries regarding requests to reprint all or part of *The Real Goods Solar Living Sourcebook*
should be addressed to New Society Publishers at the address below.

To order directly from the publishers, please contact:

Real Goods
Solar Living Center
13771 So. Highway 101
Hopland, CA 95449 USA
www.realgoods.com
Toll-free (North America): 1-800-919-2400

OR:

New Society Publishers
P.O. Box 189
Gabriola Island, BC V0R 1X0 Canada
www.newsociety.com
Toll-free (North America): 1-800-567-6772

Real Goods has checked all of the systems described in this book to the best of its ability; however, we can take no responsibility whatsoever regarding the suitability of systems chosen or installed by the reader. Due to the variability of local conditions, materials, skills, site, and so forth, New Society Publishers, John Schaeffer, and Real Goods assume no responsibility for personal injury, property damage, or loss from actions inspired by information in this book. Always consult the manufacturer, applicable building codes, and the National Electrical Code™ before installing or operating home energy systems. For systems that will be connected to the utility grid, always check with your local utility first. When in doubt, ask for advice; recommendations in this book are no substitute for the directives of equipment manufacturers or federal, state, and local regulatory agencies.

New Society Publishers' mission is to publish books that contribute in fundamental ways to building an ecologically sustainable and just society, and to do so with the least possible impact on the environment, in a manner that models this vision. We are committed to doing this not just through education, but through action. This book is one step toward ending global deforestation and climate change. It is printed on Forest Stewardship Council-certified acid-free paper that is **100% post-consumer recycled** (100% old growth forest-free), processed chlorine free, and printed with vegetable-based, low-VOC inks, with covers produced using FSC-certified stock. Additionally, New Society purchases carbon offsets based on an annual audit, operating with a carbon-neutral footprint. For further information, or to browse our full list of books and purchase securely, visit our website at: www.newsociety.com

Contents

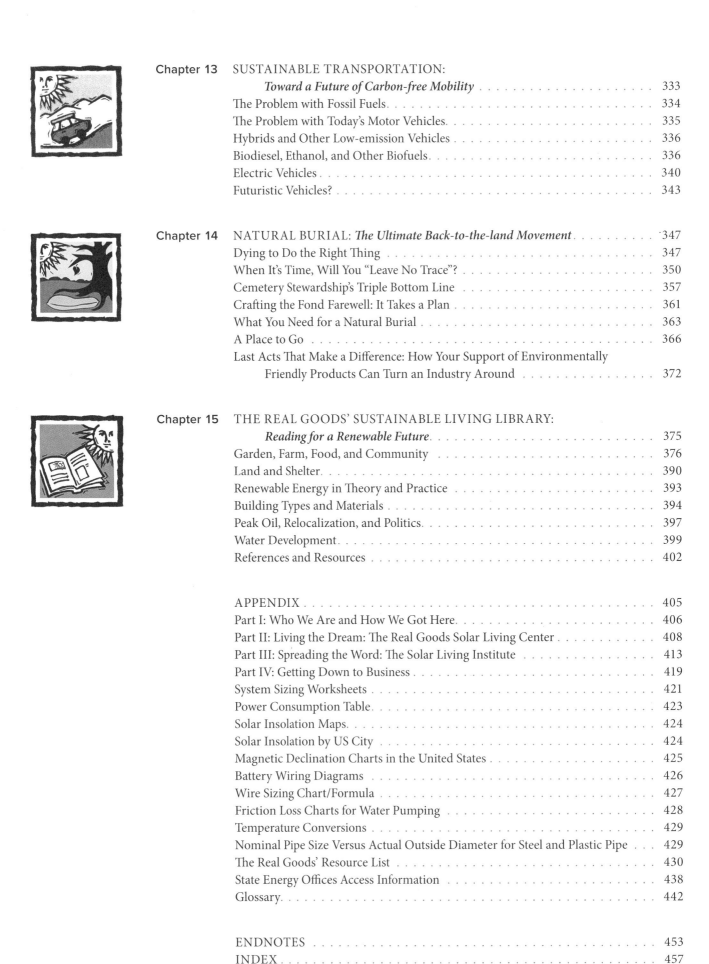

Acknowledgments

WHILE IT'S IMPOSSIBLE to adequately thank everyone who has contributed to our 14th edition *Solar Living Sourcebook* and to Real Goods' extraordinary 36-year odyssey in general, several people stand out above the crowd. In the 36 years and 14 editions of this *Sourcebook*, more than 700,000 copies have been sold in 44 English-speaking countries. For this edition, thanks go first to Alan Berolzheimer, our managing editor, who toiled tirelessly over many months to manage the new chapters as well as the old while putting his all into the entire book. His masterful pen and mind often performed miracles of weaving straw draft copy into gold nuggets of final prose. Hearty thanks go to several people who helped get our chapters new to this 14th edition launched: to Daniel Lerch for his revision of the Relocalization chapter, to David Arkin and Amanda Knowles for a thorough rewrite of Land and Shelter, to Brad Lancaster for generous contributions to the updated Water Development chapter, and to Art Ludwig for revising Composting Toilets and Greywater Systems. Special thanks to Rachel Kaplan for her work on the brand new Urban Homesteading chapter. Thanks to Bob Ramlow, who has helped tremendously on the last several *Sourcebooks* with the Solar Hot Water chapter. I also want to thank Benjamin Fahrer for his work on the Permaculture section and to Jim Fullmer for his updates on Biodynamics. Raphael Schiffman helped us tremendously with much needed updating to Chapters 3 and 4, really the renewable energy heart of this book. Of course I thank both Steve Heckeroth and Dave Blume for their passionate defense of both electric vehicles and alcohol as a vehicle fuel. The book wouldn't have been complete without the final "back to the land" Natural Burial updates from Cynthia Beal. And thanks to Lindsay Wood for her last-minute updates about efficient lighting and LEDs. Thanks also go to Tony Novelli for helping out with the resource list and with Chapter 2.

Thanks also to the folks at New Society Publishers for doing such an amazing job of distribution with our 30th anniversary edition and for encouraging us to continue to keep the *Sourcebook* updated. And for *Sourcebooks* past, there are many to thank who have nurtured the book over the years and some of whose work still remains current to this day. I especially thank Doug Pratt, Stephen Morris, and Jeff Oldham for their inspiration and dedication to Real Goods and the entire *Sourcebook* project over the years. Thanks also go of course to Real Goods' current CEO, Kam Mofid, without whose encouragement this book would not have been possible, and to Melissa Watson, who has taken over management from me of the retail store and our webstore (realgoods.com) so that I could work on this *Sourcebook*. And on a personal note, I thank my wife, Nantzy, for her frequent insight and inspiration for this and most all of my writing, and my children, Sara, Ashley, and Cy, for their help over the years. Finally, for their unwavering financial support, I thank the original 10,000 Real Goods shareowners from the 1990s for directly bringing the Solar Living Center into fruition, and the directors and staff of the Solar Living Institute for nurturing that original vision to the present day.

The Nonprofit Real Goods Solar Living Institute

The Solar Living Institute is a 501-C-3 nonprofit organization that promotes inspirational environmental education through hundreds of hands-on workshops; SolFest—an annual renewable energy educational event now in its 16th year with displays and interactive demonstrations at our solar-powered 12-acre permaculture oasis called the Solar Living Center in Hopland, California; and numerous children's educational programs. More information about the Institute can be found on pages 413–18 or at its website (solarliving.org) or by calling 707-472-2450. The Institute has an innovative membership program and encourages tax-deductible donations to further its programs.

FOR THE EARTH

May we preserve and nurture it in our every action.

A Message from the RGS Energy CEO

HUMANS TEND TO THINK of themselves as smarter than animals, yet in some ways one has to question that assumption. Animals only take what they need from the planet, yet we, the supposedly superior species, take far more than we need. It is both ironic and tragic to see how much damage our "progress" has done to our only home, our beautiful yet fragile planet. If a paradigm shift in how we think about growth and progress doesn't occur in the near future, for the first time in human history, we will leave our children and grandchildren a planet in far worse shape than the one we inherited. We have an unquestionable obligation to protect our air, water, and other natural resources, and to preserve the habitat of other species that have an absolute right to coexist with us. Their health and survival will be ours, and the reverse is also true.

Visionaries like John Schaeffer, the founder of Real Goods Solar, have been working in a very pragmatic way to help people think differently and realize that simple yet impactful options are available to live in harmony with nature and to reduce our collective carbon footprint. The information and practical "how to's" of the *Solar Living Sourcebook* reflect the cumulative efforts of many people whose intention has always been to do something good—real good—for the world. John, who I had the pleasure of first meeting in 2010, has always walked the talk. He lives in a sustainably built, off-grid home, practices organic farming, and encourages others to do the same.

In 1978, John started Real Goods Solar, and became one of the first people in the country to sell solar systems to the public and educate them about the benefits of this wonderful clean and renewable source of energy. What started as a small operation selling solar panels to off-gridders in the hills has become one of the nation's pioneering and largest solar companies. In 2014, Real Goods Solar was rebranded as RGS Energy to highlight our vision to make solar a critical part of the country's energy mix. We have

already deployed tens of thousands of solar systems and serve commercial, residential, and utility customers. In the process, we have not only provided economic benefits to our customers by reducing their total energy costs, but in my view equally important, we have made a meaningful contribution to reducing greenhouse gas emissions and our society's reliance on dirty and toxic fuels such as coal.

We need a sea change, and as you know, rivers are made up of individual drops all flowing in the same direction. So whether you are an individual or a family considering off-grid living, a person buying an electric vehicle, a family interested in urban homesteading, a small or large business installing solar, or a utility with the vision to invest in solar rather than coal or nuclear, remember that our choices impact everyone on the planet. Every time we make environmentally responsible choices such as using solar to power our homes, using fewer synthetic chemicals in our lives, and making food choices that are healthy, humane, and environmentally prudent, we make a real positive difference to our planet in all its diversity. As more and more people awaken and challenge the status quo, we will end up with a large river of change—a river that will heal our scarred planet.

The *Solar Living Sourcebook* and RGS Energy offer real solutions to some very big problems. We need more people and more leaders like John Schaeffer, who are willing to set an example for the better and truly walk the talk. We have to learn to live smarter, to use less, to give back to the land, and to protect what nature has so generously supplied. I hope the *Solar Living Sourcebook* inspires you to make positive changes. All of us need to tread more lightly on this magnificent planet. I hope you will join us in becoming part of the solution.

Kam Mofid
CEO, RGS Energy

Foreword

by Bill McKibben

I'M NOT ALWAYS the most hopeful person on the planet—back in 1989 I wrote the first book about global warming for a general audience, and it came with the cheerful title *The End of Nature*. Since that time, we've watched the Arctic melt, and the oceans acidify. We've had all 10—all 20—of the warmest years ever recorded. We've seen deep and horrific droughts, and storms of unprecedented power. Hell, we've watched the New York City subway system fill up with the Atlantic Ocean. We've left behind the Holocene—the 10,000-year stretch of benign climatic stability that underwrote the rise of human civilization. So perhaps a certain amount of gloom is in order.

But the book you hold in your hands is the antidote to much of that despair. Written by the folks who sold the very first PV panel in the US back in 1978, it shows exactly how far we've come toward the solutions that we need. Because 25 years ago, solar power was still *hard*. You needed to really know what you were doing. You needed some serious cash, and maybe a hairshirt. In the intervening years, the price of solar panels has crashed—down 99% over the last 40 years. More to the point, the knowledge base has gone straight up. Now, most of the time, the utilities will know what you're talking about when you want to connect. They may not like it—and in some states, at the behest of the Koch Brothers, they're doing their best to make it harder. But they'll fail: The technology is too good, the desire too strong.

There are other deeply positive trends that you can sense in these pages too. Take transportation. A quarter century ago, we were still in the complete thrall of the private automobile, the bigger the better. But that's starting to shift. We've seen a wave of hybrid cars, and now electric hybrids, and straight plug-ins. (If you've got those solar panels up top, you're running your car off

the sun—it's like a plant!) Biking has come of age, with commuting lanes in any city worth the name (though no American burgs have come close to, say, Copenhagen, where 40% of commuters now pedal to their jobs). As a result, young Americans are far less interested in cars then they used to be—the age at which they get their first driver's license has gone steadily up, and the percentage desiring to own a car has plummeted. Most of them want to live in Brooklyn, where only a fool would own a car anyway.

Or think about food. A quarter century ago, only a few folks—Wendell Berry, say—were carrying the flag for local growing. But for the last 15 years, the local farmer's market has been the fastest-growing part of the food economy. We have ten times the number of local breweries (there's a good sign!), and we have scads of young people who spend their summers Woofing (that there's even a word to mean volunteering on organic farms should tell you something). And the bottom line: A couple of years ago the USDA reported that, for the first time in 150 years, the number of farms in the US is going up, not down. The biggest demographic trend in the history of our country has bottomed out and turned slowly around. It's got a long ways to go (there are still more prisoners than farmers in this country), but at least the trajectory is right.

All these trajectories are right—and there are plenty more, as the chapters in this Real Goods *Sourcebook* make clear. From garage to dining room, from cradle to green grave, we're pioneering new/old and most of all *sensible* ways of doing things that let us see a future not only possible but beautiful. Imagine a world where most energy didn't come from a few huge power plants and utilities, but from a million solar panels on a million rooftops all tied into the grid. Imagine a

> All these trajectories are right—and there are plenty more, as the chapters in this Real Goods *Sourcebook* make clear. From garage to dining room, from cradle to green grave, we're pioneering new/old and most of all *sensible* ways of doing things that let us see a future not only possible but beautiful.

Simply installing your solar system and calling it a day isn't going to cut it, not by itself. It's a good start, but it's too firmly in the American tradition of just taking care of yourself. This kind of individualism is a great thing—but climate change is above all a structural, systemic problem.

To counter all that fossil fuel industry money, we need a strong movement. And the good news is, that movement is building. We need you very badly to spend Saturday morning up on the roof putting in your new solar system. And then we need you to climb down the ladder and spend the occasional Saturday afternoon with as many of your neighbors as you can round up, trying to change the nation's power structure just as strenuously as you've been working to change your household power system.

future where food didn't come from the distant corporate plantation but from your neighbors. You don't have to imagine as hard as you used to.

I've stayed positive as long as it's humanly possible for me. A little reality check is in order now. All these things are coming—they're just coming too slowly. If we had a hundred years to deal with climate change, we'd be going at just the right pace: After all, humans and their institutions do better with slow change than with fast. In an ideal world, you'd put a solar hot water heater on your roof (why, by the way, does anyone *not* have one of these—the payback couldn't be much quicker), and then your brother-in-law would see it and he'd put one up, and…. Over a hundred years, we'd get where we'd need to be, with the kind of gentle transition that's ideal.

But physics calls the tune here, and it's not giving us a hundred years. It may not be giving us another decade: Remember, the Arctic has already melted, which is not a good sign (you really don't want to be, you know, *breaking* the major physical features of your home planet). If the scientists who study this greatest crisis we've ever faced agree on one thing above all, it's that we need fast action. As in, preferably yesterday, but certainly in the next couple of years.

Which means that simply installing your solar system and calling it a day isn't going to cut it, not by itself. It's a good start, but it's too firmly in the American tradition of just taking care of yourself. This kind of individualism is a great thing—but climate change is above all a structural, systemic problem. You can be pretty sure that putting in the best possible light bulbs isn't going to solve it. Not unless you can convince almost everyone else to do it too, and right away.

There are ways to make that happen. Almost every economist—left, right, and center—has said we need to put a price on carbon to reflect the damage it does in the atmosphere. Once we've done that, market forces can begin to go to work; every time we pay the electric bill, we'll be reminded to get the PV up on the roof, because our artificially cheap fossil fuel will be a thing of the past. And we can even do it without bankrupting people: As long as you're Googling, look up "fee and dividend" for details of the scheme that will tax the hell out of carbon but return all the money to consumers directly. A check every

month—80% of us will come out ahead, with the cash in our pockets to, you know, buy more solar panels.

But the simple fact that economists and scientists have told us that we should take these steps does not mean they will in fact be taken. So far Washington has produced a perfect 20-year bipartisan record of accomplishing absolutely nothing on climate change. And that's for a simple reason: The fossil fuel industry is the most profitable enterprise humans have ever invented. Exxon made more money last year than any company in the history of money. And in our semi-corrupt political system, that money allows them to buy enough influence to keep the status quo going, even as it's changing the pH of the ocean. I mean, Chevron gave the largest campaign contribution in the post-Citizens United era two weeks before the last federal election—and it was successful, helping keep control of the House of Representatives in the hands of climate deniers.

To counter all that money, we need a strong movement. And the good news is, that movement is building. At 350.org, the global climate campaign, we've held 20,000 rallies in every country but North Korea. Opposition to the Keystone XL tar sands pipeline has led to the largest civil disobedience actions about anything in 30 years. The drive to get colleges, churches, and cities to sell their shares in fossil fuel stocks has become the fastest growing divestment campaign in history. We're not winning yet—the US continues to produce more oil and gas every year, and our coal exports are at an all-time high. But we're starting to make progress.

Which means we need you very badly to spend Saturday morning up on the roof putting in your new solar system. And then we need you to climb down the ladder and spend the occasional Saturday afternoon with as many of your neighbors as you can round up, trying to change the nation's power structure just as strenuously as you've been working to change your household power system. Screwing in a new light bulb really is great; and screwing in a new senator is pretty darned good too.

There's no guarantee we can win this fight—the science is fairly dark. But one thing is certain: This is the fight of our lifetimes. And we need to battle it on every front, home and away!

BILL MCKIBBEN *is Schumann Distinguished Scholar at Middlebury College, founder of 350.org, and 2013 winner of the Gandhi Prize. He enjoys taking slightly-longer-than-absolutely-necessary showers thanks to the solar hot water heater on his roof.*

Introduction

by John Schaeffer

Welcome to the 14th edition of the *Solar Living Sourcebook*! As I look back to our humble beginnings at Real Goods in 1978, it's clear that we've made significant progress toward the regenerative and sustainable goals that we set for ourselves nearly 37 years ago. Progress in the solar sector is both dramatic and quantifiable. That first year, we sold a 9-watt PV module for $900, or $100 per watt. Today's prices are well below $1 per watt, a decrease of over 90%!

I've mused about the "dawn of the solar age" for 37 years, and finally, it's true. The "too-cheap-to-meter" nuclear age is on its way to the scrap heap. Utilities have discovered that nuclear power is nowhere near as cost-effective as new solar installations. Last year in the US, a record 4.7 gigawatts of solar was installed, bringing the total to over 10 gigawatts. Each of those gigawatts can power 164,000 homes, making more than 1.6 million homes solar powered. Every four minutes, another American home or business goes solar, and it's predicted that the pace will accelerate to one every 90 seconds by 2025. While impressive, this still represents less than 1% solar electric generation in our country. Meanwhile, utilities are steadily reducing the number of operating nuclear power plants to fewer than 100 for the first time in 20 years. Nine planned upgrades were recently canceled because the investments are no longer economically justifiable.

2013 was another record-shattering year for solar in the United States. According to GTM Research and the Solar Energy Industries Association's (SEIA), PV installations increased 41% over 2012 to reach 4,751 megawatts (MW). Additionally, the cost to install solar fell 15%. Moore's Law (for every doubling of the source of supply, the price declines 15%) proved true again. By the end of 2013, more than 440,000 operating solar electric systems in the US generated well over 10,000 megawatts (10 gigawatts), offering the first real glimpse of mainstream status. The combination of rapid customer adoption, grassroots support, improved financing terms, and public market successes represent clear gains for solar with both the general population and the investment community.

Japan and Germany have, for the last several decades, been the innovators in solar. Germany truly *has* paved the way. Currently 59% of that country's energy comes from renewable sources, 11.2% from solar. In 2013, however, the US surpassed Germany in PV deployed, so, finally, we're truly back in the solar game. It is telling that if we filled the 25,000 square mile Mojave Desert with solar, we could produce 6 times the total electricity demand of the country. (See page 89 for an illustration of how a solarized area less than 100 miles square could provide all the energy needs of the US.)

Solar is proving to be a major job creator. 2013 saw tens of thousands of new American jobs connected with the solar industry, which pumped tens of billions of dollars into the US economy. In fact, more solar has been installed in the US in the last 18 months than in the previous 30 years. 143,000 people are employed in the solar industry, up 20% from a year earlier and 10 times the national average job growth rate. That's a remarkable record of achievement.

When I started Real Goods in 1978, our clientele was a cadre of young and idealistic hippies living in the woods of Mendocino County, California. They were refugees from major urban centers looking for a simpler and more meaningful existence. Many got light from kerosene, heat from wood, food from the garden, social contact from friends and family, and entertainment from books. There were no computers, cell phones, Internet, Google, Facebook, or YouTube. Jimmy Carter was president, Jerry Brown was governor (for the first time), and optimism abounded that

> 2013 was another record-shattering year for solar in the United States. According to GTM Research and the Solar Energy Industries Association's (SEIA), PV installations increased 41% over 2012 to reach 4,751 megawatts (MW). Additionally, the cost to install solar fell 15%.

with the strength of our convictions, we would soon overcome the misguided practices of our over-logging, over-consuming, polluting culture. Climate change, calculations of parts per million of CO_2, oil and natural gas fracking technologies, and the Keystone XL Pipeline were still in the future. So was oil depletion. I wrote an editorial in our 1979 catalog that stated, "According to the US Congress's own Office of Technology Assessment, all known oil reserves will be exhausted by 2038." At the time, 60 years seemed like plenty of time to wean ourselves from fossil fuels! But, from the present perspective, time is clearly no longer on our side. Reaching the benchmark 400 ppm of CO_2 in 2013 makes it imperative that we accelerate the pace to correct our fossil fuel addiction. Sadly, it may already be too late.

This 14th edition of the *Solar Living Sourcebook* is dedicated to supporting the complete lifestyle change necessary to give our planet the best possible path toward the maintenance of a climate and ecosystems that have allowed humanity and other species to flourish. While we still call it the *"Solar" Living Sourcebook*, you'll find that this book is not limited to the details of solar technologies, but instead engages with all facets of sustainable, resilient, and regenerative living.

Our 14th edition begins with a thorough update of the **Relocalization** chapter that debuted in our 30th anniversary (13th) edition. There, we introduced the "Transition" movement, showing practical ways for individuals and communities to return to a regenerative and sustainable economy. The end of cheap, abundant fossil fuels, and the disruption of long-stable climatological and biological systems—the results of rampant industrialization—are altering life on Earth. When the relocalization movement began at the start of the 21st century, it was focused on concerns about "Peak Oil." Since the amount of oil on the planet is finite, and consumption is rising, at some point in the near future, less will be available and prices

will rise accordingly. Peak oil consciousness has evolved from the theory that oil would be gone within a few decades to the more contemporary view that oil won't ever completely run out, but it will become increasingly too expensive to extract. Indeed, the easily developed oil is nearly gone now. What remains will be extracted at a progressively slower rate and will cost much more.

Our present human economy is unsustainable, because it relies on non-renewable raw material sources whose development, production, and consumption create pollution that causes negative "feedbacks" that impair ecosystems and disrupt the climate. A sustainable economy needs to run on income from solar energy while not degrading ecosystems by creating wastes or mining nutrients. Relocalization seeks to preserve the "natural capital" of the Earth by acknowledging that our well-being is derived from the ecological and geological richness of the planet. With roots extending back to the back-to-the-land movement that gave birth to Real Goods in 1978, relocalization encompasses the best of environmental protection, sustainable living, regenerative design, natural building, urban homesteading, Slow Food, and voluntary simplicity. This is a new way of saying "thinking globally while acting locally." It is, at once, a strategic response to peak oil, climate change, and the overshoot of Earth's physical limits. It honors, encourages, and nurtures local businesses, farmer's markets, energy production, and community involvement while rejecting the malign aspects of globalization and our current fossil fuel-based economy. Relocalization and the Transition movement are perfect preludes to the nuts-and-bolts education found in the balance of the *Sourcebook*.

Then, it's right into the essence of the *Sourcebook*. We explore the concept of "home" in the **Land and Shelter** chapter. It all begins and ends with land. Our land and homes are not only the cornerstones of our existence but also the site of

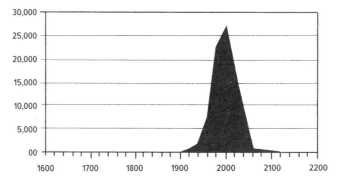

World oil production from 1600 to 2200, history and projection, in millions of barrels per year. (Source: C. J. Campbell)

our largest carbon-footprint impact. The choices we make about where we choose to settle, how we treat the land, what we create for shelter, and how those homes function are fundamental to any notion of sustainability or regeneration.

Shelter is, quite literally, rooted in the land. Our homes are built upon the topsoil that nourishes the plants that make all life possible. The materials we use to build them come from the land. By displacing organisms and extracting resources for our shelters, the decisions made in the building process have profound impact. Through intelligent placement and smart design, we can create shelter—and entire communities—that honor and respect the Earth and its ecosystems. If our goal is to create carbon-neutral homes that minimize their contribution to climate change, the choice of where and how to create shelter becomes critical.

Last year Americans built around 900,000 new homes, each one requiring massive amounts of energy and materials. Wood is still the material of choice for the majority of new houses. In 2012, the housing industry consumed about 32 million cubic meters of lumber in residential construction and remodeling. If cut into conventional 2×4s, that lumber would stretch for half a million miles, circling the Earth 20 times—a mind-numbing illustration of the impact of the building industry on the planet's forests. Besides the number of homes built, the size of homes also pushes the demand for wood. After a short-lived drop-off during the economic downturn that began in 2008, the average house size has rebounded to 2,500 square feet. According to the US Department of Energy, America's homes consume about 22% of the nation's fossil fuel energy, which, in turn, accounts for one-fifth of America's enormous annual release of CO_2.

While humans have accomplished amazing feats and created astonishing civilizations, like all species we depend upon nature for the resources that make our lives and our economies possible. The source of all the goods and services we consume, nature is also the sink for our wastes. We must recognize our limited ability to control nature and learn to live cooperatively with nature to achieve harmony and balance on Earth.

As with all previous editions of the *Sourcebook*, the guts of the book are in Chapters 3 and 4: **Sunshine to Electricity** and **Panel to Plug**. These are the nuts and bolts of renewable energy: photovoltaics, wind turbines, and hydroelectric turbines, along with all the necessary peripherals for living off the grid or on the grid with solar or other forms of renewable energy. Renewable energy is the heart of Real Goods. Dramatic shifts have occurred in the solar sphere since we published our 30th-anniversary *Sourcebook* in 2008. The combination of rapid customer adoption, widespread grassroots support, and the welcome appearance of innovative financial leasing mechanisms has made going solar almost a no-brainer. This has not been overlooked by the public capital markets, whose successes have propelled the industry to the forefront of our economy.

While the solar market has grown annually by nearly 50% since 2008, prices have plummeted by 50%! In 2008, Real Goods sold grid-tied solar systems for around $8 per watt installed; now, the same system sells for less than $4 per watt. Nationwide, solar installations have increased 16-fold from 298 megawatts to 4,751 megawatts. Another change is that "third party owned" (TPO) has emerged as the dominant method for homeowners to install solar. Today, through a lease or a power purchase agreement, as many as three-fourths of all new systems will be third party owned. For the homeowner, this usually means no money down and a significant discount on your monthly utility bill. Cash purchases are becoming a relic of the past, even though PV is still a much better investment than most financial instruments, with annual returns often exceeding 15%. If you live off the grid, like I do, Chapters 3 and 4 will guide you through every aspect of your home energy system, from conception through purchase, installation, and annual maintenance. I still refer to my *Sourcebook* whenever I'm working with my batteries or my monitoring system, or whenever I need a refresher course on the finer points of solar and micro-hydro.

With fresh memories of natural disasters, ranging from Hurricane Katrina to the nuclear meltdown at Japan's Fukushima Daichi power plant, to the accelerating pace of floods, tornadoes, and tsunamis, we've expanded our chapter on **Emergency Preparedness** (Chapter 5). The world can be a scary and hazardous place with power grid failures, disastrous storms, wildfires, fuel price spikes, and terrorist attacks. We all harbor deep fears that our society has become vulnerable to infrastructure disruption. Who can doubt that factors as diverse as growing population pressures, agonizing poverty, economic globalization, political conflict, climate change, and additional natural disasters will continue to place enormous strains on the networks that supply us with food, water, power, and other necessities? Many of us will experience these disruptions and breakdowns in the foreseeable future. You will rest easier if you've taken steps to protect your

In 2012, the housing industry consumed about 32 million cubic meters of lumber in residential construction and remodeling. If cut into conventional 2×4s, that lumber would stretch for half a million miles, circling the Earth 20 times—a mind-numbing illustration of the impact of the building industry on the planet's forests.

The combination of rapid customer adoption, widespread grassroots support, and the welcome appearance of innovative financial leasing mechanisms has made going solar almost a no-brainer.

family from events that threaten to undermine your ability to meet basic needs. Be prepared.

Following the chapters on Renewable Energy and Emergency Preparedness comes Chapter 6 on **Energy Conservation**. Every dollar spent on conservation translates into $3 to $5 savings on solar system costs—another no-brainer. Conservation reduces greenhouse gas emissions, slows the depletion of natural resources, decreases environmental pollution, takes strain off the planet's organic life-support systems, and saves a lot of money. Who can afford *not* to conserve energy? The further beauty of conservation is that regardless of what skeptics say, conservation does *not* mean sacrifice. Many European and Scandinavian societies enjoy a comparable standard of living to the United States, but on a much tighter energy budget. Build smaller rather than larger. Make sure your building envelope is tight. Use passive solar strategies to minimize heating and cooling loads. Install low-flow showerheads, low-flush or composting toilets, greywater systems, and compact fluorescent or LED lights. Buy the most efficient appliances possible for your needs and budget.

Conservation is the cheapest, most cost-effective way to "produce" energy. Years ago renewable energy guru Amory Lovins introduced us to the concept of "negawatts," or watts we never need to use. But conservation does not come intuitively. We were raised on a steady diet of cheap fossil fuel energy during the 20th century, especially since World War II. Our society is enormously wasteful of energy. Not only are fossil fuel supplies finite and dwindling, but we now recognize their dire ecological impact on Earth's climate. Energy conservation is the starting point for a sustainable future, *before* implementing the transition to solar and other forms of renewable energy.

Following Energy Conservation is Chapter 7, **Water Development**. Two age-old sayings play preeminently in this chapter: "Water is life," and "You buy the water, and the land comes free." Water is the single most important factor in any homestead. Without a dependable water source, no place is home very long. Unfortunately, today many people literally *are* buying water, often unnecessarily and at outrageous expense. Some are swayed by multimillion-dollar corporate ad campaigns selling bottled water; others are victims of the privatization of previously public water supplies. Globally, *nearly 800 million* people still lack access to clean water. Sadly, as supplies become polluted, privatized, or acquired by for-profit enterprises, that number will steeply rise.

Once we have figured out how to develop our water, the next step is to heat it, the subject of Chapter 8, **Water Heating**. Most of us take hot water at the turn of a tap handle for granted. It makes modern life possible. Yet many people do not realize the total costs of this convenience. The average household spends an astonishing 20%–40% of its energy budget on water heating! Those energy dollars are typically dedicated to an appliance that has a life expectancy of only 10–15 years and wastes 20% or more of the energy it consumes. Efficiency improvements to your water heater will reduce overall energy consumption, lower your carbon footprint, and lessen the environmental impact of your home. There are better, cheaper, and more durable ways to get hot water than using fossil-fueled water heaters—in particular, solar hot water. This chapter describes common water heater types, examines the good and bad points of each, and offers suggestions for efficiency improvements. It also provides a comprehensive review of solar hot water systems that underscores their environmental and economic benefits.

Chapter 9 is on **Water and Air Purification**. If you are concerned about the foods you eat and what you put into your body, you want to be equally conscientious about the quality of the air you breathe and the water you drink. Like food, air and water can be the vehicles into your body for the nasty contaminants in our environment. Most people are not aware of the extent to which domestic water and inside air can be polluted. This chapter examines the bad stuff that may be lurking in your air and water, the conditions that contribute to unhealthy levels of contaminants in the living space, and what you can do to alleviate these problems. The Environmental Protection Agency (EPA) states that no matter where you live in the United States, some toxic substances will be found in the groundwater. Indeed, the agency estimates that one in five Americans, supplied by one-quarter of the nation's drinking water systems, consume tap water that violates safety standards under the *Clean Water Act*. Even substances added to our drinking water to protect us, like chlorine, which is still legal in the US while long banned in Europe, can form toxic compounds.

Chapter 10, **Composting Toilets and Greywater Systems**, is especially relevant to the prevailing drought conditions caused by climate change. Both technologies offer outside-the-box ways to save significant amounts of water. Composting toilets, while traditionally employed in country cabins and cottages, are beginning to come into more frequent use in more suburban settings. In our collective rush to sanitize everyhing, we discard a potentially valuable and money-saving resource by designing expensive and energy-intensive disposal systems that pollute surface and groundwater. Conventional plumbing systems mix a few pounds of valuable nutrients and a few micrograms of potentially dangerous pathogens with hundreds of gallons of very lightly polluted greywater from our sinks, showers, tubs, and washers. This chapter takes a fresh look at how to incorporate non-traditional technologies into our homes to convert waste into a valuable resource.

Chapter 11, **Regenerative Homesteading and Farming**, is close to my heart as my wife, Nantzy, and I run a biodynamic farm in Hopland, California, where we grow olives, grapes, fruit trees, vegetables, and lavender. **Biodynamics** is emerging in much the same way as the organic movement in the late 1970s and early 1980s. Conceived in 1924 by Dr. Rudolf Steiner (also the founder of the Waldorf education movement), biodynamic agriculture shares the original foundation of organic and sustainable agriculture. Steiner outlined his principles in a series of lectures to European farmers alarmed at both the growing use of synthetic chemical fertilizers and pesticides, and the corresponding decline in crop and animal vitality.

The biodynamic program focuses on the biological systems of agriculture and advocates making soil amendments directly on the farm, rather than importing them. Importing materials to an organic agricultural system, says Steiner, introduces the same problems as synthetic industrial products. Both require additional natural resources to mine, refine, and transport a myriad of products, a practice that puts pressure on natural resources and systems wherever these materials are mined or harvested. The goal of biodynamics is to be as regenerative as possible by developing inputs on site from the living dynamics of the agricultural system itself.

Permaculture, also featured in Chapter 11, is much more than a style of gardening. It is a fully integrated design system and philosophy in which diverse techniques create a food production system that emulates the natural world. Permaculture design assembles conceptual, material, and strategic components in patterns that provide mutually beneficial, regenerative, and secure places for all forms of life. The permaculture ethos comes from the teachings of indigenous cultures and from patterns found in nature. It is an approach to gardening and a way to manage land as well as a way to create shelter. Beyond these arenas, permaculture design concepts can

be applied to all economic and social aspects of society.

Over the last 10,000 years, our species has developed cultures that are the most destructive our planet has ever seen. Based on a short-sighted extractive process, the industrial capitalistic complex has left us in a terrible ecological mess. Our soils are depleted, water and air are polluted, and natural resources are peaking in their supply vs. demand. Limiting permaculture to simply a way of gardening would be like limiting the concept of energy production to only solar power. The concept is much broader, invoking all aspects of a healthy food system and a holistic way of thinking and living. Permaculture is truly about design, connectivity, and relationships.

In conjunction with Permaculture, Paul Stamets is one of the people we think is likely to change the social-ecological paradigm in the 21st century. Paul is the founder of Fungi Perfecti (fungi.com) and Host Defense Organic Mushrooms (hostdefense.com), and among his many accomplishments as a scientist and author, including working with the Centers for Disease Control and the National Institutes of Health, he has pioneered countless techniques in the field of edible and functional food mushroom cultivation. His groundbreaking work on fungi, toxic waste remediation, cancer, and most recently the Colony Collapse Disorder that is devastating bees around the world connects models of sustainable economics and holistic design with strategies for healing the Earth. As Paul puts it: "*We are now fully engaged in the 6th Major Extinction ('6 X') on planet Earth. Our biosphere is quickly changing, eroding the life-support systems that have allowed humans to ascend. Unless we put into action policies and technologies that can cause a course correction in the very near future, species diversity will continue to plummet, with humans not only being the primary cause, but one of the victims. What can we do? Fungi, particularly mushrooms, offer some powerful, practical solutions, which can be put into practice now.*" The central premise of his research is that habitats have immune systems, just like people, and mushrooms are the cellular bridges between the two. Our close evolutionary relationship to fungi can be the basis for novel pairings that lead to greater sustainability and immune enhancement. The concept of mycelium as "nature's internet" is a transcendent metaphor that ties together all of the topics covered in this book. I urge you to check out Paul's amazing work, through his TED Talk, "How Mushrooms Can Help Save the World" (rated in the top 10 of all TED talks), his books, and his websites.

Biodynamic agriculture focuses on the biological systems of agriculture and advocates making soil amendments directly on the farm, rather than importing them. Permaculture design— much more than a style of gardening— assembles conceptual, material, and strategic components in patterns that provide mutually beneficial, regenerative, and secure places for all forms of life.

Totally new in this 14th edition of the *Sourcebook* is our extensive chapter on Urban Homesteading. Taking the concepts of homesteading deep into the urban environment can and will make a big difference to wide-ranging communities, and will assist in redesigning our cities on a new template based on nature's bounty and resilience.

Lower-carbon and more energy-efficient alternatives to fossil fuels for automotive transportation are finally here. Powering vehicles with electricity or biofuels is now realistic, affordable, and increasingly popular.

Totally new in this 14th edition of the *Sourcebook* is our extensive chapter on **Urban Homesteading**, a trend that has strongly emerged in just the last five years. Urban homesteading is happening in small and large cities across the country, with practitioners relearning heirloom skills that have been largely abandoned in our relentless decades-long march toward convenience. Urban homesteading values thrift and community self-reliance in our homes, while repudiating the cultural forces of speed, need, and greed. It's part of an emerging global movement working for change that is rooted in respect for indigenous peoples and their values, while seeking peace and reconciliation at every level of community. Urban homesteading provides an opportunity to rewrite the story of our relationship to the Earth in the places where most of us live, and allows the possibility of remaking culture around an ethic of care and stewardship for our home base. Taking the concepts of homesteading deep into the urban environment can and will make a big difference to wide-ranging communities, and will assist in redesigning our cities on a new template based on nature's bounty and resilience.

All the systems that sustain us—food, water, shelter, medicine, family, and community—are at risk from the ongoing disintegration of life brought about by global capitalism's disrespect for natural limits. It's time for us to redesign our cities around an ethic of care and remake local systems on the model of the Earth itself—adaptive, lush with diversity, and fertile with possibility. Urban homesteading offers urban folks a strategy for maximizing interdependence, community resilience, and a sense of sufficiency in living locally. Practically speaking, our Urban Homesteading chapter gives you the tools you'll need to live a sustainable urban lifestyle that includes gardening, seed banking, composting, mushroom cultivation, orchards, and energy-efficient green homes. It also will inspire you to integrate animals such as chickens, rabbits, ducks, bees, quail, and even goats into your urban homestead. Further, it delves into harvesting and drying techniques, fermentation, rainwater catchment, greywater systems, recycling and upcycling, and even "humanure."

Chapter 13, **Sustainable Transportation,** tackles another issue of momentous importance. Driving is probably the aspect of our personal carbon emissions that stares us in the face most obviously and consistently, and often seems the most daunting to change. But in recent years, we've witnessed a sea change in transportation technology. In America personal mobility is practically considered a civil right. The key conceptual shift is to understand that transportation should be about access to people and goods, rather than freedom of mobility. Urban design and transportation policy should favor people and healthy communities over motorized vehicles. However, with more than 300 million gas-guzzling cars and trucks filling roadways in the US, it's clear we need to transform the core concepts behind providing fuel and systems of fuel production while these vehicles still rule our streets. Lower-carbon and more energy-efficient alternatives to fossil fuels for automotive transportation are finally here. Powering vehicles with electricity or biofuels is now realistic, affordable, and increasingly popular.

In this chapter, we've asked long-time Real Goods associates and experts in their respective fields—David Blume (biofuels) and Steve Heckeroth (electric vehicles)—to update the state of affairs with available alternatives. They have very different perspectives about the relative merits of biofuels and electricity as the best choice for addressing climate change in the short term and building a world of sustainable transportation in the longer term—and they're both skeptical about the promise and utility of fuel cells. We've let David and Steve each have his say and will let you, the reader and decision maker in your own life, assess the options that might work for you. We've had several alternative fuel "smackdowns" between David and Steve at our annual SolFest celebrations, so we thought it would be appropriate to air it out here in the *Sourcebook* as well. *Any* shift away from fossil fuels that results in a reduction of greenhouse gas emissions is positive. Ultimately, building sustainable transportation systems will require bold public policy decisions, investment in the infrastructure, and incentives that encourage and assist consumers to kick the fossil fuel habit.

The 14th edition of the *Solar Living Sourcebook* comes to a close with a chapter on a concept whose time has come. Aptly subtitled "The Ultimate Back-to-the-Land Movement," **Natural Burial** is a fitting conclusion to an environmentally responsible life and a righteous send-off to an afterlife, no matter what your beliefs. Think about what's buried along with our loved ones in American cemeteries every year: more than 800,000 pounds of embalming fluid, over 180 million pounds of steel, more than 5 million pounds of copper and bronze, and over 30 million board-feet of wood. Contrast this to the UK, where there's a burgeoning new movement in which people are burying loved ones in biodegradable containers—without toxic embalming fluid or synthetics—and returning bodies to the Earth to compost into soil nutrients with a forest of trees marking the spot. We must question the waste management behavior of our society and the wisdom of leaving toxic burial chemicals and other synthetic substances in the ground (and the atmosphere as a result of cremation) for future generations to clean up. As 80 million American baby boomers cross the finish line over the next 25 years (myself included!), the natural burial movement will gain momentum. Because many of us will be "dying to do the right thing," it makes sense to conclude our *Sourcebook* with this topic.

But knowledge is even more powerful than death, so the *Sourcebook* once again wraps up with our comprehensive **Sustainable Living Library**. This final chapter highlights many of our all-time best sellers, mixed in with the best new book offerings for those who want to learn still more about living right. We're proud to offer even more resources to help fulfill your dreams.

When we opened our first Real Goods store in Willits, California, in 1978, our mission was to demonstrate and provide renewable energy alternatives—and it still is. After 37 years, we are better positioned than ever to help you transform your lifestyle in healthy, regenerative, and fulfilling directions, whether your goal is to buy land and build a totally self-sufficient solar home, or to reduce your carbon footprint with energy efficiency, or dabble in urban homesteading with a few chickens and a biodynamic garden, or figure out the best alternative fuel for your next car. Over the years, we've assembled an unbeatable team of renewable energy experts with hundreds of years of combined experience in living the sustainable lifestyle described in this book. Our Real Goods residential and commercial solar divisions (RGS Energy) specialize in renewable energy design and installation, often made possible with no money down and an electric bill that is far below what you currently pay. Our Real Goods eCommerce division (realgoods.com) is online and tree-free, featuring the latest products for energy conservation, healthy living, renewable energy, and environmental education, as well as the most comprehensive sustainable living library on the planet.

We are headquartered at the Real Goods Solar Living Center in Hopland, California, our 12-acre permaculture oasis where products, ideas, and concepts come alive every day—not only in the interactive displays onsite but in the 200,000 people who annually visit (more than four million visitors since our opening 20 years ago). The Solar Living Center is operated by the nonprofit Solar Living Institute (SLI, solarliving.org), which nurtures and provides stewardship to the site while offering classes and even some professional accreditation on renewable energy, green building, permaculture, urban homesteading, and other sustainable living topics. The SLI's mission is to provide inspirational, environmental education. If you haven't visited northern California's #1 tourist attraction, we invite you to see the reality of sustainability. The Solar Living Center is, of course, 100% solar powered.

I've been a passionate adventurer in the solar industry and the sustainability movement my whole life. I try hard to walk my talk. My wife, Nantzy, and I live in an off-the-grid home (see page 70) built of recycled and green materials, powered by solar (passive and active) and hydroelectric energy, with gorgeous biodynamic gardens and fruit orchards that provide most of our food, a 15-acre biodynamic olive orchard, an 8-acre biodynamic vineyard, and a dozen beehives. I'm fortunate to benefit from the fruits of all our collective labors. As the solar industry continues to grow and mature, and as our cultural consciousness evolves, I remain hopeful that, once and for all, we will get things right in our homes, communities, country, and on our planet. Instead of forever being blamed for the excesses that put our planet on the destructive path, perhaps we can be viewed as the generation that rose above comfort and decadence to turn things around. We are living on borrowed time, but we have a chance, maybe our last, to embrace a vision of a fulfilling and sustainable future.

For the Earth,

John Schaeffer
Founder, Real Goods
and Solar Living Institute

"The Ultimate Back-to-the-Land Movement," Natural Burial is a fitting conclusion to an environmentally responsible life and a righteous send-off to an afterlife, no matter what your beliefs.

The Solar Living Center is operated by the nonprofit Solar Living Institute (solarliving.org), which nurtures and provides stewardship to the site while offering classes and even some professional accreditation on renewable energy, green building, permaculture, urban homesteading, and other sustainable living topics. The SLI's mission is to provide inspirational, environmental education.

CHAPTER 1

Relocalization

A Strategic Response to Peak Oil and Climate Change

As we approach the middle of the second decade of the 21st century, peak oil and global climate change continue to loom as the two most critical issues of our time. The end of the age of cheap and abundant fossil fuels on one hand and the unprecedented disruption of long-stable climatological and biological systems on the other are unquestionably altering life on Earth as we know it. This is not an ideological statement. It is grounded in the overwhelming weight of scientific fact and the laws of physics and ecology as we understand them. Even the last few remaining scientist skeptics—even the last holdouts among the oil and gas companies themselves!—now grudgingly acknowledge the reality of these earthshaking, interlocking trends. We begin the book by exploring some promising strategies for grappling with these profound issues. Many thanks to Daniel Lerch of the Post-Carbon Institute, who tackled the revision of this chapter.

The greatest hope rests in the ability to honestly accept the reality of a situation and then make the best of it.

People may be scared or shocked by predictions of ensuing environmental and social chaos driven by the end of cheap fossil fuels and the decline of the planet's ecological systems. But while awareness and concern about these issues are growing, many people still remain indifferent. How individuals respond emotionally to facts and deductions is important, but if they are unable and unwilling to accept what is true because it makes them feel bad, positive change is not possible. The greatest hope rests in the ability to honestly accept the reality of a situation and then make the best of it.

The world officially woke up to the looming challenge of climate change at the 1992 Rio Earth Summit. Then in late 1999 at the World Trade Organization (WTO) protests in Seattle, the world woke up to another looming (and controversial) challenge—economic globalization. Both trends were driven in large part by fossil fuels: climate change, by overconsumption of coal and oil; globalization, by the dropping cost of transportation and manufacturing enabled by abundant and affordable oil, coal, and natural gas. Suddenly, people who had been separately concerned about the environment, society's oil addiction, and the concentration of economic power faced an interconnected triple threat of global energy, economic, and environmental crises.

The period around the turn of the 21st cen-

tury proved to be a crucible for new approaches to these vexing challenges. Influential publications exploring the potential of local economies and local action had begun to appear.[1] The November 1999 WTO protests that brought together "Teamsters and Turtles" (labor activists and environmental activists) were quickly followed by Y2K, which prompted not only scattered doomsday panic but also real community concerns around local infrastructure and provisioning—ideas that, publicly, hadn't been much discussed since the oil crises of the 1970s. A few months later, the 30th anniversary of Earth Day in April 2000 turned into a rallying point for environmentalists eager to breathe new life into the second

One aspect of relocalization is buying locally grown produce, such as the fruits, vegetables, and other products available at Rosaly's Garden, an organic farm in Peterborough, New Hampshire.

Photo © Ann Card

half of the "think globally, act locally" mantra. In the summer and fall, Ralph Nader's passionately supported but ultimately doomed candidacy for President on the Green Party ticket gave an unexpected voice to voters disenchanted with the pro-corporate and pro-globalization policies of both the Democratic and Republican parties. And finally, the Supreme Court's awarding of the presidency to George W. Bush in December 2000 firmly closed the door on many people's hopes for progress on the world's sustainability crisis from the US federal government.

Out of those years emerged a scattered but quickly growing grassroots movement of people and organizations focused on local sustainability, some of whom started using the terms "localization" or "relocalization" to describe what they were doing. Where a decade earlier the call was for individuals to make small changes that collectively could impact global issues (see, for example, the famous 1989 book *50 Simple Things You Can Do to Save the Earth*), this new activism was decidedly community focused. Groups organized community gardens, clamored for bicycle lanes, set up car-sharing clubs, launched local currencies, and pushed for their local governments to adopt climate action plans. By 2004, with the release of the movie *The End of Suburbia* and James Howard Kunstler's book *The Long Emergency*, concern about peak oil was added to the mix. The movement started developing a national and even international identity with the launch of the Relocalization Network, the predominant pre-Facebook online meeting place supporting relocalization initiatives and ideas.[2]

Of course, the movement drew from a rich history of activism related to environmental and social concerns. "Relocalization" may have been a new term, but the concept and the activities it encompassed had deep roots. Its precursors include: thinkers like E. F. Schumacher, Ted Trainer, Garrett Hardin, and Wendell Berry;[3] social trends like the conservation movement, the back-to-the-land movement, the voluntary simplicity movement, and the slow food movement; practices like organic gardening, biodynamics,

placemaking, natural building, and permaculture; concepts like ecological footprint, import substitution, new urbanism, and ecocities; and centuries-old American traditions of individual and community self-sufficiency. In general, the common themes included the decentralization of political and economic structures; lower material consumption and pollution; a focus on the quality of relationships, culture, and the environment as sources of fulfillment; and downscaling of infrastructure development.

The movement has since grown and evolved in myriad ways. The Transition Towns concept, developed in the British Isles in the mid-2000s, gave rise to the global Transition Network; in 2009, the Relocalization Network was folded into the Transition movement, and in the United States is now coordinated by Transition US. Similarly minded groups also formed and spread (and persisted, or faded away), including peak oil awareness meet ups, Resilience Circles, and Local Living Economy groups.[4] Crafts and skills related to relocalization, like urban farming and DIY (do-it-yourself), have been embraced by popular culture. The number of people—and books, conferences, websites, videos, etc.—involved in relocalization activities has truly grown exponentially.

Before describing some of the details of relocalization, let's examine its basic premises. We believe that these premises are sound, being grounded in good science and common sense. By contrast, the assumptions underlying most of the economic and social models that have led to our current environmental and resource predicaments are essentially unsound rationalizations to justify short-term, often individual, interests. Our society's obsession with growth and gain have blinded us to the real common good, the needs of future generations, and the welfare of nonhuman life on the planet. Changing these paradigms and reorienting the trajectory of our society is a major undertaking, but all of us—as individuals and working together collectively—*do* have the power to make a positive contribution.

Ecological Economics

During the era of cheap energy, the study of economics became divorced from an understanding of how human systems are connected to ecological systems. Not surprisingly, the nearly free energy available from fossil fuels, and the rapid technological advances they fostered, made people in modern industrialized societies believe they were no longer constrained by tangibles like food, energy, water, and the weather. But the hubris of our recent past is being revealed, and many are searching for a more honest and realistic reckoning of humanity's place on Earth.

A helpful place to look is the discipline called ecological economics.[5] A conceptual model based on ecological economics is useful both to understand the current economic system and its vulnerabilities and to guide the development of a sustainable alternative.

Mainstream economic thinking usually distorts or fails to fully understand the fundamental interconnectedness of "the economy" and "the environment." Only in recent decades have economists begun to consider the environmental or ecological dimensions of human productive activity. But even when economists do take account of these relationships, their formulations are typically partial or misguided. For example, wealthy and environmentally responsible countries are sometimes touted as examples of how economic growth and stewardship of the planet go hand in hand. But while local measures of air quality, forest cover, and water cleanliness may be high, the raw materials and goods that the wealthy countries consume still have an environmental impact—in the poorer countries where those materials are extracted and those goods are manufactured. The damage—in addition to the jobs—has simply been outsourced.[6]

In the ecological economics model, the Human Economy is a subset of the Earth System, and therefore the *scale* of the human economy is ultimately limited. The human economy depends upon the *throughput* of materials from and back into the Earth system. Just pick up any trinket in your possession and ask, What is it made of? Where did these materials come from? How much energy was used and what happens to the waste products?[7] Limits to the size of the human economy are determined by three related factors: 1) the capacity for the Earth system to supply inputs to the human economy (sources), 2) the capacity of the Earth system to tolerate and process wastes from the human economy (sinks), and 3) the negative impacts on the human economy and the resources it relies on (feedbacks) caused by too much pollution.

For example, mining coal makes available a "source" of energy for industry that produces pollution, including sulfur dioxide that causes acid

Ecological Economics

Earth Systems

Human Economy

Population · Industry

Sources: raw material inputs

Sinks: waste stream

Feedbacks

The ecological economics model of the relationship between the human economy and the Earth system highlights the importance of source, sinks, feedbacks, and scale.

rain. Too much acid rain degrades built infrastructure and overwhelms the capacity of natural "sinks," such as forests, killing them or slowing their growth. The loss of ecosystems also creates new costs to society for ecological services that were previously accomplished "free of charge" through ecological processes. Clean air and water, stable climates, and species interactions that moderate outbreaks of disease are all compromised when damage is done to ecosystems. The human economy then invests in expensive technologies to try and compensate for this damage, such as pollution control devices, flood control walls and canals, pesticides, medicines, and more.

The current human economy is clearly unsustainable because it relies heavily on nonrenewable raw material sources. These are by definition finite, and using them produces tremendous pollution that leads to many negative "feedbacks" that impair ecosystems and disrupt climate. A sustainable economy would need to run on the income from solar energy and not degrade ecosystems through the buildup of wastes or the mining of nutrients.

Relocalization is based on an ethic of protecting the Earth system and its "natural capital," knowing that despite human cleverness, our well-being is fundamentally derived from the ecological and geological richness of Earth.

> A sustainable economy would need to run on the income from solar energy and not degrade ecosystems through the buildup of wastes or the mining of nutrients.

Overshoot

If the scale of the human economy (the inner circle within the ecological economics model) is too large relative to the Earth system, the human economy is in a state of *overshoot*. This means that the environmental load of humanity on the planet is greater than the long-term ability of the planet to support it. Overshoot means we are *above carrying capacity*. This environmental load will eventually be reduced through declines in some combination of population, resource consumption, and pollution. Either we manage to reduce our environmental load, or resource constraints and pollution will limit it for us—with unpleasant and potentially catastrophic consequences.[8]

The concept of overshoot can be confusing. You may ask, How can a population go beyond the carrying capacity of the environment to support it? Won't a population simply increase until it reaches carrying capacity, and then stabilize? Isn't the human population projected to stabilize in this century? Sophisticated modeling of population, resource, and consumption dynamics provides answers to these questions that persuasively suggest the reality of overshoot.

Population overshoot may happen for several different reasons: 1) resource windfall and drawdown, 2) release from negative species interactions, 3) demographic momentum, and 4) fluctuating carrying capacity. These mechanisms of overshoot are not exclusive, and in fact they can feed positively on one another. Here is an example of how these mechanisms have interacted in modern human history.

In the middle of the 19th century, people discovered a dense and versatile energy source in fossil fuels, especially petroleum. The use of fossil energy freed up other resources, such as land and labor. Without the need to feed draft animals to power equipment, more land was available to grow food for humans (i.e., resource windfall and drawdown). With fossil fuel-powered equipment, fewer humans were needed for manual labor, enabling extended educational opportunities and a

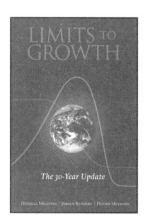

The 30-year updated edition of *Limits to Growth*.

shift of resources into fields such as public health and medicine. Increased societal attention to health and medicine, and corresponding technologies like vaccines, antibiotics, and sanitation, resulted in increased human life expectancy (i.e., reduced negative species interactions). A rapid increase in the human population increased the number of fertile women of childbearing age, leading to an even larger population (i.e., demographic momentum). As this population became very large, it began to impact the natural world around it substantially. Toxic emissions built up that harmed the basic life-support systems humans depend on, eventually making it more and more difficult to provide essentials, such as food (i.e., fluctuating carrying capacity).

Experts in the field of human demography project that the human population will stabilize around the middle of the 21st century.[9] Most people accept this analysis without knowing the underlying assumptions. Unfortunately, most studies of human population are akin to most studies of the human economy. The broader environment is not factored in to models of growth. If you have ever asked yourself, How are we going to feed 9 billion people when the soils are eroding and the aquifers are being depleted and the climate is changing and the deserts are expanding and oil and natural gas supplies are dwindling?—then you have stumbled upon this disconnect between most human population models and the physical world. Biologists studying any population would include those environmental factors in their models, whereas human demographers do not.

However, models do exist that contextualize the human population and our well-being within a dynamic study of resource availability, pollution levels, and even climate change and the fate of ecosystems. The classic example is the World3 model developed by the authors of *The Limits to Growth*, where the baseline scenario shows human population declining after 2020.[10] Another model is GUMBO, from the University of Vermont's Gund Institute of Ecological Economics.[11] These models are not perfect, but they at least begin with the right premises and tell us what aspects of human civilization are likely pushing the boundaries of, or already exceeding, the physical and ecological capacities of Earth.

Relocalization starts from the premise that the world is a finite place and that humanity is in a state of overshoot. Perpetual growth of the economy and the population is neither possible nor desirable. It is wise to start planning now for a world with less available energy, not more.

Human demographic models of population show a plateau this century (solid line is approximate historic and demographic projected), whereas systems models show a decline (gray line). The difference exists because human demographic models do not include negative feedbacks from either resource scarcity or pollution, whereas systems models do.

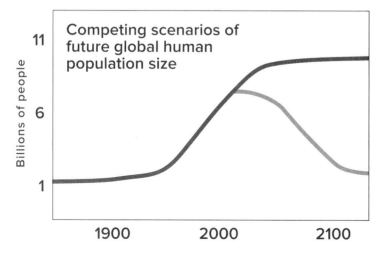

Competing scenarios of future global human population size

Billions of people

11

6

1

1900   2000   2100

Peak Oil and Implications for a Transportation-dependent Economy

To a great extent, the Relocalization movement was sparked by concerns about "peak oil."[12] The concept of peak oil is simply that we can expect global oil production to follow the same pattern that's been observed countless times in individual oil fields and regions: Production climbs, peaks, and falls in a rough bell curve. Oil production has this pattern because the largest and easiest oil deposits tend to be discovered and tapped first. As the easy oil runs out, a point of all-time maximum production is reached, followed by perpetual decline as production shifts to the smaller, less profitable deposits.

With any particular oil field or region, post-peak declines are simply made up for by other fields or regions and oil continues flowing to the global market. The worry with global peak oil, therefore, was that there is no other place to turn for additional cheap oil: Declining global supply would force the price of oil to start increasing, with potentially disastrous results for the world economy. Predictions in the mid-2000s for the future of oil were all over the map: The most pessimistic "peakists" warned of economic crashes and social chaos, while the most optimistic "cornucopians" claimed that markets and innovation would keep cheap oil flowing no matter what.

In reality, global production of cheap oil hit a plateau in 2005—and the nearly ten years that followed have been a whirlwind of economic, technological, and social changes that both fulfilled and disproved parts of just about every prediction. Oil prices indeed rose: from the old "normal" of $10–$40 per barrel to a new "normal" of $80–$120 per barrel. And some economic and social pain has been quite severe: The United States experienced its worst economy since the Great Depression, some European countries are in or near financial collapse, and related economic shocks have played a major role in the latest wave of revolutionary unrest to sweep the Middle East. On the other hand, overall global oil production has managed to keep growing, thanks indeed to technological developments but more so to the fact that a lot of previously unprofitable oil reserves (like deepwater and tight oil) are now quite profitable at prices over $80 a barrel.

Whether the world has technically experienced "peak oil" yet or not is a matter of interpretation that is largely irrelevant.[13] The economic and social unrest predicted by peak oil have already arrived.

In the ecological economics model, peak oil is a "source" issue. Several source problems face the human economy, including peak natural gas and peak water.[14] Greater expansion of the human economy requires greater inputs, and aside from the ecosystem services provided by nature, oil is probably the single most important economic resource on the planet. Oil is critical for at least two reasons: energy density and versatility.

The energy output of a single person doing manual labor, averaged over a period of days, is equivalent to about 200–300 British Thermal Units (BTUs) per hour. A single gallon of gasoline contains about 150,000 BTUs of potential energy, roughly equivalent to 500–750 hours of hard human labor.[15] The energy density of oil

Whether the world has technically experienced "peak oil" yet or not is a matter of interpretation that is largely irrelevant. The economic and social unrest predicted by peak oil have already arrived.

has not simply permitted a life of leisure and travel for those with access to it—it has in fact greatly expanded the short-term carrying capacity of the human population. By harnessing the energy of oil (and other fossil fuels), our species has been able to outcompete others for space and resources. The expansion of industrial agriculture and "green revolution" technologies are based on oil and natural gas feedstocks and energy. Construction of large dams, water diversion systems, and pumps for groundwater and water delivery to fields and cities depends upon plentiful fuel. Land, water, and other resources that in the past had been available to a diversity of species are being funneled toward the appetite of only one species—hence the biodiversity crisis.

Oil is versatile because it is a liquid, making it easier to extract and transport than coal and natural gas. Oil is more readily available as a fuel for a global market because it can be put into pipelines and tankers without requiring special treatment. Natural gas, by contrast, needs to be cooled and pressurized for tanker travel, and coal needs to be pulverized into slurry to be piped or put onto freight cars or barges for long-distance transport.

Because oil can be delivered anywhere so efficiently, modern transportation systems have become reliant on it. Some buses and cars use natural gas. Some trains run on electricity. But the vast majority of energy used for transportation worldwide—more than 95%—comes from oil as gasoline, diesel, or kerosene (jet fuel).[16] Consequently, modern economies are extremely vulnerable to shortages in transportation fuels. The relative stability of the oil market in previous decades led to the development of "just in time" delivery of products and to commercial linkages across the globe. Local and regional warehouses are uncommon now, with stores and businesses relying on frequent shipments to maintain a low overhead. Before the era of cheap transportation, each town and city had a full complement of craftspeople who relied on each other. Today, businesses are connected through vast transportation networks, with a manufacturing company in California, for example, relying on components shipped in from Asia and Europe.

The food economy is perhaps the finest example of the insecurity that is now bred into normal societal infrastructures. Markets selling food are typically restocked daily with only a few days' supply available in the store. This fact leads many people concerned about peak oil to reason: no fuel, no trucks; no trucks, no food. The shifts in agricultural practices that have occurred in the past 30 or 40 years make it difficult to quickly switch to a less transportation-intensive food system. Many agricultural regions are overly specialized to serve global markets. For example, a place where granaries, dairies, vegetable farms, and ranches coexisted 50 years ago is now dominated by premium wine grapes.[17]

These developments have been possible only because cheap oil has allowed us to overcome the limitations of local ecologies. And because oil possesses a unique combination of attributes, finding a suitable and equally effective substitute is no easy task—and perhaps is impossible.[18]

All proposed "substitutes" for cheap oil appear to fail the test of Energy Returned on Energy Invested (EROEI).[19] For an energy source to be useful to society, it must deliver more energy than it takes to find, harvest, and distribute it. Our economies have become addicted to energy sources with EROEIs of from 100:1 to 20:1, whereas biofuels, tar sands, and many renewable energy technologies range from about 10:1 to 1:1 or less. If a fuel has an EROEI of 1:1, it is essentially useless, because as much energy goes into producing the fuel as the fuel delivers. A complex and sustainable society will probably require substantial EROEI ratios, such as 5:1 or greater. Energy policies need to be devised based on sound EROEI analyses, which are currently difficult to find.

In the US, a high-EROEI energy source permits about 1% of the population to feed the other 99%. In places without access to fossil fuels, such as Afghanistan, more than 90% of the working population is engaged in growing food. Agriculture is, in essence, a means of capturing solar energy through investment in planting, maintenance, and harvesting. While the Afghan agricultural system looks inefficient from a labor point of view, it is actually far more efficient from an EROEI perspective than US agriculture. The extensive use of fossil fuels in industrialized food systems makes them energy sinks. Highly industrialized agriculture requires about 10 times more energy to grow, harvest, process, and distribute the food than is contained in the food itself—an EROEI of 1:10. Such a system is clearly unsustainable and begs for relocalization. With investment today in the right kind of equipment and training, industrialized nations could transition away from fossil fuel-dependent agriculture and grow food with far less than 90% of their working population engaged in farming. According to some estimates, perhaps a third of the population would suffice.[20]

Highly industrialized agriculture requires about 10 times more energy to grow, harvest, process, and distribute the food than is contained in the food itself.

Climate Change and the Need to Eliminate Fossil Fuels

While peak oil is a source problem, climate change is a sink problem.

During the most recent ages of geologic history, Earth has cycled between ice ages and intervening warm periods. These cycles are primarily driven by orbital variations, both with respect to the angle of Earth's tilt toward the sun and the shape of Earth's orbit around the sun.[21] Carbon dioxide (CO_2) fluctuated as a result of how ecosystems responded to changes in Earth's temperature, and changes in temperature then amplified those ecosystem changes. In systems theory, which is the guiding paradigm for ecological economics and computer modeling, this process is known as a positive feedback loop.

Currently, CO_2 and other greenhouse gas concentrations are rising not because of orbital changes but from the use of fossil fuels. The pre-industrial level of CO_2 in Earth's atmosphere was 280 parts per million (ppm); now it's about 400 ppm. Consider that 100 ppm of CO_2 is what separated the Ice Age from the warm, stable climate of the past several thousand years, and that the corresponding temperature transition took about a thousand years. Today, global average temperatures are rising about 100 times faster than during transitions out of ice ages. In fact, the current rate of change in the chemistry of Earth's atmosphere and oceans is comparable to only a few previous mass-extinction episodes over the past several hundred million years that appear to be related to radical, rapid climate change.[22]

The rate of change is perhaps more important to the climate system and life on Earth than is the amount of change. A slow rate of change is akin to gently applying the brakes to stop at a light, while a fast rate of change is akin to hitting a brick wall. Both take the vehicle and a passenger from 60 to 0 miles per hour, only one is faster.

Nobody really knows what this means for the climate system, the pH of the oceans, the physiology of plant growth, and other planetary processes. Policymakers ask scientists how much pollution can be tolerated before "dangerous interference" occurs. Unfortunately, answering how much is too much is not possible, and in all probability, we have already passed some very dangerous thresholds that will become apparent only as the future unfolds.

There are many reasons why a precise answer to "how much is too much" is not possible. Consider that for any factor built into a model, scientists 1) work with what they know, 2) try to incorporate plausible ranges for what they don't know, and 3) obviously exclude what they don't know they don't know. Some would argue that because we can't be sure climate models are correct, we should do nothing. Would "do nothing" skeptics be as cavalier about uncertain dangers if the food their children ate had *possibly* been contaminated by a deadly poison? What you don't know can kill you. Given the stakes, many advocates of energy policies leading to a curtailment of greenhouse gas emissions take a precautionary

Although Americans make up less than 5% of the world's population, we produce about 16% of all greenhouse gas emissions directly, and a significant share indirectly through our consumption of goods manufactured in other countries.

Photo © jupiterimages, comstock.com

A systems view of how ecosystems and biological, hydrological, and climatological processes interact on a planetary scale is sensible. Therefore it is not a stretch to imagine that the changes being wrought by greenhouse gas emissions could, relatively suddenly, inflict even more massive damage to Earth than we are already seeing.

stance.[23] After all, if the US is so concerned about security that it is willing to spend over $650 billion dollars a year on the military, what is it worth to help secure our climate? Regardless, there is no longer any real scientific debate on the basic facts of climate change. A 2013 study of 21 years of peer-reviewed scientific papers on climate change found that over 97% of scientists endorsed the consensus view that climate change is caused by humans and is cause for serious concern.[24]

Climate systems are too complex to allow us to model many of the details of change. For example, models can't scale down to the future climate of a single town, which makes it difficult, perhaps, for local officials to understand the implications of global models. Nor can models usually identify critical thresholds in a complex system with much accuracy. Systems can remain remarkably stable over long periods under stress until something snaps, like a balloon expanding until it pops. The Earth System has been remarkably tolerant of the stresses it is under, but when something finally gives, it will probably be "loud." It is very possible that current models actually underestimate the true threats of climate change.

Those of us living in the United States have a special responsibility to deal with climate change: Although Americans make up less than 5% of the world's population, we produce about 16% of all greenhouse gas emissions directly, and a significant share indirectly through our consumption of goods manufactured in other countries. A constellation of interrelated climate change pressure points examined by Professor John Schellnhuber, former chief environmental advisor to the German government and currently director of the Potsdam Institute for Climate Impact Research, reveals both the complexity of the problem and the shockingly realistic possibility of dramatic environmental damage.[25] Schellnhuber has identified 12 global ecological "tipping points," weak links that could, under the impact

of global warming, trigger the catastrophic collapse of some of Earth's critical ecosystems. Because each of these phenomena operates on a planetary scale, disruption of one influences the behavior of others in chains of positive feedback loops. The dozen fragile systems and/or places in Schellnhuber's model are:

- Amazon Rain Forest
- North Atlantic Current
- Greenland Ice Sheet
- Ozone Hole
- Antarctic Circumpolar Current
- Sahara Desert
- Tibetan Plateau
- Asian Monsoon
- Methane Clathrates within Siberian permafrost and ocean sediments
- Salinity Valves in the oceans, especially the Mediterranean Sea
- El Niño
- West Antarctic Ice Sheet

The ways in which these tipping points interact with one another are too numerous and complicated to detail here, but a couple of examples will give a good idea of what's at stake. As the Greenland ice sheet melts, for example, it discharges massive volumes of freshwater into the sea where the vast oceanic river known as the North Atlantic Current delivers warmth to the European continent via a mechanism called thermohaline circulation (THC). That new freshwater dilutes the salt content of the North Atlantic's surface waters; and with enough freshwater, the THC could be disrupted, making parts of Europe considerably colder than they are now. As another example, global warming is expected to increase rainfall along the southern edge of the Sahara, which means more plants would grow and the desert would shrink. But windblown dust from the Sahara seeds the ocean with nutrients that support the phytoplankton population, which is the foundation of the entire oceanic food chain. More rain would mean less dust, and thus less food for the phytoplankton, and thus decreasing phytoplankton populations, and thus less food for fish.... You get the picture.

Of course, these and similar scenarios involving fragile environmental tipping points are somewhat hypothetical, especially when spun out to what some might see as worst-case scenarios. But the point is that the evidence that global warming and large-scale climate change are occurring is real, if not overwhelming, to those who will see it. And a systems view of how ecosystems and biological, hydrological, and climatological

processes interact on a planetary scale is sensible. Therefore it is not a stretch to imagine that the changes being wrought by greenhouse gas emissions could, relatively suddenly, inflict even more massive damage to Earth than we are already seeing. Indeed, it only makes sense to take Schellnhuber's warnings seriously.

And here's one more sobering consideration. Although climate models have limits, they also do an incredible job of accurately modeling the *past* climate. For example, when comparing images from weather satellites to the most advanced climate models, you can see how well models match the actual formation and movement of storm clouds around the globe.

One of the tests climate modelers perform to decide whether human-induced changes in the atmosphere are causing climate change is to run climate models for the 20th century *as if* we hadn't burned so much fossil fuel. The rise in global temperatures and the shifts in rainfall patterns seen during the 20th century can be accurately modeled only when fossil fuel-induced greenhouse gas emissions are included. Natural variations in solar radiation and the shape of Earth's orbit around the sun do not account for recent climate change. Climate change is our problem.

While we can't know future threats precisely, scientists agree that creating a carbon-cycle-neutral economy should be the dominant task occupying our minds. This is exactly what relocalization aims to do.

Relocalization: A Strategic Response to Overshoot

Economic and population growth were made possible by the synergies permitted by cheap energy. The limits of productivity in one locality could be overcome by importing something that was produced in excess elsewhere. A global economy emerged, propelled by an imperative that each place seek its comparative advantage and specialize in the marketplace. The spread of "free trade" agreements is further indication that most economists, policymakers, and political leaders see only the benefits of globalized commerce and ignore or minimize the long-term liabilities.

One particular flawed assumption behind globalization is especially glaring, i.e., that transportation costs will always be low, both in terms of fuel availability and the environmental problems associated with their use.[26] If that assumption is false—and certainly peak oil and climate change make it appear false—then localities should not be specializing to trade globally. Take the example of California wine country again. That place grows far more grapes than the local population can eat, but it lacks just about every other kind of food production in sufficient quantity. As long as the region can sell its wine to a global market and buy the other stuff people need, this situation seems reasonable. But a peak oil perspective reveals the region's vulnerability, and a climate change perspective calls this entire socioeconomic system irresponsible.

Relocalization advocates rebuilding more balanced local economies that emphasize securing basic needs. ***Local food, energy, and water systems are perhaps the most critical to build.***[27] The movement toward relocalizing food networks is perhaps the most advanced today. A book by Sandor Ellix Katz, *The Revolution Will Not Be Microwaved*, documented the growth of early relocalization initiatives that worked to restore traditional food production and distribution methods and revive local economies. (Katz's books on fermentation are available at realgoods.com.) Katz's analysis speaks directly to the issues we've been discussing and the promise of relocalization:

> Food-related political activism…seeks to revive local food production and exchange and to redevelop community food sovereignty. There is no sacrifice required for this agenda because, generally speaking, the

Relocalization advocates rebuilding more balanced local economies that emphasize securing basic needs.

food closest at hand is the freshest, most delicious, and most nutritious. This revolution will not be genetically engineered, pumped up with hormones, covered in pesticides, individually wrapped, or microwaved. This is a revolution of the everyday, and it's already happening. It's a practice more of us can build into our mundane daily realities and into a grassroots groundswell. This revolution is wholesome, nurturing, and sensual. This revolution reinvigorates local economies. This revolution rescues traditional foods that are in danger of extinction and revives skills that will enable people to survive the inevitable collapse of the unsustainable, globalized, industrial food system.[28]

In the absence of reliable trade partners, whether from peak oil, natural disaster, or political instability, a local or regional economy that at least takes care to produce the essentials will have a true comparative advantage. Relocalization will promote local and regional stability. Because it reduces the distances that goods travel between production and consumption, it will also significantly reduce pollution and greenhouse gas emissions. Ideally, relocalization is grounded in the principles of ecological economics and is developed around renewable energy inputs and cycling of nutrients. The more we can do, the greater the positive impact it will have on reducing the potential consequences of these impending crises.

> A local economy that takes care of its basic needs is also a very interesting place to be.

Approaching Social Change

The problems we face tempt many to drop out of society as much as possible and live a simple life, semi-isolated from the horrors "out there." However appealing this may be, global traumas will most likely catch up with everyone. Reversing course and implementing a complete overhaul of our collective lives requires massive cooperative action.

History shows some instances when societies have responded wisely and plenty of other instances when they didn't change in time.[29] There's no guarantee that relocalization will be successful, but we can hope that our work will improve

Members of the Peterborough, New Hampshire, community garden gather to share the bounty.

Photo © Ann Card

the odds. And the journey has many inherent benefits in any event.

When approaching others, it's useful to keep an old sales adage in mind: Sell the benefits, not the features. Truthfully, nobody really knows "how" to relocalize economies in any intimate detail. Many examples of the components of a local, sustainable economy can be found, but nowhere can we point to an example of a place where it has all been put together harmoniously. This is something we will have to learn how to do together. So don't get too hung up on the "features" of a local economy beyond some broad principles and working examples. In the meantime, sell the benefits to enroll people in the vision.[30]

The benefits of a local, sustainable economy would be extensive, beyond the environmental pluses already noted. Such an economy is more responsible, secure, and potentially more "free" in some ways. With greater self-reliance comes greater political autonomy and less vulnerability to instabilities elsewhere. A local economy that takes care of its basic needs is also a very interesting place to be. The diversity of goods, services, and skills required is much greater than what many in the US are accustomed to in their communities. Such a place to live would be attractive to all generations, as each individual would be more likely to find a role suited to his or her talents and interests. The stability of a locally focused economy enhances community bonds. Instead of a hypermobile society in which anonymity is prevalent, people would have the time to get to know each other and work together based on mutual understanding.[31] This social lubricant lowers what economists call transaction costs, which can be very important for getting work done together efficiently and responding cohesively during crises. It is hard for people nowadays to "love thy neighbor," when in many cases they wouldn't even recognize their neighbor.[32] (For an in-depth discussion of the urban homesteading movement, which seeks to achieve the goals outlined in this paragraph, see chapter 12.)

Beyond the personal level, the challenge is to engage our neighbors and communities in the project of shifting public investment strategies and creating laws that can lead to significant behavioral changes at the societal level. Changing a behavior required to get by on a daily basis, such as driving a car, requires that the built environment make alternatives relatively convenient. That means that tax dollars going toward highway projects and airport expansion need to go instead toward a locally scaled, non-fossil fuel-dependent transportation system. (For more information about sustainable transportation, see chapter 13.)

Activists will often hear this response: "Sounds great, but we have no money." In the United States, however, vast material resources are devoted to expanding freeways, building cities in deserts, generating electrical power using coal and natural gas, and producing military hardware. In truth, shifting public investment resources is mainly a matter of priorities. But that requires involvement in one of the most difficult social environments of all: politics. Bumping up against institutional norms may sound daunting, but the good news is that reality is conflicting with dominant belief systems. This conflict sets up an opportunity to help the disillusioned or confused by offering a coherent explanation of what is happening, and pathways to realign their thinking with the new realities.

Being politically active doesn't require running for office or joining a political party, though those are fine options. Because relatively few Americans are actually civically engaged, a small group of well-organized and thoughtful people can wield great power. After all, this is what professional lobbyists do.

The message that we need to develop local economies is an easy sell because most people readily understand the advantages of greater self-reliance and strengthened communities. Job loss trends related to the dissipation of local manufacturing and agriculture and the disruptions brought on by the economic crisis of 2007–2009 are sore spots that lead many to question the wisdom of globalization. In this context, people are receptive to cogent arguments that reveal the insecurity and environmental problems of our dependence on fossil fuels and other imports, and they are more willing to see the vision of an interesting and beautiful alternative.

Shifting established patterns of behavior and generating the same sense of urgency many activists feel is more difficult. Two parallel strategies can help us to move forward.

The first is a strategy of earnest but careful dialog. Developing a relationship with leaders in your community based on mutual respect and trust is critical. Establishing such personal connections will improve the likelihood that decisions will be made from the perspectives and priorities of relocalization. However, be aware that because the message of overshoot challenges cherished assumptions and intrudes on many people's comfort zones, it can be a somewhat awkward dance getting to know someone and communicating your purpose.

Understanding the art of conversation is a useful skill when navigating interpersonal relationships. Being able to listen carefully to understand others' perspectives, and finding avenues of shared concern, is a great way to start. Build from an initial foundation of respect. Pointing fingers and demanding that people "get on board" is less effective than asking friends to assist in dealing with a mutual problem. Strive for understanding and agreement, but learn how to live with differences and to tolerate tension and conflict. We don't always get along with those we care most about, but we still try to stick together.

The second strategy is modeling through tangible examples the kinds of changes being advocated. Those of us "ahead of the curve," so to speak, will need to pull up our sleeves and create some of the alternatives we're talking about. Examples abound, including local monetary systems, renewable energy devices, community gardens and farms, farmer's markets, bike clubs, and small businesses or nonprofit organizations helping people do all of the above and more. Ideally, as these activities demonstrate success, gain credibility, and become more cost-effective, more and more people will rally to support them.[33]

If you want to work on peak oil, climate change, or relocalization issues in your own community, there's probably already an existing group you can join: you'll find thousands listed at resilience.org/groups. And if there's nothing that appeals to you, start your own group! Most important, don't be paralyzed by indecision and fear. Doing nothing is a capitulation to disaster, while doing something is empowering and potentially transformational.

Those of us "ahead of the curve," so to speak, will need to pull up our sleeves and create some of the alternatives we're talking about.

Pioneering Solar Down on the Farm

Left: The Decaters plow their fields with "original" solar farming technology—draft horses that eat grass fed by the sun. **Above:** Tended by apprentices, compost piles at Live Power Community Farm recycle onsite waste. **Below:** John Schaeffer visits with Steve and Gloria Decater down on the farm.

LIVE POWER COMMUNITY FARM has been generating its electricity via some form of solar energy for decades. Now the farm is set to add another leg to its legendary system.

Talk about your renewable energy pioneers. For the past 40 years, Stephen and Gloria Decater have been powering Live Power Community Farm, their 40-acre Demeter-certified biodynamic operation in Covelo, California, with some form of solar energy.

Whether it be from the draft horses that plow their fields ("Horses are eating grass and forage, and their energy source is the sun, so, in effect, horses are solar powered," Stephen says) or from photovoltaics (PV), the Decaters have been committed to solar energy for nearly as long as the term *solar energy* has been around.

"Our goal has always been to produce food from solar energy rather than fossil fuel. We designed the farm on that principle," Stephen says.

Right now the farm has completed its third leg of solar power development. In addition to their draft horses, the Decaters currently have a 28 kW PV system that provides much of the farm's power and water pumping capacity. For help designing

and installing the third phase, they contacted Real Goods. "The system is getting pretty involved," Stephen says. "The people at Real Goods were extremely knowledgeable in helping us understand the next leg of it."

Powering Community Supported Agriculture

If there's a quintessential statement to be made about reverence for the Earth, it's being made every day at Live Power Community Farm. The farm was one of the first Community Supported Agriculture (CSA) projects in California. CSAs now number between 6,000 and 6,500 in the United States and are based on an economic paradigm vastly different from traditional market-based systems.

Essentially, CSA is a social and economic contract between growers and consumers. "For us, CSA means 100% community based. Members support our annual operating budget, and all the food we grow is distributed to members," Gloria says. "It's an associative economy rather than a market economy. The association is between growers, consumers, and the Earth."

Tech specs	
Solar system size	28 kW after new panels installed
Est. average annual savings	$1,600+
Solar panels	32 Siemens SR 100; 72 Shell SP 140; 30 Sharp 165 W modules; 36 Trina TSM-305PD14 modules
Inverters	SMA SB300 for Siemens panels; SMA SB4000 and SMA SB6000 for Shell panels; SMA SB6000U 6 kW inverter for Sharp panels; SMA SBT1000LUS-12 for Trina panels

CHAPTER 2

Land and Shelter

Designing and Building Your Green Dream Home

IT ALL BEGINS AND ENDS WITH LAND. The choices we make about where we choose to settle, how we treat the land upon which we live, and how we create our sheltering places are absolutely fundamental to any concept of sustainability or regeneration.

Human shelter, of course, is rooted in the land. We build our homes upon the thin layer of topsoil that nourishes the plants that make all life on Earth possible. And all the materials we use to build our homes come from the land. Wood, metals, glass, and even synthetic materials such as plastics are harvested from the world's forests and fields or deep within the Earth's crust.

Because our homes take up space that other organisms once occupied, because the resources needed to build and maintain our shelters come from the Earth in vast quantities, and because there are so many of our kind on the planet, few human activities so profoundly influence Earth and its ecosystems as creating shelter. Through intelligent placement and smart design, we can create shelter—even entire communities—that honor and respect the Earth and its ecosystems. In order to curb global climate change, our goal is to create carbon-neutral homes and lifestyles, and the choice of where and how to create shelter is an important one.

> Few human activities so profoundly influence Earth and its ecosystems as creating shelter.

The original "Land and Shelter" chapter from a decade ago was authored by several in-house Real Goods staffers. For our 30th-anniversary edition, major new portions were contributed by sustainable design consultant Dan Chiras, and this latest 14th edition has been updated by Amanda Knowles and David Arkin of Arkin Tilt Architects.

The passive solar home of residential renewable energy and green building author and consultant Dan Chiras in Evergreen, Colorado.

Photo: Dan Chiras

The Impacts of Modern Building

Although humans have accomplished amazing feats and created astonishing civilizations, we mustn't lose track of the fact that, like all other species, we depend on nature for the resources that make our lives and our economies possible.

For years, the culture of consumption—predicated on convenience, comfort, and immediate personal gratification—has damaged the Earth's systems. The elements that make up our climate are complex, and the impact on them of the unsustainable methods used to produce, harvest, and maintain the materials modern culture demands is showing. Undeniable shifts in weather patterns are bringing strong storms to communities unprepared for them, and droughts to regions accustomed to plentiful rain.

These and other ecological catastrophes are turning our culture's attention to the importance of living a sustainable and regenerative life. We must understand the relationship between a healthy environment and strong societies and economies, and use societal and economic pressures to swing us toward a more sustainable civilization. This requires recognizing that we are an integral part of nature, not apart from it.

Although humans have accomplished amazing feats and created astonishing civilizations, we mustn't lose track of the fact that, like all other species, we depend on nature for the resources that make our lives and our economies possible. In short, nature is the source of all the goods and services we consume, and it is also the sink for all of our wastes. Intricately connected as we are, we must recognize that we cannot control nature. The concrete levees of New Orleans, engineered to maintain a single course for the Mississippi River, gave way in the face of swelling waters caused by Hurricane Katrina in 2005.

Water flooded into the city, depositing built-up pollutants and destroying neighborhoods in the process. In catastrophes like this, we witness how technologies fail when they cannot react to the complex and flowing systems they are meant to contain. Instead of designing and implementing more restrictive systems, we as a species must learn to cooperate with nature, and in doing so, rediscover how to live in balance on the Earth.

Our home planet is a resilient one, but not all of the systems within it are. Climates change, but today's more sudden shift will leave a mark on the species, ecologies, and landscapes on Earth. We must take responsibility for our part, and move forward with the goal of living sustainably.

Although most *Sourcebook* readers understand the fundamental incompatibility of modern human culture and sustaining a healthy environment, many only barely grasp the full potential of a response in the building sector—how profoundly we can change our future by building sustainably. Fortunately for all of us, many people are exploring innovative ways of building shelter, using techniques and materials that could, if widely used in the future, help steer society onto a sustainable path.

When most of us think about environmental damage that threatens our long-term future, one of the last things we think about is our homes. Our homes are supposed to be places of refuge, sanctuaries in a maddeningly hectic world. We think that greenhouse gases, pollution, and other significant environmental threats come not from

The Ninth Ward in New Orleans, four years after Hurricane Katrina hit land. Very few structures had been rebuilt, and city streets were not being maintained.

Amanda Knowles

All photos by Dan Chiras.

our homes but from factories, power plants, cars and trucks, smelters, corporate farms, and other activities. Unfortunately, shelter is anything but environmentally benign; in fact, shelter poses a major environmental threat. Environmental damage can arise from two aspects of shelter: construction and day-to-day living. Both require investments of energy and natural resources, both are responsible for outputs of waste, and both contribute significantly to local, regional, and global decline in environmental quality. Let's take a closer look at this bold allegation.

In 2013, Americans started construction on more than 924,000 new homes. Each house requires a significant amount of energy and materials. Consider, for example, the amount of wood that goes into a typical new house.

Wood is used to build the majority of new houses in America. It is a primary component of walls, floors, ceilings, and roofs. It is used to build decks, cabinets, shelves, fences, and furniture. In 2012, with the housing market in a slow recovery, the US still consumed about 79 million cubic meters of lumber. If cut into conventional 2 × 4s, that lumber would stretch for 9.5 million miles, circling the Earth over 380 times. Add to that the 32 million cubic feet of wood panel products consumed in 2012, and you start to understand the impact that the building industry has on the Earth's forests (fpl.fs.fed.us/documnts/fplrn/fpl_r n0328.pdf).

Besides the number of homes built, the size of homes is also pushing the demand for wood. New homes built in 2005 averaged 2,434 square feet, up from 2,200 square feet just five years earlier. After a drop-off during the economic downturn of 2008, the average house size has rebounded, returning to 2,500 square feet in 2012 (census.gov /construction/chars/highlights.html).

Although wood is a renewable resource and forests can be managed sustainably, humankind's insatiable demand for this material is leading to ever-greater loss of native forests. Deforestation can lead to serious soil erosion that damages nearby rivers, streams, and lakes. Deforestation also significantly decreases the planet's ability to

absorb the primary greenhouse gas, carbon dioxide (CO_2). The existing forests in the US alone remove approximately 1.5 billion pounds of CO_2 from the atmosphere, off-setting around 10% of the annual US production of the greenhouse gas (continuingeducation.construction.com/article .php?L=5&C=645&P=3).

Deforestation is also responsible for the loss of valuable wildlife habitat that, in turn, leads to species extinction. Nowhere are these problems more acute than in the tropical rain forests of the world. In this narrow band of land that straddles Earth's equator, forests continue to be leveled in mind-boggling fashion to provide wood and wood products for a fast-growing human population and the rapacious appetite of a continually expanding global economy. These and an assortment of additional impacts are expected to only worsen in the near and long-term future as this growth continues. Utilizing wood resources that bear the FSC (Forestry Stewardship Council) stamp of approval ensures that the wood one does use isn't exacerbating these negative impacts.

In addition to wood, house building consumes a huge amount of minerals and metals, among them concrete, granite, other kinds of stone, aluminum, copper, and steel. Supplying the demand for these materials requires extensive mining. Construction also requires a huge input of energy—electricity, oil, natural gas, and gasoline—to manufacture materials, assemble products, transport products from the factory to the building

Wood is a renewable yet endangered resource. Huge amounts of wood are used to frame walls, floors, and roofs in new construction and renovation.

In 2012, with the housing market in a slow recovery, the US still consumed about 79 million cubic meters of lumber. If cut into conventional 2×4s, that lumber would stretch for 9.5 million miles, circling the Earth over 380 times.

Courtesy of Re-New Wood, Inc.

site, and, finally, to assemble them into a sturdy, weatherproof structure. This embodied energy in each and every piece of a building is often forgotten once it is in place and serving a purpose—and then forgotten again when its useful life is over, and it is thrown away.

Toxic chemicals found in numerous building materials can poison the occupants of a home. Certain manufactured wood products such as plywood, particleboard, and oriented strand board (OSB, also known as wafer board or chip board) contain formaldehyde resins that are used to bind the wood fibers together. (Formaldehyde is the chemical biologists use to preserve biological specimens.) Studies show that these materials release formaldehyde and other toxic chemicals into room air long after a building is completed.

Potentially toxic chemicals are also found in furniture made from particleboard, furnishings such as carpeting and curtains, as well as paints, stains, fireproofing chemicals, and finishes. Appliances like water heaters and furnaces can also release toxins (in this case, carbon monoxide). Our homes are meant to provide life safety and shelter, yet we fill them with materials that poison the air we breathe.

Building a house creates enormous quantities of solid waste. Scrap wood, glass, drywall, stone, concrete block, packaging materials, and other waste is generated at building sites in astronomi-

cal amounts. In fact, a typical new stick-frame house produces three to seven tons of solid waste. Nationwide, waste from new construction and remodeling constitutes approximately 25%–30% of the municipal solid waste stream. In regions where home building is occurring at a rapid rate, the percentage can be much higher. Hauling this waste to the landfill consumes valuable resources, especially energy, and adds to the cost of building. By comparison, consider the construction of the Real Goods Solar Living Center in 1995, which sent less than 3% of building materials to the landfill! (Learn more about the Solar Living Center on pages 408–12.)

But while construction accounts for an enormous amount of resource consumption, it is responsible for only part of the environmental impact of modern housing. Many impacts continue long after a home is finished: the occupation or habitation impacts, including travel to and from our homes. To keep our homes running, we use enormous quantities of electricity, oil, natural gas, and water. Food and countless household items also stream in. Let's take energy as an example.

According to the US Department of Energy, America's homes consume about 22% of the nation's fossil fuel energy. This energy is used to heat, cool, light, and ventilate our homes and to power a growing assortment of appliances and electronic devices meant to make our lives more

Our homes are meant to provide life safety and shelter, yet we fill them with materials that poison the air we breathe.

Net-zero Energy Home Mandate (California by 2020)

A net-zero energy home is one that is designed, modeled, and constructed to produce as much energy as it consumes on an annual basis. Creating a tight, efficient building envelope and then installing energy-producing systems (typically PV) to meet the annual energy demand of the home is the common means of achieving this goal. In a grid-intertie setup, most utilities have an "annual true-up," whereby one aims for balance on an annual basis, putting kilowatt-hours "in the bank" during the higher-production summer months, and then drawing these down during the winter.

Of course, if one is living off the grid, the home needs to be designed with and feature energy systems that keep it comfortable and powered at any moment during the year. This can require some supplement of heat or electricity from a generator during the coldest, darkest months, making a true net-zero balance more difficult to achieve without some form of offsets. This is also true when one is running vehicles that burn fossil fuels. It is difficult for anyone to claim that they are living a true fossil fuel-free lifestyle, given its use in creating much of our modern world (including this book).

According to the Net-Zero Energy Home Coalition, globally buildings account for more than 40% of primary energy use. Our built environment, consisting of houses, buildings, and the communities they form, account for approximately 41% of total US primary energy consumption, and 50% of all energy consumed in Canada.

Achieving a net-zero energy home is as much a matter of choice and behavior as it is the design and construction of the home and energy systems. As Andy Shapiro of Energy Balance Inc. once put it, "There's no such thing as a net-zero house, only net-zero families. Occupant choice matters hugely." What many families discover is that the choice to consciously use less energy often yields healthier, more satisfying lifestyles.

A home with passive and active solutions for sustainable living in northern California.

Edward Caldwell Photography

comfortable, enjoyable, and convenient. Residential energy demand, in turn, is responsible for approximately one-fifth of America's enormous annual release of CO_2. Add to this the impact of commercial buildings, which consume about 18% of the nation's energy, and the overall impact that the building sector could have on reducing carbon emissions is undeniable. Some municipalities are taking action. In California, all new homes built after 2020 must be net-zero, consuming no more energy than they are responsible for producing.

Net-zero policies can begin to curb the increase in carbon emissions, but the existing building stock if not improved will continue to add carbon to our atmosphere at high levels. While the causes can be argued, climate change is no longer a theory—we see the impacts of shifting global weather patterns in the superstorms, floods, and droughts of recent years. The question now is how we will react, and if we can see a way toward a future of a more responsive and responsible built environment. (See Chapter 1, "Relocalization: A Strategic Response to Peak Oil and Climate Change.")

Our homes and lifestyles are also a source of tremendous amounts of waste, including food scraps, empty cans and bottles, junk mail, packaging materials, and yard waste. To get an idea of the volume generated each year, take a look at the garbage cans lined up on the street of a typical suburb on garbage day and multiply that by the 115 million homes throughout the country and multiply that by 52 weeks in a year.

Less obvious are the billions of gallons of sewage, containing human excrement, poten-

tially hazardous cleaning chemicals, and excreted medications, which have a deleterious effect on fish and other aquatic organisms. Each day, these billions of gallons of sewage flow to wastewater treatment plants that remove as much potentially harmful materials as they can before the water is released back into waterways where it is used for crop irrigation, industrial production, and drinking water, and where it also serves as habitat for many aquatic species. Modern treatment plants can return sewage to clean, drinkable water—but instead of recycling this water back into the supply, it is returned to rivers, after which it must be treated again before being used for drinking water.

Another way to think about the environmental impact of our homes is to realize that they are like patients in an intensive care unit. They're supplied by electrical, gas, and water lines. Wastes are removed by sewage lines and garbage companies. Just like an intensive care patient, our homes are highly vulnerable to disruptions in the flow of inputs and outputs: Cut off the supply of resources such as electricity, even for a short period, and they cease to function.

In sum, then, when most of us think about our homes, the last thing that comes to mind is resource depletion or pollution. We don't think of our homes as a major source of greenhouse gases that contribute to global warming. Nor do we think about the loss of wildlife habitat, species extinction, dangerous exposure to toxic chemicals, and our extreme dependence on outside resources. We think of our homes as sanctuaries, not sources of personal and environmental damage.

Recognizing that our climate is shifting, and that the infrastructure we depend on may not be up to the challenge, we should be considering how to build or improve our homes so that they can be self-reliant.

The bad news is that our homes are anything but environmentally friendly. The good news is that they don't have to be.

Houses can be built or improved to operate at a fraction of the impact of a conventional modern home. They can be designed and built in ways that require significantly less of Earth's natural resources. They can be healthful places to live. If we're smart, our homes can work for us. They can heat and cool themselves, and they can even be a source of energy—not just for us, but for our neighbors as well, generating tax-free income by reducing monthly fuel bills. Our homes can be built to survive catastrophes, and to be independent from the grid for short or long periods of time. Shelter can be part of the solution to creating a sustainable future, if only we rethink our strategies.

Fortunately, many architects, builders, and consultants are rethinking their approach to building. They are embracing the idea of "green building," which has assumed the status of a social movement. Green building means producing houses that satisfy a triple bottom line. They are good for people, good for the economy, and good for the environment. It is one strategy that will help us build a sustainable future.

Passive Survivability

The causes of climate change are now unambiguous, with at least 97% of scientists in agreement that it is manmade, and it is impossible to deny the impacts increased carbon dioxide and other greenhouse gases are having on our planet. While the climate has been known to undergo serious shifts throughout the Earth's history, we have few recorded examples of shifts in weather patterns as drastic as those being experienced today. In the United States, changes in Arctic temperatures are shifting the jet stream, creating dry, warm, drought-causing winters on the West Coast and drastically cold temperatures through the Midwest and East Coast. The shift is allowing strong summer storms to travel further north along the eastern seaboard, such as when Hurricane Sandy devastated many coastal communities in 2012.

All of this means that communities can expect to see weather patterns unlike those they have previously prepared for. Our infrastructure may not be up to the test when the next superstorm rolls through, or drought sets in. Shortly after Hurricane Katrina hit the Gulf Coast, the concept of Passive Survivability began to gain traction within the building community.

Passive Survivability recognizes the weaknesses in our built environment—that without the support of utilities such as city water, sewer, and electricity, many homes and buildings would no longer function as shelter for their occupants. Depending on the local climate and conditions, homes would overheat, or freeze. They would no longer carry away the toxic waste created by humans, nor provide us with clean drinkable water or safe food storage.

Recognizing that our climate is shifting, and that the infrastructure we depend on may not be up to the challenge, we should be considering how to build or improve our homes so that they can be self-reliant. We cannot control the waves of storms or droughts that may be coming our way. In fact, attempting to do so may lead to more catastrophic damage when those systems fail. Implicit in this is the idea that we must find a way to strike a balance with the local ecologies in which we live.

Local/Locale

The solution to creating a functional home is not the same across all regions and climates. Our response must be nuanced, and focused on the specific location and ecology in which we are working. Whether you are building new, or improving an existing house, consider your own hometown. Is it in a temperate climate, with four definable seasons, like the Mid-Atlantic? Or, are you living in a desert locale, where temperature

The polar jet stream meandering across North America. Climate change is causing greater north-south fluctuations in the jet stream, in turn leading to weather patterns holding over a region, causing prolonged floods, droughts, and heat or cold waves.

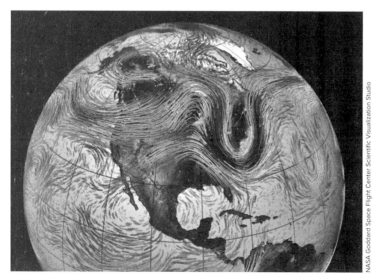

NASA Goddard Space Flight Center Scientific Visualization Studio

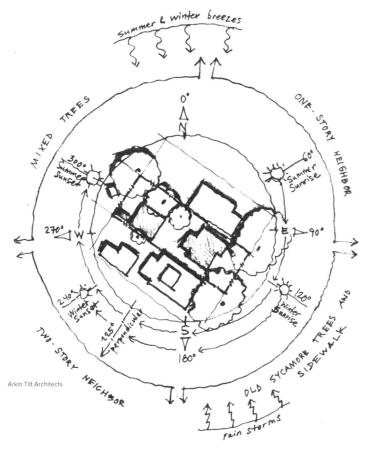

Arkin Tilt Architects

Diagram showing local site conditions in a dense neighborhood.

swings are more drastic across a typical day, but steadier over the course of the year? Your region may have an abundance of freshwater sources, or it may lack the raw materials needed to easily build and maintain a shelter. By taking stock of what is available to you, and what is in demand, you can start to identify what elements you may want your home to be self-sufficient in.

Now look closer at the specific site of your home. Is it sheltered by hills to the north? Exposed to heavy winds and storms rolling in from the west? Do you have running water nearby? A sunny hillside where you could collect energy? Are you in a dense neighborhood, where breezes are funneled onto the street? Or are your neighbors further away, allowing for more access to light and air? Your microclimate, or the condi-

tions created by the forms and flora surrounding you, will also have a big impact on what your home can provide for you, and what you might struggle to get, or learn to do without.

Once you've taken an inventory of the place and conditions, you can begin to strategize how to create a responsive home for your site. The solution need not be daunting. Simple space planning and low-maintenance systems can take a home from grid-dependent to resilient. Depending on your budget and abilities, you will find a balance between high- and low-tech solutions. We'll get into more details about building, heating and cooling techniques, and water and energy systems. But first, it is important to think more holistically about design and planning strategies for your home.

Strategies for a Responsive Home

By framing your home as not just the shelter you live in, but the community you are a part of, the definition of the support systems you depend on will shift. A good first step for planning your home is a diagram, showing the various inputs

and outputs that your home will need or produce. This diagram will be influenced by the study of the local ecology and microclimates that you have already undertaken. Important inputs/outputs will vary with locale, but are likely to include

Diagram looking at the inputs needed and outputs created in a small residence.

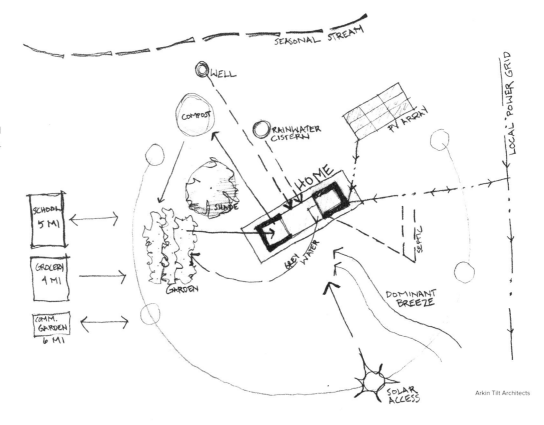

Arkin Tilt Architects

water (fresh and waste), power, heat sources and excesses, air or breezes, food, compost, and data or information.

If you live or are building out in an undeveloped area, you must identify the natural systems around you that will allow you to survive in that place. You will need to identify a local water source, likely by digging a well, and a plan for how to deal with the wastewater your home produces. You'll need to tap into an energy source with solar panels, or wind turbines. In a city or suburban neighborhood, your home will run on municipal water, power, and electricity. You should think about where your food will come from, and what you can do with the compost or excess you may end up with, and how you can source food if stores are not open. In either case, look for weaknesses in your system. Are there any necessities for which your diagram only identifies one source? By knowing where your home lacks resilience, you can plan for the inevitable day when that input/output no longer functions.

Consider a strong hurricane passing through the Gulf Coast in the depth of summer. The storm system itself will cause untold damage (Hurricane Katrina's damage totaled approximately $81 billion). But what happens after the storm is gone? Homes that were dependent on municipal systems are suddenly left with no water or power. Food supplies dwindle as shipments to stores are

delayed. The summer heat returns in full force, and the population left behind no longer has functioning shelters to protect them.

What if this community had resiliency built into the system? If individual homes or neighborhoods had secondary sources of power, or had been collecting and storing fresh water in cisterns, the damage from the storm would not cause the economic impact we are used to seeing fester for years after a disaster strikes. In a city, a communal attitude toward disaster preparedness means that not every home needs backup systems. Neighbors can depend on each other, allowing them to more quickly rebound and rebuild.

Building Your Community

The first step to building responsibly is choosing to build in a community or on a site that can support the addition of your home. Living in a densely populated area has benefits for both you and the larger environment. Containing human habitat is one way of protecting delicate ecologies beyond the city's boundaries.

Urban neighborhoods are likely to have a wealth of services nearby that will improve your lifestyle—grocery stores, farmer's markets, community centers, transportation systems, schools, parks, and public forums all help us to forge ties with our fellow residents. By connecting with

Amanda Knowles

Functioning community garden in Albany, California.

each other, we ease the creation of communal backup systems that can support us in our times of need. Of course not all communities are set up with an eye toward Passive Survivability. If the place you have chosen to live is lacking in some facet, consider getting involved and creating that service for yourself and others. Perhaps your town could use a food bank to help struggling families, or maybe it needs improvements to city water treatment and storage facilities in case of a prolonged drought. Your city, like your own home, should be resilient, and it is only the residents who can make it that way. Our new chapter about Urban Homesteading discusses many kinds of individual and collective actions for sustainable living in city environments; see chapter 12, pages 305–32.

Whether your home is remote or in a bustling city, local regulations and rating systems can point you in the right direction when you're deciding how to invest time and money in improving your home. Get in touch with your local building department, and learn about the requirements you need to meet in order to build. By doing so, you can get an idea of what are the biggest pressure points in your community, be it energy usage, water quality, wetland protection, or water generation.

It's possible that your building department will require both new construction and renovation projects to meet certain levels in a rating system, such as the US Green Building Council's LEED[1] system or Build It Green's Greenpoint Rated[2] program. Each of these programs, and the many others out there, touch on a variety of issues—material selection, energy use reduction and production, water systems, interior air quality—and rate a project based on how well it meets certain requirements. Setting a ratings goal for your home can be a good way of ensuring you make responsible and sustainable choices as you move through construction.

Not long ago, no common definition existed for a high-performance green building. Yet today more than 100,000 housing units are LEED (Leadership in Energy and Environmental Design) registered (usgbc.org/articles/infographic-leed-world). By 2013, projects in North America representing 658 million gross square meters were LEED registered or certified. The green building market now exceeds $36 billion, which represents 22% of the overall building market. This is projected to increase to $114 billion by 2016.[3] The growth of green building can be traced to increased awareness created by programs like LEED, and regulations set in place by local municipalities.

Of course, many of the lifestyle choices we make beyond the homes we build will impact the environment around us. Consider the importance of supporting sustainable agricultures (see Chapter 11, pages 291–303), or responsible municipal energy production.

Design Tools for Your Home

When planning a house project, it can be helpful to visualize your project before construction begins. Doing so can help you to understand the solar condition on site, look at air flows, consider various orientations, and even perform complex

Not long ago, no common definition existed for a high-performance green building. Yet today more than 100,000 housing units are LEED (Leadership in Energy and Environmental Design) registered.

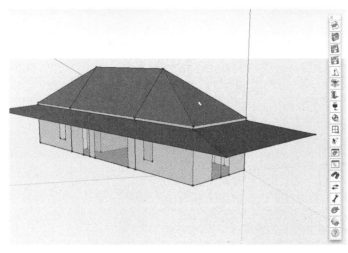

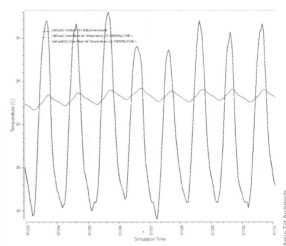

Arkin Tilt Architects

Model of a small residence built with OpenStudio in Sketch-Up, and graphic results produced showing internal temperatures versus outdoor dry bulb temperature.

thermal studies. A green building professional can help to bring these tools into your process, but you can certainly tackle the basics yourself.

Free programs like Sketch-Up,[4] from Trimble, have simple modeling tools, geo-location abilities, and solar modeling functions. This allows you to build a digital representation of your home and its site, and explore orientations, window placements, and shading systems to maximize your home's passive solar function. We'll describe more about Passive Solar further on in this chapter.

Another free program to explore comes from the US Department of Energy: EnergyPlus.[5] This program, along with a Sketch-Up plug-in called OpenStudio, allow you to build basic models of your home, assign material properties to the walls, roofs, and floor, and then run complex thermal tests to explore heat gain, light levels, and energy loads. This program takes some training and exploration, but can help you to answer basic questions about the performance of your home,

and to make decisions about where improvements can have the biggest impact. There are numerous other thermal and fluid dynamic modeling softwares available on the market, varying in complexity and required training (EcoTect, IES, Tas, and Maxwell). If maximizing your home's energy performance is a top priority for you, it may be worthwhile to engage an energy modeling specialist to help.

If the idea of using a digital modeling pro-

Solar Pathfinders are used to map the solar access at a particular site. The glass cover reflects back the sky dome, so that it is visible over the diagram of the path the sun takes through the sky.

Solar properties of a physical model of a residence being testing with a Pocket Heliodon.

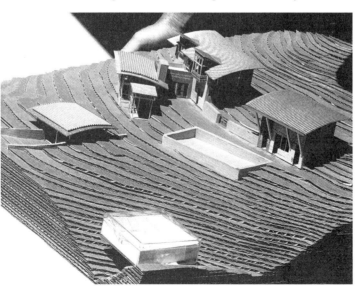

Arkin Tilt Architects

Arkin Tilt Architects

REAL GOODS

Tools & Calculators

BEES provides environmental and economic performance data, including *climate change* impact, for 230 building products using the *life-cycle assessment* (LCA) approach.

Athena Institute's **EcoCalculator** provides LCA results, including climate impact, from hundreds of common building assemblies.

Athena Institute's **Impact Estimator** is a more detailed and flexible version of the Athena EcoCalculator.

Lawrence Berkeley National Laboratory's

B-Path is an LCA tool for structural materials in commercial buildings.

Rocky Mountain Institute's **Green Footstep** considers the carbon impacts of construction including landscape impacts, along with operational energy, to help project teams set comprehensive carbon reduction goals.

Build Carbon Neutral is a simple carbon footprint calculator for buildings.

GaBi Envision enables manufacturers to integrate LCA into product development.

---

gram is too daunting, you can still explore your home in 3-D before construction begins! Consider building yourself a scale model of your home and the site. Using tools and supplies available at craft and model stores, you can create an accurate depiction of your project. And, because sunlight is scale-less, you can take that model onsite and look at the solar exposure in real time. Tools like a pocket heliodon allow you to recreate an entire year's solar patterns simply by manipulating your model while it is in direct sun.

Even those of us who are intimidated by modeling our home can better understand the solar exposure on our site through the use of a Solar Pathfinder.[6] This tool is placed onsite, and allows the user to see how trees, ridgelines, and other structures will limit your site's solar exposure throughout the entire year. There are also now applications such as the Helios Sun Position Calculator, or SunFollower, available for various mobile operating systems. These tools use augmented reality through your phone's camera to trace the path of the sun through the sky. Take a look at what tools are available, and start exploring possibilities.

Renovations and Development

Renovating an existing home, or creating a second unit in or adjacent to a home, is a great way to build green and live sustainably. These kinds of projects support density, thereby minimizing disruption of functioning habitats. Residents of dense neighborhoods are less likely to drive to their jobs, grocery stores, or schools. They have access to shared community resources, like public parks and transportation systems.

Renovations obviously use fewer virgin materials than a new construction project. Because you're starting with an existing structure, you can focus on improving its function. Each renovation project will present its own set of challenges. Take the time at the beginning of the process to lay out your construction goals.

Renovations are often focused on improving the function and flow of the spaces in your home. Just as important is taking the opportunity to update and upgrade the mechanical systems, thermal performance, and material properties of the house.

Consider writing a how-to guide to your home. Break down the various functions it provides—heating/cooling, light, air, energy, and water—and describe in detail how each works. Include what can be done to make each system function as efficiently and effectively as possible. Refine the guide over time. Perhaps at first you can only say that you know you should close your window shutters on hot, sunny days. But, as you spend time in the house, you learn that the shutters have the biggest impact on hot days in the shoulder seasons (fall and spring), and that you have to pull them closed early in the day in order to have an impact on the internal temperature.

Use the how-to guide to direct your investments in your home. Defining the systems your home uses will make it hard to ignore which ones aren't functioning well. Perhaps you find that your heating system doesn't seem to efficiently heat the rooms with the original windows in them, and it's time to upgrade to new insulated windows. Your guide might make note that on rainy days your

> Residents of dense neighborhoods are less likely to drive to their jobs, grocery stores, or schools. They have access to shared community resources, like public parks and transportation systems.

Amanda Knowles

Insulation comes in a variety of forms and for a variety of applications. Take care to select the best option for your installation.

It's not just insulation that makes a difference in your home's energy performance—creating a strong building envelope is just as important. Your building envelope defines the point at which air and water are unable to penetrate further into the building.

basement gets moist, causing musty air and possible mold growth, and that perhaps the water-proofing and drainage under the house needs to be redone. As you improve your home, update your guide and keep track of the conditions you are experiencing. Staying aware of your home's passive and active systems well help you to optimize them.

Another method is to simply identify outdated technologies or materials still in use in your home. If you don't already have one, getting a report by a home inspector can help to locate the biggest existing issues. Once you know what you'd like to improve, work with building consultants or designers to explore which new materials or systems make the most sense in your application. Finally, make a plan for what to do with the waste materials that will be removed in the course of the work. Contact local scrapyards, recycling centers, or salvaged goods stores to give items a new productive life.

Build It Tight and Ventilate It Right

Older neighborhoods have a lot of attractive values—varied character, period details, and established communities. They are also full of leaky, poorly insulated homes with single-pane windows and drafty spaces. Depending on when and how the house was built, it could have old insulation in the walls, or it could have been built with no insulation at all. Some homes have insulation in the walls, but don't have any insulation under the floor or in the roof—meaning drafty attics and basements allow air to seep into the house.

If you're planning on renovating anyway, you have a great opportunity to improve the existing insulation levels. Assuming you live in a wood-framed home, new blown-in cellulose, fiberglass batts, or rigid foam insulation can be added to your exterior walls with ease. It will be important to consider the existing depth of your walls, and what level of insulation (or R-value) you'll be able

to achieve with different insulations in that depth. You'll need to aim for different R-values in your walls, roofs, and under the floor, which will vary based on your climate. The EnergyPlus program[7] provides simple guidelines for improving insulation. More substantial levels can be achieved by adding a wrap of insulation; this can be rigid foam or rock wool, or a more drastic transformation such as a wrap of straw bales or other natural materials.

It's not just insulation that makes a difference in your home's energy performance—creating a strong building envelope is just as important. Your building envelope defines the point at which air and water are unable to penetrate further into the building. This line of defense is created by properly sheathing the building structure with plywood, which provides a continuous surface over which building paper and moisture barriers can be applied.

There are many possible weak points in a home's building envelope: penetrations for mechanical and plumbing systems, crawl spaces under occupied floors, drafty attics, and poorly installed windows and doors, to name a few. Weaknesses in your building envelope lead to unwanted exterior air infiltration, or loss of internal warmed or cooled air. A good designer and contractor can help to develop careful air-sealing details and construction sequences to protect these locations.

Properly insulating and sealing your exist-

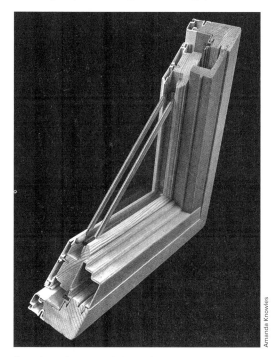

Amanda Knowles

A modern aluminum-clad wood window, with a double pane of glass.

ing attic or crawl space can be a quick and easy project that will drastically improve your home's energy performance. With the use of insulation, drywall, expanding spray foam for small cracks, and a little know-how, your home can go from drafty and cold to sealed and warm over the course of a few days.

Even if your home is well insulated and sealed, old windows and doors can still harm the overall energy efficiency. On average, homes lose 30% of their heat or air conditioning energy through their windows. Old windows are often single pane, don't seal well when closed, and have no thermal break in their frame. They transfer heat through radiation (through the panes of glass), conduction (across the materials of the frame), convection (as air moves along the interior or exterior surfaces of the windows), and through direct air leakage.

New windows, while not perfect, have methods for combating all of these heat transfers. They have double panes of glass, filled with argon or other low-conductance gases. Window frames are built to seal tight when closed, helping to prevent air leakage. Low-E coatings on the glass reflect long-wavelength (infrared, or heat) energy from the sun, while allowing in short-wavelength energy (light). Good frames, with thermal breaks for aluminum windows, well-built wood frames, or hollow vinyl frames all prevent heat loss through conductance.

One option to consider is having your home scored on the HERS (Home Energy Rater) Index. This program tests a home for insulation levels in walls, floors, ceilings, roofs, attics, foundations and crawl spaces; well-sealed windows, doors, and penetrations; and tests efficiencies of your HVAC and water heating systems. You can do this before you start improving your home to help identify where you might focus your efforts. Once upgrades are complete, a HERS Index score can command a higher property value, and help you to anticipate energy bills. To learn more about the program, or to get in touch with a local rater, visit the HERS Index website, hersindex.com

Renovations can go way beyond improving the building envelope. You may consider updating your water fixtures to low-flow fixtures, or a new mechanical system that lowers your energy use, or you may replant your yard with native species and food-producing crops. We will explore many of these strategies in the context of a new home, but remember that any of them can be applied to a renovation project as well.

Building a New Home

In some circumstances, the right option may be to build a new home. This opens up many possibilities for a homeowner. Since you're starting without an existing structure, you can make responsible decisions each step of the way that lead you and your home onto a more sustainable path.

This opportunity also brings some added responsibilities. As mentioned earlier, in California, all new homes built after 2019 will have to be net-zero, or produce as much energy as they consume. Regardless of whether or not your local municipality has any such policy, you should understand the inputs and outputs of your home and its system. Decisions ranging from the site you choose to the mechanical systems you build will all have an impact on your carbon footprint. In order to remind yourself of this throughout the design, construction, and operation of your new home, you should set goals for yourself from the beginning that force you to confront the embodied energy in the decisions you make.

Carbon Accounting and Sequestering

As our awareness of the impacts of human activity in the form of greenhouse gas emissions increases, we are obligated to account for and offset these impacts in our own lifestyles and construction activities. Information on measuring these impacts is becoming more and more widely available.

An organization called Build Carbon Neutral has created a carbon calculator that helps one measure these impacts in relatively simple terms, which in most cases is adequate. If you desire more detailed accounting, this information is out there; for example, BuildingGreen.com and Architecture2030.org provide ample sources and links for finer-grained accounting of one's impact.

Sequestering carbon is done when building materials—through their growth and/or manufacturing—are taking CO_2 molecules out of the atmosphere and locking them away within the construction. Most building materials emit far more carbon during their extraction or harvesting, processing, and delivery than they sequester. A few materials such as wood or earth building systems come close to sequestering more than they emit, and even fewer—straw bale, bamboo, and a couple other natural building systems—actually sequester more carbon than they emit,

As our awareness of the impacts of human activity in the form of greenhouse gas emissions increases, we are obligated to account for and offset these impacts in our own lifestyles and construction activities.

Tiny House Movement

To design a home is a practice of aspiration. We dream of a home that supports a lifestyle that provides us with all of the things and places we've imagined. But could we aspire to living a more economical life? Is there something to be said for a monastic existence?

"In a bare existence, economy is necessary for survival. But it is also, in any existence, an ethical act that regrets the taking; imposing itself as a respectful, if insufficient, act of atonement." *Replacement*, W. G. Clark

Out of respect for the land we use, and the ecologies we impact, can we imagine a dream home that has a smaller impact? Would it allow more of the land we fell in love with at first to remain?

"Build as Little as Possible" is a principal worth taking to heart. When wondering how to build a home that costs less money, and consumes less energy, perhaps the easiest but not often obvious answer is to build less house!

A small home means less material, shorter construction time and disruption onsite, and a lifetime of less energy use. Tumbleweed Tiny House Company, of Sonoma, California, sells house plans and ready-built homes that range from 117 to 874 square feet, or just a fraction of the average new house size in the US (tumbleweedhouses.com). They do this by creating multipurpose areas, finding usable space in what is normally considered "left over," and embodying a lifestyle of minimal consumption.

While a home under 200 square feet may not work for the average person or family, the lessons of efficient use of space can be applied to any new house being designed. One way to create livability without sacrificing too much space is to think of outdoor spaces—porches, courtyards, and similar areas—as rooms. One can expand living into these rooms during times of the year when it is comfortable to be outdoors, and then shrink back into a smaller area that is easier to keep warm or cool during more extreme weather. With thoughtful shaping and orientation, these outdoor rooms can be shaded or sunny, protected from the wind, or connected to breezes, depending on one's microclimate.

By the way, if you want a real-life experience in a Tiny House, the Solar Living Institute offers one via Airbnb at the Real Goods Solar Living Center, where you can stay for a night for a very reasonable fee.

Courtesy of Tumbleweed Tiny Houses

The Elm, a tiny house model from Tumbleweed Tiny Homes. This 117-square-foot home has a bedroom, sleeping loft, one bath, and a great room.

thus helping to achieve construction that is closer to carbon neutral and potentially carbon sequestering.

Choosing Land

Choosing land is one of the most important challenges facing those who want to build a house. For most people, land is selected by proximity factors, for example, how close it is to work, stores, schools, recreational opportunities, doctors, and services such as fire protection. In rural areas, proximity to the electrical grid is an important factor for some. In urban areas, proximity to mass transit may be important.

Although proximity factors may seem to have no bearing on sustainability, they are relevant. They involve more than mere lifestyle and convenience choices. Choosing a site near a major transit line or a place of work, for instance, helps reduce travel time. Fewer miles on the road each day means fewer gallons of gas each week, less money spent on gas, and less air pollution and carbon emissions. Proximity might also permit you to commute by bicycle, a lifestyle change that has many health benefits. Reducing commuting time also means more time spent with loved ones.

We tend to associate buying land with rural living, but if you're interested in an urban location, you can search for a vacant lot to build on. You'll be amazed at how many lots and homes are available when you are open to considering those that need work, and you just might find one that suits your needs and goals.

If you dream of escaping to the country, choose your land wisely, and think in the long term. Watch out for easements, zoning restrictions, and covenants in rural subdivisions. Check into future land development plans in the area. You don't want to build a house only to find that a landfill is slated to go in next door, next year. You want to be sure that you'll be able to enjoy your home for many years to come, and that it will serve you through multiple phases of your life. After all, building a new home with the plan of being there for many years is both the sustainable and the logical thing to do.

Wherever you're looking for land, don't forget to shop for the many "free services" a piece of property can offer, such as a good source of water, hydroelectric potential, natural drainage, or access to the sun.

Here are some things to look for—and avoid—when shopping for land for building a sustainable house:

1. **Choose a site with good solar access.** Sunlight can provide or contribute to your space heating and domestic hot water. It can also provide free lighting during the day and be used to generate electricity. When selecting a site, be sure that it provides good unobstructed access to the sun from 9 AM to 3 PM each day.

2. **Choose a south-sloping site for earth sheltering.** "Backing" a house into a south-facing slope (in the Northern Hemisphere) so that it is sheltered by the earth dramatically reduces energy consumption. Houses built this way stay cool in the summer and warm in the winter with little outside energy input. When combined with passive solar design, earth sheltering can virtually eliminate the need for heating and cooling equipment.

3. **Choose a site that offers unobstructed access to winds.** Wind can provide a significant amount of electrical energy in many areas. Wind can also help cool a house in the summer. If you are building in an area with a good wind resource and want to tap into wind energy, be sure to select a site that is free from potential obstructions for a wind turbine.

4. **Seek favorable microclimates.** Climate can vary dramatically from one location to another within a state, a region, and even a neighborhood. Some microclimates may be

A Humboldt County homestead with great southern exposure.

suitable for self-sufficiency; others may make it difficult to live on renewable energy.

5. **Select a dry, well-drained site.** Moisture is becoming a major issue of concern among homebuilders throughout the world. It can seep into basements and walls through the foundation. To prevent moisture problems, select a well-drained site. This will protect your house from water damage, reduce the cost of grading, and decrease construction costs.

6. **Choose a site with stable subsoils.** Some subsoils are very unstable. They expand and contract in response to changing moisture levels, which can cause a considerable amount of damage to foundations and overlying walls. Be sure to select a site in an area with stable subsoils or design your house accordingly.

7. **Avoid marshy areas.** Wetlands are a precious natural resource that support many species and provide many free services (for example, they recharge groundwater and reduce flooding after rainstorms). Building near wetlands can damage them and also expose inhabitants to pesky mosquitoes that carry dangerous diseases such as West Nile virus. Stay clear of wetlands.

8. **Select land that has a suitable site for growing food.** Gardens and orchards can provide a large portion or even all of a family's food in favorable climates with good topsoil. Growing food provides exercise, time outdoors, and substantial economic savings. It also offers many environmental benefits, especially if you grow with organic methods. Choose land that has good topsoil somewhere, or be prepared to spend years building up the organic matter in the soil. Always do a simple soil test before buying land.

9. **Select land that offers building resources.** Land can provide many building materials,

among them clay-rich subsoils for making natural plaster, wood for framing walls, and stones for building pathways or garden walls. Using materials harvested from a site dramatically decreases the energy required to build a house. Determine the type of house you want to build and then select a site that offers abundant free materials, or better yet, design the house around the site you've selected and the building materials it offers.

10. **Choose a site with a good water supply.** It's important to select land that has a reliable supply of clean water, either in the form of groundwater or a stream or spring you can tap.

11. **Don't destroy beauty in your search for it.** Be sure that the property you buy will allow you to build a house that respects the beauty of the place, instead of attempting to become the focus of it. As Christopher Alexander, co-author of *A Pattern Language*, advises, "Leave those areas that are the most precious, beautiful, comfortable, and healthy as they are, and build new structures in those parts of the site which are least pleasant now." Create beauty; don't be an agent of its destruction.

Once you have purchased land, you'll have to determine exactly where your house should be situated. It is highly advisable that you become intimately familiar with your building site. Visit it often during different seasons and different times of day. Plan on camping on the land on several different spots during different seasons to see how it feels at night and early in the morning. Study wind patterns. Determine the path of the sun. Observe the land during rainstorms and snowstorms. Study natural drainage patterns. Notice where snow accumulates in the winter. If you're building on property with a stream, look for evidence of flooding. Be on the lookout for wildlife corridors or paths. Locate the best views. Check for hot spots and cold spots that exist because of topography or natural windbreaks. Determine how you can access the site so that you do the least amount of damage.

After you are intimately familiar with the land, you can select a site for your house that affords a wonderful view; does the least damage; harvests the energy of the sun, wind, and water (hydroelectric); avoids natural drainage paths and erosion; and provides optimal year-round comfort. Selecting the wrong site is a decision that will haunt you for a lifetime.

After you've settled on a house site, you can begin to design your home. Don't make the all-too-common mistake of imposing your dream home on a site. Design with the site in mind, and remember that your design should enhance the natural beauty of a site, not become an eyesore.

Passive Solar Design: A Primer

Simply defined, passive solar is a means of providing space heating without the use, or at most with minimal use, of costly outside energy sources or mechanical systems.

Creating sustainable shelter is about homes that provide comfort during all seasons, have a small environmental footprint, and are easy on the pocketbook. Many such homes rely heavily on a design strategy known as *passive solar heating*. Simply defined, passive solar is a means of providing space heating without the use, or at most with minimal use, of costly outside energy sources or mechanical systems.

Passive solar heating is not a new idea. It's been around for centuries. It has been tried and tested by ancient cultures ranging from the Greeks, who knew to face their homes to the winter sun, to the Anasazi of the desert Southwest, who built their communities using the same passive solar design principle. Today, using passive solar techniques remains the first step in designing a sustainable shelter.

Passive solar design is intelligent, climate-sensitive design. If properly executed, it ensures that homes stay warm in the winter and cool in the summer with little, if any, outside energy or mechanical support from air conditioners or evaporative coolers. Year-round comfort happens because many of the measures that permit a house to be passively heated in the cold season, such as insulation and proper orientation, also help to passively cool it when the weather turns warmer. But that's not all.

Passive solar houses also provide a significant amount of light during the daytime, saving even more money on energy bills. This strategy, known as daylighting, creates a more healthful and productive environment and for this reason is often incorporated into offices, warehouses, schools, and factories. Daylighting has been shown to increase test scores in schoolchildren, boost productivity in the work environment, and promote a healthy circadian rhythm and sleep cycles.

Passive solar houses rely on south-facing glass that is designed to allow the low-angled winter sun to enter the building, where it is converted into heat that warms the interior through the day and into the night, while protecting the in-

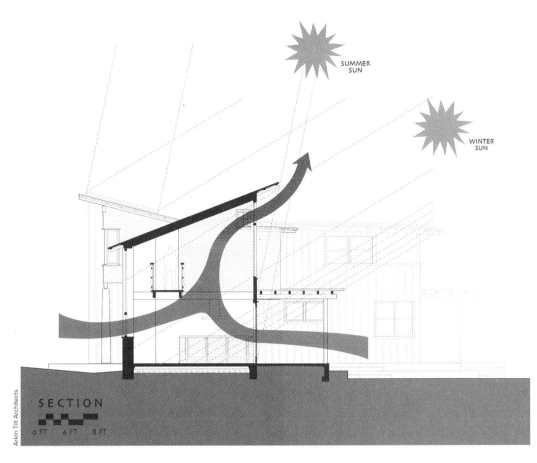

SUMMER
SUN

WINTER
SUN

SECTION

Arkin Tilt Architects

0 FT   4 FT   8 FT

Overhangs and trellises protect from the hard summer sun, while winter sun is allowed to enter the space and warm interior surfaces.

terior from direct sunlight in the warm summer months, minimizing heat gain. We like to think of passive solar as a heating system with only one moving part, the sun. With the sun moving from high to low in the sky through the year, the house itself can have a simple design and rely on standard building components such as windows, walls, floors, insulation, and overhangs. In many climate zones, a properly designed passive solar house can provide all of the heating and cooling needed. But even in those places where more expensive active techniques are needed to augment the thermal control and energy loads, passive solar design is the place to start.

It is important to remember that depending on where your home will be built, some aspects of passive solar design may be more or less appropriate. Consider your own climate, and make note of the range of temperatures you can expect to experience. If you're building in a hot and humid climate, it will be very important to reduce summer heat gain to a minimum, as the heat of the day is held in the air long after the sun sets. In a warm, dry climate, you'll want to add thermal mass that receives direct solar heat gain, which will help to heat a space after the day's heat has dissipated.

You'll also need to know the preferences of yourself and your family members. "Thermal neutral," or the temperature at which the human body does not lose or gain heat from the air around it, is what most mechanical systems strive to reach. But in reality, that neutral temperature is different for every body, based on activity levels, dress, and internal hormonal conditions. With a passive solar design, there is no switch where you can easily adjust the internal temperature. So, by knowing if you tend to "run hot or cold," you're more likely to design a home that will provide you comfort.

To begin, lets explore a variety of passive solar design principles that can be helpful no matter who you are or where you're building.

Passive Solar Design Principles

Principle 1: Select a Site with Good Solar Exposure

South-facing windows are vital to all passive solar buildings. They allow the low-angled winter sun to enter the interior of a structure. Inside, sunlight consisting of visible light, near-infrared radiation, and ultraviolet light is converted into heat, warming the interior.

It may seem silly to point out the obvious— that a solar home needs sunlight to work well— but you'd be amazed how many passive solar

Passive Solar Checklist

✔ Small is beautiful
✔ Long east-west axis
✔ South-facing glazing
✔ Carefully sized overhangs
✔ Thermal mass inside building envelope
✔ High insulation levels
✔ Radiant barriers in roof
✔ Open airways to promote internal circulation
✔ Tight construction to reduce air infiltration
✔ Air-to-air heat exchanger
✔ Best high-tech windows available
✔ Reduced glazing on north side to reduce heat loss
✔ Reduced glazing on west sides to reduce afternoon heat gain
✔ Take advantage of the daylighting
✔ Careful attention to details
✔ K. I. S. S. (aka Keep It Simple, Stupid)

> The first principle of solar house design is to select a site with unobstructed solar access, now and in the long-term future. Ideally, the south side of the house should be exposed to the sun from 9 AM to 3 PM during the heating season.

houses suffer from a chronic lack of sunlight. In many instances, they become shaded over time by trees or neighbors whose houses block the sun.

The first principle of solar house design is to select a site with unobstructed solar access, now and in the long-term future. Ideally, the south side of the house should be exposed to the sun from 9 AM to 3 PM during the heating season. In most locations, the heating season occurs in the late fall, winter, and early spring. In northern latitudes, it is even longer, around eight or nine months of the year, beginning in the early fall and ending in the late spring. In warmer areas,

like the southeastern United States, the heating season may be a brief two-month period in the "dead of winter."

Avoid wooded lots, or be prepared to remove trees on the south side of the house to open it up to the sun. And avoid lots where the house will be shaded by large obstructions, such as hills or nearby buildings that will block the low-angled winter sun.

When in doubt, visit the site in the winter, especially around December 21 (the shortest day of the year when the sun is the lowest in the sky), to be certain that the house will receive a good dose of sunshine. Refer back to the earlier section on Design Tools (page 29) to explore other methods of identifying the sun's daily and annual path across your site.

If you can afford only a small lot, select one that is deep from north to south to ensure good solar access. That way, it is less likely that a yet-to-be-built neighboring house will block the sun at some future point. In rural settings, locating your septic drainage field within the solar access zone is a good strategy to maintain good solar access, because that area will need to be kept clear of trees that could obstruct the sun.

Principle 2: Orient the Long Axis of the House from East to West

To ensure maximum solar gain, orient your house so that its long axis lies on an east-west plane. The east-west axis should be oriented so that the house points within 30 degrees (east or west) of true south (in many locations 15° east of due south is ideal, giving a bit more gain in the morning and then less in the afternoon). These steps ensure that the maximum surface area is directed south for optimal solar gain. Any side of your home can be the "solar face," though you may want to consider which rooms you would like to receive the most light as you plan your home.

When staking out a house site, bear in mind that true north and south are not the same as magnetic north and south. True north and south correspond to the lines of longitude, running from the North Pole to the South Pole. Magnetic north and south are determined by magnetic fields created by iron-containing minerals in the Earth's core, and they rarely run true north and south. In most places, however, true north and south vary slightly from magnetic north and south, from a few to as much as 10 to 15 degrees.

To determine whether and how much magnetic north and south deviate from true north and south in your location, contact a local surveyor

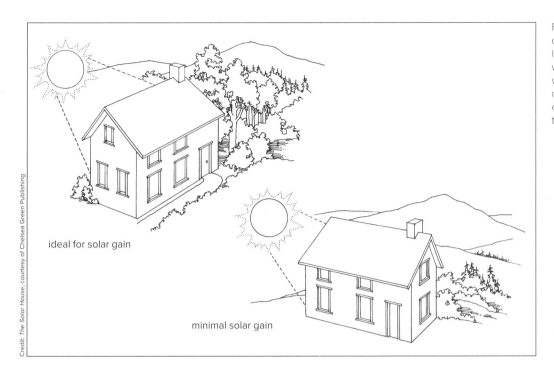

ideal for solar gain

minimal solar gain

Credit: *The Solar House*, courtesy of Chelsea Green Publishing

For ideal solar gain, orient a house so its long axis is aligned with true east and west. This places the maximum amount of south-facing wall toward true south.

and ask for the magnetic declination or simply consult our chart on page 425 of US magnetic declinations. The magnetic declination is given in degrees and direction. If, for instance, the answer is 10 degrees east, true south is 10 degrees east of magnetic south on your property.

For optimal performance of rooftop-mounted photovoltaics, it's best to orient a house exactly toward true south (in the Northern Hemisphere). Although you can orient a house so that it faces slightly east or west of true south without much reduction in solar heat gain, it's not advisable to deviate any more than you absolutely must. While rotating off of true south does not seriously impact your winter heat gain, it can greatly increase your summer heat gain. This is because your solar collecting windows are now more likely to face east or west, and can allow in morning or afternoon light, when the sun angle is lower. Depending on your home's location, excess solar gain at these times of day can result in substantial overheating and discomfort and much higher cooling costs.

Principle 3: Concentrate Windows on the South Side of the House

As should be perfectly clear by now, a passive solar house relies on south-facing windows to capture sunlight energy during the heating season. Concentrating windows on the south side of the house therefore helps ensure sufficient solar gain.

As a general rule, true passive solar houses require south-facing glass whose cumulative square footage ranges between 7% and 12% of the total area of the house. If you're designing a 1,000-square-foot house, for example, the amount of south-facing window surface should be between 70 and 120 square feet. The more heat you want, the greater the square footage of south-facing glass you need. Consider speaking with an experienced builder or designer in your area who can give you more specific advice on your likely heating needs.

Principle 4: Minimize Windows on the North, East, and West Sides

While increasing the amount of glass on the south side increases solar gain in the winter, designers typically decrease window square footage elsewhere. As a rule, north- and east-facing glass should not exceed 4% of the total square footage. West-facing glass should not exceed 2%. Limits on glazing are imposed for several reasons. North-facing glass is restricted primarily to reduce wintertime heat loss. Limits on east- and west-facing glass are imposed to reduce heat loss as well, but also to reduce summertime heat gain to prevent overheating.

These performance concerns need to be balanced with an overall attitude toward providing the best daylight for your lifestyle. If you're planning a studio or other creative space, you may want to focus your north-facing windows in that area, as indirect light is the best light for artistic ventures. If your house site has a view to the west that simply cannot be ignored, you may want to include a large west-facing window, but take care to protect it with a large overhang, or

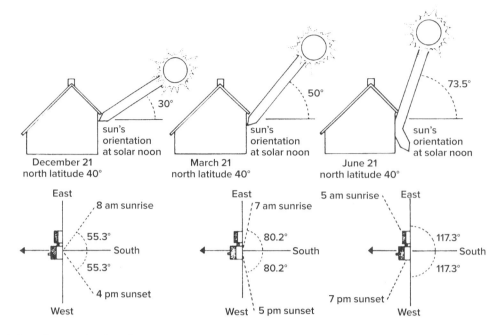

December 21 north latitude 40°	March 21 north latitude 40°	June 21 north latitude 40°

sun's orientation at solar noon — 30°

sun's orientation at solar noon — 50°

73.5° — sun's orientation at solar noon

East
8 am sunrise
55.3°
South
55.3°
4 pm sunset
West

East
7 am sunrise
80.2°
South
80.2°
5 pm sunset
West

5 am sunrise
East
117.3°
South
117.3°
7 pm sunset
West

sliding shutters for days that are too hot. Perhaps you have a rigorous yoga practice, and you can't imagine doing sun salutations without a window facing the sunrise. Allow yourself to have that east-facing window, but consider installing operable windows that can vent out unwanted heat, or separating that yoga space from the main volume of the home, to keep the heat gain at bay.

Principle 5: Install Adequate Overhangs

Overhangs, or eaves, protect the walls of a house from rain and thus ensure durability. In a solar house, however, they also regulate solar gain. To understand this phenomenon, you must first understand the annual path of the sun.

Overhangs regulate solar gain—that is, when sunlight can begin to enter a solar home and when solar gain ends each year.

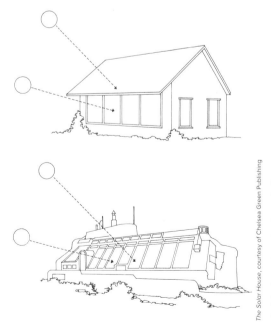

The Solar House, courtesy of Chelsea Green Publishing

As most readers know, the angle of the sun from the horizon changes throughout the year, with a higher sun angle in the summer and a lower angle in the winter. On June 21, the longest day of the year, the sun angle from the horizon (or altitude angle) is greatest. From that time until December 21, the shortest day of the year, the altitude angle decreases. After December 21, the sun angle increases again in a cycle that repeats itself year after year. What does all this have to do with eaves or overhangs on a passive solar house?

The eaves of a house are the on-off switch of a passive solar system. They determine when sunlight begins to enter the house in the fall and when it exits in the spring. In the summer, the eaves shade walls and windows from the high-angled sun, facilitating passive cooling. As the sun descends in the sky after the summer solstice, sunlight is gradually able to enter the house below the overhang. As the sun falls lower and lower in the sky, more and more sunlight enters, creating more heat just when it is needed most, during the winter heating season. As the heating season winds down, however, the sun is on the rise again, providing less and less heat just as the days grow longer and warmer. By June 21, the sun is at its peak again, beating down on the roof and overhangs, unable to penetrate the windows.

Overhangs on solar houses must be custom tailored to one's heating needs. In some locations, the shoulder seasons of spring and fall are the most likely to cause overheating. At this midpoint in the sun's path, deeper overhangs, or even shutters, may be needed to protect the interior from heat gain. On the other hand, if the windows are too protected, the house may not heat up soon enough and the heating system may shut off too soon as well.

Overhangs are most important on the south side of a passive solar house. Contrary to popular belief, eaves provide very little shade on the east and west sides of a house because the angle of the sun in the morning and late afternoon is fairly low. That doesn't mean that you should skimp on this detail, however. Overhangs may not help to protect the walls and windows from sun, but they do protect walls from rain.

In most northern locations, a 2-foot overhang on south-facing walls works well. That will shade an 8- to 9-foot wall in most locations during most of the summer and ensures solar heating as the heating season begins. In southern locations, less heat is required and a longer overhang is advisable. Generally, the warmer and sunnier the climate, the greater the overhang. Log on to the website susdesign.com/tools.php for help designing overhangs for a house in your region.

Be careful to choose a house design with few projections such as decks and porches on the south side that could impair the function of the solar collecting windows. Porches on the east and west sides of a house, however, are often beneficial, as they shade windows and exterior walls from the hot summer sun in the mornings and afternoons. They also provide a great space to watch the sunrise, or sunset!

Principle 6: Incorporate Sufficient Thermal Mass

Thermal mass consists of heat-absorbing material in walls, floors, and ceilings. Tile on a concrete slab, for example, serves as thermal mass, as do walls and floors made of earthen material. Even framing and drywall in a conventional house absorb heat from solar energy.

Thermal mass performs three important functions: 1) It converts sunlight into heat; 2) It radiates heat into the room of a house to provide warmth; and 3) It stores excess heat for later use at night or during long, cloudy periods.

Thermal mass is essential to all passive solar houses, regardless of type or design. For sun-tempered houses, those with less than 7% solar glazing, incidental mass found in the home is generally sufficient to accommodate the heat produced by the sun. When south-facing glass falls within the 7%–12% range, additional mass is required to prevent overheating. Concrete slabs, adobe floors, masonry walls, and planters all work well.

Thermal mass is costly and tricky. It is tricky because placement of thermal mass is directed by two contradictory principles. For optimal performance, mass should be located in direct contact

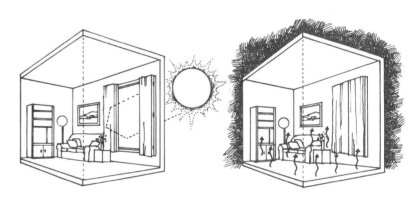

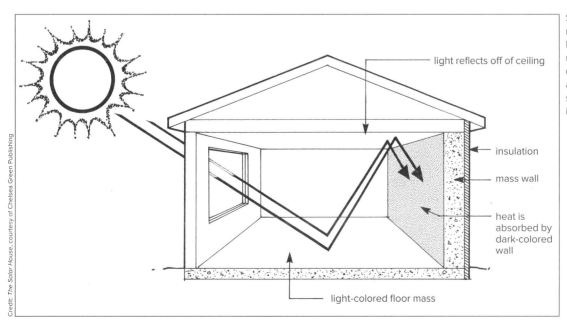

Sunlight can reflect off of light-colored mass onto darker, heat-absorbing mass situated deeper in the house.

light reflects off of ceiling

insulation

mass wall

heat is absorbed by dark-colored wall

light-colored floor mass

Credit: *The Solar House*, courtesy of Chelsea Green Publishing

The Solar Slab: Thermal Mass

Thermal mass is vital to the function of a passive solar home, and its absence is often the Achilles' heel of passive solar houses. As noted in the accompanying text, thermal mass is tricky. It needs to be in direct contact with the sun as much as possible and needs to be widely dispersed throughout a building for maximum comfort. Doing so is not easy.

One solution to the thermal mass dilemma that faces most designers is the Solar Slab, a technique invented by James Kachadorian and detailed in his book, *The Passive Solar House.*

As shown in the accompanying drawing, a Solar Slab is a concrete slab poured over cement blocks. The blocks are laid in parallel rows that are oriented north-south. This ensures that the cavities in each block line up to create a continuous path for air to move from the north side of the house to the south side. How is air propelled through the sub-slab channels?

According to Kachadorian, air movement through the sub-slab air channels is powered as a result of natural airflow—more specifically, a natural convection current. As you can see, sunlight streaming through south-facing windows warms air in the vicinity. Because heat causes air to expand, warm air along the south side of the house rises. Cool air flows in to replace it. That cool air comes out of the ducts under the slab.

Warm air from the south side then moves slowly along the ceiling to the back of the house. As it flows northward through the interior of the house, it is cooled. Along the north wall, heavier cooler air sinks. It then enters the vents on the north side of the house that connect to the sub-slab channels. The movement of air into the system at this end and the movement of air out of the channels along the south side are the drivers for the natural convection loop, also known as a thermosiphon. As the air flows through the channels in the cement blocks, it gives off additional heat.

Heat absorbed by the blocks then migrates upward to warm the slab. The slab stores heat and radiates it into the room at night or on cold days.

As illustrated, the natural convection cycle is energized by incoming solar radiation. However elegant the system appears, though, it often needs a little boost in the form of fans to increase airflow through the slab.

Concerns have been raised about the potential for moisture to build up in the block cavities and promote the growth of mold and mildew. Accordingly, some builders use plastic pipes instead of cement blocks. They're very likely most important in areas that experience higher levels of humidity, according to Kachadorian. Although such modifications may be required, the Solar Slab remains a simple and effective means of storing solar heat in a location that permits the widest dispersal.

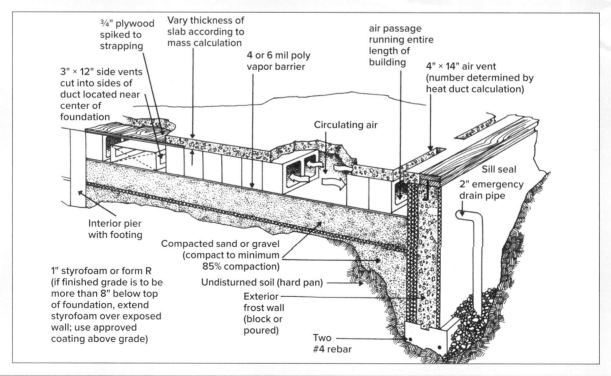

¾" plywood spiked to strapping

Vary thickness of slab according to mass calculation

4 or 6 mil poly vapor barrier

air passage running entire length of building

4" × 14" air vent (number determined by heat duct calculation)

3" × 12" side vents cut into sides of duct located near center of foundation

Circulating air

Sill seal

2" emergency drain pipe

Interior pier with footing

Compacted sand or gravel (compact to minimum 85% compaction)

1" styrofoam or form R (if finished grade is to be more than 8" below top of foundation, extend styrofoam over exposed wall; use approved coating above grade)

Undisturbed soil (hard pan)

Exterior frost wall (block or poured)

Two #4 rebar

with incoming solar radiation. For greatest comfort, however, thermal mass should be distributed throughout the house for even heat. That way, occupants are surrounded by heat at night and on cold winter days as the thermal mass relinquishes its stored solar energy. To achieve these contradictory goals, designers can use a lighter-colored floor mass near the windows to bounce light into the back of the home where it strikes darker-colored wall mass. Earthen wall systems and thicker coats of plaster can achieve a distributed mass effect, where heat can be built up across a space. Another innovative way to deposit sunlight energy into thermal mass is the Solar Slab, described in the accompanying sidebar.

Principle 7: Insulate, Insulate, and Insulate!

Solar energy is a diffuse form of energy. It's not concentrated like gasoline or coal. Even so, small amounts of heat can be used to warm a home, even in the chilliest climate zones. To do so, you need insulation to conserve the heat you collect.

In a well-designed solar house, insulation will form a complete and uninterrupted layer that courses through the walls and ceilings, under floors, around foundations, and over windows and skylights. Well-insulated homes are able to preserve the heat they collect or produce in the winter, and maintain the cool internal temperatures in the summer.

Many passive solar houses built in the 1970s and early 1980s gathered a significant amount of heat during the day but lost much of that heat through leaky, poorly insulated walls. Modern passive solar designers don't make that mistake; in fact, smart designers tend to superinsulate, knowing that the better a home retains heat in the winter, the better it will perform and the more comfortable it will be.

A passive solar house requires superior levels of insulation in ceilings, walls, floors, and foundations to conserve hard-earned solar energy and maintain comfortable interior temperatures despite wide swings in outdoor temperature. Ron Judkoff, of the National Renewable Energy Lab, recommends insulating at least to the level prescribed by the International Energy Conservation Code for residential structures—but preferably beyond them. For initial guidance you can refer to the government's ENERGY STAR program and their prescribed levels of insulation for various climates. (See energystar.gov/index.cfm?c =home_sealing.hm_improvement_insulation _table.) Even so, many energy-smart builders prefer to insulate at levels that exceed these recommendations, often by 30%.

General insulation guidelines for superefficient passive solar houses are R-30 walls and R-60 roofs in temperate climates, and R-40 walls and R-80 roofs in extremely hot or cold climates, according to Ken Olson and Joe Schwartz, authors of "A Home Sweet Solar Home: A Passive Solar Design Primer," published in *Home Power* magazine.

When designing passive solar houses in sun-challenged climates, remember that adding insulation can offset the lack of sunshine. Even in good solar regions, in cases where builders can't add as much south-facing glass as is required for optimal solar gain, say for aesthetic reasons, extra insulation may offset the reduced solar gain.

Bear in mind that insulation has to be installed correctly, too. Pay careful attention to detail, or hire pros that know how to do the job right. For more information on insulating techniques ranging from standard batt insulation to high-tech passive homes to natural building techniques like straw bale, refer to the Building Materials section (page 51).

Although wall, roof, floor, and foundation insulation are important, don't forget to install well-insulated windows. Insulated windows are especially important in north-, east-, and west-facing walls. To make a house even more energy efficient, you can cover windows at night with insulated shades or rigid thermal shutters.

Principle 8: Protect Insulation from Moisture

Most commonly used forms of insulation lose most of their thermal resistance (their ability to block heat flow) when they contain a small amount of moisture. Even a tiny increase in moisture level can reduce the R-value of insulation by half. So it is important to protect insulation from moisture.

In conventional stick-frame construction, builders frequently install a plastic vapor barrier on the warm side of the wall to retard the movement of moisture into and through walls. (In cold climates, the vapor barrier is installed on the inside of the wall; in warm, humid climates, it is installed on the outside.) Placing a plastic house wrap on the exterior sheathing may provide additional protection. (Natural building materials typically don't require moisture barriers—see page 32.)

A passive solar house requires superior levels of insulation in ceilings, walls, floors, and foundations to conserve hard-earned solar energy and maintain comfortable interior temperatures despite wide swings in outdoor temperature.

While vapor barriers are a good idea, they are not as important as many would believe. That's because most moisture enters walls elsewhere. Very little actually penetrates a wall surface under most circumstances. The main source of moisture in walls is through improper flashing around doors, windows, and roofs. The second major source is through penetrations in walls, for example, electrical outlets, light switches, recessed lighting, and other electrical and plumbing penetrations. When building a new house or energy-retrofitting an existing one, pay close attention to these areas to help protect the insulation in your walls and ceilings from moisture. Careful flashing and caulking of penetrations can pay a huge dividend.

Principle 9: Build It Tight, Ventilate It Right

Energy-efficient passive solar houses are built airtight to prevent moisture from entering the walls, but also to prevent infiltration and exfiltration. Infiltration, the movement of air into a building, and exfiltration, the movement of air in the opposite direction, can rob a home of heat in the winter. On windy winter days, for instance, cold air may leak into a house through openings in the building envelope, creating uncomfortable drafts. On days when the wind isn't blowing, warm interior air may escape through leaks in the building envelope, reducing your heating efficiency and comfort levels.

Competent contractors seal all cracks in a building envelope around windows and doors and wherever else they occur, in order to reduce infiltration and exfiltration. Designers may also include entryways and mudrooms that can be closed off from the main living areas, especially in cold climates. In the winter, sealed entryways (sometimes referred to as airlocks or vestibules)

prevent cold air from rushing in when the outside door is opened.

Another effective way of reducing air infiltration is earth sheltering, where the home is built partially into or beneath the earth (usually into a hillside). This technique helps to create a much more airtight house because it dramatically reduces the exterior wall space exposed to the elements. In addition, it helps to reduce heat loss through exterior walls and roofs because the earth surrounding the house stays at a constant temperature below the frost line, usually around 50°F. This consistent and relatively warm thermal blanket around the earth-sheltered portion of a house helps to keep it warm in the winter and cool in the summer.

But how airtight should a house be? An airtight house should permit 0.35–0.5 total air changes per hour. That's sufficient to permit fresh air inside and purge stale air. If your house is tighter, you may need to install a whole-house ventilation system with a heat-recovery ventilator. Heat-recovery ventilators remove stale air and replace it with fresh outside air, and thus help to maintain indoor air quality while reducing heat loss in the winter and heat gain in the summer. For more guidance, refer to the Passive House standards described below (page 52).

Good insulation, protecting the insulation from moisture, and effective sealing—Principles 7, 8, and 9—all make a residence more energy efficient. Energy efficiency is integral to passive solar design, almost as important as the sunlight that provides the energy to heat a house. Without efficiency measures, passive solar design won't work. Efficiency pays huge dividends in the summer months as well by helping to keep a home comfortable when outside temperatures are in the 90s.

Principle 10: Design Your House with Solar Space Planning

Take the time to do some space planning in relation to light access and optimal thermal conditions. Consider each room in your home, the activities you will likely perform there, and what kind of light and temperature is appropriate for that activity. You may know that you like to sleep in a cooler room, but that you often feel too cold in your office, where you end up sitting still for longer periods of time. By organizing your home in relation to each space's access to heat and sunlight, you can avoid over- or under-heating spaces.

Most passive solar houses are rectangular. This shape is thermally superior to others for sev-

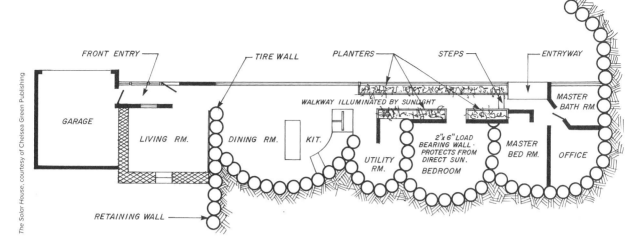

The Solar House, courtesy of Chelsea Green Publishing

FRONT ENTRY   TIRE WALL   PLANTERS   STEPS   ENTRYWAY

GARAGE

LIVING RM.   DINING RM.   KIT.

WALKWAY ILLUMINATED BY SUNLIGHT

RETAINING WALL

UTILITY RM.

2" x 6" LOAD BEARING WALL PROTECTS FROM DIRECT SUN.
BEDROOM

MASTER BED RM.

MASTER BATH RM.

OFFICE

eral reasons. First, rectangular designs result in the largest possible amount of exterior wall space and window area on the south side of the building. Second, rectangular designs permit sunlight to penetrate more deeply into the interior of a building than square floor plans. In fact, if the house is narrow enough, a rectangular design can ensure that the sun heats each room independently. This, in turn, eliminates the need for fans and ducts to move warm air from one part of the house to another. Third, rectangular floor plans minimize the exposure of east and west walls to summer sun, thereby reducing unwanted heat gain, a feature that is especially important in hot climates.

If a rectangular design is not possible or desirable, it is best to place rooms that require less heat, such as bedrooms—or rooms that have their own source of heat, such as kitchens and utility rooms—on the north side of the house. Place frequently used rooms, such as home offices and living rooms, on the south side where they can be warmed by sunlight. Dining rooms are often best placed on the south side, too.

Unless you live in an area that is starved for sunlight—regions solar designers refer to as the nation's gloom belt—it is also wise to create sun-free zones where family members can watch TV, work on computers, or escape for a midday nap. Kitchens are often placed away from the direct sunlight on the south, as the extra heat can make meal preparation rather uncomfortable.

Creating a design that permits sunlight to enter a home yet doesn't render living space useless

This floor plan of Dan Chiras's passive solar rammed earth tire/straw bale house shows various design features that reduce sun drenching and create usable living space. The front air lock (an attached sunspace) shields the living room; a wide hallway down the south side protects the kitchen and living room; partition walls shield bedrooms; and the bathroom shields the office.

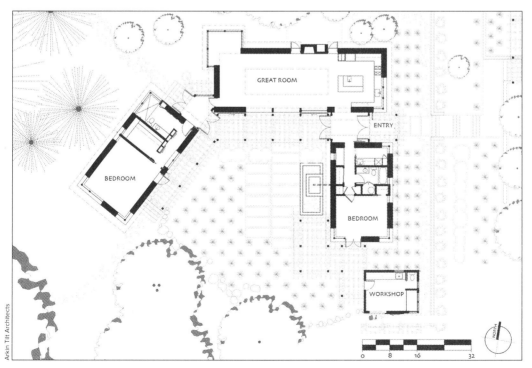

Arkin Tilt Architects

GREAT ROOM

ENTRY

BEDROOM

BEDROOM

WORKSHOP

NORTH

0   8   16   32

A passive solar home should organize spaces based on their need to access the light and heat that the sun provides.

on sunny days can be an enormous challenge. But by recognizing which spaces will benefit from bright light, and which spaces will want to be protected from it, you can find a good balance that supports your daily habits while still heating the home.

Principle 11: Provide Efficient Backup Heat

Although passive solar houses often receive most of their heat from the sun, most of them still require backup heat for long stretches of cloudy weather. A furnace, boiler, or wall heater meets a family's needs during long, cold, cloudy periods when all the solar energy stored in the thermal mass has been dissipated. In most locales, backup heating systems are required by code. To meet code, they must be automatic—that is, able to function on their own, turning on and off as needed when a family is not home. Before choosing a backup heat source, you should research what is required where you are building. For some owner-built residences, restrictions are less stringent, and it's permissible for wood stoves to be the only source of heat.

Remember, however, that because most of your home's heat will come from the sun, a backup heating system must generally be downsized, often substantially. Downsizing—or more accurately, rightsizing—means installing a smaller-than-usual heater, wood stove, or boiler. This reduces the initial costs, thus offsetting higher costs elsewhere, such as for extra insulation. It also saves fuel and money over the long haul because an oversized heating system tends to operate inefficiently. (The same pertains to mechanical backup cooling systems.)

When selecting a backup heating (or cooling) system, it's best to purchase one that is efficient and nonpolluting. Not only does this help to make a passive solar house more environmentally friendly, it cuts fuel bills and may qualify for tax incentives from various governmental agencies or rebates from your local utilities. Among the most environmentally friendly backup heating systems are solar hot water, masonry heaters, clean-burning wood stoves, and heat pumps.

Radiant floor heating has become popular over the last several decades and is an excellent backup for passive solar houses. Typically powered by propane or natural gas, these systems use tubes to circulate hot fluid through concrete floor slabs or under conventional wood floors. They keep a home at a very constant temperature throughout the winter. These systems are often more pricey than a typical heating system—remember that if your home was designed well, you'll rarely be using the backup heat, so you may choose to invest your budget elsewhere.

Passive Solar Design Options

Keeping the preceding background information in mind, let's look more closely at various passive solar design options. Passive solar design falls into two broad categories: sun-tempered and true passive solar. Those who seek greater energy independence build true passive solar houses. This type of house can reduce heating and cooling requirements by 30%–80% or more. But the simplicity and cost-effectiveness of sun-tempered design is attractive to those who are renovating or adding to an existing home. Each of the following strategies may be more or less appropriate depending on your site's climate and your personal preferences.

Sun-tempered Design. This is the simplest type of passive solar heating. It is achieved by placing the house on a lot so that the long axis is oriented along a line running east and west. The number of windows on the south side of the house (in the Northern Hemisphere) is increased to capture solar energy. Slightly higher than normal levels of insulation in walls, ceilings, and floors are usually installed. Unlike more advanced designs, no additional thermal mass is necessary. A sun-tempered house functions well without extra thermal mass, relying only on the "free" or "incidental" mass in framing, floors, walls, surfaces, and furnishings.

Sun-tempered designs are simple to conceptualize and build. The small effort required to convert an ordinary house into a sun-tempered house pays fairly large benefits. It can reduce heating and cooling requirements by 20% to

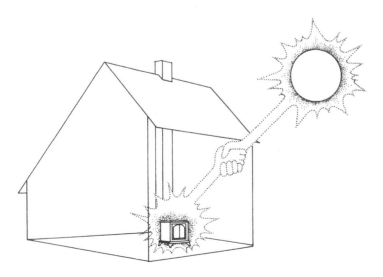

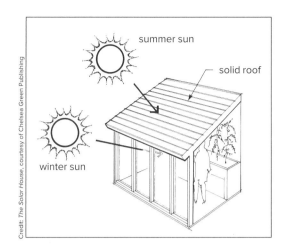

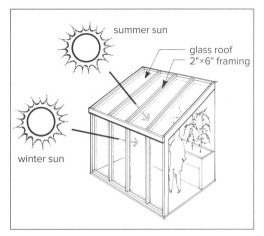

Credit: *The Solar House*, courtesy of Chelsea Green Publishing

30%. The amount of savings depends on many factors, such as local conditions (sunlight availability and ambient temperature), house design, and construction choices (including insulation).

While the energy savings in a sun-tempered house are modest compared with a true passive design, the cost is equally modest. As a rule, building a sun-tempered house doesn't cost a nickel more than any other house—but a homeowner will be rewarded with a substantial decrease in annual heating bills for the life of the house.

Direct-gain Systems. The most popular passive solar design is called direct gain. It is probably what you think of as the classic solar house with lots of south-facing windows that allow the space to be heated directly by sunlight. Excess heat is stored in thermal mass within the living space and released into the interior at night or on cold days—anytime the indoor air temperature falls below the surface temperature of the thermal mass.

A direct-gain house must be carefully thought-out, however. If not designed correctly, sunlight can overheat and damage carpeting and furnishings and generally make the living space very uncomfortable.

Attached Sunspaces. Adding an attached sunspace or solar greenhouse is a common way to retrofit a house, though this strategy can also be used in new construction. Built on the south side of a house, an attached sunspace collects heat from sunlight that streams in through the glazing, which is transferred to the house through a door or window in the wall between the sunspace and the house. Fans are sometimes used to facilitate the flow of warm air into the living space. This concept is sometimes referred to as *isolated gain passive solar* because the attached sunspace is a

solar collector that is isolated from the main living space, meaning the occupant can control how much of that gain is allowed into the home.

Although sunspaces are relatively easy to attach to an existing home or incorporate into new construction, they must be designed carefully, and likely won't be usable year-round. All-glass attached sunspaces, for example, tend to overheat in the spring, summer, and early fall, sometimes even in the winter. They also tend to get very cold at night. For most applications, an attached sunspace with a solid roof works best. The south-facing windows suffice to heat the structure during the day.

Thermal Storage Walls. This kind of passive solar design is also known as a *Trombe wall* (pronounced TROM) in honor of the French engineer who came up with the idea. The strategy consists of a mass wall, such as rammed earth, adobe, poured concrete, or cement block, located on the south side of a house (just the opposite in the Southern Hemisphere). The surface exposed to the sun is typically painted black, and glass is placed 1 to 6 inches from the wall, creating a small air space.

In a Trombe wall system, sunlight penetrates the glass, strikes the mass wall, and heats up its surface. The heat then slowly migrates into and through the thermal mass, moving from the warm surface to the cooler interior of the home. By the time the sun sets, the heat absorbed by the wall begins to radiate into the adjoining room, providing gentle radiant heat.

For daytime heating, some builders put vents in the wall. As illustrated in the accompanying drawing, cool air enters the lower vents and is warmed by sunlight. As it heats up, the air expands, becomes lighter, and rises in the cavity between the glass and the surface of the mass wall, exiting through the upper vents. More cool

Sun-tempered designs are simple to conceptualize and build. The small effort required to convert an ordinary house into a sun-tempered house pays fairly large benefits. It can reduce heating and cooling requirements by 20% to 30%.

Thermal storage (Trombe) walls use mass and convection to heat and cool a space.

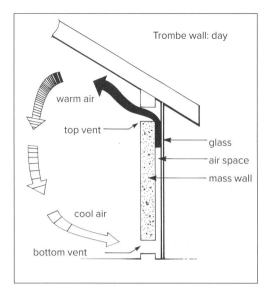

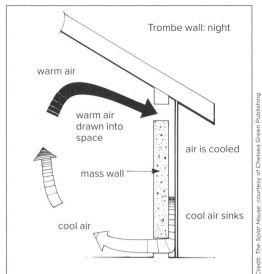

Credit: *The Solar House*, courtesy of Chelsea Green Publishing

Contrary to popular belief, passive solar can work well in most climates. Over the lifetime of your home, passive solar design could save you thousands, more likely tens of thousands of dollars in fuel bills.

air flows in through the bottom vents, creating a thermal convection loop.

A thermal storage wall heats a home indirectly, and this is considered a form of *indirect gain*. So that you're not living behind solid concrete walls, many designers install windows in their thermal storage walls to provide light during the day and a view of the outside world. Because this is a double-wall system, there are cost implications to a thermal storage wall. But for some climates, their efficiency and controlled heat gain is worth the extra cost.

Designers often combine passive solar features. For example, they might choose to heat some rooms via direct gain and others, such as bedrooms or a home office, using thermal storage walls. Why? A thermal storage wall reduces the amount of direct sunlight in a room, a big plus for spaces that need more even lighting, such as offices or studios. A bedroom with a thermal storage wall may also be darker at night, ensuring a better night's rest.

Choose your system or combination of systems carefully. Be sure to read more about each design, each of which offers a unique set of advantages and disadvantages.

Passive Solar Design Challenges

Passive solar is simple in theory. Find a sunny lot, orient the house to true south, concentrate windows on the south side, provide overhang, insulate well, and add a sufficient amount of thermal mass. After the house is done, sit back and let the sun shine in and enjoy the free solar heat while your neighbors pay extravagant fuel bills to maintain some degree of comfort.

But even though the rules of passive design are elementary, designing a passive solar house

that works well year-round can be a challenge. Many houses can't be oriented to true south for one reason or another. Floor plans may create obstacles to optimal solar gain. Cabinets in kitchens on the south side, for instance, may reduce solar gain. Even entryways, if designed improperly, can block solar gain. Trees may shade the site. Budgetary constraints may make certain desirable design features impractical.

In our experience, very few passive designs offer optimal solar gain. Most involve compromises, which may lead to substantial decreases in energy performance and comfort if they aren't considered and counteracted in some way. If you design correctly, however, passive solar will result in a lifetime of low-cost comfort.

Contrary to popular belief, passive solar can work well in most climates. Don't exclude passive solar from your plans simply because your house site is not as sunny as that of a friend who lives in Colorado, New Mexico, or California. Plan to obtain as much free solar energy as possible.

Even if you live in a sun-challenged area, such as the northeastern United States, solar energy can provide a significant amount of heat to an energy-efficient household. A new house in that region can take the shape of the traditional saltbox home, with its tall south face and minimal overhangs. These homes opened up to the winter light, while using a low roofline on the north to protect from heat loss and winter winds. With a passive solar design in the northeast, you may acquire most of your heat early and late in the heating season, but the gains can be impressive.

You will be amazed at how little passive solar design actually costs. In fact, if you plan well, a passive solar house may not cost a penny more than a conventional house. Proper orientation and concentrating the windows on the south side, for example, won't add to the cost of your new home. Adding insulation beyond levels required by code will increase the price a bit, but higher costs on insulation are often offset by installing a smaller heating and cooling system. You will certainly save huge amounts of money over the lifetime of your home in lower energy bills that will pay back any additional initial investment many times over.

In recent years, passive solar house design has become even more cost competitive with conventional construction, because many municipalities have increased energy-efficiency requirements. As a result, measures such as energy-efficient windows that were once unique to passive solar houses are now required by code. Bottom line: It isn't going to cost you an arm and a leg to employ passive solar. And over the lifetime of your home, passive solar design could save you thousands, more likely tens of thousands of dollars in fuel bills.

Passive Ventilation and Active Living

Depending on your climate, making a comfortable home can be as simple as opening your windows. As a culture, we have gotten used to controlling the conditions around us by pushing a button on a thermostat. But, before mechanical cooling and heating was introduced, our daily lives were full of rituals that took advantage of the heat and coolness around us.

Consider an Italian villa on a hot summer day. The walls of the villa are built of thick masonry construction, with windows set deep into wall. The depth of the walls itself protects the interior from direct solar gain. But, in addition to the walls, the villa is outfitted with shutters at the exterior, and curtains at the interior. In the morning, the shutters are open, and the curtains are drawn, allowing in some early morning light, to take the overnight chill out of the air. Before the heat of the day builds up, the shutters are closed at the exterior of the wall, cutting off solar gain from the interior. Slowly the interior warms, through internal gains such as cooking, and the heat coming off of the occupants' bodies. In the evening, when the temperature outside starts to drop again, the drapes are pulled across the opening, trapping the warmed air inside, and keeping the interior comfortable over the cool night.

This complex interaction with the infrastructure of the home is common in many historic typologies, across cultures. However, once mechanical systems took over, attention was turned to maximizing interior space, or opening onto views. The home became less of a shelter, and more of a showpiece.

Passive Ventilation is a simple technique that can be applied to almost any home by any homeowner. It requires an understanding of the climate you live in, and of the temperature preferences of your family. Consider getting a weather station, and making note of temperatures, humidity, and solar gain throughout the year. By learning the patterns of the weather around you, you will be more capable of tapping into them when you are in need of warmth, or looking to purge excess

An integrated passive and active heating/ cooling system.

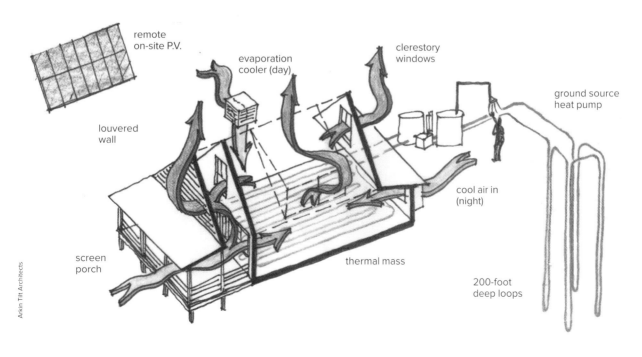

remote on-site P.V.

evaporation cooler (day)

clerestory windows

ground source heat pump

louvered wall

cool air in (night)

screen porch

thermal mass

200-foot deep loops

Arkin Tilt Architects

heat. To take advantage of your climate, there are some basic principles of air movement to understand.

Principle 1: Warm Air Rises

As air heats up, the molecules begin to move faster, and spread out more, making the air less dense. Warm air will rise to the top of whatever it is contained by. This can cause issues when you are trying to heat a space—and it is why heat is most efficiently added to a space from below.

In a still space, air will separate out based on temperature, with cool air sitting at the floor and the warmest air floating by the ceiling. This delineation based on temperature is known as the Stack Effect.

Principle 2: Air Moves toward Low Pressure

The Bernoulli Principle states simply that as pressure decreases, the movement of a liquid increases. This can be applied to the flow of air through a space. Consider a home that is a simple box, oriented north-south. This home is experiencing a strong wind coming from the west. The west wall of the house is under high pressure from the air blowing at it, and the air is forced to split, and move to either side, or over the home. As the air flows across the house, and out over the

east side, the east side experiences low pressure (or, air is pulled away from it). If an occupant of that home were to open a window on the east side, air would flow out of the house, through that window.

As the shape of the home becomes more complex, so does ascertaining where pockets of low pressure are forming. Taking advantage of these conditions requires being aware of both interior and exterior conditions.

Principle 3: Hot Air Creates Low Pressure

If you're looking to evacuate hot air from a space, the easiest solution is to allow it out at the top of a space. The hot air will seek out higher ground, and the air in the rest of the space will move toward the opening.

A device called a solar chimney is a common element in homes in hot and arid climates. These chimneys are exposed to the hot southern sun, which heats the material of the chimney and the air within it. As the air heats and exhausts from the chimney, the air within the occupied space below is pulled into the chimney, and cooler air from a lower crawl space or below grade can replace the air in the occupied space. As it is warmed by internal gains, it then moves toward the chimney, and the exhaust continues, without mechanical aid.

Principle 4: Well-placed Windows Facilitate Air Movement

A complex solar chimney is not required to take advantage of passive ventilation. A simple window can do wonders when it is opening onto a low-pressure pocket, or is placed high in a room.

Interior transom windows allow air movement along the ceiling of your home, and can get hot air from the interior of the house out toward the edges. From there, high windows or clerestories can be opened to vent the air to the outside. Windows at standard heights can also play an important role. When placed in the warmest room of your home (perhaps the kitchen, or a room with lots of southern windows), the same principles can apply.

Important to any successful passive ventilation strategy is a deep understanding of your local climate, prevailing winds, internal heat sources, and cool air supply. It also requires an active participant—occupants must know when to open or close windows, shutters, drapes, and skylights in order to create the conditions needed for comfortable internal temperatures. Occupants must be open to a larger swing in numerical temperature, and instead be satisfied with a deeper con-

Solar chimneys use heat from the sun to create an updraft, venting hot air out of attached spaces. In hot and dry climates, the function can be reversed by releasing small droplets of water into the air near the top of the tower. These droplets evaporate, cooling the air, which causes it to drop down into the spaces. More hot air is then pulled in from above.

Arkin Tilt Architects

nection to their condition and awareness of the climate surrounding them. By really engaging your home to control the temperature, rather than pushing a button, residents gain an understanding of temperature swings, and conversely, find comfort.

Building Technologies

Earlier, when discussing renovations, we considered various techniques for improving the insulation values and air tightness of an existing home. When you're starting from scratch, you can be sure to design and construct your home with optimal insulation values and healthy materials.

Depending on your budget, and how adventurous you are, your building options range from high-quality but standard wood-frame construction, to prefabricated and modular systems, to natural building materials such as straw bale or earth, to high-tech modern building techniques such as Passive House. The focus of each of these options is to provide an optimal building envelope that creates an energy-efficient home that doesn't require much effort to heat or cool. By building right in the first place, you can expect to see savings long into the future as your home operates better than the conventional homes around it.

Building with Wood

Wood is a versatile material that is used to build houses throughout the world. Most new houses in more developed countries are made from wood harvested from the world's forests. It is often shipped hundreds or thousands of miles to house sites. Even homes built primarily of other materials, such as concrete or earth, often require a significant amount of wood to build roofs, floors, interior walls, cabinets, and decks.

Stick-frame Houses

Although wood-frame houses are often criticized in natural building circles, they can provide many years of service if they are well designed and well built. Roofs and foundations require special consideration. The problem is that some builders cut corners to save money up front. Cost-cutting measures may lead to potentially devastating problems, including mold buildup and rot, that dramatically reduce the useful life of a stick-frame house.

Wood does offer many advantages. Stick-frame houses, erected by the millions each year, are the most popular type of construction in the United States, and in many other developed countries. Wood is widely available and well understood, and stick-frame homes can be re-

modeled with relative ease. Wood is a renewable resource and can be grown and harvested sustainably. In addition, there are many sources of recycled and repurposed wood products that work as well as virgin timber and can be more attractive and pleasing.

Unfortunately, sustainable wood production is just beginning to emerge in many parts of the world. Although more and more wood is grown and harvested intelligently, abuses continue that cause significant damage to forests and the surrounding environment.

If you are thinking about building a house using wood, you can take many steps to reduce the environmental impact. One way is to harvest wood locally on your land or on a woodlot nearby (with permission, of course!). Locally harvested wood can be used as is, peeled, or dried, with appropriate engineering. If you prefer a standard wood-framed system, the wood can be milled into usable lumber onsite using a portable sawmill like a "mobile dimension saw," or hauled to a local sawmill for processing.

Another way to reduce the impact of wood is to purchase reclaimed lumber—that is, lumber salvaged from demolition sites or old redwoods submerged for years in a river (called "buckskins"). Salvaged lumber is widely available in

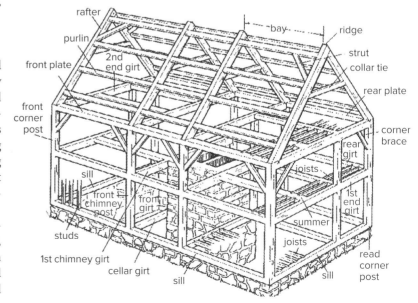

Framing nomenclature

many parts of the country. Be aware that most municipalities will require that the locally harvested or salvaged wood is graded and stamped in order to ensure it has the structural value required for the construction, and that stamping fees can be substantial.

You may also want to use engineered lumber—sheathing and dimensional lumber made from wood chips that are glued together. Engineered lumber provides sufficient strength and uses more fiber from each tree than does dimensional lumber. In other words, there's less waste. In addition, in many instances, engineered lumber is manufactured from smaller-diameter trees. When selecting engineered lumber, consider the binders or adhesives used in the member's assembly—some contain formaldehyde and other chemicals that may off-gas in your home.

Other Wood Techniques

Post-and-beam construction is a building technology that uses wood for framing and is popular among the natural building community, especially straw bale builders. Post-and-beam, which has been around for centuries, produces amazingly beautiful houses. In the hands of a skilled craftsman, timber framing results in a work of art. Bear in mind, however, that it requires heavy lifting, as timbers weigh a great deal. Timber framing also requires special tools; some builders perform all the work using hand tools. Some also fabricate the frame indoors during the off-season and then haul it to a site for quick and efficient assembly.

Using logs for the structure of a house is another technique that is prevalent among natural builders. Log house construction begins with a solid foundation to elevate the logs off the ground. Logs are debarked and then laid down one at a time. In olden days, logs were joined only by interlocking corner notches. The spaces between adjacent logs were filled with a material known as *chinking* in an attempt to create an airtight and energy-efficient structure. Mud was used in early days; a cement mix is now more common.

Unfortunately, say critics, timber frame and log construction are fairly wood-intensive. They often require fairly large timbers that come from much older and much higher-quality trees that are increasingly more difficult to find and often quite costly. They may need to be shipped hundreds or thousands of miles. These building techniques may also be dependent on the thermal mass of the wood itself for temperature control, unless an extra layer of insulation is added over the structure.

Natural Building

Despite the ubiquity of wood construction today, earthen materials have been used far longer and continue to be used to build shelter throughout the world. Today, perhaps a quarter the world's population lives in homes built from earth. Many of them are beautiful and durable structures, and many have been continuously occupied for hundreds of years. Grasses and other fibrous materials have also played a huge role in building and are being used more and more by a growing legion of builders who want to create locally sourced, low-impact, high-performance, and energy-efficient buildings.

Straw Bale Construction

Straw bale building got its start in the late 1800s in the windswept Sand Hills area of Nebraska. Because the soil was sandy, sod homes were out of the question. Because trees were rare, stick-frame homes were beyond the reach of the early settlers.

Shortly after the invention of mechanical baling equipment in the late 1800s, however, early settlers began to experiment with hay bales, using them to create walls of temporary housing. Much to their surprise, their temporary homes turned out to be exceedingly comfortable, both in the winter and the summer. Many settlers plastered the walls and forsook their dreams of building more expensive stick-frame houses like their eastern relatives had. As testimony to the durability of the early straw bale structures, a number are still standing today in good shape.

Straw bale building experienced resurgence in the 1970s, and continues to gain popularity due to the fact that straw bale homes are energy efficient and suitable for a wide range of climates, even humid ones. Although initially touted as a low-cost building method, experience shows that, in most cases, it isn't necessarily cheaper than standard construction, especially if you hire outside help. If you hire a contractor to build a straw bale house, expect to pay as much as for a conventional spec or custom-built house, perhaps more. Although straw is less expensive than wood, building the walls of a house represents only 15% to 20% of the total cost of construction. Slight savings on wall material are often offset by higher costs of plastering.

Straw is the stem of harvested grains such as wheat, barley, oats, and rice. Harvesting machines remove the seeds and spit out straw, which is baled and used as animal bedding or left on the fields and either plowed under or burned to return the nutrients to the soil. Straw is a waste product produced throughout the world and is

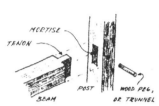

Mortise and tenon joint. From *A Shelter Sketchbook*. Used with permission.

Straw is popular for many reasons. It can reduce wood use and produces superinsulated walls. It is inexpensive and fairly easy to work with. With careful attention to detail and an attractive stucco or plaster treatment, straw bales can create a house that is extremely attractive and durable.

A straw-bale and wood-framed home in northern California.

Edward Caldwell Photography

available in abundance anywhere grain crops are grown, meaning you can expect to find your building materials locally.

Straw is often confused with another agricultural product, hay. Hay consists of meadow grasses that are harvested green to retain as many nutrients as possible. It is used primarily as feed for livestock. Homes should not be built using hay bales because they could attract rats and mice. The high nitrogen-to-carbon ratio of hay also makes it more susceptible to composting, and can promote mold growth. In contrast, straw has no nutritional value and is quite unattractive to rodents, except as a place to nest.

In this technique, straw bales are used to build exterior walls. Most commonly, they are stacked within a post-and-beam frame to provide insulation in a technique known as the in-fill method. But sometimes no framing is used, and straw bales form the walls and support the roof and upper stories. Such structures are known as load-bearing straw bale houses.

Straw is popular for many reasons. It can reduce wood use and produces superinsulated walls (in some cases R-50 or higher). It is inexpensive and fairly easy to work with. With careful attention to detail and an attractive stucco or plaster treatment, straw bales can create a house that is extremely attractive and durable. Moreover, as recent tests show, straw bale is quite fire-resistant. One test blasted a plastered bale wall with 1,700-degree heat for two hours—and the wall survived, performing much better than a standard wood-framed wall (treehugger.com/sustainable-product-design/straw-bale-construction-passes-the-fire-test.html). Even unplastered walls burn very slowly because the density of the bales means that little oxygen is available to support combustion.

Although the idea of building the walls of a house out of straw bales may seem absurd to the uninitiated, these homes perform very well in a variety of climates. Like any building technique, however, straw bale must be detailed well and sealed correctly. To endure weather over the long haul, straw bale houses must be equipped with good roofs with adequate overhang and solid foundations that prevent moisture from seeping

Bales in place in a post-and-beam wall system. Metal lath and lime plaster will be added over the bales.

Arkin Tilt Architects

Straw bales can be used in construction in two ways—as infill in a post-and-beam system, or as the structural wall itself.

into the walls. Builders must pay careful attention to details such as flashing around windows and doors and on roofs. They must carefully seal all penetrations in the building envelope to prevent moisture from entering the walls. They must also apply wall finishes (like earthen or lime plasters) that permit the escape of moisture that can creep into bale walls. If careful attention is paid to preventing moisture incursion, the walls will be free of mold and decay.

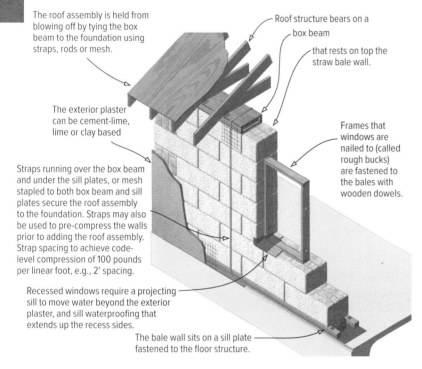

The roof assembly is held from blowing off by tying the box beam to the foundation using straps, rods or mesh.

Roof structure bears on a box beam

that rests on top the straw bale wall.

The exterior plaster can be cement-lime, lime or clay based

Straps running over the box beam and under the sill plates, or mesh stapled to both box beam and sill plates secure the roof assembly to the foundation. Straps may also be used to pre-compress the walls prior to adding the roof assembly. Strap spacing to achieve code-level compression of 100 pounds per linear foot, e.g., 2' spacing.

Frames that windows are nailed to (called rough bucks) are fastened to the bales with wooden dowels.

Recessed windows require a projecting sill to move water beyond the exterior plaster, and sill waterproofing that extends up the recess sides.

The bale wall sits on a sill plate fastened to the floor structure.

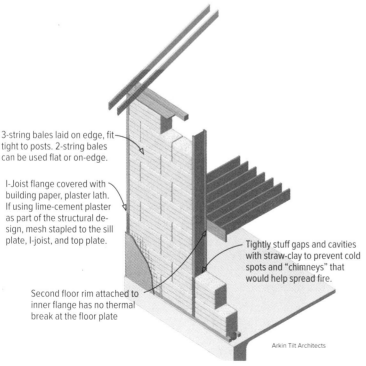

3-string bales laid on edge, fit tight to posts. 2-string bales can be used flat or on-edge.

I-Joist flange covered with building paper, plaster lath. If using lime-cement plaster as part of the structural design, mesh stapled to the sill plate, I-joist, and top plate.

Second floor rim attached to inner flange has no thermal break at the floor plate

Tightly stuff gaps and cavities with straw-clay to prevent cold spots and "chimneys" that would help spread fire.

Arkin Tilt Architects

Straw bale building results in an energy-efficient structure. The R-value of straw bale construction has been tested and retested over the years, with varying results from R-50 to R-19. It is generally accepted today that a plaster and bale wall has a value of about R-30, which is still considerably higher than a 2 × 6 framed wall insulated with wet-blown cellulose or fiberglass batts. And R-value is just one measure of thermal performance. With straw bale, the real benefits come from the impacts of the thermal mass exposed to the interior of the space. The plaster at the interior tends to stay at the ambient temperature of the surrounding air, meaning the wall will not feel cool or hot to the touch, which is a major factor in thermal comfort for occupants.

Using straw bales to build houses also helps clean up the air. For years, farmers in California, Oregon, and other states burned straw in the fields after the grain harvest. This practice contributes significantly to air pollution, as each ton of straw contains approximately 880 lbs. of carbon, all of which recombines with oxygen during burning, being released as CO_2 (naturalmatters.net/article-view.asp?article=4577). This fact has lead to the practice being banned in many places. If the straw is baled and used to build houses, however, this waste product can be turned into a building material, and its carbon is sequestered.

In spite of these environmentally friendly qualities, however, we must not forget that huge quantities of water, fertilizer, and, in many cases, pesticides and herbicides are used to grow cereal crops. If at all possible, use bales from organic farms.

Many straw bale builders and designers promote the technique because homeowners can easily get involved. While the foundations, detailing, framing, and mechanical system installations may be outside of their ability, stacking straw bales can be done by just about anyone. With a large crew of enthusiastic helpers, placing the bales in your walls can occur over the course of just a few days! This requires a well-prepared building site, and will not end with a home ready for occupation. Plastering, finishes, and systems often still need to be installed. But the opportunity to engage in the construction of your own home is very appealing to many people.

When all is said and done, straw bale building can produce dramatic results. The walls are impressively thick, with great insulation value, and can range in form from clean and modern lines to the rounded walls with beveled window wells and arched doorways.

If you want to drastically reduce your heating and cooling bills, reduce stress on the world's forests caused by overharvesting, and help build a sustainable future, straw bale building may be right for you. Don't forget to design for passive solar, insulate ceilings well, install high-quality windows, and make sure the house is airtight. Real Goods offers numerous books on the subject as well as the hardware and specialized tools that make straw bale building easier; check out realgoods.com.

Earth

Earth construction is one of the oldest natural building techniques in use today. It has been used to build houses and other structures such as churches for centuries throughout the world. So durable is it that a great many ancient earth or adobe structures remain standing today in China, the Middle East, northern Africa, South America, Central America, and the United States. Many of these buildings are still occupied.

Historically, most adobe and earth building has occurred in warmer regions of the world—for example, the Middle East, Central and South America, and the southwestern United States. In these regions, adobe homes are been built without insulation, so that the performance depends on the building's mass to provide a "thermal flywheel" that moderates temperature swings and ensures comfort. However, this isn't appropriate in most climates. In areas where temperatures can drop significantly, adobe homes can perform well only if the walls are well insulated. Too much thermal mass in a home can create other issues—if the walls continue to soak up hot or cool air produced by your mechanical system, the space can become impossible to maintain at

The Straw Bale House, courtesy of Chelsea Green Publishing

a comfortable temperature. But when used intelligently, earth walls can be excellent performers and truly beautiful.

There are many techniques for using earth to build a home. They range in complexity, cost, and appearance. Some are suited for being integrated into homes with other building technologies, while some are stand-alone systems. We'll look at a series of techniques below, and review their basic principles and assembly.

Adobe buildings are made from bricks fashioned from local subsoils containing a mixture of clay and sand, with chopped straw or horse manure added to increase their strength. This material is wetted, mixed, and then poured into wooden forms the shape and size of the desired bricks. In arid climates, the forms are often left to bake in the sun. After the bricks have dried a bit, the forms are removed and the blocks are left in the sun to complete the process. The forms

Straw bale interior finished in the traditional Southwestern style.

Despite the ubiquity of wood construction today, earthen materials have been used far longer and continue to be used to build shelter throughout the world.

Eric Millette

Interior of a straw bale residence.

Making adobe bricks.

Below: A rammed earth passive solar home in Buena Vista, Colorado.
Bottom: Rammed earth requires serious formwork during construction.

designers, and contractors such as David Easton, Rick Joy, and Ron Rael.

In this technique, a mixture of clay and sand is compacted (rammed) in forms, like those used to create concrete walls. Some rammed earth builders add cement to their mix while others use sand stabilized with a small amount of cement, rather than clay.

Temporary wooden or steel forms are erected on the foundation. Moistened subsoil containing approximately 70% sand and 30% clay is shoveled into the forms, about 6–8 inches at a time. The mix is then compacted, typically with a pneumatic tamping device powered by a compressor. Once compacted, additional soil is added, tamped down, and the process continues until the forms are filled. When a section is completed, the forms can be removed. The result is a solid block of earthen material, typically 12–18 inches thick. Exterior surfaces of rammed earth walls may be plastered to protect them from the weather or left as is in areas where rain is scarce. Interior surfaces are typically left unplastered so the occupants of the home can enjoy the raw natural beauty of this material.

David Easton, a California builder and author of *The Rammed Earth House*, thinks of rammed earth construction as a means of creating "instant rock." "To me," he says, "rammed earth construction is magic…watching soil become stone beneath your feet, and knowing that, when the forms are removed, a well-built wall will be here that will survive the test of centuries." (This book is available at realgoods.com.)

To reduce costs, some builders are now using a modified rammed earth technique known as **PISE**, which stands for Pneumatically Impacted Stabilized Earth. This technique requires much less formwork and no tamping. Rather, the earthen mix material is sprayed from the outside, using a gunite spray rig against a plywood form, to a thickness of 18–24 inches. In other words, the wall is built out laterally instead of vertically, and much more quickly. In seismically active regions, the walls are reinforced with rebar for protection. The Real Goods store at the Solar Living Center is built of 600 rice straw bales all covered with PISE, applied by a machine similar to the one that affixes gunite to a swimming pool. After 20 years, the PISE material remains strong and the straw bale building still performs flawlessly.

Earthship, the invention of architect and builder Michael Reynolds, is a technique in which automobile tires are stacked and filled with earth. Americans throw away almost 300 million automobile tires each year. Although more and more

are reused to make more blocks. To build walls, adobe bricks are laid in a running bond (overlapping pattern), just like modern bricks or stones in a stone wall, directly on a foundation. A mortar made of the same dirt is used to hold the bricks in place. When completed, adobe walls are often coated with an earthen plaster that protects them from the weather, especially wind and rain.

Another ancient natural building technique, **rammed earth**, is growing in popularity in the US thanks to the pioneering work of architects,

Dan Chiras

The Earthship is made from used automobile tires, collects rainwater from its roof, and heats and cools itself naturally.

tires are recycled into useful products, millions of tires are still simply discarded. Today, however, a small percentage of used tires is being salvaged to build houses. This technique, known as **rammed earth tire construction**, is not entirely natural, but it has many redeeming qualities worth serious consideration.

Rammed earth tire houses are typically built into south-facing slopes for earth sheltering and passive solar design, so excavation must precede building. Once the site is ready, the tires are laid on the compacted subsoil. Tires are typically laid in large U-shapes, one for each room of the house. Next, a small piece of cardboard is placed over the center hole at the bottom to prevent dirt that is shoveled in from escaping. One worker begins shoveling dirt from the site into the tires while another compacts the dirt using a sledgehammer or a pneumatic tamping device. After extensive pounding or tamping, the tire is fully compacted and easily weighs 300 pounds. The tire rows are set in a running bond pattern, typically six to eight rows high. After the wall is finished, a roof is attached. The tire walls are then plastered with mud or stuccoed with cement.

Other innovative earth techniques lend themselves to being built by the homeowner or with unskilled labor. **Cob** construction uses a mixture of sand, clay, and straw, and is applied directly to the foundation, and massaged into shape. Cob walls are strong and thick, up to 36 inches in some homes, and are as solid as a rock. To protect against weather, cob walls are either whitewashed, lime plastered, or coated with an earth plaster.

Earthbag construction attempts to democratize rammed earth building. It was pioneered by Iranian-born architect Nader Khalili and popularized and further perfected by two maverick builders in Moab, Utah, Doni Kiffmeyer and Kaki Hunter. Rather than using wooden or steel or tire forms, earthbaggers use polypropylene bags. These widely available bags are filled with slightly moistened dirt from the site—typically subsoil containing a mixture of sand and clay. The bags are filled on the foundation one by one and laid flat into a course, which is then tamped. Tamping flattens and compresses the earthen material within the bags, transforming them into solid blocks. Before the next course is laid, two strands of four-point barbed wire are laid down on top of the bags. The wire provides continuous tensile strength around the building, and the barbed points dig into the bags above and below and "Velcro" the courses together. When completed, the walls are plastered, preferably

Left: Becky Bee; right: Robert Bolman

Cob lends itself to organic forms. The house to the left is finished with lime plaster.

with earthen or lime-sand plaster. Both adhere admirably well to the earthbags without stucco netting or diamond lath.

Cob and earthbag are both good choices for owner-builders, as they are easy to master, though labor-intensive. These styles of construction lend themselves to sensuous curved walls, arches, and niches, and are delightful places to live that evoke serene feelings.

Cast Earth was invented in 1993 by Harris Lowenhaupt, a Phoenix-based metallurgist. These houses are made from a slurry consisting of water, soil, and 10%–15% heated (calcined) gypsum. The slurry is poured into forms set on a concrete foundation (as in rammed earth construction). The slurry sets up fairly quickly after being poured, thanks to the gypsum. Once the forms are removed, the exterior walls are typically coated with plaster or stucco for protection and aesthetics. Cast earth is one of the quickest and most expensive ways of building a natural house. If you want a cast earth home, you will need to contract with a certified builder, as it is a proprietary process.

Straw-clay is a relative newcomer to the natural building field in North America, although it has been used in Europe (especially Germany) for at least 500 years. Reintroduced by New Mexico-based architect and builder Robert LaPorte, straw-clay is made from straw with a little clay slip, a mixture of clay and water that serves as a binder. Clay slip is poured onto the straw and mixed in until it slightly wets the surface of all the straw. The mixture is then packed as in-fill into 2-foot-high wooden forms attached to the wall framing, either by lumber or posts. The slightly moist straw-clay mix is then tamped by hand. Bamboo reinforcing is installed to ensure lateral stability, usually attached to holes drilled in the framing members every few feet along the height of the wall. When the walls have dried, they can be coated with plaster.

Because it provides a fair amount of insulation, straw-clay is an excellent choice for those interested in building an energy-efficient, passive solar house. Likewise, when interior surfaces of exterior walls are coated with a 1- to 2-inch layer of earthen or lime-sand plaster, they provide a fairly good source of thermal mass. Along with straw bale, straw-clay homes are suitable for a wide range of climates.

However, building with straw-clay is a fairly slow and labor-intensive process. The drying time is the slowest of all the natural wall systems— about a week per inch of wall thickness—so they should be built when a lot of good drying weather is expected. Generally, walls do not exceed a foot in thickness; otherwise they dry so slowly that mold becomes a concern.

Building techniques that use earth produce beautiful, solid, fireproof walls. These massive walls can withstand extremes of weather, including hurricanes and tornados. Techniques like adobe and rammed earth stand up to wind much like concrete or concrete block walls and far better than the walls of stick-frame homes. Other types, like cob, require more protection from the elements. They are ideal for passive solar heating and cooling in desert climates, and can work well in colder climates as well so long as exterior insulation is applied to prevent heat loss. These homes stay unbelievably cool in the summer and warm in the winter.

Earth homes have some disadvantages as well. They tend to be highly labor-intensive, and require heavy equipment to transfer the mix or materials onto the foundation. While some of these techniques can be taken on by homeowners, others require special skills and specific tools. Hiring a qualified contractor can mean labor costs will be substantial, which could eat up any savings found on material costs.

Hybrid Natural Houses

There was a time when a straw bale homebuilder built entirely with straw bales and a cob builder built exclusively out of cob. As more and more natural builders have learned about the benefits of the wide array of natural building materials, however, the field has opened up considerably. Most natural builders today utilize two or more natural materials in any given project. They may, for instance, build a straw bale house with rammed earth interior walls for thermal mass. They might build a cob bench and shed. They might even build interior straw-clay walls in places where acoustic insulation is required. Houses may be placed on a stone or rammed earth tire foundation. Many builders use cob and straw-clay to fill in unusual places, like nooks and crannies that are hard to fill with straw bales. They may use cob to sculpt window and door openings as well as shelves and even freeform fireplaces. Finally, many natural builders use earthen plasters or an earthen plaster base coat with a lime-sand plaster finish coat on their creations.

In many cases, they even use conventional building materials and techniques like stick framing for interior walls, roofs, and floors. They may integrate products such as wet modules, which are a complete kitchen or bath that is built in a factory, then shipped and craned into place

> Building techniques that use earth produce beautiful, solid, fireproof walls. They are ideal for passive solar heating and cooling in desert climates, and can work well in colder climates as well so long as exterior insulation is applied to prevent heat loss.

within a natural building. These modules reduce waste and increase the speed of construction. Natural homes may take advantage of high-tech mechanical systems, such as heat recovery systems or radiant heating, in order to add to the home's energy efficiency. In addition, manufacturers continue to explore simpler and more cost-effective ways to utilize natural building materials. Watershed Blocks, a product developed by David Easton, uses native soils to create an "earth masonry unit," which can be built into walls with the same techniques and tools as the very affordable concrete masonry unit (or CMU).

All this is to say that contemporary natural builders frequently mix and match natural, conventional, and high-tech building techniques to achieve the best results. The result is structures that are beautiful, healthy, and efficient.

"By viewing building as a process of combining different, but complementary, materials rather than adhering to a particular building system," note Bill and Athena Steen, two extraordinary veteran natural builders, "we have given ourselves the freedom to create structures that respond to a wide variety of contexts and circumstances. They can be elegant or simple, quick or detailed, inexpensive or costly, and probably most important, they can be built from predominantly local materials in whatever combination best matches the local climate."

Utilizing a number of materials does create challenges, so be sure to heed the advice of two veteran straw bale builders, Matts Myhrman and S. O. MacDonald, authors of another classic book, *Build It with Bales*: "By its nature, a hybrid structure often requires extra thought during the design process. Draw it, model it, get a 'second opinion,' and still expect to have to think on your feet once you get started."

Insulated Concrete Forms

As just noted, in the process of hybridization that currently characterizes the natural building movement, builders sometimes use green products that, while not natural, offer substantial benefits. One of them is the insulated concrete form.

Insulated concrete forms, or ICFs, are hollow blocks of varying length made from rigid foam insulation (beadboard, also known as expanded polystyrene). One common brand, known as Rastra, is made of 85% recycled polystyrene (Styrofoam) and 15% cement, making it very environmentally friendly. Another, known as Durisol, is made of recycled wood and cement.

ICFs are used to build foundations and exterior walls, and often interior walls as well. They

Watershed Blocks made with various locally sourced soils.

David Easton

are internally reinforced with rebar to increase tensile strength and then filled with concrete.

Unlike wood or steel or concrete forms used to build walls and foundations, ICFs are lightweight and easy to assemble. They pretty much snap in place like pieces of a three-dimensional puzzle. To prevent blowout when concrete is poured into them, manufacturers install plastic or steel cross bridges that attach one side of the form to the other. Products like Durisol and Rastra, however, are more like concrete blocks in their structure. After concrete is poured, the ICFs are left in place. The foam sandwich produces a foundation or external wall with an insulation value of around R-35, depending on the width and the manufacturer. Tech Block, a relative newcomer, boasts an R-value of over 45!

Most natural builders today utilize two or more natural materials in any given project. They may, for instance, build a straw bale house with rammed earth interior walls for thermal mass.

Interlocking insulated concrete forms (here made by American Polysteel) are placed, reinforced with rebar, then filled with concrete. They're great for foundations and walls, providing superior insulation.

American Polysteel

Sweetwater Springs: A Hybrid Solar Green House

Sweetwater Springs Ranch is an off-grid straw bale and Watershed Block (earth masonry unit) home in Sonoma County, California. Located in a sensitive watershed area, the design and construction of the home took great measures to ensure a harmonious fit with minimal impact on the landscape. Here we highlight some of the more important features, with hopes they provide some inspiration or insight into your own building project. The house is for a couple with two young twin girls; the design was by Arkin Tilt Architects, with Nadia Khan assisting early in the design and then Amanda Knowles serving as project manager; Darryl Berlin designed and managed construction of the infrastructure improvements; Jeff Oldham of Regenerative SOLutions designed and installed the renewable energy system; and Earthtone Construction of Sebastopol was the general contractor.

Siting: The 170-acre parcel offered several possible building sites, but with the use of a Solar Pathfinder and other site analysis aspects as discussed in this chapter, we quickly narrowed in on a south-sloping open hillside that was dubbed the "Big Sky" site. Other sites under consideration were eliminated for various reasons: One possible site was across a year-round creek and would have required an expensive bridge, and also need extensive environmental study and approvals that would have delayed the project substantially. Another site was closer to the main road and near a lovely rock formation that would have been a delightful home site—however, when a rattlesnake joined one of our meetings, we concluded that this was already his (and likely many of his cousins) home site, and the location was removed from the list. The final home site allowed us to route the driveway access further from the creek, and eliminate a ranch road that had been channeling rainwater runoff into the creek. The homeowners have also been participating in a salmon restoration effort located along their portion of the creek.

Passive Solar: The "Big Sky" site was so dubbed due to the ample access to sun throughout the year, thus ensuring passive solar heating during the winter months. When the sun is lowest in the sky, it still clears the ridge across the creek in this spot because the creek makes a sharp turn to the south. The homestead is designed as a series of east-west running bars, roughly paralleling the topography and stepping up the hill. These ensure direct sunlight into all of the main living spaces during the cooler times of the year. Overhangs and exterior shade fins keep the higher summer sun out of the windows during the warmer times of the year. Nearly

Arkin Tilt Architects

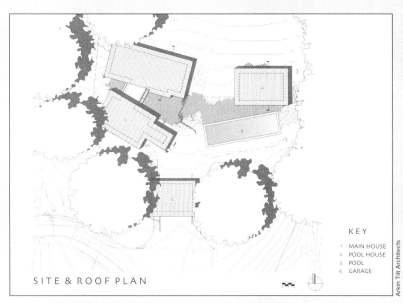

SITE & ROOF PLAN

KEY

1 MAIN HOUSE
2 POOL HOUSE
3 POOL
4 GARAGE

Arkin Tilt Architects

Arkin Tilt Architects

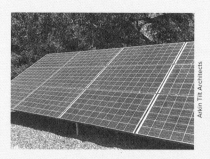

Arkin Tilt Architects

all of the items listed on the Passive Solar Checklist are implemented in the design of this house.

Active Solar: The electrical power is provided by photovoltaics along with an LP backup generator. Micro-hydro was explored but found to not be cost-effective given the ample sunlight and anticipated very infrequent times when the generator might be needed. The 8.64 kW PV array features thirty-six 240-watt Trina panels, ground mounted just east of the home site. (Trina Solar, a Chinese company, was recently named the "Greenest Manufacturer on the Planet" by a German certification group.) An 8 kW Outback Radiant inverter is configured for 120/230 V tied to a 1490 Ah 48 V battery array (71.8 kW). One of the unique features of this setup is the way it is set to auto-start if one of three different conditions are met: 1) if the batteries reach 50% for 24 hours, or 2) 30% for two hours, or 3) 20% for 2 minutes. It will also shut down all power if the generator doesn't start, in order to protect the batteries. The project also features solar-direct pool filter pumping powered by four 405-watt panels. Solar hot water collectors are located on the highest of the roofs, reducing the domestic water heating loads significantly, approaching zero additional energy needed in the summer months.

Straw Bale and Watershed Block and Site-milled Timber: The project is a hybrid of materials, each

performing a function to which it is best suited. For the foundation retaining walls, we chose block due to the remote location, aiming to limit the amount of delivery trips for concrete. During development of the design, David Easton's Rammed Earth Works company built their first home out of Watershed Block, an Earth Masonry Unit (EMU) made in a hydraulic press and to standard concrete block specifications; we quickly volunteered to be the second house. While more expensive, the blocks do not need to be subsequently finished with stucco or another material, as they are beautiful unto themselves. The walls that are out of the ground are in-filled with rice straw bales, coated inside and out with lime plaster. These have at least an R-value of R-30—but more significantly, the 12-hour time lag of thermal transfer is perfectly suited to the large diurnal temperature swings of northern California. Lastly, the project features a good amount of timber milled at the site. Actually, it was material milled on a neighbor's site, which had cured for a year, and in turn the owners are letting the neighbor mill a comparable amount of timber from their site, selectively and sustainably harvested.

Living Roof and Curved Metal Roofs: Literally rounding out the project are the roofs, which feature a deeply corrugated zinc-alume-coated steel. It is curved to a 65-foot radius, which is close to the limit it can be bent without requiring special factory shaping. The living/dining room space—located to the south of the bedrooms—would have presented quite a glare problem if finished in steel, so this roof instead features a curved living roof planted with sedums and other drought-tolerant varietals. The roof deck over the kitchen presents an easy access for occasional weeding and maintenance of the roof. There is a similar roof deck over a portion of the pool house, the roof of which shades the upper studio space. Carefully sized overhangs ensure thermal comfort throughout the year, with the exposed structure of these sized as heavy timbers to meet fire safety regulations.

ICFs significantly reduce both the amount of concrete needed to build foundations and exterior walls and the amount of lumber used throughout a structure. Although expanded polystyrene is made from a chemical extracted from oil, bear in mind that the foam is 98% air bubbles. (Air bubbles block heat flow through walls.) Also, no ozone-depleting chemicals are used in the production of expanded polystyrene.

Although ICFs are, for the most part, made from virgin materials derived from oil, there's a notable exception: the Rastra block. Rastra blocks, which Real Goods founder John Schaeffer used to build his house (see page 71), are made from a small amount of cement (about 15%) and 100% recycled expanded polystyrene (EPS) foam (about 85%). Like other ICFs, Rastra blocks are lightweight and easy to handle—they can even be cut with a saw. Another huge advantage to Rastra, then, is that it lends itself beautifully to being shaped and sculpted into just about any design.

According to *Environmental Building News*, Rastra "is insect-proof, creates a very strong wall, and is extremely fire-resistant.... structural testing found a Rastra wall to be seven times better under earthquake-type stresses than a wood-framed shear wall, according to marketing director Richard Wilcox" (July/August 1996).

Passive House

Originating in Europe, the Passive House movement was founded in extreme climates where a tight, super-insulated building envelope in conjunction with solar heat gain have combined to create dwellings that use remarkably little energy.

In order to create a passive house envelope, attention must be paid to details where there is a potential for thermal bridging, or air leakage.

Tight seals around penetrations in roofs, walls, and ceilings can help to reduce heat transfer through air movement. A mineral wool insulation wrap on the exterior of the home (under your finish siding) minimizes thermal bridges that allow heat to escape.

Successful building envelopes depend on multiple layers of defense, creating resiliency within the walls of your home. Building a Passive House in this way requires a level of expertise, and may require putting together a team of designers and builders who know the specific requirements and are capable of producing results.

The Passive House Institute US defines a Passive House as one that has the following performance characteristics:

- Airtight building shell ≤ **0.6 ACH [air changes/hour] @ 50 pascal pressure**, measured by blower-door test
- Annual heat requirement ≤ **15 kWh/m²/ year (4.75 kBtu/sq ft/yr)**
- Primary Energy ≤ **120 kWh/m2/year (38.1 kBtu/sq ft/yr)**

In addition, the following are recommendations, varying with climate:

- Window U-value ≤ **0.8 W/m²/K**
- Ventilation system with heat recovery with ≥ **75% efficiency with low electric consumption @ 0.45 Wh/m³**
- Thermal Bridge Free Construction ≤ **0.01 W/mK**

The Passive House Planning Package (PHPP) is a tool to help you design and analyze the performance of a design relative to the performance characteristics outlined above. Even if you do not pursue Passive House Certification, or even achieve the performance levels, applying these techniques can help you make significant progress toward creating a net-zero energy home.

Green Building Materials

One element of green building that receives a great deal of attention, and rightly so, is the use of environmentally friendly and people-friendly building materials, commonly referred to as *green building materials*. The term defies easy definition because it has multiple dimensions. In general, building materials are called *green* because they are better for the environment. For example, they may be produced from recycled materials, produced locally, minimize waste or harmful by-products, be durable and long lasting, or have the ability to be recycled at the end of their life as a building product.

Rigid insulation installed below a fiber cement rain screen.

Arkin Tilt Architects

The Source

Let's start by considering where the material comes from. Many products available today come with a recycled content, such as crushed glass tiles. In a case like this, less virgin material is used in the production, and a former waste product is given a new life. Another example is ICFs or Rastra blocks, made of 85% recycled Styrofoam and 15% cement. Other building materials are considered green because they are made from renewable resources that are harvested sustainably. Flooring made from sustainably grown and harvested lumber or bamboo is a good example.

The Process

Most of the time, what you put in your home is not a raw material that was just harvested from the land. Instead, it has gone through an intensive production process, where it is cut, crushed or broken down, mixed with other additives, and reformed into a final product. All of this takes tools, energy, and labor, and can generate waste by-products like polluted water, off-cuts, and fumes. Efficient manufacturing is a mark of a green building material. Companies strive for production that requires much less energy and fewer raw materials than for a conventional counterpart. Examples include engineered lumber such as glue lams and wafer board or wooden I-joists. Engineered lumber is more efficient than conventional wood products because floor and ceiling joists made from it use much less wood than a solid 2×10 framing timber. Moreover, you don't have to cut down an old-growth tree to create a 2×10 engineered wood joist. Engineered lumber is typically made from small, fast-growing trees.

Another group of green building products of considerable importance is the low- or no-VOC (volatile organic compound) paints, stains, finishes, and adhesives. These products are made without toxic resins that can cause cancer or other health problems. They're safer for the factory workers who produce them, the subcontractors who install them, and the family who lives in the house. In other words, they're healthier for all who come in contact with them.

The Performace

Some building materials are considered green because they're durable. A durable form of siding like cement-fiberboard outlasts less durable products such as wood shingles, resulting in significant savings in energy and materials over the lifetime of a house. If a durable product is made from environmentally friendly materials, for instance, recycled waste, it offers even greater benefits. For example, shingles made from recycled tires are class-A fire rated and expected to last 50 years. Siding made from cement and recycled wood fiber (wood waste) is a good example of a green building material that offers at least two significant advantages over less environmentally friendly materials. It will very likely outlast many homeowners, and because it is durable, it saves on costly maintenance.

The End Game

Some building materials are green because they can be recycled once their useful lifespan is over, for example, aluminum roofing shingles. If that product itself is made from recycled materials, the benefit is multiplied.

The Market

Ideally, green building products should be manufactured by companies that embrace environmental stewardship and resource efficiency,

> Green building materials may be produced from recycled materials, produced locally, minimize waste or harmful by-products, be durable and long lasting, or have the ability to be recycled at the end of their life as a building product.

Green building materials help builders create more energy-efficient and environmentally friendly homes. Hardie plank (far left and left), made from recycled wood fiber and cement, creates a durable siding that will outlast cheaper alternatives. Structural insulated panels (right), used to build energy-efficient walls, roofs, and floors, reduce labor and lumber use. Wooden I-beams (far right), for framing floors and roofs, use substantially less wood than solid dimensional lumber and help protect old-growth trees.

Dan Chiras

support the local economy and community, and demonstrate a strong commitment to protecting workers from exposure to toxic substances. Pollution prevention programs are an important sign of a company's environmental responsibility and its desire to safeguard the health of its workers. Energy-efficient operations, factory recycling programs, and the use of recycled waste or sustainably harvested wood are additional signs of a company's commitment to the environment.

The production of green building materials has expanded tremendously in recent years. In fact, these days nearly every material or appliance or piece of furniture or furnishing has at least one green alternative. The number of retail sources has also skyrocketed. Online mail-order companies that can drop ship hundreds of products to a building site are now operating nationwide. Some products, such as recycled plastic decking and certified lumber, can often be purchased at local building supply stores. As green building continues to grow, you can expect wider availability. With so many options, and various ways to define "green," it is important for each builder or homeowner to consider his or her own goals in choosing a sustainable material, and to weight the costs and benefits of those selections.

Surely you're asking yourself, "But what about the cost of green building materials? Aren't they terribly expensive?"

The good news about green building materials is that many of them are cost competitive with conventional materials. Other green products may cost more—sometimes a lot more—but they provide peace of mind and satisfaction that you're doing the right thing for the planet as a home-owner, including helping this critically important industry gain a foothold in the marketplace. In time, as more and more builders go green and the demand for green building materials rises, prices of materials that cost more than conventional materials could start to tumble. And don't forget, a product that is healthier may be expensive, but then again, so is cancer.

Choosing a green building material is not always easy. Many products will require making trade-offs; for example, they may be produced from a renewable resource that is sustainably harvested halfway around the world. To ship them to your building site requires an enormous amount of energy. Should you use a product like this?

That's your decision. You may have to compromise on some energy or cost factor, but you may still decide that the choice of a certain green building product makes sense anyway, even if isn't quite perfect. By using these environmentally friendly products and helping to stimulate markets needed to drive this important endeavor, you can become an agent of change.

To achieve the greatest benefit, you probably want to choose green building materials that offer the greatest gains. You may also want to concentrate on big-ticket items—that is, products that are used in greatest quantity, such as framing lumber, insulation, roofing materials, floor coverings, wall finishes, and foundation materials. Resources like the *GreenSpec Directory*, published by Building Green and available online at build inggreen.com/menus, list, describe, and provide contact information for thousands of green building products.

Characteristics of Green Building Material

- Produced by socially and environmentally responsible companies
- Produced sustainably—harvested, extracted, processed, and transported efficiently and cleanly
- Low embodied energy
- Locally produced
- Made from recycled waste
- Made from natural or renewable materials
- Durable
- Recyclable or reusable
- Nontoxic
- Efficient in their use of resources
- Nonpolluting

The Healthy Home

Green building materials also help create a healthier indoor environment. As noted above, some products can actually improve indoor air quality, making our homes more healthful places in which to live.

Healthy home construction is essential today if you are going to build to more exacting energy standards—that is, better insulated and much more tightly sealed. Although tight construction is a good idea from an energy standpoint, it can create significant issues. Many modern building materials are manufactured with potentially toxic chemicals. Oriented strand board (OSB), for example, which is used to sheath walls, floors, and roofs, is an engineered wood product held together by a resin that contains formaldehyde, which has been shown to be hazardous to human health. In addition, sealing your home tight can trap moisture from internal sources, such as showers, creating potentially harmful mold.

Combine air tightness with mold and toxic chemicals and you could be creating a health nightmare. Occupants may have only minor complaints, such as headaches or allergies, or they may suffer more serious illnesses such as chronic bronchitis and asthma. Unhealthy indoor air can even lead to debilitating illnesses such as multiple chemical sensitivity and cancer.

Using healthy green building materials is a good idea but won't eliminate all health problems created by indoor air pollution. That's because indoor air pollution comes from many sources besides building materials. Prominent among those are the many combustion technologies we rely on, such as gas stoves, ovens, water heaters, and furnaces. Wood stoves and fireplaces also release potentially harmful pollutants, such as carbon monoxide, into the air inside a home. But that's not all.

Conventional cleaning products and disinfectants, household pesticides, personal care products, and pets also generate indoor air pollution. That shouldn't be too surprising, since you're probably aware that these items contain volatile chemicals. But indoor air pollutants also come from sources you'd never suspect, such as carpeting, window coverings, furniture and cabinetry made from particleboard, and the stains and finishes that coat their surfaces. Vinyl floor coverings and wallpaper may also release potentially toxic chemicals. These and other household furnishings may release dangerous substances for six months to a year after application or installation, sometimes longer. The more airtight a home is, the more problems these sources create.

To create a healthy home, it's best to eliminate as many of these sources as possible. Choose healthy alternatives, for example, nontoxic cleaning agents, formaldehyde-free building materials, and water heaters with closed-combustion chambers. Take care in selecting furnishings and finishes that you add to your home. It is easy to source paints and stains with low VOCs. Since January 2014, you can also find upholstered furniture without flame retardants, which have been shown to cause cancer. Look for furniture labeled with the new regulation TB117-2013, or stuffed with natural fillings such as down or wool. These will not have been treated with the ineffective and harmful fire retardants. (For more information see greensciencepolicy.org/faq/.)

Eliminating sources of pollution is the first line of defense against indoor air pollution. The second line of attack is ventilation. An airtight energy-efficient home requires adequate ventilation. Make sure your home has operable windows so that, in the summer, you can open them and purge air of toxic pollutants. When the house is closed up, however, you must provide mechanical ventilation.

These days, many contractors attentive to energy efficiency install heat-recovery ventilators in the houses they build. These devices continuously remove stale indoor air when a house is closed up and replace it with fresh outdoor air. The opposing air streams pass through a heat exchanger, a device that transfers heat from the warm outgoing air to the incoming cool air, thereby saving energy. There's no sense in losing all that hard-earned heating or air conditioning!

For more information about indoor air quality and cleansing or purification techniques, see Chapter 9, "Water and Air Purification," pages 267–77.

> Eliminating sources of pollution is the first line of defense against indoor air pollution. The second line of attack is ventilation. An airtight energy-efficient home requires adequate ventilation.

High operable windows used to flush out the heat of the day during the cool nights in northern California.

Edward Caldwell Photography

Sustainable Choices

We've discussed a multitude of ways that you can design and plan a home that is responsible and resilient. Regardless, it is still dependent on outside systems to provide water to your fixtures and energy to your lights. In order to really be a good member of your community, you should consider how to make the most of the resources you need to consume to survive.

Even the best passive designs still use some form of energy to turn on lights at night, cook food, and heat the interior during the coldest seasons. To learn more about how to produce your own energy, and make the most of what you consume, see Chapters 3 and 4 on renewable energy, 6 on energy conservation, and 8 on water heating.

Water is another resource that is often taken for granted. The average American family uses 400 gallons of water a day! The water goes to indoor uses such as showers, toilets and sinks, and also to watering lawns and gardens, or filling swimming pools. There are simple and cost-effective techniques to reducing our water usage; see Chapters 7 and 10 for more information.

Your Home, Your Choice

> We've discussed a multitude of ways that you can design and plan a home that is responsible and resilient. In order to really be a good member of your community, you should consider how to make the most of the resources you need to consume to survive.

Building or renovating a home is not just an academic exercise. It is a real-life process, with multiple inputs and outputs, pros and cons to be weighed, and restrictions to be considered. You must acknowledge the neighborhood and micro-climate you are building within, your regional climate, and the limits it presents. You can't ignore your own lifestyle, preferences, family size, and budgetary concerns. The reality is you won't be able to find a solution that solves every problem.

But, you can do your best, and ask the same of those around you. You can make decisions about what elements are most important to you, where you want to get involved, and where you'd like others to share the load. You can recognize the cost and benefit of each decision, and weigh it against your overall impact. You can find a solution, and a home, that doesn't only meet your needs, but inspires you to live a fuller and more productive and peaceful life.

Once you have your home, you can share it with others, record your story to be passed on, and engage more deeply in the community around you. After all, once you get past all of the technical work, the monetary concerns, and the global impact, we are still talking about home. Make it one worth living in!

Words of Wisdom from Real Goods Staffers and Colleagues

Many of the staffers, colleagues, and friends at Real Goods have bought their own land, built their own houses, and acquired some good down-to-earth advice from the school of hard knocks in the process. The following tips are a distillation of lessons they and others have learned along the way. We hope that some of the mistakes we have made and some of the advice we offer can help you build your dream home with less stress and fewer errors, while saving you some money.

On Where to Build

"Don't buy too far out in the boonies. Elbowroom is great, but there are practical limits. My property was five miles out a steep, rough dirt road. That's fine if you're independently wealthy and don't have kids or friends. The difficult commute was my primary reason for eventually selling the property."
—Doug Pratt

"Permaculture teaches us to use whatever we do to the land as a healing action. Leave the beautiful lookout as just that—don't build your house on it. Build where the land needs the most help and make that place beautiful. Also, notice the flows of water, wind, wildlife, and sun across the land. Enhance the performance of your spaces by aligning them with the natural flows."
—Tony Novelli

"Don't build your driveways on north slopes that don't get at least 1–2 hours a day of sun in the winter. Ice or snow can stay there for weeks. The steeper the drive, the more important it is! We learned this the hard way: This winter we could not get our car out even with chains for 2 weeks. It was 4 days before the truck could get out in a 4×4 *and* chains."
—Jeff Oldham

"When looking for land, consider in-town lots. You'll ride your bike more, shop more locally, and help strengthen the web of your community. After building my first straw bale home in a beautiful but remote location, I love the options I have living in town and not needing my car all the time. My life is far more sustainable now."

—Laura Bartels

On Being Realistic with Time and Money Plans

"Everyone thinks they have to push the green envelope with every decision and ends up with compromises in the wrong places and budgets and time frames blown. Know that the choices you make cumulatively represent positive impacts, and be smart about where to use mainstream approaches and where to go totally eco-conscious."

—Tony Novelli

"Be patient. You're going to be living in your house hopefully for many decades. If you can do it right by spending a few extra months, stop rushing, and pay attention to detail."

—John Schaeffer

"If you're going to use an alternative building approach, start with a shed. Learn the techniques, and weigh the efforts. You might decide that building a house of straw bale is a better investment of your energy than trying to build an entire house out of cob."

—Tony Novelli

"If I had to do it over again, I'd realize that it takes three times as long and costs three times as much as expected. I'd have my finances together so it could be built within a year's time instead of 15 years!"

—Debbie Robertson

"Think concentrically: Put the most attention, detail, and quality materials closest to you on a daily basis. Only use wood where it's visible and its beauty and functionality can be appreciated."

—Tony Novelli

"Once you realize you've made a mistake, your first loss is your best loss."

—Stephen Morris

"Pay as you go."

—Scott Nearing, via Stephen Morris

"There's an old saying: "Don't use up all your friends at the bale-raising; save a bunch for the plaster parties!""

—Tony Novelli

On Energy and Size

"Renewable energy makes lifestyle support possible, even miles from the nearest utility pole. The abundant solar resource is thousands of times more plentiful than required for continuing the enterprise of modern civilization, and Real Goods offers the appropriate technology to harvest, store, and convert it. 'That which is not sustainable will not be sustained'—the sooner we transition to sustainable production and consumption, the better. It's never too late to do the right thing."

—Erik Frye

"Small is beautiful in more ways than you think. Five hundred fewer square feet at \$100–\$150 per square foot is \$50,000–\$75,000 that you can invest in higher-quality and more sustainable materials, higher-performance systems, and better design. The smaller house also makes it easier to achieve zero-energy with renewable energy systems. Whatever you build, you're going to pay for initially and then pay for over and over as you pay taxes on it and clean, maintain, repair, and heat and cool it. Building a smaller house will have a smaller ecological footprint—an even bigger benefit."

—David Eisenberg

"Know your climate, and the microclimates on your site. Simple decisions on orientation, ceiling height, window placement, thermal mass, and overall aesthetics can be inexpensive to free during the design process, saving oodles of energy later. One example is that most areas of the country have extended periods of the year where outdoor living is comfortable. Cooking outside in the summer reduces cooling loads inside, and covered porch areas can reduce total indoor conditioned space."

—Tony Novelli

"Before you even think about installing a solar electric system or a wind machine or a solar hot-water system, be sure to make your home as energy-efficient as possible. Each dollar you spend on energy-efficiency measures will save \$3–\$5 in renewable energy systems cost. Remember: Efficiency first!"

—Dan Chiras

"If building new, consider wiring part of your house to carry the most basic loads: Air handling, refrigeration, water pumping, a bit of lighting. Put the main loads in another box so you can size a backup system efficiently for the smaller load center."

—Tony Novelli

"There is no such thing as too much mass or insulation." —Jeff Oldham

On When to Move In

"Get a cheap, portable living space initially. I had a refurbished school bus that allowed me to move onto the property with a minimum of development. This saved rent and allowed me to check out solar access and weather patterns before choosing a building site and designing a house. A house trailer can do the same thing. Don't move in until it's finished. It's real tempting to move in once the walls and roof are up. Resist if at all possible." —Doug Pratt

"I lived in a school bus until it was finished. It was good to be able to move into the shade in the summer and into the sun in the winter. Buses are cheaper than trailers and come with a motor and charging system. One of my biggest mistakes was moving into the house before it was 100% completed. Finish it before you move in, or you never will." —Jeff Oldham

On General Building Plans

"Keep a notebook of all the houses and spaces you feel the most comfortable in. Take photos, make sketches, and write descriptively about each and what appeals to you. Interfacing with a designer becomes much easier when he or she gets a clear picture of what you want." —Tony Novelli

"Take an honest assessment of your skills. Decide up front which tasks you can accomplish through your own efforts and which will require assistance. Assign a dollar amount and time required for each. Establish a working budget, which you should update throughout the project." —Robert Klayman

"Question Authority. Since most of us think we know best, debate yourself on all plans and try to find flaws or a better way. You'll soon notice that you will continually evolve the need into something far more elegant and affordable." —Jeff Oldham

"Hire a carpenter and become his apprentice." —Terry Hamor

"If you're hiring a contractor, get him or her involved from the beginning of the design phase. Get his/her buy-in for the project before you start. Always find a contractor you can trust and pay him/her time and materials instead of doing

a contract. With a contract, one of you loses every time—with time and materials, it's a win-win." —John Schaeffer

"Examine carefully your contractor's work and his or her relationships with clients. Choose professionalism over friendliness. Notice how organized his or her toolboxes and materials are." —Tony Novelli

"Take a permaculture course. Nothing will teach you better about how we live in zones and are well served by designing carefully the zones we spend our time in, inside and outside our homes." —Tony Novelli

"Complete your plan (for a house or garden), then cut by half and cut by half again." —Stephen Morris

"Build your power shed first for an off-grid home and install your solar, wind, or hydro. THEN build out the site using your RE system instead of a noisy and expensive to use genset. You're going to buy the energy system anyway, so do it first and in the end have a brand-new backup genset." —Jeff Oldham

"Use passive solar design. This was one of the things I did right. My house was cool in the summer and warm in the winter, and with an intelligent design, yours can be too. Avoid the temptation to overglaze on the south side. You'll end up with a house that's too warm in the winter and cools off too quickly at night. Buy quality stuff. Cheap equipment will drive you crazy with frustration and eat up your valuable time." —Doug Pratt

"Avoid west-facing windows and control solar gain with exterior shutters/blinds to stop the heat *before* it gets inside your windows and home." —Jeff Oldham

"Money is no object when it comes to insulation! Don't use aluminum door or window frames. Aluminum is too good a conductor of heat. Use wood, vinyl, or fiberglass instead. Pay attention to little details. It's not a matter of which roofing or siding system you choose; it's the small details like flashing, corner joints, and drainage runoffs that really count." —Jeff Oldham

"Design a compost bucket into your kitchen counters that's flush with the top. Design recycling chutes into the kitchen walls with 6-inch PVC so you can use it for bottles and cans—it makes recycling fun for kids and a breeze to keep clean and organized." —John Schaeffer

"Design your wood storage with the ability to load from the outside and retrieve from the inside. Build a laundry chute and a dumbwaiter. Have a small utility bathroom accessible from a back door. Plan for, or install, a dishwasher." —Debbie Robertson

"Spark arrestors, spark arrestors, spark arrestors! Stay on top of wood stoves, mowers, weed eaters, chain saws, and all things that can toss a spark. Keep them clean and intact—these are common causes of wildfires." —Jeff Oldham

"The house that George and I built on Low Gap was the first off-the-grid home in California to be funded by Cal-Vet. My advice is to never take "NO" for an answer from any lender. We were originally turned down by Cal-Vet, so we wrote up an appeal and presented it to the State Department of Veteran Affairs in Sacramento and won the appeal!" —Jeanne Kennedy

"Learn the Snow Pants Rule. If something (such as going outside to play in the snow) is worth doing, it better justify the time and effort (of bundling up, putting on boots, etc.) of getting there." —Stephen Morris

SunHawk is located next to a 10-acre foot pond that provides passive cooling for the frequent 100°F+ summer days.

Mark Boyer

Spirit of the Sun

Between the two of them, John Schaeffer and Nantzy Hensley have 50 years of experience researching green building, off-the-grid living, and permaculture. John founded Real Goods, the nation's first solar retail business, in 1978, now in Hopland, California. Nantzy has lived off the grid since 1973 and joined Real Goods in 1989. The couple has realized their dream of employing their collective wisdom to create an energy-independent, nontoxic, environmentally gentle home that promotes sustainability—while also being tastefully beautiful and soul soothing.

They were right on track with that vision as construction on their 2,900-square-foot roundhouse—oriented to the cardinal directions and patterned after a red-tailed hawk ready to take flight—began in early 2001. The house is set on 320 acres that are richly landscaped with organic and biodynamic gardens, orchards, two large ponds for recreation and wildlife habitat, as well as a grotto with a waterfall, and that overlook the Hopland Valley.

In 2002, John and Nantzy moved into a barn on their property where they could watch the building progress. That summer, they noticed several large black birds pecking mercilessly at the windows. How cute, they thought, until they realized that, weeks later, the birds—ravens—continued to attack the house's 69 windows with a vengeance.

"We went into major raven research mode," Nantzy says. "We talked to biologists, ornithologists, and shamans. People told us to dance around in circles with corn, build altars, give them offerings. We eventually learned that when the birds are nesting, they're very territorial. They saw their reflections and tried to scare off those 'other ravens.'"

"A shaman told us the ravens were upset because we were calling the house SunHawk—ravens apparently have a natural animosity toward hawks," John says. "So we made an altar, and for 30 days we brought them tobacco, fish, and meat." In marked contrast to locals who thought a shotgun was the answer to their raven problem, John and

Nantzy's attitude was that the birds had inhabited this piece of land first and that they, as the human "intruders," should strive to live in harmony with the ravens. That open-mindedness translated into every step of building their ecologically friendly home.

Better Living Through Technology
John and Nantzy moved into SunHawk on the summer solstice in 2003 and have now been living there for over a decade. Built from Rastra blocks (aka Insulated Concrete Forms or ICFs), which are made from 85% recycled polystyrene beads and 15% cement, SunHawk is a masterful example of sustainability and regeneration. Nantzy spent months researching building materials and appliances. Her finds include recycled-tire roof shingles and repurposed granite countertops from a Berkeley cafe. Roof decking, fascia, barge rafters, and beams were made from reclaimed redwood, Douglas fir from an area winery, a reclaimed walnut tree that had to come down for a driveway at the nearby Solar Living

1930s recycled irrigation pipe form the gazebo top on SunHawk's outside patio space.

Center, and other recycled and reclaimed materials from a vinegar plant, warehouse, and converted orchard.

John and Nantzy's house is, of course, entirely off the grid. (How could you expect less from a renewable energy pioneer?) A 17-kilowatt solar system—the PVs recycled from a Real Goods installation in Belize blown down in a hurricane—provides ample power in summer months, and a hydroelectric turbine produces as much as 36 kWh per day from a seasonal creek that runs through the property from December through June. "Our micro-hydro system provides 36 kilowatt-hours per day—almost twice the national average for electricity use," John points out. "And it cost only $2,500—it's way more cost-effective than the electric company or any other source of electricity. It's a powerful feeling knowing we'll never have to pay an electric bill for the rest of our lives."

John is most proud of the home's innovative heating and cooling systems, which work so well largely because of the passive heating and cooling design. Cooling the house without air conditioning is no small feat in Hopland, where temperatures consistently rise to three digits or even to 110°F several days in the summer. SunHawk is cooled by earth-cooling tubes, two 150-foot-long pipes (12-inch culverts) that draw outside air 10 feet below the surface of the ground, where it is cooled naturally. The culverts empty into a 9-foot-deep rock storage chamber below the central portion of the house. Two solar-powered fans pull cool air into the central rock core, which remains at 67°F. From there, it travels by convection to the rest of the house. Even on the hottest day, the home's interior has never exceeded 76°F. The home is heated primarily by radiant floor tubing powered by

Stone "tree" in the living room from local rocks.

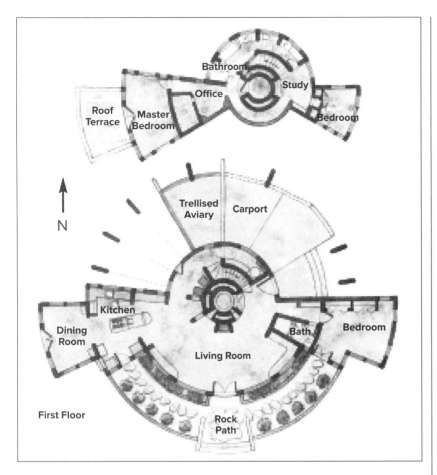

First Floor

Master bedroom.

rooftop solar hot water panels and water heated by the excess voltage from the solar and hydroelectric systems.

Living off the grid entails absolutely no sacrifice, John quickly points out. In fact, there are unexpected bonuses. "We got a wireless Internet system that is two to three times as fast as DSL, and our cell phones have 5-bar reception. The only real maintenance on the solar system is watering the batteries every six weeks," he says. "The fun part is we get to use our house as a laboratory for the technology and products we order for Real Goods."

Now This Is Permaculture

After John and Nantzy bought their acreage in 1998, they spent numerous nights camping in various places around their property to find the ideal spot for their house—in the end, exactly the location where Nantzy's intuition had initially told her it should be.

Building the Real Goods Solar Living Center in Hopland had taught the couple the importance of landscaping as an integral part of designing a homestead, so they made that a priority, even before finding a designer for the house. "We knew early on that this would be much more than just a house-building project," John says.

Their first major project after grading the road was to dig a 10-acre-foot pond flanked by a 30-foot-wide grotto overflowing with waterfalls from the property's three natural springs. Three and a half acres of lush permaculture landscaping include over 50 fruit trees, an acre of olive trees, a large vegetable and flower garden, native grasses, a coastal redwood grove, Mediterranean foliage, lavenders, and a corridor of swamp cypress by the pond, which attracts a variety of wildlife including herons, egrets, ducks, coots, swallows, otters, and giant bullfrogs. To complement the landscaping, the couple added 14 acres of olive trees from which they yield over 100 gallons of biodynamic extra virgin olive oil every

year, and an 8-acre biodynamic vineyard of Rhone varietal grapes. In addition they added a "lavender labyrinth" modeled after a 14th-century French cathedral's labyrinth that they harvest every July to make lavender oil and lavender hydrosol. The orchards and the pond are key to the home's comfort; prevailing winds from the northwest flow across them and bring evaporative cooling inside.

Four Rastra window boxes on the home's south side provide enough growing area for the family's kitchen and herb gardens. Vegetable beds are located 6 feet from the kitchen door so John and Nantzy can step outside in any kind of weather to harvest arugula, cilantro, and exotic lettuces. A few years ago, as part of a Real Goods Solar Living Center permaculture workshop, students spent four days building an herb spiral on the home's north side and installing a composting system, worm bin, and drip irrigation systems to water the vegetation.

Letting the Hawk Soar

With the major landscaping in place, John and Nantzy faced the task—and privilege—of creating a house that could live up to their ideals. "A home is far more than a shelter," John says. "It's an expression of our values and commitment, and it enables us to put our convictions into action. We wanted

A Conversation with the Homeowners

What do you love most about this house?
Nantzy: I love the feeling of permanence and belonging. European houses are frequently homes to families for a thousand years. I see no reason ours won't be here to see our newly planted redwood trees turn 1,000 years old.
John: There's nothing better than walking with bare feet on the warm floors on a cold winter day and curling up in the cushions in the south-facing window seats.

What's your favorite room?
Nantzy: In the dining room, we can sit at the table and enjoy 360-degree views of the pond and the wildlife. We keep a bird guide and binoculars on the dining room table.
John: I love having a great "dishwashing" window in the kitchen, where I can look out over the valley while I work. And, of course, there's the tower where we hold great solstice rituals.

How does the home perform? How comfortable is the house throughout the seasons?
John: The house has exceeded our expectations. The telling moment was a Solar Living Institute board meeting in July. The meeting took place in the midst of the worst heat wave I've ever experienced in the 35 years I've lived in Mendocino County. The temperature topped out at 110°F on Friday, 112°F on Saturday, and 110°F on Sunday and never fell below 85°F, even at night.

In the board meeting on the 112°F Saturday, 12 people met in our dining room. We were all totally comfortable, and the temperature never got over 76°F inside! The passive cooling measures (high levels of insulation) and the earth-cooling tubes worked admirably. All we used was a small fan in the meeting room.

In the winter, the passive solar and radiant floor provide all the heat we need. We have found that our wood stove (Tulikivi) is oversized, as the R-65 insulation on the roof and R-35 Rastra walls ensure that a minimum amount of radiant floor heat keeps us toasty, even on 25°F winter days. The wood stove is more of a fireplace with a glass door and is wonderfully cozy but not really needed to heat the house.

What would you do differently during the construction phase?
Nantzy: I would have created a hazardous materials collection station when we began building and introduced all the subcontractors to it. You don't want even nontoxic paints and sealers ending up in the landfill or your future garden.
John: I would be more patient. When you're going to live in a home for decades, who cares if it takes an extra six months to complete? I'd spend more time and think out every detail. It's easier to change things before the house is done.

What advice can you offer new homebuilders?
Nantzy: Include the outside areas in your planning. We just recently poured the sidewalks around the house, so we had a year of mud and dust. Don't wait until the house is done—do it at the same time.
John: Erect your solar system before you start construction. That way, you build with solar energy and don't have to listen to and smell generators—and you save money in the long run.

Now that you have lived in the house for a few years, do you think you would change anything about the energy design?
John: The earth-cooling tubes work great in the immediate vicinity of the central core of the house, and it's a great place to hang out and play drums when it's really hot outside, but we've found that the insulation and passive cooling account for 95% of the cooling and the cooling tubes probably only 5%.

The only thing I can think of would be to extend pipes carrying cool air (from the earth-cooling tube system) throughout the house to cool the upstairs better. On the hottest of days, the temperature downstairs climbs to 76°–78°F, but upstairs can sometimes reach 84°F. We have found that window blinds significantly help to cool down the upstairs on the west side. I think awnings or overhangs on the west side of the house would have helped reduce the late-afternoon heat gain in the summer.

All cabinets are made from recycled walnut from a tree that came down at the Solar Living Center.

ours not just to promote the principles of sustainability but to engender restoration and regeneration of the environment, while also nourishing the spirit."

They searched until they found architect Craig Henritzy of Berkeley, California, who understood that vision. However, they were taken aback when he showed them a set of plans for a Rastra house based loosely on the California Native American roundhouse (in the style of the indigenous Pomo Indians) and symbolic of the hawk—which he declared John and Nantzy's

"house totem." "We thought it was visionary and unique but a real challenge to pull off in a practical sense," John admits. Being open-minded, John and Nantzy visited another roundhouse Henritzy had designed in Napa, and they knew they'd found their home. "That house was unlike any other we'd ever seen," John says. "It felt Native American yet 21st century—both ancient and futuristic."

"For Native Americans, the hawk symbolizes 'vision,' which has been important in John and Nantzy's work," Henritzy explains. "Also, in the green architecture field, we often use just shed-type designs, which I feel has limited the progress of alternative designs being accepted in the housing market. This design explores passive solar with a geometry that celebrates the sun's cycles and playfully and beautifully assumes a hawk shape."

Because Rastra's polystyrene beads give it a fluid quality, it's easily cut and sculpted, making the hawk shape and orientation with the cardinal directions possible. Sunlight falling on a solar calendar running from north to south on the living room floor marks the passing seasons. On the winter solstice, sunbeams stream through a stained-glass hawk above the south-facing French doors, causing the bird to "fly" across the floor from west to east. At exactly solar noon, the sun illuminates a slate hawk in the floor in front of the living room wood stove.

The equinoxes and solstices are marked by the stained-glass hawk aligning with two ceramic hawks inside.

"I appreciate always knowing the position of the sun," John says. And that, Nantzy adds, is really just a fringe benefit of a good passive solar design. "The very basic, most important thing is good southern exposure—taking advantage of sun and light," she points out. "What I love most is that the sun comes in at the right time and doesn't come in at the wrong time."

Adapted from *Natural Home* magazine, November/December 2004, by Robin Griggs Lawrence, photography by Barbara Bourne.

REAL GOODS

CHAPTER 3

Sunshine to Electricity

Creating Renewable Energy with Solar, Wind, and Hydroelectric Power

THE MOST ABUNDANT SOURCE OF ENERGY AVAILABLE TO US HERE ON EARTH IS THE SUN. There's certainly no better source of power than a nuclear fusion reactor, at a safe distance, providing virtually unlimited energy. Ultimately, almost all energy on Earth comes from the sun. Some is harvested directly by plants, trees, and of course solar panels, also known as photovoltaics; much is used indirectly in the form of wood, coal, or oil; and a tiny bit is supplied by the nonsolar sources, geothermal and nuclear power. Solar panels receive this energy directly. Both wind and hydro power sources use solar energy indirectly, for without the changes in water and air temperature caused by the sun, there would be no air or water movement, and thus no energy to harvest. Even the coal and petroleum resources that we're so busy burning up now represent stored solar energy from the distant past. But we could do a lot more with current solar energy than we do: Each minute, enough sunlight falls on Earth to meet the energy demands of the world for a whole year. Best of all, with this energy source there are no hidden costs and no borrowing or dumping on our children's future. As sure as the sun will rise tomorrow, solar energy will be available to provide more than enough energy for generations after generations to come.

Solar energy can be directly harnessed in a variety of ways. One of the oldest and most efficient uses of solar is heating domestic water for showering, dishwashing, or space heating. At the end of the 19th century, solar hot water panels were an integral part of 80% of homes in southern California and Florida until gas companies, sensing a serious low-cost threat to their businesses, started offering free gas water heaters and installation. (See pages 252–66 for the full story of solar hot water heating.) Another way to utilize the sun's energy besides direct water heating is to convert it to electricity, which is the subject of this chapter. Photovoltaics, or PV, is one of the cleanest and most direct ways to harvest the sun's energy, and it has declined in price by more than half in just the last decade.

> Each minute, enough sunlight falls on Earth to meet the energy demands of the world for a whole year.

What Are Photovoltaic Cells?

Photovoltaic cells were developed at Bell Laboratories in the early 1950s as a spinoff of transistor technology. This became necessary because the utility companies could not figure out how to extend their wires into space when the early US vs. USSR space race got under way. Very thin layers of pure silicon, called "wafers," are impregnated

In 1954, Bell Telephone systems announced the invention of the Bell solar battery, a "forward step in putting the energy of the sun to practical use."

with tiny amounts of other elements such as boron and phosphorous, in a process called "doping." When exposed to sunlight, small amounts of electricity are produced. They were mainly a laboratory curiosity until the advent of spaceflight in the 1950s, when they were found to be an efficient and long-lived, although staggeringly expensive, power source for satellites. For electricity in space, there is still no match for PV. Since the early 1960s, PV cells have slowly but steadily come down from prices of over $40,000 per watt to $100 per watt when Real Goods sold the very first PV at retail in the US in 1978 to current retail prices of $1 per watt or even less in very large quantities. To give you an idea of the potential of solar energy, and the scale of deployment necessary, we could equal the entire electric production of the United States with photovoltaic power plants using about 10,000 square miles (100 miles by 100 miles), or less than 12% of the state of Nevada. (See the map on page 89 for details.) The only catch would be transmission: getting the electricity where it needs to go. As the true environmental and social costs of coal and petroleum become more apparent, PV is growing faster than any other sector to meet the challenges of our energy needs. The solar industry today is the fastest-growing sector in the US economy. And with a growing consensus that worldwide petroleum sources are past peak (see chapter 1), we are rapidly running out of "cheap" oil, making renewable energy more and more practical and essential.

Sixty Years of Photovoltaics

Solar Energy Still Strikes a Powerful Chord

Hearing the hit songs of Rosemary Clooney or Perry Como was an ordinary occurrence in the mid-1950s. But it turned extraordinary on Sunday, April 25, 1954.

That was the day that Bell Laboratories executives electrified members of the press with music broadcast from a transistor radio—powered by the first silicon solar cell. Bell called its invention "the first successful device to convert useful amounts of the Sun's energy directly and efficiently into electricity."

The discovery was, indeed, music to people's ears. The *New York Times* heralded it as "the beginning of a new era, leading eventually to the realization of one of mankind's most cherished dreams—the harnessing of the almost limitless energy of the Sun for the uses of civilization."

Single Cell Gives Birth to Millions

This 60-year-old prophecy has, in many ways, come to fruition, with total global installed PV rapidly approaching 100,000 megawatts or 100 gigawatts (GW)—or 100 billion watts of solar electric modules now powering satellites, telescopes, homes, businesses, utility grids, water pumps, and even Antarctic research stations. In the past few years, more PV was installed than any other source of energy! Solar power has officially arrived.

Today's solar cells are a vast improvement on the original technology, with typical residential and commercial solar panels achieving 15% to 20% efficiency compared with the 6% efficiency of the original Bell Labs panels. Silicon is still used—though in slightly different forms—to achieve higher efficiency rates as well as lower costs. The current world record for individual solar cell efficiency was recently set at an astounding 44.7%.

The rapid increase in efficiency is only part of the equation in the success of photovoltaics. The industry has grown more than 20% compounded per year since 1980 and continues at breakneck pace, growing 67% in 2010. A four-thousand-fold price drop since 1954, along with tax incentives and rebates, has made the move to solar power economically feasible for many homeowners and businesses, with typical payback periods of less than 3 years in markets with higher energy costs like Hawaii, and still less than 10 years in markets where utility energy remains relatively inexpensive. PV seems to follow the same economic rules as other new technologies, like computers and cell telephony—that is, Moore's Law that for every doubling of production, the price drops by 15%–20%. PV is now competitive with fossil fuel-generated power company electricity even without state or utility rebates or incentives. (The 30% federal investment tax credit [ITC] remains in effect for the system owner until at least 2016 and helps to reduce the installed cost significantly.) This phenomenon, where the cost of solar power is more or less equivalent to the cost of utility power, is known as "grid parity" and is considered to be the holy grail of the solar industry.

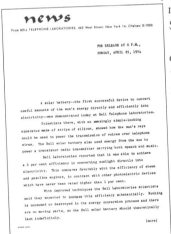

New Bell Solar Battery Converts Sun's Rays into Electricity

Where Will Solar Energy Be After Its First 100 Years—in 2054?

For over 36 years, Real Goods has pioneered the solar industry, and we couldn't have imagined the explosive growth we've seen in the recent past. We're thrilled to see what the future holds. As Margaret Mead said so eloquently, "Never doubt that a small group of thoughtful, committed citizens can change the world. Indeed, it is the only thing that ever has." In fact, the solar industry is no longer a "small group" but now a worldwide network that has built up too much momentum to ever slow down.

Since we sold the very first PV panel in the US in 1978, Real Goods, as of this writing, has now solarized more than 22,500 homes and businesses amounting to over 235 megawatts (MW) of solar between our store, website, catalogs, and our commercial and residential installation service, RGS Energy. What's even more exciting to consider is that we're just getting started—less than 0.02% of the roof space with solar potential has been used thus far! The United States still has less than 1% of homes using PV vs. Germany's over 10%. We can't wait to see what will happen as the number of solarized buildings continues to grow over the next 50 years. You can count on Real Goods and RGS Energy to continue leading the way.

A Brief Technical Explanation

Despite the huge leaps and bounds in the PV industry, the basics of how the technology works have remained tried and true. A single PV cell is a thin semiconductor sandwich, with a layer of highly purified silicon. The silicon has been slightly doped with boron on one side and phosphorus on the other side. Doping produces either a surplus or a deficit of electrons, depending on which side we're looking at. Electronics-savvy folks will recognize these as P- and N-layers, the same as transistors use. When our sandwich is bombarded by sunlight, photons knock off some of the excess electrons. This creates a voltage difference between the two sides of the wafer, as the excess electrons try to migrate to the deficit side. In silicon, this voltage difference is just under half a volt. Metallic contacts are made to both sides of the wafer. If an external circuit is attached to the contacts, the electrons find it easier to take the long way around through our metallic conductors than to struggle through the thin silicon layer. We have a complete circuit, and a current flows. The PV cell acts like an electron pump, as there is no storage capacity in a PV cell. Each cell makes just under half a volt regardless of size. The

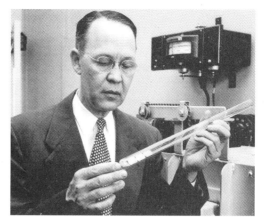

amount of current is determined by the number of electrons that the solar photons knock off. We can get more electrons by using bigger cells, or by using more efficient cells, or by exposing our cells to more intense sunlight. There are practical limits, however, to size, efficiency, and how much sunlight a cell can tolerate. Panel efficiency for flat plate collectors has been holding steady at ~15–16%, and near 20% for higher-end products used in residential and commercial markets. Another improvement has been wattage density. Just a few years ago, a 39" × 65" panel was rated at 216 watts; we now have 265-watt panels with the same dimensions.

Since 0.5-volt solar panels won't often do us much good, we usually assemble a number of PV cells in series for higher voltage output. A PV "module" or "panel" consists of many cells wired in series to produce a higher voltage. Lower-wattage modules consisting of about 36 cells in series are typically used for small off-grid and remote power applications, since this arrangement delivers power at 17 to 18 volts, a handy level for 12-volt battery charging. In today's residential and commercial PV market, 24-volt modules consisting of 72 cells have become the standard. The module is encapsulated with tempered glass (or some other transparent material) on the front surface and with a protective and waterproof material on the back surface. The edges are sealed for weatherproofing, and there is often an aluminum frame holding everything together in a mountable unit. Wire leads providing electrical connections are usually found on the module's back, commonly referred to as a J-Box. Truly weatherproof encapsulation was a problem with early module assemblies; however, we have not seen any encapsulation or weatherproofing problems with glass-faced modules in many years.

Common applications use a number of connected PV modules, called an "array." A PV array consists of a number of individual PV modules that have been wired together in series and/or parallel to deliver the voltage and amperage appropriate for the application. An array can be as small as a single pair of modules, used for lighting a street sign at night, or large enough to cover acres, used for a megawatt-size utility-scale solar power plant.

PV costs are now economically competitive with fossil fuel energy sources, making them the clear choice for both remote *and* grid-intertie power. For decades, they have been routinely used for roadside emergency phones and most temporary construction signs, where the cost and trouble of bringing in utility power outweighs the higher initial expense of PV, and where mobile generator sets present significant fueling and maintenance trouble. It's hard to find new gate-opener hardware that isn't solar powered. Solar with battery backup has proven to be a far more reliable power source, and it's usually easier to obtain at the gate site. More than 180,000 homes in the United States, largely but not always in rural sites, depend on PVs as a primary power source. This figure is growing rapidly since homeowners and developers are aware of how clean, reliable, and maintenance-free this power source is. We see how deeply our current fossil fuel-based energy practices are borrowing from our children.

A Technical Step Back in Time

Before the invention of the silicon solar cell, scientists were skeptical about the success of solar as a renewable energy source. Bell Labs' Daryl Chapin, assigned to research wind, thermoelectric, and solar energy, found that existing selenium solar cells could not generate enough power. They were able to muster only 5 watts per square meter—converting less than 0.5% of incoming sunlight into electricity.

The Solar Cell That Almost Never Was

Chapin's investigation may have ended there if not for colleagues Calvin Fuller and Gerald Pearson, who discovered a way to transform silicon into a superior conductor of electricity. Chapin was encouraged to find that the silicon solar cell was five times more efficient than selenium. He theorized that an ideal silicon solar cell could convert 23% of incoming solar energy into electricity. But he set his sights on an amount that would rank solar energy as a primary power source: 6% efficiency.

Nuclear Cell Trumps Solar Scientists

While Chapin, Pearson, and Fuller faced challenges in increasing the efficiency of the silicon cell, archrival RCA Laboratories announced its invention of the atomic battery—a nuclear-powered silicon cell. Running on photons emitted from Strontium-90, the battery was touted as having the potential to run hearing aids and wristwatches for a lifetime.

That turned up the heat for Bell's solar scientists, who were making strides of their own by identifying the ideal location for the P-N junction—the foundation of any semiconductor. The Bell trio found that shifting the junction to the surface of the cell enhanced conductivity, so they experimented with substances that could permanently fix the P-N junction at the top of the cell.

The three inventors of the first silicon solar cell, Gerald Pearson, Daryl Chapin, and Calvin Fuller, examine their cells at their Bell lab.

When Fuller added arsenic to the silicon and coated it with an ultrathin layer of boron, it allowed the team to make the electrical contacts they had hoped for. Cells were built using this mixture until one performed at the benchmark efficiency of 6%.

Solar Bursts the Atomic Bubble

After Bell released the invention to the public on April 25, 1954, one journalist noted, "Linked together electrically, the Bell solar cells deliver power from the Sun at the rate of 50 watts per square yard, while the atomic cell recently announced by the RCA Corporation merely delivers a millionth of a watt" over the same area. The solar revolution was born.

Special thanks to John Perlin, author of *From Space to Earth: The Story of Solar Electricity.*

Worldwide, there is now enough PV operating to meet the household electricity needs of nearly 70 million people at the European level of use. Because they don't rely on miles of exposed wires, residential PV systems are more efficient and reliable in terms of power delivery than utilities, particularly when the weather gets nasty. PV modules have no moving parts; degrade very, very slowly; require only an annual hosing off for maintenance; and boast a lifespan that isn't fully known yet but will be measured in multiple decades. We frequently have happy Real Goods customers come into our store at the Solar Living Center in Hopland, California, who are still using systems they originally installed in the late 1970s, replacing only the batteries and inverter, while the panels still provide clean and reliable energy. Standard factory PV warranties offer 25-year performance guarantees. What's more, many of these 25-year warranties come with insurance backed by multiple insurers, providing even more peace of mind to the consumer that even 25 years down the line, even with the original manufacturer gone, she'll be taken care of in the unlikely event of a defect or a manufacturer bankruptcy. Compare this with any other power generation technology or consumer goods. What other product exists that not only pays for itself, but carries that kind of guarantee?

Construction Types

There are four commercial production technologies commonly used to make PV cells. Newer technologies are currently being developed in the lab, and we're looking forward to seeing what impacts they will have in the PV marketplace.

Single- or Monocrystalline

Far less common in today's market, this is the oldest and most expensive production technique, but it's also the most efficient sunlight conversion technology commercially available. Complete modules have sunlight-to-wire output efficiency averages of about 13%–15%. Efficiencies up to 22% have been achieved in the lab, but these are single cells using highly exotic components that cannot economically be used in commercial production.

Boules (large cylindrical cylinders) of pure single-crystal silicon are grown in an oven, then sliced into wafers, doped, and assembled. This is the same process used in manufacturing transistors and integrated circuits, so it is well developed, efficient, and clean. Degradation is very slow with this technology, typically 0.25%–0.5% per year. Silicon crystals are characteristically blue, and single-crystalline cells look like deep

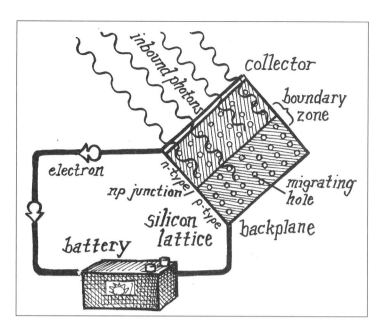

blue glass. Examples include Sunpower, Solar World, and Canadian Solar single-crystalline products.

Polycrystalline or Multicrystalline

Tried and true, this is the most common technology in today's PV market. In this production process, pure molten silicon is cast into cylinders, then sliced into wafers off the large block of multicrystalline silicon. Polycrystalline cells are slightly lower in conversion efficiency compared with single crystalline cells, but the manufacturing process is less exacting, so costs are a bit lower. Module efficiency averages about 15%–16%, sunlight to wire. Degradation is very slow and gradual, similar to that of single-crystal. Crystals measure approximately 1 centimeter (two-fifths of an inch) thick, and the multicrystal patterns can be clearly seen in the cell's deep blue surface. Doping and assembly are the same as for single-crystal modules. Today there are hundreds of multi- and polycrystalline panel manufacturers, including LG, First Solar, Rena Sola, Canadian Solar, Sanyo, and Kyocera.

String Ribbon

Overshadowed by the massive growth seen in the mono- and polycrystalline market, this technology currently sits on the sidelines of the PV industry, in part because the primary patent holder of the technology, Evergreen Solar, filed for bankruptcy and sold off its string ribbon intellectual property rights in the process. We anxiously await string ribbon's return to the market and are interested to see what form it may take and what applications develop. This clever technique is a refinement of polycrystalline production. A pair of strings are drawn up through molten silicon, pulling up between them a thin film of silicon like a soap bubble. It cools and crystallizes, and you've got ready-to-dope wafers. The ribbon width and thickness can be controlled, so there's far less slicing, dicing, or waste, and production costs are lower. Sunlight-to-wire conversion efficiency is as high as 13%. Degradation is the same as for ordinary slice-and-dice polycrystal.

Amorphous or Thin-film

Rarely if ever used in the residential market, thin-film panels have been relegated to larger-scale commercial projects where, due to large size and low efficiency, plenty of space must be available for the array. In this process, silicon material is vaporized and deposited on glass or stainless steel. The three mainstream thin-film solar cell technologies are cadmium telluride, amorphous silicon, and copper indium gallium selenide, or CIGS. All have the advantage of being able to be deposited on flexible substrate materials, producing highly flexible and lightweight solar panels. These production technologies cost less than any other method, but the cells are less efficient, so more surface area is required. Early methods in the 1980s produced a product that faded up to 50% in output over the first three to five years. Present-day thin-film technology has dramatically reduced power fading, although it's still a long-term uncertainty, making large-scale deployment a higher risk for investors and therefore less common. Uni-Solar has a "within 20% of rated power at 20 years" warranty, which relieves much nervousness, but we honestly don't know how these cells will fare with time. Sunlight-to-wire efficiency averages about 5%–7%. These cells are often almost black in color. Unlike other modules, if glass is used on amorphous modules, it is not tempered, so breakage is more of a problem; tempered glass can't be used with this high-temperature deposition process. If the deposition layer is stainless steel and a flexible glazing is used, the resulting modules will be somewhat flexible. While rare in residential applications, because they are flexible or in some cases even foldable or rollable, these modules have found a place in marine, RV, and military applications. We're seeing fewer examples of thin-film technology as time goes on and the efficiency of flat plate collectors improves. Uni-Solar makes flexible, unbreakable modules.

Emerging PV Cell Technologies in Pre-industrial Development

Light-absorbing Dyes or Dye-sensitized Solar Cells (DSSC): Developed at the University of California, Berkeley, in 1988, part of the thin-film family, this low-cost technology utilizes a semiconductor formed between a photosensitized anode and an electrolyte. Features and benefits include simple manufacturing using roll-printing techniques, and being semi-flexible and semi-transparent, allowing for versatile applications. Hurdles this technology will need to overcome before becoming more commonplace are eliminating some of the more expensive materials needed, and problems with chemical stability. Many observers predict that DSSC will be a significant part of the solar technology mix within the next 5 to 10 years.

Quantum Dot Solar Cells (QDSC): Currently undergoing widespread research in the field, this technology uses "quantum dots" as the photo-

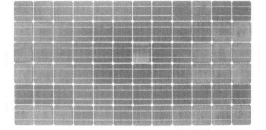

Single-crystalline module.

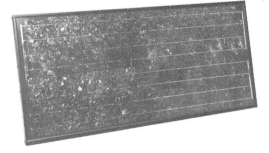
Multicrystalline module.

String ribbon module.

Thin-film module.

REAL GOODS

voltaic absorbing material. Quantum dots have band gaps that are tunable to a wide range of energy levels, making them attractive for multi-junction cell technologies where a variety of materials are used to improve efficiency by harvesting energy from selected frequencies of the solar spectrum. They are used in "solar paint" applications, which are exciting and promising yet have a long way to go, with efficiencies currently around 1%. Developments in the nanotechnology sector have paved the way for research in QDSC technology, and we expect to hear more news of this emerging technology.

Organic/Polymer Solar Cells: By far the most novel of emerging solar technologies, organic or polymer solar cells are built from extremely small thin films (typically 100 nanometers) made of organic or polymer semiconductors and small-molecule compounds. These cells can be processed from liquid solutions, holding the potential for relatively inexpensive roll-to-roll printing at a large scale. Benefits include applications with as high as 70% transparency used to produce power-generating windows. Efficiencies as high as 3% have been measured in the lab, and research is ongoing, making Organic/Polymer Solar Cells another exciting technology we'll be keeping our eye on!

Putting It All Together

The PV industry at large has standardized on 24-volt modules for both grid-tied and off-grid battery systems. We still see smaller 12-volt panels used for remote power applications with relatively low power needs. Whether 12 or 24 volts, this moderately low voltage is relatively safe to work with. Under circumstances where modules are connected in parallel, this isn't enough voltage to puncture the skin's surface and pass through your body. While not impossible, it's pretty difficult to hurt yourself on such low voltage. Still, whenever working with electricity, make sure you take the necessary safety precautions. Batteries, however, which store enormous quantities of accumulated energy, can be very dangerous if mishandled or miswired. Please see Chapter 4, which discusses batteries and safety equipment, for more information.

Multiple modules can be wired in parallel or series to achieve any desired voltage output. As systems get bigger, we usually run collection and storage at higher DC voltages because transmission is easier. We're also seeing more equipment components that accept higher input voltages and step the output voltage down to whatever

is appropriate for the application. Small systems processing up to about 2,000 watt-hours are fine at 12 volts at each part of the system. Systems processing 2,000–7,000 watt-hours will function better at 24 volts, even if the panels are at a higher voltage, and systems running more than 7,000 watt-hours should probably be running at 48 volts or AC coupled (see Chapter 4). These are guidelines, not hard and fast rules! The modular design of PV panels and the versatility of today's system components allow systems to grow and change as system needs change. Modules from different manufacturers, different wattages, and various ages can be intermixed with no problems, so long as all equipment input and output voltage and amperage levels remain within manufacturer specifications. The modular nature of PV system components allows you to buy what you can afford today and expand tomorrow. There's no need to wait to begin building the PV system of your dreams!

One of the most frequently asked questions we get at Real Goods is, "Shouldn't I wait to buy until this great new technology is ready for prime time or until the price drops?" The answer is unequivocally "no," because in the 36 years we've

---

What's a Watt? An Amp? A Volt?

A watt (W) is a standard metric measurement of electrical power. It is a rate of doing work.

A watt-hour (Wh) is a unit of energy measuring the total amount of work done during a period of time. (This is the measurement that utility companies make to charge us for the electricity we consume, only they do it 1,000 watts at a time, in kilowatt-hours or kWh.)

An amp (A) is a unit measuring the amount of electrical current passing a point on a circuit. It is the rate of flow of electrons through a conductor such as copper wire: 1 Amp = 6.28 billion billion electrons moving past a point in one second. (Amps are analogous to the flow rate in a water pipe.)

A volt (V) is a unit measuring the potential difference in electrical force, or pressure, between two points on a circuit. This force on the electrons in a wire causes the current to flow. (Volts are analogous to water pressure in a pipe.)

In summary, a watt measures power, or the rate of doing work, and a watt-hour measures energy, or the amount of work done. Watts can be calculated if you know the voltage and the amperage: **Watts = Volts×Amps** (this is known as Ohm's Law). More pressure or more flow means more power.

been selling PV, everything is more or less the same, just slightly more efficient. And if you had waited, you would have either wasted all that solar power or gone without any power in off-grid applications.

Efficiency

By scientific definition, the sun delivers 1,000 watts (1 kilowatt) per square meter at noon on a clear day at sea level. This is defined as "peak sun" or "full sun" and is the benchmark by which modules are rated and compared. That is certainly a nice round figure, but it is not what most of us actually find on our rooftop where we would like to install a PV system. Dust, water vapor, air pollution, seasonal variations, altitude, and temperature all affect how much solar energy your modules actually receive and how much electricity they produce. For instance, the 1991 eruption of Mt. Pinatubo in the Philippines reduced available sunlight worldwide by 10%–20% for a couple of years. It is reasonable to assume that most sites will actually average about 85% of full sun, unless they are over 7,000 feet in elevation, in which case they'll probably receive more than 100% of full sun. PV system design and performance depend on all of these factors.

PV modules do not convert 100% of the energy that strikes them into electricity (we wish!). Current commercial technology averages about 16%–20% conversion efficiency for single- and multicrystalline modules, depending on panel quality and price. Conversion rates slightly over 47% have been achieved in the laboratory by using experimental cells made with esoteric and rare elements. But these elements are far too ex-

pensive to ever see commercial production, so we do not foresee these levels of efficiency entering the residential or commercial market any time soon. Conversion efficiency for commercial single- and multicrystalline modules is not expected to improve significantly; this is a mature technology. There's better hope for increased efficiency with amorphous technology, and much research is currently underway, as discussed above.

How Long Do PV Modules Last?

PV modules last a long, long time, especially compared to other technologies. How long, we honestly don't yet know, as the oldest terrestrial modules are barely 50 years old and still going strong. In decades-long tests, the fully developed technology of single- and polycrystalline modules has shown to degrade at fairly steady rates of 0.25%–0.5% per year. First-generation amorphous modules degraded faster, but there are so many new wrinkles and improvements in amorphous production that we can't draw any blanket generalizations for this module type. The best amorphous products now seem to closely match the degradation of single-crystalline products, but there is little long-term data. Most full-size modules carry 25-year warranties, reflecting their manufacturers' faith in the durability of these products. The performance guarantees that are a common part of most module warranties are for 80% production at year 25. Also, many of these warranties are linear, meaning the manufacturer is willing to replace the panel if the rate of degradation ever begins to speed up. PV technology is closely related to transistor technology. Based on our experience with transistors, which just fade

Photovoltaic prices have decreased dramatically since 1955. Between 1999 and around 2010, we saw dramatic price decreases of 10% to 15% per year. Prices are still decreasing, but not as dramatically. With hard costs of the major system components leveling off, the PV industry has begun focusing more effort on reducing the soft costs involved with residential PV installation, including permitting, utility interconnection paperwork, and inspection of PV systems from the building department.

The Swanson Effect
Price of crystalline silicon photovoltaic cells, $ per watt

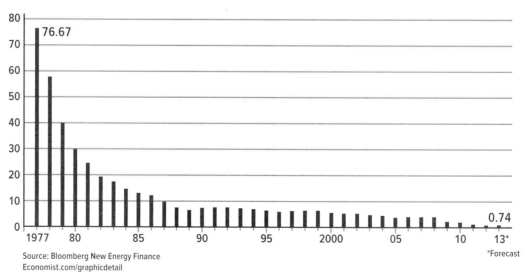

Source: Bloomberg New Energy Finance
Economist.com/graphicdetail

REAL GOODS

away after 20 years of constant use, most manufacturers have been confidently predicting 20-year or longer lifespans. However, keep in mind that PV modules are seeing only six to eight hours of active use per day, so we may find that lifespans of 60–80 years are normal. Cells that were put into the truly nasty environment of space in the late 1960s are still in use. The bottom line? PV modules have been performing as expected for over 35 years, and we haven't been measuring their performance long enough to know when they'll stop working. Only time will tell.

Payback Time for Photovoltaic Manufacturing Energy Investment

In the early years of the PV industry, there was a nasty rumor circulating that PV modules would never produce as much power over their lifetimes as it took to manufacture them, something called "embodied energy." Since then, we've seen most panels offsetting their embodied energy in only 1 to 2 years, depending on the manufacturing process and installation environment. However, during the early years of development, when transistors were a novelty, and handmade PV modules costing as much as $40,000 per watt were being used exclusively for spacecraft, it was true that it took more energy to manufacture them than they would produce. The National Renewable Energy Laboratory has done some impartial studies on payback time (see the results at nrel.gov/docs/fy04osti/35489.pdf). It concludes that modules installed under average US conditions reach energy payback in 3 to 4 years, depending on construction type. The aluminum frame all by itself can account for six months to one year of that time. Quicker energy paybacks, down to 1 to 2 years, are now more common, as more "solar grade" silicon feedstock becomes available and as modules have become more standardized across the industry.

Maintenance

One of the most common questions homeowners have about PV installations is maintenance, and the simple truth is that there really isn't any required. Because they have no moving parts, they are virtually maintenance free. Since dust or pollen will affect module performance, you want to keep them clean. In dry environments, the occasional spraying from a garden hose is sufficient—no need to get on the roof either, if you can spray them from the ground, that will do. Do not hose them off when they're hot, since uneven thermal shock could theoretically break the glass. Wash them in the morning or evening. For PV mainte-

nance, that's it. There is not much data out there on how much panel cleaning will improve system performance, or what the value of keeping them clean is, but as the industry has grown, we've even seen module cleaning services begin to sprout up, and we leave it to them to prove themselves as a worthwhile service. If indeed the value of the added energy produced by a squeaky-clean PV system justifies the cleaning services, more power to the module cleaners! For example, the 132 kW solar array at our Solar Living Center in Hopland gets hosed down once a year in July, and typically we see a 10% improvement in power output in the following month—so it's a task definitely worth doing!

Charge Control for Battery-based PV Systems

Controls for PV systems are usually simple and have remained so in recent years. When the battery reaches a full-charge voltage, the charging current can be turned off and simply "open-circuited" or

One of the most common questions homeowners have about PV installations is maintenance, and the simple truth is that there really isn't any required. Because they have no moving parts, they are virtually maintenance free.

A Mercifully Brief Glossary of PV System Terminology

AC (alternating current): This refers to the standard utility-supplied power, which alternates its direction of flow 60 times per second (60 hz), and for normal household use has a voltage of approximately 120 or 240 (in the USA). AC is easy to transmit over long distances, but it is impossible to store. Most household appliances require this kind of electricity.

DC (direct current): This is electricity that flows in one direction only. PV modules, small wind turbines, and small hydroelectric turbines produce DC power, and batteries of all kinds store it. Appliances that operate on DC very rarely will operate directly on AC, and vice versa. Conversion devices are necessary.

Inverter: An electronic device that converts (transforms) the low-voltage DC power we can store in batteries to conventional 120- or 240-volt AC power as needed by lights and appliances. This makes it possible to utilize the lower-cost (and often higher-quality) mass-produced appliances made for the conventional grid-supplied market. Inverters are available in a wide range of wattage capabilities. We commonly deal with inverters that have a capacity of anywhere between 150 and 10,000 watts.

PV Module: A solar panel that makes electricity when exposed to direct sunlight. PV is shorthand for photovoltaic. We call these panels PV modules to differentiate them from solar hot water panels or collectors, which are a completely different technology and are often what folks think of when we say "solar panel." PV modules do not make hot water.

The typical American home consumes about 20–30 kilowatt-hours daily, compared to about a third of that in European countries and a tenth or less in many developing countries. Supplying this demand with PV-generated electricity can be costly, but it makes perfect sense from an investment standpoint.

directed elsewhere, to a water heating element, for example. Open-circuited PV module voltage, when the controller stops charging batteries, rises 5–10 volts and stabilizes harmlessly. It does no harm to the modules to sit at open-circuit voltages, but they aren't doing any work for you either, hence the value of diverting the power to a water heater or another worthwhile application. When the battery voltage drops to a certain set-point, the charging circuit is closed and the modules go back to charging. With older or less expensive solid-state PWM (Pulse Width Modulated) controllers, this opening and closing of the circuit happens so rapidly that you'll simply see a stable voltage.

The most common PV control technology today is Maximum Power Point Tracking, or MPPT. These sophisticated solid-state controllers allow the modules to run at whatever voltage produces the maximum wattage. This is usually a higher voltage than batteries will tolerate. The extra voltage is down-converted to amperage the batteries can digest comfortably. MPPT controls extract an average of about 15% to 30% more energy from your PV array, depending on the time of year. They do their best work in the wintertime when most residential systems need all the help they can get. Most controllers offer a few other bells and whistles, like nighttime disconnect and LED indicator lights. See the Controls and Monitors section in Chapter 4 for a complete discussion of controllers.

Photovoltaic Summary

Advantages
1. No moving parts
2. Ultralow maintenance (Hose 'em off!)
3. Extremely long life
4. Noncorroding parts
5. Easy installation
6. Modular design
7. Universal application
8. Safe low-voltage output
9. Simple controls
10. Economic payback

Disadvantages
1. Initial cost to own equipment
2. Works only in direct sunlight
3. Sensitive to shading
4. Lowest output during shortest days
5. Low-voltage output difficult to transmit

Powering Down

One could argue that the downside to all this good news is that it costs money to install a PV system. Compared to plugging into the grid and paying for what you use in kilowatt-hours, you need to spend money up front on equipment and installation. After decades of cheap, plentiful utility power, we've turned into a nation of power hogs. The typical American home consumes about 20–30 kilowatt-hours daily, compared to about a third of that in European countries and a tenth or less in many developing countries. Supplying this demand with PV-generated electricity can be costly to do out of pocket; however, it makes perfect sense from an investment standpoint. Fortunately, at the same time that PV-generated power started to become affordable and useful, conservation technologies for electricity started to become popular, and given the steadily rising cost of utility power, even necessary. Compact fluorescent and LED lighting are good examples of conservation technologies that were once rare, and are now commonplace. The two emerging technologies dovetail together beautifully. Every kilowatt-hour you can trim off your projected power use in a stand-alone (off-grid) PV-based system will have a huge impact in reducing your initial setup cost.

Using a bit of intelligence and care in your lighting and appliance selection will allow you all the conveniences of the typical 20-kWh-per-day California home, while consuming less than 10 kWh per day. With this kind of careful analysis applied to electrical use, most of the full-size home PV electrical systems we design come in between $10,000 and $20,000, depending on the number of people and intended lifestyle. Simple weekend cabins with a couple of lights and a boom box or iPod setup can be worked up for $1,000 or less. With the renewable energy rebates and incentives available in an increasing number of states, grid-tied PV can be very cost-effective. Typical payback times in California run 4–7 years depending on utility energy prices (a 15%–24% return on investment!). Commercial paybacks with tax incentives typically take half that time.

Other chapters in this *Sourcebook* present extensive discussions of electricity conservation, for both off- and on-grid (utility power) scenarios, and Real Goods offers many of the lights and appliances discussed. We strongly recommend reading these sections before beginning to size your system. We are not proposing any substantial lifestyle changes, just the application of appropriate technology and common sense. Stay away from 240-volt watt hogs, electric space heaters, cordless electric appliances, standard incandescent light bulbs, instant-on TVs, and monster side-by-side refrigerators, and our friendly technical staff can work out the rest with you.

PV System Leases and Power Purchase Agreements

We've spent time discussing the costs, benefits, payback, and return on investment (ROI) of PV systems purchased up front, and for many of our customers able to make the investment in their own energy independence, this is the best option for them. But many of us are already strapped financially with little disposable income, and rising utility costs are not helping change the situation. How would you like to pay $0 to install a solar system on your rooftop and begin spending a lot less on electricity than you currently pay your utility company? Sound too good to be true? Well, ladies and gentlemen, believe it: Solar energy at no out-of-pocket expense is here now! One of the biggest innovations in the grid-tied PV industry in the last few years came not in the form of a new technology or efficiency breakthrough, but in the way we finance and pay for systems to be installed. They're called leases and power purchase agreements, or PPAs for short, and they are one of the reasons we've seen such a rapid deployment of residential solar systems recently. Most of you are probably too young to remember how hard it was to own a car in the 1950s, until the leasing companies came around to make cars affordable for many more Americans. What's happened in the last few years with PV leasing is analogous.

Leases and PPAs are quite simple and work like this: A third party finances the installation of your PV system, and you purchase the electricity that it produces from the system owner. The third party maintains ownership of the equipment, takes care of monitoring and maintenance, while you simply enjoy the benefits: clean, renewable energy that costs substantially less than you would otherwise spend purchasing the power from your utility. Most leases and PPAs are for an initial term of 20 years, so you're able to lock in the rate you pay for solar electricity and protect yourself from utility rate increases (which have averaged over 7% escalation annually over the last 25 years) for the next two decades! If you decide to move, the agreement is transferrable to the new homeowner, and who wouldn't want to buy a house that comes with lower utility bills! Time flies when you're not spending a lot on energy

What happens at the end of a solar lease or PPA? Most of these agreements offer a few options at the end of the initial 20-year term. One option is to purchase the system at fair market value. This option is also available to you throughout the term of the agreement, so if you ever decide you want to own the system and the energy produced, you can. The second option is to simply extend the agreement for a second term, and continue purchasing inexpensive clean energy. The last option is for the system owner to remove the equipment. Given the likelihood that the equipment won't be worth a whole lot after being removed 20 years down the road, we anticipate that many third-party system owners will make it worthwhile to the homeowner to keep the system, and many lease and PPA agreements contain a clause allowing the system to remain onsite in an "as is—where is basis," meaning you'll just get to keep it.

Solar Lease or PPA?

What's the difference between a lease and a PPA? In a solar lease agreement, you make a fixed monthly payment for solar energy provided by the system regardless of the actual kilowatt-hours produced each month. That is to say, your solar energy bill is the same in the summer as it is in the winter, even though your system production is different. With a solar PPA, you would pay monthly for each kilowatt-hour generated that month, so your solar energy bill changes throughout the year to reflect the actual amount of energy produced by the system. In either case, lease or PPA, any energy usage beyond what was produced by the system is provided by your utility company. If the system produces more than needed in the home, the excess kilowatt-hours are credited to your account for future use

Another difference between leases and PPAs is the way system performance above or below the expected production is handled. In a lease, you would be reimbursed for any kilowatt-hours that were not produced if the system underperformed, and if the system produces more than expected, you essentially get that energy for free, as your monthly payment is fixed for the term of the agreement. In a PPA, since you're buying each kilowatt-hour that is produced, if the system underperformed, you would pay a lower than expected solar bill that month, or if the system produced more than expected, your expected solar bill would match accordingly. While a well-designed and installed PV system should perform as expected, equipment failure, however unlikely, can occur, and as we'll see in the next section, there are a number of real-world factors that, if not accounted for, will impact the PV system production.

> Solar energy at no out-of-pocket expense is here now! One of the biggest innovations in the grid-tied PV industry in the last few years came in the way we finance and pay for systems to be installed. They're called leases and Power Purchase Agreements.

Community-owned Solar

Another breakthrough in the grid-tied PV industry that has more to do with financing and logistics than technology is community-owned solar, sometimes also known as solar gardens. The concept is straightforward. You own panels and enjoy the benefits, but they're not installed on your roof! They're installed in a nearby commercial solar array, and you purchase the right number of panels to meet your energy needs. Through metering technologies and agreements with your utility, you have "virtual ownership" of the power your panels produced remotely. This model has huge implications for the industry, because none of the site-specific requirements for grid-tied PV systems apply. All the people who couldn't take advantage of the technology because their roof is shaded, obstructed, or otherwise not suited for a PV system, can now enjoy the benefits with community-owned solar. Real Goods installed the first solar garden to go up in Colorado, and we couldn't be more proud! As of this writing, many more community-owned solar projects are in the works, and RGS Energy, the installation arm of Real Goods Solar, is working on completing them.

PV Performance in the Real World

We live in the real world, and unfortunately the ideal conditions for PV modules to perform are not always found on your roof 365 days of the year. There are a few things to watch out for. First off, be aware that wattage ratings on PV modules are given under ideal laboratory conditions that are rarely found outside the lab, and if they are, the conditions are fleeting. A 250-watt-rated panel will range in output above and below its rating, depending on module temperature. Assuming you can avoid or eliminate shadows, the most important factors that affect module performance out in the real world are percentage of full sun, soiling from dust or pollen, and operating temperature.

Shadows

Short of dropping it on the hard ground from the rooftop, hard shadows are the worst possible thing for a PV module. Even a tiny amount of shading dramatically affects module output. Electron flow is like water flow: It flows from high voltage to low voltage. Normally the module is high voltage, and the battery, load, or grid-tied inverter is lower voltage. A shaded portion of the module drops to very low voltage. Electrons from other portions of the module and even from other modules in the array will find it easier to flow into the low-voltage shaded area than into the battery. These electrons just end up making heat and are lost to us. A little shading goes a long way. This is why bird droppings are a bad thing on your PV module. A fist-size shadow at the right place with respect to the cell wiring arrangement will effectively shut off a PV module. Always avoid installing your modules where they will be shaded during the prime midday generating time from 10 AM to 3 PM. Early or late in the day, when the sun is at extreme angles and has a lot more atmosphere to contend with, little power is being generated anyway, so don't sweat shadows then. Sailors may find shadows unavoidable at times, but just keep them clear as much as practical.

Full Sun

As mentioned above, most of us have less than 100% full sun at our disposal. If you are not getting full, bright, shadow-free sunlight, then your PV output will be reduced, but don't worry, this is the reality for most of us. If you are not getting bright enough sunlight to cast fairly sharp-edged shadows, then you do not have enough sunlight to harvest much useful electricity. Most of us actually receive 80%–85% of a "full sun" (defined as 1,000 watts per square meter) on a clear sunny day. High altitudes and desert locations have better sunlight availability. On the high desert plateaus, 105%–110% of full sun is normal. They don't call it the "sunbelt" for nothing.

Temperature

The power output from all PV module types decreases at higher temperatures. This is not a serious consideration until ambient temperatures climb above 80°F, but that's not uncommon in full sun. The backs of modules should be as well-ventilated as practical. This is given special consideration when selecting the method of mounting the panels. Always allow some air space behind the modules if you want decent output in hot weather. On the positive side of this same issue, all modules increase output at lower temperatures, as in the wintertime, when most residential applications can use a boost. We have seen cases when modules were producing 30%–40% over specs on a clear, cold winter morning with a fresh reflective snow cover and hungry batteries. Because we rely on our PV power year-round, we take the extreme minimum and maximum temperatures we can expect year-round into consideration. In the last several years, PV modules with black back sheets and black frames have become more popular because they are very attractive on a home roof. But bear in mind that for efficiency purposes the white back sheet is preferable, as it reflects the heat rather than absorbing it, meaning you get more power output.

As a general rule of thumb, we usually derate official manufacturer-specified "nameplate" PV module output by about 25%–30% (grid-tied systems) to 40%–50% (off-grid, battery-based systems) for the real world. For panel-direct systems (where the modules are connected directly to the pump without any batteries), derate by 20%, or even by 30% for really hot summer climates if you want to make sure the pump will run strongly in hot weather.

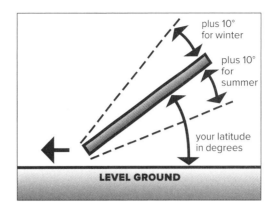

Proper PV mounting angle.

Module Mounting

Modules will catch the maximum sunlight, and therefore have the maximum output, when they are perpendicular (at right angles) to the sun. This means that tracking the sun across the sky from east to west will give you more power output. But tracking mounts are expensive and prone to mechanical and/or electrical problems, and PV prices have been coming down significantly. We used to make extra effort in parts and labor to achieve the optimal module orientation, but the cost benefit is no longer there, and you're better off just adding a couple of more panels to make up the difference. Even if you've got a summer-time high-power application, like water pumping, tracking mounts don't make a good investment anymore, and your money is better spent on a larger array. We've also seen almost every tracking mount on the market fail at some point, something nobody wants to deal with.

PV systems are most productive if the modules are approximately perpendicular to the sun

at solar noon, the most energy-rich time of day for a PV module. The best year-round angle for your modules is approximately equal to your latitude. Because the angle of the sun changes seasonally, in an off-grid situation, you may want to adjust the angle of your mounting rack seasonally. In the winter, modules should be at the angle of your latitude plus approximately 15 degrees; in the summer, your latitude minus a 15-degree angle is ideal. On a practical level, many residential grid-tied systems will have power to burn in the summer and are net metered so kilowatt-hours will roll forward, so seasonal adjustments are no longer necessary.

Generally speaking, most PV arrays end up on fixed mounts of some type. Tracking mounts are rarely used for residential systems but have found their place on some larger-scale commercial projects that factor in the costs of ongoing maintenance and repair. Today, water-pumping arrays are the only small-scale application of tracking mounts.

System Examples

Following are several examples of photovoltaic-based electrical systems, starting from simple and working up to complex. All the systems that use batteries can also accept power input from wind or hydro sources as a supplement or as the primary power source. PV-based systems constitute better than 95% of Real Goods' renewable energy system sales through our store, website, and residential installation division, RGS Energy, so the focus here will be mostly on them as opposed to wind and microhydro systems.

A Simple Solar Pumping System

PV direct, in which a module is connected directly to DC motor, is the most basic form of PV system. In this simple system, all energy produced

by the PV module goes directly to the water pump. No electrical energy is stored; it's used immediately. Water delivered to the raised storage tank is your stored energy. The brighter the sun, the faster the pump will run. PV-direct is the most efficient way to utilize PV energy. Eliminating all other system components, including the electrochemical conversion of the battery, saves about 20%–25% of the energy, a very significant chunk! However, PV-direct systems work only with DC motors that can use the variable power output of the PV module, and of course, this simple system works only when the sun shines. Another common example of a PV-direct system used today is the solar attic fan. Sold as a single unit installed on the roof, a DC fan helps to expel hot air from an overheated attic, reducing summer air conditioning loads.

A PV-direct system has one component you won't find in other systems. The PV-direct controller, or linear current booster (LCB), is unique to systems without batteries. This solid-state device will down-convert excess voltage into amperage that will keep the pump running under low-light conditions when it would otherwise stall. An LCB can boost pump output by as much as 40%, depending on climate and load conditions. We usually recommend them for PV-direct pumping systems.

For more information about solar pumping, see Chapter 7, "Water Development, pages 231–41."

A Grid-tied PV System Without Batteries

This scenario—usually called grid-tied PV—utilizes net metering to account for energy production vs. consumption, and is the simplest and most cost-effective way to connect PV modules to regular utility power. Grid-tied PV is by far the most common use of PV modules in today's residential and commercial market. All incoming PV-generated electrons are converted to household AC power by the grid-tied inverter and delivered to the main household via a circuit breaker in the house's main service panel, where it displaces an equal number of utility-generated electrons. When a grid-tied PV system is installed, the utility will replace your existing meter, which only counts kilowatt-hours in one direction, with one that counts both imported and exported power. As utilities update outdated metering equipment, many of the newer "smart meters" being installed are already PV system ready, and will function as net meters when PV is installed. All the electricity generated by the PV system is power you didn't have to buy from the utility company. If the incoming PV power exceeds what your house can use at the moment, the excess electrons will be forced out through your meter, turning it backward. If the PV power is insufficient, that shortfall is automatically and seamlessly made up by utility power. It's like water seeking its own level (except it's really fast water!). When your intertie system is pushing excess power out through the meter, the utility is paying you regular electric rates for your excess power. You sell power to the utility during the daytime; it sells power back to you at night. This is what is meant by the term "net metering"—any electricity your PV system produces that you don't use on the spot is still yours to use later. This treats the utility grid like a big 100%-efficient battery: Whatever you put in, you can take out. However, if utility power fails, even if it's sunny, your PV system will be shut off for the safety of utility workers. Many solar customers are surprised when they discover that if the electric utility has a power outage, they have no power from their solar system. And for them, it is always possible to purchase battery backup to have reliable power even when the grid goes down.

These grid-tied systems use a minimum of hardware: a power source (the PV modules), an inverter, a circuit breaker, and some wiring to connect everything. See below for information on grid-tied systems with batteries.

A Small Cabin PV System with Batteries

Most off-grid PV systems are designed to store some of the collected energy for later use. This allows you to run lights and entertainment

PV-direct water pumping.

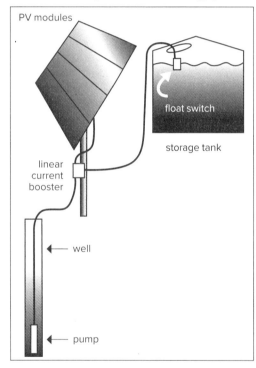

A utility intertie without batteries.

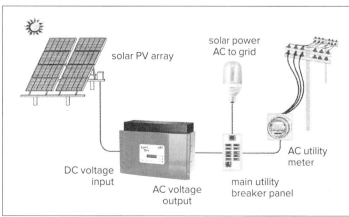

How Much PV Area Would It Take to Equal US Electric Production?

Here's the simple answer:

A solar electric array, using today's off-the-shelf technology, sited in sunny, largely empty Nevada, that's big enough to deliver all the electricity the US currently uses, would cover a square almost exactly 87 miles per side. This statement gives us a conceptual way to wrap our heads around the potential of PV technologies to provide us with clean renewable energy. Of course, the solution isn't quite this simple.... Transmitting power from Nevada to everywhere it's needed wouldn't be practical, due to voltage drop and sizing conductors over such long distances. If you envision our huge array in Nevada, all we need to do now is split it up among sunny rooftops across the county, make up for the differences in full sun at these locations, and viola! We've done it! What better solution to our energy problems than to produce solar power right at home where it is needed?

Here's the proof in more detail:

According to the Energy Information Administration of the US Department of Energy, eia.gov/totalenergy /data/monthly/pdf/sec7_3.pdf, the US produced 4,048 billion kilowatt-hours of electricity in 2012. Note that this is "production," not "use." Transmission inefficiencies and other losses are covered.

We'll want our PV modules in a sunny area to make the best of our investment. Nevada, thanks to climate and military/government activities, has a great deal of almost empty and very sunny land. So looking at the National Solar Radiation Data Base for Tonopah, Nevada, rredc.nrel.gov/solar/pubs/redbook/, a flat-plate collector on a fixed mount facing south at a fixed tilt equal to the latitude, 38.07° in this case, saw a yearly average of 6.1 hours of "full-sun" per day in the years 1961 through 1990. A "full sun" is defined as 1,000 watts per square meter.

For PV modules, we'll use the large Solar World 250-watt module, which the California Energy Commission (energy.ca.gov/ greengrid/certified_pv_modules .html) rates at 223.3 watts output, based on lab-tested performance. 222.3 watts times 6.1 hours equals 1,356 watt-hours or 1.356 kilowatt-hours per day per module at our Tonopah site. At 65 × 39 inches, this module presents 17.6 square feet of surface area. We'll allow some space between rows of modules for maintenance access and for sloping wintertime sun, so let's say that each module will need 23 square feet.

Conversion from PV module DC output to conventional AC power isn't perfectly efficient. Looking at the real-world performance figures from the California Energy Commission (energy.ca.gov/greengrid/certified _inverters.html), we see that the Power One PVI-Central 250 kW model is rated at 97% efficiency. We'll probably

be using larger inverters, but this is a typical efficiency for large intertie inverters. We'd better also deduct about 10% for whatever other losses might occur—dirty modules, etc. So our 1.356 kWh per module per day becomes 1.183 kWh by the time it hits the AC grid.

A square mile (5,280 × 5,280 feet) equals 27,878,400 square feet. Divided by 23 square feet per module, we can fit 1,212,104 modules per square mile. At 1.183 kilowatt-hours per module per day, our square mile will deliver 1,433,919 kWh per day on average, or 523,380,467 kWh per year. Back to our goal of 4,038,000,000,000 kWh divided by 523,380,467 kWh per year per square mile, it looks like we need about 7,715 square miles of surface to meet the electrical needs of the United States.

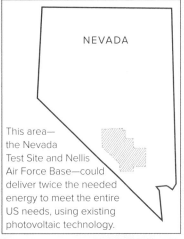

This area— the Nevada Test Site and Nellis Air Force Base—could deliver twice the needed energy to meet the entire US needs, using existing photovoltaic technology.

That's a square area about 87 miles on a side. This is about 48% of the approximately 16,000 square miles currently occupied by the Nevada Test Site and the surrounding Nellis Air Force Base.

What about the cost/benefit of such a project? Let's look at the cost of an array that would produce just one-quarter of the required electricity that is produced in the US. Currently, small commercial or residential systems cost about $3.50–$4.50 per peak watt including typical government and utility incentives. If economies of scale, advances in efficiencies, and government subsidies are considered, the job might get done for $1.50/watt (just guessing here, but you get the idea). Therefore, our quarter-sized 584,000 megawatt array would cost a cool $876 million. This certainly is a lot of money; but then the potential benefits can be enormous. For one, think of the jobs created. The Blue Green Alliance (bluegreenalliance.org) 2007 UC Berkeley report says that "renewable power production is labor-intensive.... Solar PV creates 20 manufacturing and 13 installation/maintenance jobs per MW." However, worldwide PV production in 2012 was only 38.5 GW, according to European Commission report in 2013. So we're on our way, but we've still got a ways to go to meet this potential demand.

As a practical measure, PV power production happens during the daytime, and so long as we use lights at night, we will continue to use substantial power at night. Also, out in the desert, solar thermal collection may be a more efficient power generation technology. But however you run the energy collection system, large solar-electric farms on what is otherwise fairly useless desert land could add substantially to the electrical independence and security of any country. The existing infrastructure of coal, nuclear, and hydro power plants could continue to provide reliable power at night, but non-renewable resource use and carbon dioxide production would be greatly reduced.

Batteries are the most cost-effective energy storage technology available so far, but batteries are a mixed blessing.

Small cabin systems to run a few lights and an appliance or two can start at under $1,000.

equipment at night, or to temporarily run an appliance that takes more energy than the PV system is delivering. Batteries are the most cost-effective energy storage technology available so far, but batteries are a mixed blessing. The electrical/chemical conversion process isn't 100% efficient, so you have to put back about 20% more energy than you took out of the battery, and the storage capacity is finite. Batteries have not seen the same improvements over the years as PV modules have, but in recent years, battery technology, specifically lithium ion batteries, has seen increasingly rapid development, spurred by the growth in the electric vehicle (EV) market. As the EV market grows and matures, the PV industry is sure to reap the benefits of the improvements, and to a greater extent, the reductions in cost, of batteries. Batteries are like energy buckets; they can get only so full and can empty just so far. A charge controller becomes a necessary part of your system to prevent over- (and sometimes under-) charging. Batteries can also be dangerous. Although the lower battery voltage is generally safer to work around than conventional AC house current, it is capable of truly awe-inspiring power discharges if accidentally short-circuited. So overcurrent protection devices (OCPDs) like fuses and safety equipment also become necessary whenever you add batteries to a system. Fusing ensures that no youngster, probing with a screwdriver into some unfortunate place, can burn the house down. And finally, a monitoring system that displays the battery's approximate state of charge is essential for

reliable performance and long battery life. You could do without monitoring—just as you could drive a car without any gauges, warning lights, or speedometer—but this doesn't encourage system reliability or longevity. Today's off-grid monitoring systems have become quite sophisticated, often integrating data from PV modules, charge controller, batteries, and inverter into one platform, usually allowing for web-based or computer-based monitoring that can be checked remotely with a smart phone. This is a huge benefit to both the PV system owner and the installer. What used to be a 6-hour round trip service call can now be handled remotely, saving both parties time and money.

The Real Goods 750-watt Weekender Kit is an example of a small cabin or RV weekend retreat system (see realgoods.com). It has all the basic components of an off-grid residential power system: a power source (the PV module), a storage system (the deep-cycle battery), a controller to prevent overcharging, safety equipment (fuses), and monitoring equipment. The Weekender Kit is supplied as a simple DC-only system. It will run 12-volt DC equipment, such as RV lights and appliances. An optional inverter can be added at any time to provide conventional AC power, and that takes us to our next example.

A Full-size Household System

Let's look at an example of a full-size residential system to support a family of three or more. The power source, storage, control, monitoring, and safety components have all been increased in size from the small cabin system, but most important, we've added an inverter for conventional AC power output. The majority of household electrical needs are run by the inverter, allowing conventional household wiring and a greater selection in lighting and appliance choices. We've found that when the number of lights gets above five, AC power is much easier to wire, plus fixtures and lamps cost significantly less due to mass production, so the inverter pays for itself in appliance savings.

Often, with larger systems like this, we combine and preassemble all the safety, control, and inverter functions using an engineered, UL-approved power center. This isn't a necessity, but we've found that most folks appreciate the tidy appearance, fast installation, UL approval, and ease of future upgrades that preassembled power centers bring to the system.

Because family sizes, lifestyles, local climates, and available budgets vary widely, the size and components that make up a larger residential

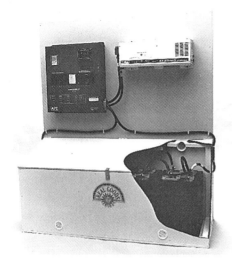

The typical Real Goods full-time system takes only 4'x8' of floor space inside your utility room.

system are customized for each individual application. System sizing is based on the customer's estimate of needs and an interview with one of our friendly technical staff. Estimating energy needs is arguably the most important step in PV system design, and usually the least interesting. It's a lot more exciting to compare PV panel sizes and racking products, and learn about the feature-rich charge controllers and inverters now available, than it is to list how many minutes a week you'll be running your toaster. Almost all future design considerations will be based off of the estimated energy needs, so be sure not to rush through this step or downplay its importance. The last thing you want is a system that won't meet your needs, or worse, spending way too much on a system you'll never take full advantage of. See the "System Sizing" hints and worksheets in Chapter 4 or in the Appendix.

A Grid-tied System with Battery Backup

In this type of grid-tied system, the customer has both a renewable energy system and conventional utility-supplied grid power. Any renewable energy beyond what is needed to run the household and maintain full charge on an emergency backup battery bank is fed back into the utility grid, earning the customer credit via the net meter to use the energy later at no additional charge. If household power requirements exceed the PV input, e.g., at night or on a cloudy day, the shortfall is automatically and seamlessly made up by the grid. If the grid power fails, power will be drawn instantly from the backup batteries to support the household. With today's inverters, switching time in case of grid failure is so fast your home computer may not even notice. This is

the primary difference between grid-tied systems with and without batteries. Batteries will allow continued operation if the utility fails. They'll provide backup power for your essential loads and will allow you to store and use any incoming PV energy.

Since the utility grid at large is for the most part reliable, and because batteries are very expensive, we do not recommend grid-tied systems with battery backup unless your utility is unreliable or you're on the edge of the grid, experiencing frequent and long-lasting outages. Otherwise, you're not getting any worthwhile return on investment for batteries that are rarely, if ever, used. An efficient generator (ideally running off biofuels!) is the best bet for emergency preparedness. That being said, you can always add batteries later, due to the modular nature of PV systems, so if in the future the utility grid became less reliable, you could add batteries at that time.

A number of federal and state programs exist to hasten this emerging technology, and an increasing number of them have real dollars to spend! These dollars usually appear as refunds, incentives, or tax credits to the consumer—that's you. Programs and available funds vary with time and state. For the latest information, call your State Energy Office, listed in the Appendix, or check the Database of State Incentives for Renewable Energy on the Internet at dsireusa.org.

System Sizing

We've found from experience that there's no such thing as "one size fits all" when it comes to energy systems. Everyone's needs, expectations, budget, site, and climate are individual, and your power system, in order to function reliably, must be designed with these individual factors in mind. Our friendly and helpful technical staff, with over 75,000 solar systems under its collective belt, has become rather good at this. We don't charge for this personal service, so long as you purchase your system components from us. We do need to know what makes your house, site, and lifestyle unique. So filling out the household electrical demands portion of our sizing worksheets is the first very necessary step, usually followed by a phone call (whenever possible) or email and a customized system quote. Worksheets, wattage charts, and other helpful information for system sizing are included at the end of Chapter 4, our "panel to plug" chapter—which also covers batteries, safety equipment, controls, monitors, and all the other bits and pieces you need to know a little about to assemble a safe, reliable renewable energy system.

Handy Formulas for Estimating Household Renewable Energy Installations

[Photovoltaic (PV) array size (watts)] × [solar radiation (hours/day)] × [system efficiency] = [system output (watt-hours/day)]

Off-grid Solar

[Average daily electric usage (watt-hours/day)] ÷ [solar radiation (hours/day)] ÷ [65% off-grid system efficiency] = [PV watts required]

Ballpark estimate:

[PV array size (W)] × 3 = [Output (Wh/day)]
[Output (Wh/day)] × ⅓ = [PV array size (W)]

Let's say we want to produce 10,000 watt hours per day in an area with an average of 4.5 peak sun hours per day. We would need an approximately 3.4 kW array (10,000 watt hours / 4.5 peak sun hours / .65 = 3,418 watts). We could get our ballpark estimate by taking our 3,400-watt array × 3 = 10,256 watt hours, or taking our desired output per day of 10,000 watt hours × ⅓ = 3,333 watts. There are lots of variables at play, so sometimes using a ballpark estimate will serve you well, since you can always adjust your energy usage habits depending on the energy you have available. If you're working with a set budget and know how large an array you can afford to buy, you'll also know how many kWh per day you can expect to generate without too much head scratching.

On-grid Solar

[Average daily electric usage (kilowatt-hours/day)] ÷ [solar radiation (hours/day)] ÷ [77% on-grid system efficiency] = [PV kilowatts required]

Ballpark estimate:

[PV array size (kW)] × 4 = [Output (kWh/day)]
[Output (kWh/day)] × ¼ = [PV array size (kW)]
1 kW = 75 sq. ft. of PV panels

1 MW (system rating) of PV energy powers 130 homes at the US average of 31 kWh/day (220 homes in California at 18 kWh/day average)

1 MW (system rating) of wind energy powers 250 average US homes (450 homes in California)

Battery Bank Sizing

[kWh/day] × [3–5 days of storage] × 3 = [kWh size for battery bank]

Charge Controller Sizing

[PV short-circuit current amps] × 1.56 = [Total amp size]

Fusing/Breakers Sizing

[Short-circuit current amps] × 1.56 = [Fuse/breaker amp size]

A full-size household system has all these parts.

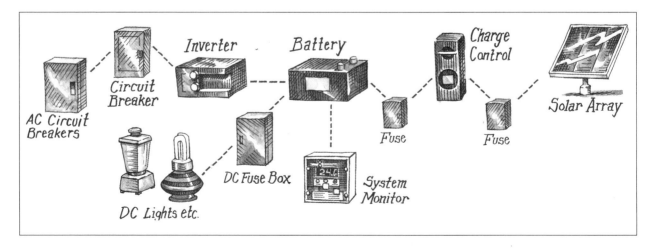

What's It Going to Cost Me to Go Solar?

Three easy steps to get a ballpark calculation for grid-tied systems:*

1. Find your daily utility usage by dividing the kilowatt-hours (kWh) used on an average month's utility bill by 30.
2. Divide that number by 5 (the average number of peak sun hours in the US) and multiply by 1.43 to account for system losses. This is the size of the solar system, in kilowatts, that you will need for taking care of 100% of your electrical needs.
3. Multiply that number by $3.75 ($3.75/watt installed) for a good ballpark idea of the gross installed cost.

Can state rebate incentives take a chunk out of that price? Go to dsireusa.org to find out what grants or incentives are available in your state. For instance, in California, you can multiply your gross installed cost by 70 to account for rebates and tax credits. In New York or New Jersey, multiply by 0.6.

What ongoing savings can I expect? Whatever you're now paying the utility for electricity will change to $0 (service charges will still apply).

Call our techs at 800-919-2400 for more information on how solar can work for your house.

* For off-grid systems, roll up an estimated watt-hour calculation using our system sizing worksheet on pages 127–128 or 421–22.

PV Racking

A PV racking structure will secure your modules to either a roof or ground mount, both keeping them safe from wind damage and allowing some cooling air behind them. PV mounting hardware can be as small as one module for an RV or big enough to carry thousands of modules for a large MW utility intertie system. PV systems are most productive if the modules are approximately perpendicular to the sun at solar noon, the most energy-rich time of the day for a PV module. If you live in the Northern Hemisphere, you need to point your modules roughly south or slightly west. The best year-round angle for your modules is approximately equal to your latitude. For better winter performance, raise that angle about 10°; for better summer performance, lower that angle about 10°. Given the minor increase in production vs. the cost of added racking and labor to tilt modules to the optimal angle, for typical residen-

tial grid-tied systems, flush-mounted racking is the best choice.

Off-the-grid systems should probably have the modules oriented for best wintertime performance, as this is typically when they are most challenged for power delivery. Grid-tied systems are usually set up for best summer performance to match summer air-conditioning demand. In addition, most utilities allow credits to be rolled over from one month to the next (net metering). We'll make the most of those long summer days to deliver the maximum kilowatt-hours for the year.

You can change the tilt angle of your array seasonally as the sun angle changes, but on a practical level, many residential systems will have power to burn in the summer. Most folks have found seasonal adjustments to be unnecessary.

Tracking mounts aim the PV array directly at the sun and follow it across the sky every day. In the early days of PV, Real Goods enthusiastically promoted trackers. But times change. The electrical and/or mechanical complexity of tracking mounts—specifically moving parts—assures you of ongoing maintenance chores (ask us how we know…), and the falling cost of PV modules makes trackers less attractive when you consider simply using more panels. For most systems now, an extra module or two on a simple fixed mount is a much better investment in the long run.

On our website (realgoodscom), you'll find numerous mounting structures, each with its own particular niche in an independent energy system. We'll try to explain the advantages and

plus 10° for winter

plus 10° for summer

your latitude in degrees

LEVEL GROUND

Proper PV mounting angle.

The best year-round angle for your modules is approximately equal to your latitude. For better winter performance, raise that angle about 10°; for better summer performance, lower that angle about 10°.

disadvantages of each style to help you decide if a certain PV racking product is the best selection for your system.

In ascending order of complexity, your choices are:

- RV mounts
- Home-built mounts
- Fixed-roof or ground mounts
- Pole-top fixed mounts
- Passive trackers
- Active trackers

Check out realgoods.com for more information on racking products.

RV Racking

Because of wind resistance and never knowing which direction the RV will be facing next, most RV owners simply attach the module(s) flat on the roof. RV mounts raise the module an inch or two off the roof for cooling. They can be used for small home systems as well when looking for a simple solution. Easy to use and inexpensive, most of them are made of aluminum for corrosion resistance. Obviously, they're built to survive high wind speeds. These are a good choice for systems with a module or two. For larger systems, the fixed or pole-top racks are usually more cost effective.

DIY Racking

Want to do it yourself? No problem. Small fixed racks are pretty easy to put together. Anodized aluminum or galvanized steel are the preferred materials due to corrosion resistance, but mild steel can be used just as well, so long as you're willing to touch up the paint occasionally. Unistrut and other slotted steel angle stock are available in galvanized form at most hardware and home-supply stores and are exceptionally easy to work with. Wood is not recommended, because your PV modules will last longer than any exposed wood. Even treated wood won't hold up well when exposed to the weather for over 40 years. Make sure that no mounting parts will cast shadows on the modules. Adjustable tilt is nice for seasonal angle adjustments, but most residential systems have power to spare in the summer, and seasonal adjustments are usually abandoned after a few years.

Flush Roof Mount or Fixed Ground Mounts

This is easily the most popular mounting structure style we sell. These mounts all use the highly adaptable SolarMount extruded aluminum rails as their base. This à la carte mounting system

offers the basic rails in various lengths and strengths (Lite or Standard depending on the desired maximum span between attachments). Buy enough rails to fit your particular array, then add optional roof standoffs and/or telescoping back legs for seasonal tilt and variable roof pitches. This mounting style can be used for flush roof arrays, low-profile roof arrays with a modest amount of tilt, high-profile arrays on flat roofs that stand way up, ground mounts in either low or high profile, or even flipped over on south-facing walls. Use concrete footings or a concrete pad for ground mounts. The racks are designed to withstand wind speeds up to 100 miles per hour or more. They don't track the sun, so there's nothing to wear out or otherwise need attention. Getting snow off of them is sometimes troublesome; it can pile up at the base, something to consider when determining the front edge height of the array; 2 to 3 feet is typical. Ground mounting can leave the modules vulnerable to grass growing up in front, or to rocks kicked up by mowers. A couple of feet or so of elevation or a concrete pad is a good idea for ground mounts. Note that attaching these racks to the roof will require roof penetrations every 4–6 feet. For example, a 3 kW PV array might need 20–24 attachment points, so the skill of the installer is important here to ensure a leak-free roof, aside from selecting a reliable flashing. It's also a good idea to stagger your attachments to distribute the loaded weight of the array evenly between all rafters or trusses.

For larger flat roof-mounted systems (usually on commercial buildings), several ballast-type racks are available these days. If roof penetrations are undesirable, then securing the mounting hardware with heavy weights (such as concrete blocks or containers of sand) or the weight of the hardware—ballasts—is the answer. The disadvantage of this type of mounting is usually the high cost, even though installation is not too difficult. Sometimes structures cannot handle the weight a ballasted system would add, so a combination of ballasts and attachments are used at the right balance to minimize the attachments needed. High wind or seismic areas will probably also require at least a minimum number of mechanical attachments to the roof anyway.

Finally, a mention of BIPV (Building Integrated PV) products should be included here, even though these roof PV modules don't require mounting hardware at all. Intended primarily for new construction, these modules actually are part of the roof covering. They provide an architecturally pleasing look that blends in seamlessly with the rest of the home or building. One type is the frameless roof-integrated module, such as Solar Shingles, which look much like a typical composition shingle and blend seamlessly with the surrounding roof. Another type is more like a standard framed polycrystalline module but is sized and installed like a roof tile. The main advantage of BIPV products is that they are unobtrusive and usually look pretty good if aesthetics are important. Disadvantages include their higher cost and low efficiency due to higher operating temperatures (lack of cooling air circulation). Troubleshooting can also be difficult, depending on the quantity and accessibility of junction boxes and electrical connections. For more information about BIPV and ballast mounts, be sure to call one of our Solar Technicians.

Pole-top Racking

A popular and cost-effective choice, pole-top mounts are designed to withstand winds up to 80 mph and in some cases up to 120 mph. The UniRac rails for the larger pole-top arrays are a heavier-duty version of the standard SolarMount rails (see realgoods.com). This mounting style is a good choice for snowy climates, because it keeps the front edge of the panels high off the ground clear of snow pileup, and with nothing underneath it, snow tends to slide right off.

For small or remote systems, pole-top mounts are the least expensive and simplest choice. We almost always use these for one- or two-module pumping systems. Tilt and direction can be easily adjusted. Site preparation is easy, just get your steel pipe cemented in straight. The pole is common schedule 40 steel pipe, which is not included (pick it up locally to save on freight). Make sure that your pole is tall enough to allow about one-third burial depth and still clear livestock, snow, or weeds. Ten feet total for pole length is usually sufficient, but always consult your product installation manual. Taller poles are sometimes used for theft deterrence. Pole diameter depends on the specific mount and array size. Pole sizes listed are for "nominal pipe size." For instance, what the plumbing industry calls "4-inch" is actually 4½-inch outside diameter. When a mount says it fits "4-inch," it's actually expecting a 4½-inch-diameter pipe.

Passive Trackers

Tracking mounts will follow the sun from east to west across the sky, increasing the daily power output of the modules, particularly in summer and in southern latitudes. Trackers are most often used on water-pumping systems with peak demands in summer. See the sidebar "To Track or Not to Track" for a discussion on when tracking mounts are appropriate.

Passive trackers follow the sun from east to west using just the heat of the sun and gravity. No source of electricity is needed—a simple, effective, and brilliant design solution. The north-south tilt axis is seasonally adjustable manually. Maintenance consists of two squirts with a grease gun once every year.

Tracking will boost daily output by about 30% in the summer and 10%–15% in the winter.

Do Tilt Angle or Orientation Matter?

Much has been made in years past of PV tracking and tilt angle, and early adopters obsessed over having just the right angle for their panels. You would think that it's nearly a life-or-death matter to point your PV modules *directly* at the sun during all times of the day. We're here to shake up this belief after more than 36 years of experience. Tilt angle and orientation make a lot less difference than one would think.

From the Sandia National Labs comes this very interesting chart, which details tilt angle vs. compass orientation, and the resulting effect on yearly power production. South-facing, at a 30° angle (7:12 roof pitch) delivered the most energy. They labeled that point 100%. All other orientations and tilt angles are expressed as a percentage of that number. Note that we can face SE or SW, a full 45° off due south, and lose only 4%. Our tilt angle can be 15° off, and we lose only 3%. In reality, we can face due east or west and lose only 12%! Facing panels any further north than 90° or 270° was unheard of at one point, but now this can be done justifiably in grid-tied systems when considering the high cost of utility power and the low cost of solar energy. A couple of bird droppings could cost you more energy than having "the right" angled roof. And in fact, homes with time-of-use rate schedules can even be better off with panels facing westward rather than directly south, since that maximizes production at "on peak" times.

Roof Slope and Orientation (Northern California Data)

	Flat (0°)	4:12 (18.4°)	7:12 (30°)	12:12 (45°)	21:12 (60°)	Vertical (90°)
South	0.89	0.97	1.00	0.97	0.89	0.58
SSE, SSW	0.89	0.97	0.99	0.96	0.88	0.59
SE, SW	0.89	0.95	0.96	0.93	0.85	0.60
ESE, WSW	0.89	0.92	0.91	0.87	0.79	0.57
East, West	0.89	0.88	0.84	0.78	0.70	0.52

The moral of this story? Shadows, bird droppings, leaves, and dirt will have far more effect on your PV output than orientation. Keep your modules clean and shade free, and don't worry if they aren't perfectly perpendicular to the sun at noon every day. You'll still receive all the environmental and economic benefits.

To Track or Not to Track

Photovoltaic modules produce the most energy when situated perpendicular to the sun. A tracker is a mounting device that follows the sun from east to west and keeps the modules in the optimum position for maximum power output. At the right time of year, and in the right location, tracking can increase daily output by more than 30%. But beware of the qualifiers: Trackers are often *not* a good investment. It is necessary to take a close look at the value of the kWh increase in production from tracking compared to the price tag of the tracker.

Trackers work best during the height of summer, when the sun is making a high overhead arc. They add very little in winter unless you live in the extreme southern US, or even further south. Trackers need clear access to the sun from early in the morning until late in the afternoon. A solar window from 9 AM to 4 PM is workable; if you have greater access, more power to you (literally).

Tracking mounts are expensive, and PV power is getting cheaper. If you have a project that peaks in power use during the summer, such as water pumping or residential cooling, then tracking may be a very good choice. For many water-pumping projects, the most cost-effective way to increase daily production is to simply add a tracking mount.

If your projects peaks in power use during winter, such as powering a typical house, then tracking doesn't offer you much. In most of North America, winter tracking will add less than 15%. One of the new generation of MPPT charge controls is a much better investment in this situation. They add 15%–30% and do their best work in the winter. See the Controllers section of Chapter 4 for more info.

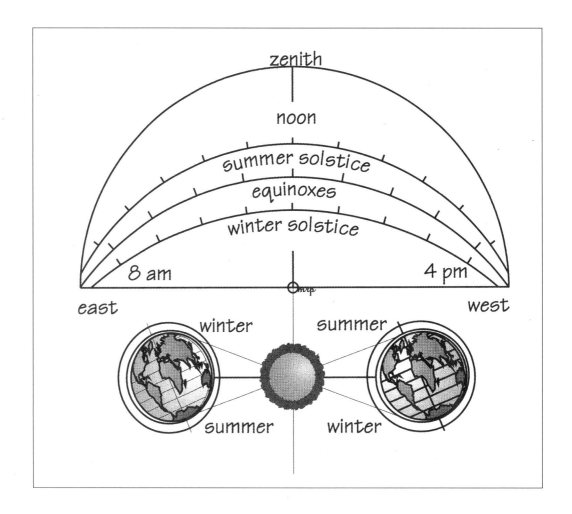

The two major problems with passive technology are wind disturbances and slow "wake-up" when cold. The tracker will go to "sleep" facing west. On a cold morning, it may take more than an hour for the tracker to warm up and roll over toward the east. In winds over 15 mph, the passive tracker may be blown off course. These trackers can withstand winds of up to 85 mph (provided you follow the manufacturer's recommendations for burying the pipe mount) but will not track at high wind speeds. If you have routine high winds, you should have a wind turbine to take advantage of those times, but that's a different subject. Passive trackers have become increasingly rare in the PV industry, and most that you see these days are stuck in the same position they were when they stopped working, usually for lack of maintenance.

Active Trackers

Active trackers use photocells, electronics, and linear actuators like those on giant old-fashioned satellite TV dishes to track the sun very accurately. A small controller bolted to the array is programmed to keep equal illumination on the photocells at the base of an obelisk. Power use is minuscule. Active trackers average slightly more energy collection per day than a passive tracker in the same location, but historically they have also averaged more mechanical and electrical problems, too. Based on our experience, the experience of hundreds of customers, and the dropping price of PV power, we no longer recommend active trackers. Their high initial cost and continuing maintenance problems just aren't worth the investment any longer. Want more power? Add some more PV. It's hard to beat the reliability of *no moving parts*. Some commercial-scale projects have chosen to use active trackers; however, they usually come with a full-time maintenance contract, and time will tell if these were wise investments or not.

Wind and Hydro Power Sources

Sunlight, wind, and falling water are the renewable energy Big Three. These are energy sources that are commonly available at a reasonable cost and complement each other seasonally in many cases. Solar, or sunlight, the single most common and most accessible renewable energy source, is well covered at the beginning of this chapter. We've found in our years of experience that wind and hydro energy sources are most often developed as a booster or bad-weather helper for a solar-based system. These hybrid systems have the advantage of being better able to cover power needs throughout the year and are less expensive than a similar capacity system using only one power source. Finding the right balance of each source depends on a number of factors, so give us a call (800-919-2400) or visit our website (realgoods.com) to help you figure it out. When a storm blows through, the solar input is lost, but a wind generator more than makes up for it, another reason to have practical dump loads like hot water heating appliances ready to take advantage. The short, rainy days of winter may limit solar gain, but the hydro system picks up from the rain and delivers steady power 24 hours a day. This is not to say that you shouldn't develop an excellent single-source power system if you've got it—like a year-round stream dropping 200 feet across your property, for instance. But good hydro sites are rare, and for most of us, we'll be further ahead if we don't put all our renewable energy eggs in one basket. Diversify!

Our experienced technical staff is well versed in supplying the energy needs of anything from a small weekend getaway cabin all the way up to an upscale state-of-the-art resort. We'll be glad to help put a system together for you. When ordering equipment from us, there is usually no charge for our friendly and personalized services. We're as excited as you are to design these types of systems.

Wind Systems

We generally advise that a good year-round wind turbine site isn't a place that you'd want to live. It takes average wind speeds of 8–9 mph and up to make a really good site. That's honestly more wind than most folks are comfortable living with. But this is where the beauty of hybrid systems comes in. Many very livable sites *do* produce 8 mph and greater wind speeds during certain times of the year or when storms are passing through. Tower height and location also make a big difference. Wind speeds average 50%–60% higher at 100 feet compared with ground level (see chart in the Wind section). Wind systems these days are almost always designed as wind/solar hybrids for year-round reliability. The only common exceptions are systems designed for utility intertie; they feed excess power back into the utility and

We've found in our years of experience that wind and hydro energy sources are most often developed as a booster or bad-weather helper for a solar-based system. These hybrid systems have the advantage of being better able to cover power needs throughout the year and are less expensive than a similar capacity system using only one power source.

turn the meter backward, and we're starting to see more grid-tied residential wind systems.

Micro-hydro Systems

For those who are lucky enough to have a good site, micro-hydro is really the renewable energy of choice. System component costs are much lower, and watts per dollar return is much greater for hydro than for any other renewable source. John Schaeffer's hydro system cost him about 1/10 of what his solar system cost for an equivalent number of watts. The key element for a good site is the vertical distance the water drops. A small amount of water dropping a large distance will produce as much energy as a large amount of water dropping a small distance. The turbine for the small amount of water is going to be smaller, lighter, easier to install, and vastly cheaper. We

offer several turbine styles for differing resources. The small Pelton wheel Harris systems are well suited for mountainous territory that can deliver some drop and high pressure to the turbine. The propeller-driven Low-Head Stream Engine is for flatter sites with less drop but more volume, and the Stream Engine, with a Turgo-type runner, falls in between. It can handle larger water volumes and make useful power from shorter vertical water-drop distances. (This turbine is available at realgoods.com.)

Read on for detailed explanations of wind generators and hydro turbines. If you need a little help and guidance putting a system together or simply upgrading, our technical staff, with decades of hands-on experience in renewable energy, will be glad to help. Call us toll-free at 800-919-2400. or visit our website: realgoods.com

Hydroelectricity

Hydropower, given the right site, can cost as little as ⅒ of a PV system of comparable output.

If you could choose any renewable energy source you wanted, hydro is the one. It also happens to be the most site-specific, unfortunately for most of us, so if you're lucky enough to have a good hydro site, consider yourself a winner of the renewable energy lottery. If you don't want to worry about a conservation-based lifestyle—always nagging your kids to turn off the lights, watching the voltmeter, basing every appliance decision on energy efficiency—then you had better settle next to a nice year-round mountain stream! Hydropower, given the right site, can cost as little as ⅒ of a PV system of comparable output. Hydropower users are often able to run energy-hog appliances that

would bankrupt a PV system owner, like large side-by-side refrigerators and electric space heaters. In fact, most hydro users have energy to spare and use it for water heating or something else practical. Hydropower will probably require more effort onsite to install, but it's well worth the initial effort. Even a modest hydro output over 24 hours a day, rain or shine, will add up to a large cumulative total. Hydro systems get by with smaller battery banks because they need to cover only the occasional heavy power surge rather than four days of cloudy weather.

Hydro turbines can be used in conjunction with any other renewable energy source, such as PV or wind, to charge a common battery bank. This is especially true in the West, where seasonal creeks with substantial drops flow only in the winter. This is when power needs are at their highest and PV input is at its lowest. Small hydro systems are well worth developing, even if used only a few months out of the year, if those months coincide with your highest power needs. So, what makes a good hydro site, and what else do you need to know?

What Is a "Good" Hydro Site?

The Columbia River in the Pacific Northwest has some really great hydro sites, but they're not exactly homestead scale (or low cost). Within the hydro industry, the kind of home-scale sites and systems we deal with are called micro-hydro. The most cost-effective hydro sites are located in the mountains. Hydropower output is determined by water's volume times its fall or drop (jargon

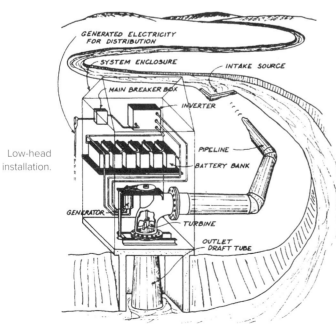

Low-head installation.

for the fall is "head"). Practically any site where water drops at least a few feet has the potential to be a great micro-hydro site. You can get approximately the same power output by running 1,000 gallons per minute through a 2-foot drop as by running 2 gallons per minute through a 1,000-foot drop. In the former scenario, where lots of water flows over a little drop, we are dealing with a low-head/high-flow situation, which is not truly a micro-hydro site. Turbines that can efficiently handle thousands of gallons are usually large, bulky, expensive, and site-specific. But if you don't need to squeeze every last available watt out of your low-head source, the Low-Head Stream Engine generator (see realgoods.com) will produce very useful amounts of power from low-head/high-flow sites, or the Turgo runner used on the Stream Engine is good at high-volume sites with 15 feet or more of head. Also keep in mind that your delivery pipe to the turbine can run quite a distance to attain enough head from start to finish. This will be necessary, but worth the effort in low-flow sites where you'll need to maximize head.

Over the years, we've learned a few things about good hydro site development. For hilly sites that can deliver a minimum of 50 feet of head, Pelton wheel turbines offer the lowest-cost generating solution. The Pelton-equipped Harris turbine is perfect for low-flow/high-head systems

(see realgoods.com). It can handle a maximum of 200 gallons per minute and requires a minimum 50-foot fall to make useful amounts of power. In general, any site with more than 100 feet of fall will make an excellent micro-hydro site, but many sites with less fall can be very productive also. The more head, the less volume will be necessary to produce a given amount of power.

A hydro system's fall doesn't need to happen all in one place. You can build a small collection dam at one end of your property and pipe the water to a lower point, collecting fall as you go. It's not unusual to use several thousand feet of pipe to collect a hundred feet of head (vertical fall).

Our Hydro Site Evaluation service will estimate output for any site, plus it will size the piping and wiring, and factor in any losses from pipe friction and wire resistance. See the example at the end of this editorial section.

What If I Have a High-flow/ Low-head Site or Want AC Output?

Typically, high-flow/low-head or AC-output hydro sites will involve engineering, custom metalwork, formed concrete, permits, and a fair amount of initial investment cash. None of this is meant to imply that there won't be a good payback, but it isn't an undertaking for the faint-of-heart or thin-of-wallet. AC generators are typically used on larger commercial systems, or on utility intertie systems. DC generators are typically used on smaller residential systems. If you have a good site and have reasonable energy needs, you're not likely to need as much juice as a larger AC generator will deliver.

DC generation systems offer several advantages for small hydro. Control is easy and cheap. The batteries we'll use to store energy allow power output surges way over what the turbine is delivering. The DC-to-AC inverters now available will deliver far cleaner and more tightly regulated AC power than a small AC hydro turbine can manage, and the inverter will cost less than a small AC control system. It's a pleasure to kick back and enjoy the time-tested reliability of a well-thought-out DC hydro setup.

For the site with a good creek but little significant fall, the Low-Head Stream Engine turbine is a far less costly alternative. With just 3 or 4 feet of fall, and some site development work, this simple turbine can provide for a modest homestead.

If you'd rather look into the typical low-head scenario, contact the DOE's Renewable Energy Clearinghouse at 800-363-3732, or energy.gov /energysaver/articles/microhydropower-systems, or check out microhydropower.net for more free

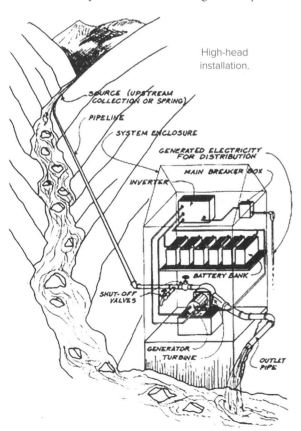

High-head installation.

SOURCE (UPSTREAM COLLECTION OR SPRING)

PIPELINE

SYSTEM ENCLOSURE

GENERATED ELECTRICITY FOR DISTRIBUTION

MAIN BREAKER BOX

INVERTER

BATTERY BANK

SHUT-OFF VALVES

GENERATOR TURBINE

OUTLET PIPE

Small hydro systems are well worth developing, even if used only a few months out of the year, if those months coincide with your hightest power needs.

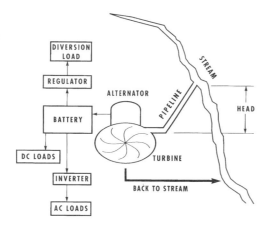

Typical micro-hydro system.

DIVERSION LOAD

REGULATOR

BATTERY

DC LOADS

INVERTER

AC LOADS

ALTERNATOR

TURBINE

PIPELINE

STREAM

HEAD

BACK TO STREAM

For the homestead with a good creek but little significant fall, the Low-Head Stream Engine turbine is a far less costly alternative.

information on low-head hydro than you ever thought was possible.

How Do Micro-hydro Systems Work?

The basic parts of micro-hydro systems are the pipeline (called the penstock in the trade), which delivers the water; the turbine, which transforms the energy of the flowing water into rotational energy; the alternator or generator, which transforms the rotational energy into electricity; the regulator or controller, which manages the hydro turbine's charging or dumps excess energy, depending on regulator style; and the conductors or wiring, which delivers the electricity. Most micro-hydro systems also use batteries, which store the low-voltage DC electricity between 12 and 48 V nominal, and usually an inverter, which converts the low-voltage DC electricity into 120 or 240 volts AC electricity to energize your load center.

Most micro-turbine systems use a small DC alternator or generator to deliver a small but steady energy flow that accumulates in a battery bank. This provides a few important advantages. The battery system allows the user to store energy and expend it, if needed, in short powerful bursts (like a washing machine starting the spin cycle). The batteries will allow us to deliver substantially more energy for short periods than a turbine is producing, as long as the battery and inverter are designed to handle the load. DC charging means that precise control of alternator speed is not needed, as is required for 60Hz AC output. This saves thousands of dollars on control equipment. And finally, with the quality of the DC-to-AC inverters now available, you'll enjoy cleaner, more tightly controlled AC power through an inverter than through a small AC turbine, ensuring that sensitive electronics won't give you a hard time due to unclean power. The bottom line is that a

DC-based system will cost far less than an AC system for most residential users, and will perform better.

DC Turbines

Several micro-hydro turbines are available with simple DC output. We currently offer the Harris and the Stream Engine (see realgoods.com).

Harris Turbine

Our most classic turbine we've used to design systems, made by micro-hydro pioneer Don Harris. The Harris turbine uses a hardened cast-silicone bronze Pelton turbine wheel mated with a low-voltage DC alternator. Pelton wheel turbines work best at higher pressures and lower volumes. Minimum site specs are about 50 feet of head. For these turbines, there is no practical upper limit, so you can take full advantage of a large mountain property. Harris offers a couple choices for the alternator. The standard Harris is based on common 1970s Motorcraft alternators with windings that are customized for each individual application. Bearings and brushes will require replacement at intervals from one to five years, depending on how hard the unit is working. These parts are commonly available at any auto parts store. Harris now comes with a standard permanent-magnet alternator that is custom-made. The PM alternators deliver more power under almost all conditions, have no brushes to wear out, and are mounted on larger, more robust bearings with two to three times the life expectancy. We suggest the PV alternator for quality and reliability.

Depending on the volume and fall supplied, Harris turbines can produce from 1 kWh (1,000 watt-hours) to 35 kWh per day, more than the average US home. Maximum alternator instantaneous output is about 2,500 watts in a 48-volt system with cooling options required. The typical American home consumes 15–25 kWh per day with no particular energy conservation, so with a good hydro site, it is fairly easy to live a conventional lifestyle.

The Harris turbine can be supplied with one, two, or four nozzles. The maximum flow rate for any single nozzle can be from 20 to 60 gallons per minute (gpm), depending on the head pressure. The turbine can handle flow rates to about 120 gpm before the sheer volume of water starts getting in its own way. Many users buy two- or four-nozzle turbines with different-sized nozzles, so that individual nozzles can be turned on and off to meet variable power needs, a huge benefit for sites that experience large seasonal swings in

Two- and four-nozzle Harris turbines. The four-nozzle is upside down to show the Pelton wheel and nozzles.

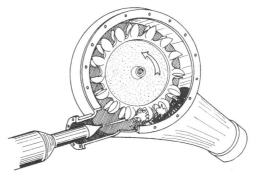

An 1880s vintage single-nozzle Pelton wheel. Only the generator technology has changed.

the amount of available water. The brass nozzles are easily replaceable because they eventually wear out, especially if there is grit in the water. They are available in 1⁄16-inch increments, from 1⁄16 inch through ½ inch. The first nozzle doesn't have a shutoff valve, while all nozzles beyond the first one are supplied with ball valves for easy visible operation.

Stream Engine

The Stream Engine turbines use a unique brushless, permanent-magnet alternator with three large sealed-shaft bearings. They are reliable and low maintenance. Permanent magnets mean there are no field brushes to wear out. Magnetic field strength is adjusted by varying the air gap between the magnet disk and the stationary alternator windings. Once the unit is set up, there is almost no routine maintenance required. Setup requires some time with this universal design, however, and involves trial and adjustment: selecting one of four alternator wiring setups and then adjusting the permanent-magnet rotor air gap until peak output is achieved. A precision shunt and digital multimeter are supplied to expedite this setup process. Output voltage can be user selected at 12, 24, or 48 volts. Provided ad-

equate site conditions, maximum instantaneous output is 800 watts for this alternator.

This low-maintenance alternator is employed on two very different turbines. The original Stream Engine uses a cast bronze Turgo-type runner wheel. This Turgo wheel can handle a bit more water volume than the Harris Pelton wheel—up to about 200 gpm before it starts choking—and starts to deliver useful output at lower 15- to 20-foot head. Nozzles are cone-shaped plastic casings; you cut them at the size desired, from ⅛ inch up to 1 inch. Two- or four-nozzle turbines are available, with replacement nozzles being a readily available bolt-in.

The Low-Head Stream Engine uses the same alternator but is packaged quite differently. It works on 2–10 feet of fall, which happens on the downstream end of the turbine for a change. Flows of 200–1,000 gpm can be accommodated through the 5-inch propeller turbine. The large draft tube on the output must be immersed in the tail water. (See the illustration.) This turbine can be more labor intensive on the site preparation side of things, but almost no maintenance or attention is required once it's installed and tuned.

HI Power Hydro

Headquartered in the coastal mountains of northern California, HI Power, manufacturer of our HI Power hydro units, brings over 30

The bottom line is that a DC-based system will cost far less than an AC system for most residential users, and will perform better.

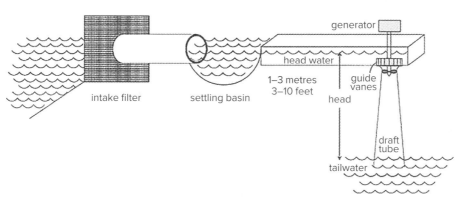

Low-Head Stream Engine installation.

A typical installation places the batteries at the house on top of the hill, where the good view is, and the turbine at the bottom of the hill, where the water ends its maximum drop. Low-voltage power is difficult to transmit if large quantities or long distances are involved.

years of experience installing and maintaining micro-hydro systems (see realgoods.com). They use brushless alternators mounted in an anodized aluminum Turgo housing for reliability and versatility. These units can generate continuous power anywhere from 30 watts all the way up to 4,000 watts, depending on your site. Output voltage is unregulated, or "wild" AC, which is then stepped down with supplied transformers and rectifiers. They can handle head ranging from 60 to 500 feet, and flows between 10 and 400 gallons per minute. The four nozzles with each unit allow for convenient adjustment for varying flows. The high-voltage models are ideal for sites where your turbine is located a long distance from where the power is needed, transmitting power anywhere from 120 to 440 volts AC, saving you a ton on wiring costs. The higher voltage is then stepped down to your battery bank voltage of 12, 24, or 48 volts nominal. They even make a grid-tied version, something we had not seen previously. Overall efficiency ranges from 30 to 60 percent, again depending on your site conditions. Not only are these hydro units among the best available today, but in addition each one is manufactured and tested using only solar and hydro power!

Power Transmission

No renewable energy system is quite perfect; there's always a small catch. One disadvantage of lower-voltage DC hydro systems is the difficulty of transmitting power from the turbine to the batteries, particularly with high-output sites. A typical installation places the batteries at the house on top of the hill, where the good view is, and the turbine at the bottom of the hill, where the water ends its maximum drop. Low-voltage power is difficult to transmit if large quantities or long distances are involved. The batteries should be as close to the turbine as is practical, but if there's more than 100 feet of distance involved, things will work better if the system voltage is 24 or even 48 volts. Transmission distances of more than 500 feet often require expensive large-gauge wire or technical tricks. Don Harris has been working with Outback Power Systems to develop a hydroelectric version of their maximum power point tracking MX60 controller, which allows inputs up to 140 volts, while feeding the batteries whatever it is that makes them happy.

One option is to set up a power shed closer to the turbine containing your battery bank, inverters, and other system components. Keeping the DC transmission short, you can send the power already at 120/240 V back to the house, thus avoiding the low-voltage transmission dilemma altogether. However, consider that this strategy requires you to build a well-suited power shed that will keep all the equipment protected from the elements. Please consult with the Real Goods technical staff about this or other transmission options.

Controllers

Hydro generators require special controllers or regulators. Controllers designed for photovoltaics may damage the hydro generator and will very likely become crispy critters themselves if used with one. You can't simply open the circuit when the batteries get full like you can with PV, otherwise you could seriously damage your generator. So long as the generator is spinning, there needs to be a place for the energy to go, a load needs to be present. Controllers for hydro systems take any power beyond what is needed to keep the batteries charged and divert it to a secondary load, usually a water- or space-heating element. So extra energy heats either domestic hot water or the house itself. These diversion controllers are also used with some wind generators and can be used for PV control as well if this is a hybrid system.

Site Evaluation

Okay, you have a fair amount of elevation drop across your property and/or enough water flow for one of the low-head turbines, so you think micro-hydro is a definite possibility and you want to see what's possible. What happens next? Time to go outside and take some measurements, then fill in the necessary information on the Hydro Site Evaluation Form. With the info on your completed form, the Real Goods technicians can calculate which turbine and options will best fill your needs, as well as what size pipe and wire and which balance-of-system components you require. Then we can quote specific power output and system costs so you can decide if hydro is worth the installation effort. If you think you have enough water to produce power, it's well worth the inquiry.

Distance Measurements

Keep the turbine and the batteries as close together as practical. As discussed earlier, longer transmission distances will get expensive. The more power you are trying to move, the more important distance becomes.

You'll need to know the distance from the proposed turbine site to the batteries (how many feet of wire) and the distance from the turbine site to the water collection point (how many feet of pipe). These distances are fairly easy to determine; just pace them off or use a tape measure. Pipe diameter and water friction will come into play as well, so don't run out and grab a bunch of PVC just yet.

Fall (Drop or Head) Measurement

Next, you'll need to know the fall from the collection point to the turbine site. This measurement is a little tougher, and some light surveying may be in order. If there is a pipeline in place already, or if you can run one temporarily and fill it with water, this part is easy. Simply install a pressure gauge at the turbine site, make sure the pipe is full of water, and turn off the water at the bottom. Read the static pressure (which means no water movement in the pipe), and multiply your reading in pounds per square inch (psi) by 2.31 to obtain the drop in feet. If the water pipe method isn't practical, you'll have to survey the drop or use a fairly accurate altimeter or GPS device—even a smartphone app would do for these purposes. If the altimeter can read ±10 feet, that's close enough. Take a hike, and record the difference in feet.

The following instructions represent the classic method of surveying. You've seen survey parties doing this, and if you've heard about them and always wanted to attend a survey party, this is your big chance to get in on the action. You'll need a carpenter's level (or a pocket sight level known as a "pea shooter"), a straight sturdy stick about eye-level tall, a brightly colored target that you'll be able to see a few hundred feet away, and a friend to carry the target and to make the procedure go faster and more accurately. (At Real Goods, we do not recommend partying alone.)

Stand the stick upright and mark it at eye level. (Five feet even is a handy mark that simplifies the mathematics, if that's close to eye level for you.) Measure and note the length of your stick from ground level to your mark. Starting at the turbine site, stand the stick upright, hold the carpenter's level at your mark, make sure it is level, then sight across it uphill toward the water source. With hand motions and body English, guide your friend until the target is placed on the ground at the same level as your sightline, then have your friend wait for you to catch up. Repeat the process, carefully keeping track of how many times you repeat. It is a good idea to draw a map to remind you of landmarks and important details along the way. If you have a target and your friend has a stick (marked at the same height,

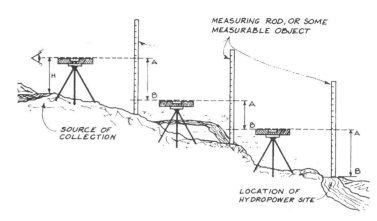

Measuring fall.

please) and level, you can leapfrog each other, which makes for a shorter party. Multiply the number of repeats between the turbine site and the water source by the length of your stick(s), and you have the vertical fall. People actually get paid to have this much fun!

Flow Measurement

Finally, you'll need to know the flow rate. If you can, block the stream and use a length of pipe to collect all the flow. Time how long it takes to fill a 5-gallon bucket. Divide 5 gallons by your fill time in seconds. Multiply by 60 to get gallons per minute. Example: The 5-gallon bucket takes 20 seconds to fill. So 5 divided by 20 = 0.25 times 60 = 15 gpm. If the flow is more than you can dam up or get into a 4-inch pipe, or if the force of the water sweeps the bucket out of your hands, forget measuring: You've got plenty!

Hydro Conclusion

Now you have all the information needed to guesstimate how much electricity your proposed hydro system will generate based on the manufacturer's output charts. This will give you an indication of whether or not your hydro site is worth developing, and if so, which turbine option is best. We've provided one output chart for reference, for the Harris turbine. To get specific information on other turbines, please call our Technical Department at 800-919-2400. If you think you have a real site, fill out the Real Goods Hydro Site Evaluation Form and send it to the Technical Department at Real Goods, or just give us a call. We will crunch the numbers to size plumbing and wiring for the least power loss at the lowest cost, and take care of all the other calculations necessary to design a working system. You'll find an example of our Hydro Survey Report on the next page, followed by the form for the info we need from you.

Now you have all the information needed to guesstimate how much electricity your proposed hydro system will generate based on the manufacturer's output charts. This will give you an indication of whether or not your hydro site is worth developing, and if so, which turbine option is best.

CALCULATION OF HYDROELECTRIC POWER POTENTIAL

Copyright © 1988 by Ross Burkhardt. All rights reserved.

ENTER HYDRO SYSTEM DATA HERE: **Customer: Meg A. Power**

Pipeline Length:	1,300 feet
Pipe Diameter:	4 inches
Available Water Flow:	100 gpm
Vertical Fall:	200 feet
Hydro to Battery Distance:	50 feet (one way)
Transmission Wire Size:	2 AWG
House Battery Voltage:	24 volts
Hydro Generation Voltage:	29 volts

Power produced at hydro:	Power delivered to house:
49.78 amps	49.78 amps
29 volts	28.20 volts
1,443.53 watts	1,403.59 watts

4-nozzle, 24 V, high-output with cooling turbine required

Pipe Calculations

Head Lost to Pipe Friction:	7.61 feet
Pressure Lost to Pipe Friction:	3.29 psi
Static Water Pressure:	86.62 psi
Dynamic Water Pressure:	83.33 psi
Static Head:	200.01 feet
Dynamic Head:	192.40 feet

Hydropower Calculations

Operating Pressure:	83.33 psi
Available Flow:	100 gpm
Watts Produced:	1,443.53 watts
Amperage Produced:	49.78 amps
Amp-Hours per Day:	1,194.65 amp-hours
Watt-Hours per Day:	34,644.83 watt-hours
Watts per Year:	12,645,362.71 watt-hours

Line Loss (using copper)

Transmission Line One-Way Length:	50 feet
Voltage:	29 volts
Amperage:	49.78 amps
Wire Size #:	2 AWG
Voltage Drop:	0.8 volts
Power Lost:	39.95 watts
Transmission Efficiency:	97.23 percent
Pelton Wheel rpm Will Be:	2,969.85 at optimum wheel efficiency

This is an estimate only! Due to factors beyond our control (construction, installation, incorrect data, etc.), we cannot guarantee that your output will match this estimate. We have been conservative with the formulas used here, and most customers call to report more output than estimated. However, be forewarned! We've done our best to estimate conservatively and accurately, but there is no guarantee that your unit will actually produce as estimated.

Real Goods Hydroelectric Site Evaluation Form

Name: _____

Address: _____

_____

Phone: _____ Date: _____

Pipe Length: _____ (from water intake to turbine site)

Pipe Diameter: _____ (only if using existing pipe)

Available Water Flow: _____ (in gallons per minute)

Fall: _____ (from water intake to turbine site)

Turbine to Battery Distance: _____ (one way, in feet)

Transmission Wire Size: _____ (only if existing wire)

House Battery Voltage: _____ (12, 24, etc.)

Alternate estimate (if you want to try different variables)

Pipe Length: _____ (from water intake to turbine site)

Pipe Diameter: _____ (only if using existing pipe)

Available Water Flow: _____ (in gallons per minute)

Fall: _____ (from water intake to turbine site)

Turbine to Battery Distance: _____ (one way, in feet)

Transmission Wire Size: _____ (only if existing wire)

House Battery Voltage: _____ (12, 24, etc.)

For a complete computer printout of your hydroelectric potential, including sizing for wiring and piping, please fill in the above information and send it to Real Goods.

Harris Turbine Output in Watts								
Feet of Head	**Gallons per Minute Flow (permanent-magnet specs in bold)**							
	3	6	10	15	20	30	50	100	200
25	–	–	–	20 **25**	30 **40**	50 **65**	115 **130**	200 **230**	–
50	–	–	35 **40**	60 **75**	80 **100**	125 **150**	230 **265**	425 **500**	520 **580**
75	–	25 **30**	60 **75**	95 **110**	130 **160**	210 **250**	350 **420**	625 **750**	850 **900**
100	–	35 **45**	80 **95**	130 **150**	200 **240**	290 **350**	500 **600**	850 **1,100**	1,300 **1,300**
200	30 **45**	100 **130**	180 **210**	260 **320**	400 **480**	580 **650**	950 **1,100**	1,500 **1,500**	–
300	70 **80**	150 **180**	275 **300**	400 **450**	550 **600**	850 **940**	1,400 **1,500**	–	–

Maximum wattage @ voltage: 12 V=750, 24 V=1,500, 48 V=2,500

A typical micro-hydro output chart, available from the manufacturer.

Wind Energy

This section was written by Mick Sagrillo of Sagrillo Power and Light. Mick has been working in the small wind energy field for over 32 years. He is coauthor of Power from the Wind: Achieving Energy Independence *(2009) and is a founding member and serves on the board of the Distributed Wind Energy Association (DWEA).*

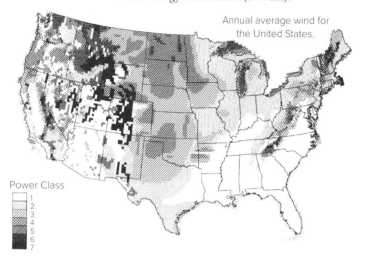

Annual average wind for the United States.

Power Class
1
2
3
4
5
6
7

Small Wind Turbines Come of Age

The debut of micro wind turbines—defined as under 1 kW—has revolutionized living off the grid. These inexpensive machines have brought wind technology within reach of almost everyone. And their increasing development has opened up new applications for wind energy previously considered off-limits, such as electric fence charging and powering remote data collection sites, once the sole domain of photovoltaics.

Micro wind turbines have been around for decades for use on sailboats, but they gained prominence in the 1990s as their broader potential for off-the-grid applications on land became more widely known. Sleek, quiet, and reliable residential micro wind turbine modules have hit the market. This article is about small wind—defined as 1 kW to 100 kW and installed on towers. Whether you're off the grid or living in town, wind power could be a good choice for you. However, it's a complicated choice that involves homework and careful assessment of multiple factors that are different from those usually considered with photovoltaics.

> Before you decide that you want to have a wind turbine, you need to answer a series of important questions. Wind is not like PV. Wind turbines are harder to site correctly, it can be difficult to get a building permit for them, and since they don't live at ground level or on your garage roof, they can be a challenge to service.

Wind Energy and Wind Turbines— What You Need to Consider to Make a Decision

You've likely been fascinated by wind turbines for many years, and now you finally have the opportunity to invest in one. An Internet search results provides a wide variety of options, but the specifications for different manufacturers and models are all over the board. How do you choose?

Before you decide that you want to have a wind turbine, you need to answer a series of important questions. Wind is not like PV. Wind turbines are harder to site correctly, it can be difficult to get a building permit for them, and since they don't live at ground level or on your garage roof, they can be a challenge to service. Let's walk through the steps you need to consider to assess whether a wind turbine is right for you.

First Things First— Do You Have Any Wind?

This is probably the most difficult question you will face, in addition to cost, and will take the most research to answer. As you go through the process outlined below, keep in mind that you are interested in *climate*, not *weather*. Weather is what you'll see this afternoon, or over the next day or two. It might be windy, but then again, it might not be. The wind seems to come and go with no apparent pattern. Weather is not a representation of your wind resource. Climate, logged over decades, will give you the seasonal and daily patterns that determine if the wind resource where you live is adequate for a wind energy system.

Wind Quantity

Any renewable energy system needs to have "fuel" for it to function as advertised. And while the solar resource for a PV or solar water system is rather easy to come by, ballparking your wind resource—your fuel of interest—takes more effort than simply observing your cat basking in the sun on the deck.

Wind is unlike any other renewable energy fuel in that the energy available to a wind turbine is a function of the cube of the fuel—the wind speed. This means that an 8 mph average annual wind speed has only half the power density of a 10 mph average. That's huge, and can make an enormous difference not only in the amount of electricity you produce but also the cost-effectiveness of your investment. So a prime question you need to answer is, "Do I have enough fuel?"

The first step in answering this question is at Wind Powering America's (WPA) website: windpoweringamerica.gov/windmaps/resident ial_scale.asp

This site has downloadable high-resolution wind resource maps for residential wind turbines for all the states. Click on your state, and you can download a map of the wind resource at 30 meters above ground, about 100 feet. This makes an excellent screening tool to see if a wind turbine might be a viable technology for your situation. Note that the maps have a 2-kilometer resolution, meaning that they can give you the average wind speed for an area about one and a quarter miles square, maybe not your exact location, but pretty darn close. But also note that the interpolated wind speed assumes an area free of ground clutter, trees, and buildings. To fine-tune the average wind speed, you need to do some more homework.

One important thing to keep in mind is that wind speed increases with height above ground. This is due to the friction between the Earth's surface, plus any trees and buildings in the area, and the moving air mass—what we call wind. Wind site assessors use wind profiles to depict this phenomenon. Increasing arrow length in the wind profile signifies a higher wind speed.

Different types of ground clutter—the terrain, buildings, trees, crops—are like different grades of sandpaper, and will result in different wind profiles because they cause more or less friction on the moving air masses. These different surface roughnesses are called wind shear coefficients, or alphas, and we'll reference them in a bit.

Wind Quality

Let's assume that you arrive at what you consider a good average annual wind speed from the WPA maps. What's next? Wind is an interesting renewable energy resource, and it can be continuously changing. The causes of these fluctuations include such things as ground drag or surface friction, turbulence, displacement height, and prevailing wind direction. Besides affecting the quantity of wind that you have—your average wind speed—these "big four" can also affect the "quality" of your fuel.

Wind turbine blades are airfoils, not unlike airplane wings. Remember the last time you flew across country and hit a pocket of turbulence? The plane was buffeted around and may have even dropped a few thousand feet in altitude. Airfoils

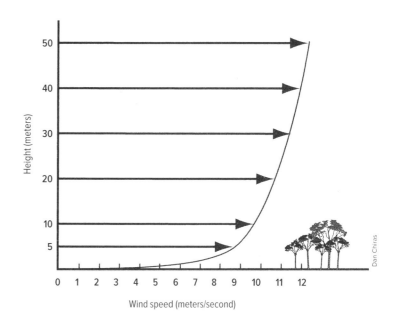

operate on a physics principle known as "lift." Lift is only created by clean, smooth, laminar airflow across an airfoil. Disrupt the airflow so that you induce chaotic air motion—that is, turbulence—across the airfoil, and lift is compromised. Turbulence on either airplane wings or wind turbine blades means they cannot do the work they were designed for. Along with greatly reduced electricity production in a turbulent location, the continuous buffeting due to turbulence will increase the wear and tear on your turbine, resulting in more maintenance and a shorter life.

The WPA wind maps assume an ideal site, quite open and clear of ground clutter. Your area's surface friction coefficient and displacement height will reduce your wind speed, while the ground clutter creates turbulence, quantified as turbulence intensity, all of which will compromise the smooth, laminar flow of your winds. You will need to adjust your wind speed down as a result of wind shear and displacement height, then reduce your expected annual kWh production due to turbulence intensity. A tutorial in this process is beyond the scope of this section. However, an excellent resource for integrating all of these factors in your average annual wind speed calculation is at smallwindtraining.org/. Click on the Site Assessor tab to find a variety of articles that will help you understand these phenomena and determine their impact on your fuel.

The 30/500 Rule

As a result of all of these compromises, wind site assessors use what we call the 30/500 rule to determine the *minimum acceptable tower height* for a wind turbine. It states that the *entire rotor* of a

Wind speed increases with height above the ground.

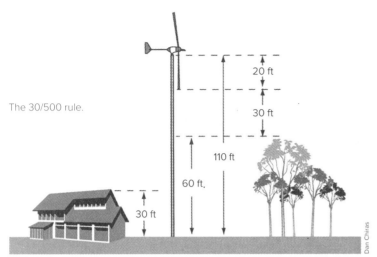

The 30/500 rule.

20 ft

30 ft

110 ft

60 ft.

30 ft

wind turbine must be *at least* 30 feet above anything within 500 feet of the tower, or the tree line in the area, whichever is higher. This rule places your wind turbine rotor sufficiently up into the wind profile while reducing fuel-robbing turbulence, so that the system will actually generate the electricity you want it to.

Keep in mind that you are installing your wind system for 20 to 30 years. "Tomorrow's" trees are going to be taller than they are today. So, when applying the 30/500 rule, you need to have a good idea of what the mature tree height will be, and use that as the basis of your tower height calculation. Remember that trees grow and towers don't, no matter how much it rains. Ignoring this

Keep in mind that you are installing your wind system for 20 to 30 years. "Tomorrow's" trees are going to be taller than they are today.

Trees grow and towers don't.

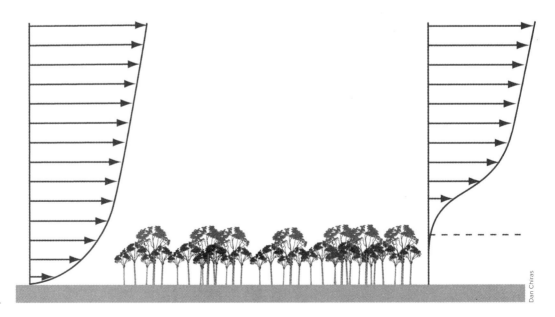

Displacement height.

fact will result in a wind system that, over time, produces less and less electricity until it simply stops generating for lack of wind.

Displacement Height

If you live in or near a wooded area, or have a tree line in your area, the wind profile is going to be bumped up to some height above the ground, making this new height the "effective ground level."

Displacement height will result in a lower wind speed than depicted on the WPA wind map. This is an excellent reason why you need to know how tall the trees are going to get over the next several decades. The amount of bump-up will depend on whether the trees are deciduous or evergreen, or some mix of the two. The articles at smallwindtraining.org/ will help you properly adjust for this height.

Prevailing Wind Direction

One of the best ways to optimize your wind speed while reducing turbulence is to site your tower upwind of any obstacles at your location toward the prevailing wind direction. The ideal reference for prevailing wind direction is the wind rose for your area. A wind rose may be available from your local weather bureau. If not, farmers or ranchers who have worked the land for decades will have a good idea of the seasonal prevailing wind directions.

Now you have an idea of the average annual wind speed for your site, which you've adjusted for surface friction due to terrain, turbulence due to ground clutter, turbulence intensity, and displacement height. You've applied the 30/500 rule, and have an idea of the minimum tower height you'll need for your wind turbine based on its rotor diameter. The next thing to investigate is tower options.

Towers

Towers are either freestanding or guyed, with two variations of each. We'll review these from the least to the most expensive.

Guyed lattice towers, the least expensive, must be climbed in order to install or visit your turbine for maintenance or repairs. Climbing a guyed tower is essentially like climbing a vertical ladder.

These towers are remarkably stable, and the tower itself has a very small footprint, only a couple of feet on a side. The guy cables, however, will stretch out in three directions to their respective anchors in the ground, about 75% of the tower height away from the tower base. These towers occupy a larger area than freestanding towers, but

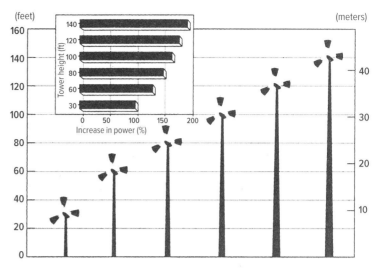

Increase in power with height above 30 feet.

Adapted from *Wind Power, Renewable Energy for Home, Farm and Business*, Chelsea Green Publishing.

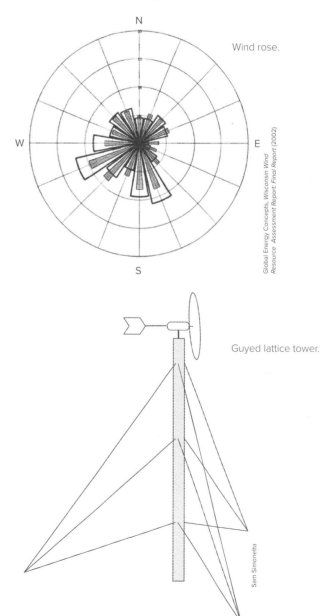

Wind rose.

Global Energy Concepts, *Wisconsin Wind Resource Assessment Report: Final Report* (2002)

Guyed lattice tower.

Sam Simonetta

Tilt-up Tower Footprint

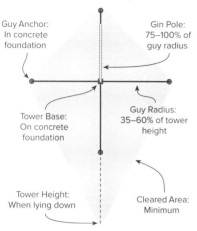

Guy Anchor: In concrete foundation

Gin Pole: 75–100% of guy radius

Tower Base: On concrete foundation

Guy Radius: 35–60% of tower height

Tower Height: When lying down

Cleared Area: Minimum

Fixed, Guyed Tower Footprint

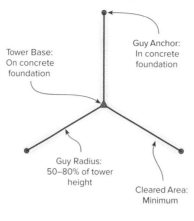

Tower Base: On concrete foundation

Guy Anchor: In concrete foundation

Guy Radius: 50–80% of tower height

Cleared Area: Minimum

Freestanding Tower Footprint

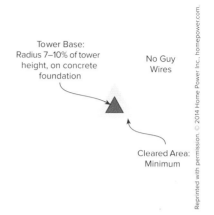

Tower Base: Radius 7–10% of tower height, on concrete foundation

No Guy Wires

Cleared Area: Minimum

Different towers present different footprints on the ground.

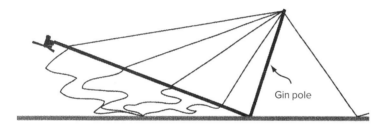

Gin pole

Guyed tilt-up towers.

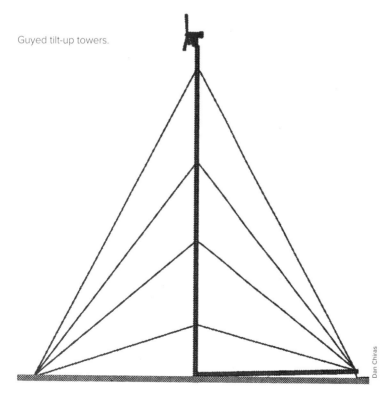

Dan Chiras

that space can still be used for gardening, livestock, yard, or work area. Guyed lattice towers are usually installed with a crane.

(One side note about living and working under a wind turbine. I am occasionally asked if I'm afraid to be under our wind turbine. My response is that if one is not confident enough to be under a wind turbine, then perhaps one should not own a wind turbine in the first place. We've lived with an 80-foot tower 34 feet from our home for 33 years, and an 84-foot tower 30 feet from the shop for 30 years, and had no incidents.)

Guyed tilt-up towers are more expensive than guyed lattice towers because a fair amount of engineering is involved to safely raise and lower the tower, as well as additional materials. Tilt-up towers are usually made of pipe or tube, and as such, are not climbable. That's also their primary advantage, because the vast majority of prospective wind turbine owners are not enamored of the idea of working at 100-plus-foot heights. Instead, the tower is lowered to the ground with a winch so the turbine can be worked on from the convenience of a lawn chair. These towers require a significant amount of open real estate to allow the guy cables to descend with the tower. While this diamond-shaped footprint can occupy a lawn or pasture, any plantings, buildings, or permanent structures will inhibit the safe lowering of a tilt-up tower. Guyed tilt-up towers are assembled on the ground, then raised and plumbed, then lowered again to install the wind turbine before the final raising. Although no climbing is involved, lowering and raising a tilt-up tower takes considerably more time than climbing a tower.

Freestanding lattice towers must be climbed to access the wind turbine. They are more expensive than the guyed towers because they are made of more steel. Their advantage is a relatively small footprint—a base of about 1/10th the tower height. Freestanding lattice towers are usually installed with a crane.

Freestanding monopole towers are sleek and much more visually appealing than lattice towers. However, these are the most expensive tower option. Monopoles have the smallest footprint of any tower type, and must be climbed to access the wind turbine. A few companies are beginning to offer hydraulic tilting options for their monopoles, but at considerable expense due to the additional engineering, materials, and required hydraulic systems.

As with any tower, you will need to acquire some very important tools to assure an incident-free experience, including a safety harness and work lanyards, plus the training to use them properly. See "Tower Climbing Safety" in Resources for more information.

Whole System Cost

When researching your dream wind turbine and what it's going to cost, it is important to keep in mind the ways that wind systems differ from other renewables. One is that a wind turbine comes complete with a dedicated controller, and if it is a grid-tied system, the inverter as well. This is because these wind system electronics are designed to control the rpm of the rotor, and mixing and matching components as you might do with a

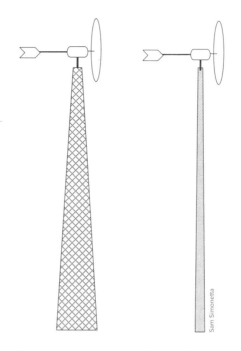

Freestanding tower.   Freestanding monopole tower.

PV system is not acceptable by the manufacturer, nor covered by their warranty.

Also, note that the cost of the *wind turbine* is a small part of the cost of a *wind system*. A turbine, controller, and inverter in a crate will not generate electricity. You also have to budget for a variety of things, including:

- the tower

Left: Installing a 1,500-watt wind turbine on a tilt-up tower using an electric winch. **Right:** Bob and Ginger Morgan's Bergey Excel being installed by a crane near Tehachapi, California.

- excavation for the tower anchors
- concrete for the tower foundation
- conduit, wire, and trenching for the wire run
- crane to install the tower and turbine
- balance of system electrical hardware
- labor to install the tower, turbine, and electrical
- sales tax
- hipping for all components
- homeowner's insurance

Some of these costs can be substantial compared to the cost of the turbine. For example, a tower for a 10 kW wind turbine might be one third to one half the price of the turbine, while the cost of a tower for a 1 kW system might be five times the cost of the turbine. However, without the tower, you will not have a wind turbine that generates any electricity.

One way to reduce costs is that, if you are handy, you may be able to do some or all of the work required to install a wind system.

Getting "Permission"

Before you plunk down a dime for a wind turbine and tower, you need to know which permits are required. The first consideration is a building permit for your tower, acquired through your local permitting or zoning authority. Zoning for wind systems is seemingly random from place to place. Local governments with experience with wind systems are usually amenable to permitting whatever tower you wish to install, provided you have tower and foundation documentation from the manufacturer to assure the authorities that the tower is an actual and properly engineered product. Most authorities are not willing to permit do-it-yourself structures.

When you apply for a building or conditional use permit, make sure that you have as much backup documentation as you can get. This should include, at a minimum, the specifications for the turbine, the tower and foundation prints, and the electrical diagrams. If there is anti-wind fervor in your area, you might also want to provide some fact sheets to quell the rampant myths. A great resource for this is the Small Wind Toolbox (renewwisconsin.org/wind/windtoolbox.htm), which has scores of fact sheets that shed light on many of the myths surrounding wind turbines. Keep in mind that most objections to wind energy emanate from the construction of utility-scale wind farms. Unfortunately, hysteria sometimes trickles down to home-sized systems, so be prepared.

If you will be installing a grid-tied system, you'll also need permission from your local utility. Contact them for their process and application so you know all their requirements as well. One of these might include sign-off from an electrical inspector. A discussion with that person to determine his or her concerns about wind systems before you begin construction is usually well worth the effort.

All of this research might take time, but without "permission" you will never be allowed to erect a wind system. You can hide a PV system, but you can't hide a wind turbine on a tower.

Finally, the Wind Turbine

If you cruise the Internet, you'll find dozens of wind turbine models available in all manner of configurations: conventional 3-bladers, vertical-axis versions, blades mounted in funnels, Savonious rotors, drag devices, and other imaginative rotor designs. If you can imagine a design, it's probably available. Wind electric generators have been around for over 85 years, and in that time, every conceivable design has been "invented." But just because you can purchase an unconventional design because it "looks neat" doesn't mean that you should. Nearly all of these designs have been rejected time and again as either inefficient, unreliable, or not an economically cost-effective way to generate electricity.

Still, there are weekly postings about the "technology breakthroughs" of unconventional designs in the media. Unfortunately, nearly all of these are press releases issued by the inventor to stir up media attention, elicit sales, and pander for investment dollars. As the saying goes: "Buyer beware!"

So, how do you separate legitimate, viable products from the half-baked ideas or over-hyped whirly-gigs? Buyers need to focus on three factors when considering possible wind turbine models: certification, swept area, and annual energy output.

Fortunately for buyers, the small wind industry took a page from the solar industry and instituted a certification process for wind turbines. The American Wind Energy Association (AWEA) 9.1 Small Wind Performance and Safety Standard was adopted in 2009. For complete certification, turbine manufacturers must submit extensive test reports for annual energy output, power curve, sound, and safety. Partial certifications are also available for each of these parameters.

Since 2009, seven wind turbines have been fully certified to AWEA 9.1 by the Small Wind Certification Council (SWCC), and another six have been certified by Intertek. Both certification

How do you separate legitimate, viable products from the half-baked ideas or over-hyped whirly-gigs? Buyers need to focus on three factors when considering possible wind turbine models: certification, swept area, and annual energy output.

agencies have a number of additional turbine models that are undergoing the detailed certification process but are not there yet. In addition, the Interstate Turbine Advisory Council (ITAC) was organized to help public benefits program managers and customers weave their way through which turbines are certified and, therefore, eligible for grant dollars. Nine wind turbine models are listed as fully certified and vetted by ITAC and meet ITAC's warranty and business performance criteria, meaning the manufacturers actually honor their warranties and have a bona fide tech support department. Collectively this represents just six companies that have gone through the certification process for their products out of the hundreds of "manufacturers" of different designs. If you buy something that is not certified, consider it a "science project." (See Resources for links to SWCC, Intertek, and ITAC.) The main thing to look for when shopping for a wind turbine is *full certification*, not just certification of the power curve or pending certification.

The second thing to look for is the swept area of the rotor. The rotor is the "collector" of a wind turbine, harvesting the kinetic energy in moving air masses and converting it into rotational momentum to turn an electric generator. As author and wind expert Paul Gipe says, "Nothing says more about a wind turbine than the size of its

rotor." For those familiar with solar collectors, this is an obvious metric. The amount of electricity that a PV array can generate is directly proportional to the area of the array exposed to sunlight. Similarly, the amount of hot water that a solar water system can produce is directly related to the size of the collectors. For whatever reason, many people don't make the same connection with wind turbines, falling instead for the "breakthrough technology" nonsense. When it comes to wind turbine rotors, size matters, just like it does for any other renewable energy collector.

Finally, shop based on the annual energy output (AEO)—after all, isn't this why you are interested in installing a wind turbine? The kilowatt-hours (kWh) that you can expect from any wind turbine are directly related to the wind speed at your site and the size of the collector you are using to harvest those winds. AEO has nothing to do with the size of the generator. Little net, small fish, big net, big fish, regardless of the size of the motor on your boat or the motor's power curve.

Unlike a PV array, which can be expanded over time, you are likely to install only one wind turbine and tower. Ever. And unless you install a heavier-duty tower than required, you will not be able to install a larger wind turbine on a tower designed for something smaller. Base your purchasing decision on how much electricity you will need or want over the life of the system—20 to 30 years for a good reliable model. If you don't have enough money, save up rather than impulsively buying some "eye candy" or a cheap substitute.

The Fine Print

Several other issues need to be considered before you make your purchase. One of these is the warranty and what it covers. A warranty is only as good as the company that manufactured the turbine. If the company disappears—all too common with start-up manufacturers—so does your warranty. Look for an established company with a good reputation for standing behind the equipment and installers. Quiz manufacturers about what is covered and what is not, including compensating the installer for unscheduled repairs. Uncompensated installers are not likely to pay much attention to your system.

My two criteria for considering any given wind system, based on feedback from long-time turbine owners, are reliability and tech support. Look for products with a solid reputation for reliability as reported by owners, not the manufacturer. Any wind turbine, no matter how good a deal it was, doesn't generate anything if it's not running. Tech support is even more critical, as

The kilowatt-hours (kWh) that you can expect from any wind turbine are directly related to the wind speed at your site and the size of the collector you are using to harvest those winds. AEO has nothing to do with the size of the generator.

Two important criteria for considering any given wind system, based on feedback from long-time turbine owners, are reliability and tech support.

eventually you will need some help from the company for maintenance or repair questions. If the feedback from owners and installers on tech support is that this department is a revolving door—or worse, nonexistent—then steer clear.

Finally, keep in mind that a wind turbine is not a PV array, no matter how it is advertised. While you can install a PV system and maybe never look at it again, if you fail to periodically visit your wind turbine, some day it will come down to visit you. Wind turbines are complicated dynamic machines with moving parts operating in one of the worst environments imaginable—atop a tall tower in the wind, sun, rain, snow, sleet, hail—you get the idea. While a certified model may be reliable, there is no such thing as a "maintenance free" wind turbine. And while "maintenance" no longer necessarily means oil changes or grease lubes, it does not exclude periodically inspecting your turbine for potential problems. In 33 years of installing and servicing dozens of models, my experience is that most catastrophic failures are not due to storms or severe winds. Instead, they happen because no one climbed the tower to inspect the turbine for loose hardware or worn and frayed components for five, or six, or even seven years. If you treated your car like that, you'd unexpectedly find yourself stranded alongside the road one day. A wind turbine logs as many operating hours in four months as the typical car does in 100,000 miles. No car owner in his or her right mind would depend on a car for 100,000 miles without periodic inspections and required maintenance.

Off-grid Systems: Go Hybrid with PV

For off-griders, wind makes the most sense in combination with a PV system. Unless you live in an area with fairly consistent winds evenly distributed throughout the year, it's best not to consider a wind-only off-grid system. Instead, design a hybrid wind/PV system. Wind and solar are remarkably compatible and interdependent resources. As the saying goes, when the sun isn't shining, the wind is blowing, and when the wind isn't blowing, the sun is shining. Wind/PV hybrid systems allow the owner to downsize both the wind turbine and the PV array. This is because the PV is sized for the summer months, when the solar resource is at its greatest, and the wind turbine is sized for the winter resource, which is usually greater than at any other time of year. Spring and fall see both generating sources contributing, and it is not unusual for an owner to reap excess electricity during these times.

In addition, since you will have two generating inputs into your battery bank, you'll also be able to downsize your storage capacity. Due to the large contribution of amps during windy weather, make sure to size the battery bank at least six times the maximum generating capacity of the wind turbine. Some sort of shunt regulator on the battery bank to dump excess electricity into a resistive load is also a must when wind is part of a hybrid system.

Wind Turbine "Problems"

The Internet is rife with reports that wind turbines generate unbearable levels of noise, create

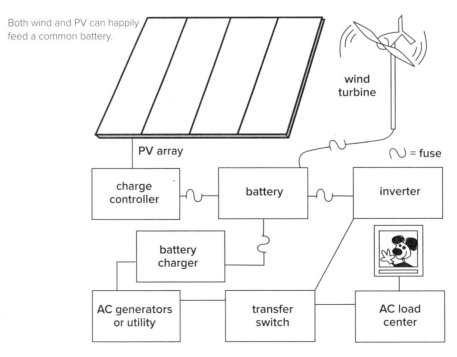

Both wind and PV can happily feed a common battery.

shadows that will cause you to become epileptic, drive down property values, and radiate infrasound that will kill you. The reality is that such reports are creative writing at its best by anti-wind groups, opposed to wind farms, trying to stop projects in their neighborhood. Most "reports" are based on little to no experience, really just misinformation propagated through the NIMBY echo chamber. While I'm not about to argue that wind farms should be allowed to locate anywhere, such claims unfortunately paint all wind installations, regardless of size, with the same brush. One can only assume that people who don't like wind energy are fearful that if they don't oppose a residential wind turbine, they may awake one day to find that a wind farm popped up overnight.

Outlandish claims need to be put into perspective relative to small wind turbines, which are defined as up to 100 kW in name-plate capacity or about a 69-foot rotor diameter. The Small Wind Toolbox (see Resources) suggests a variety of ways to respectfully address such concerns and misinformation. In reality, neighbors living next to or near small wind systems have lodged remarkably few complaints.

Probably the biggest concern is about sound, all too often repeated by someone who has read dire postings on the Internet but has no actual experience with an actual small wind turbine. Having lived with three wind turbines on our property for three decades, I can attest that such concerns have little basis in reality—for "good" wind turbines. Note the qualifier: good. I firmly believe that wind should be like PV—it should be seen and not heard. We live in a very rural area, and cherish our peace and quiet. While some wind turbines do emanate sound, they are not listed as certified turbines by the SWCC, Intertek, or ITAC. Even good turbines may make some noise during storms or power outages, when the turbine may be disconnected from the grid and running unloaded. But these incidents are infrequent, and people are typically not outside during storms. One of the best ways to gauge the suitability of a given model is to simply visit your turbine of interest and interview the owner. If you have concerned neighbors, take them with you on your field trip.

Finally, the question often arises, "Do wind turbines kill birds?" Unfortunately, the answer is "yes"—but this needs to be put into perspective. Wind farms have been documenting bird mortality for decades, and address the impact on birds by proper siting. A small wind turbine, however, is not a wind farm. Small turbines are typically installed next to a residence, farm, business, or school, locations that are already impacted by development and fragmented environments. For sites at or near wildlife areas, studies indicate that the small wind turbine has an inconsequential impact on birds. In fact, even for wind farms, the impact is relatively minor compared to collisions with buildings, windows, and automobiles, or mortality related to pesticides. On the other hand, one of the most significant causes of bird deaths, but one that most people choose to ignore, is cats. See the accompanying sidebar for a fuller discussion.

The Bottom Line About Small Wind

In the end, your decision about which wind turbine to purchase will likely revolve around cost. This is understandable. If you are already on the grid, perfectly usable electricity is available. If you're off-grid, there are competing technologies to consider. Because so many factors come into play, answering a series of questions may help you to assess how all the considerations sort out.

- What is your goal in owning a wind turbine? For electricity only? To set an example for your community? For your children and grandchildren? To leave a legacy? Each of these goals will involve different considerations.

- Are you mechanically inclined enough to install and maintain a complex piece of equipment 100 or so feet in the air? Are you

Real Goods Solar Living Center's Whisper 3000 wind turbine atop a hinged tilt-up tower in Hopland, California.

Paul Gipe

Estimated Annual Energy Output at Hub Height in Thousand kWh/y						
Avg Wind Speed (mph)	Rotor Diameter, m (ft.)					
	1 (3.3)	1.5 (4.9)	3 (9.8)	7 (23)	18 (60)	40 (130)
	thousands of kWh per year					
9	0.15	0.33	1.3	7	40	210
10	0.20	0.45	1.8	10	60	290
11	0.24	0.54	2.2	13	90	450

willing and able to climb your tower periodically for inspections and maintenance?

- If the answer is "no" to either of the above questions, are you willing to commit the dollars to hire a good installer who will attend to your wind system? Keep in mind that you are buying an entire system, not just the wind turbine. You're going to need your installer for the next several decades for tech support as well as major repairs, and maybe even inspections. Shop well for this relationship—you don't want to be left hanging when you need your installer

the most. And shop local. You might get a bargain installation quote from a company several states away, but will they think your emergency is as urgent as those of their closer customers? A local installer is far more valuable, and will be more attentive to your needs.

- Finally, even though considerable dollars might be on the table, remember to distinguish between cost and value. A reliable turbine will invariably cost more than a bargain model, but it will be a far better value. A tall tower suitable for your site will cost more than the shortest version available, but the height will seriously impact the amount of electricity you can generate. An experienced installer will cost more than someone breaking into the industry who is willing to give his services away, but that won't last long. Your up-front cost may be higher, but the value of a 20-year system far outweighs that of an orphaned wind turbine on a short tower that only runs for a year or two.

Research thoroughly, then choose well. And may the winds be with you!

Resources

Wind Powering America Residential Wind Resource Maps
windpoweringamerica.gov/windmaps/residential_scale.asp

Site Assessment Articles
smallwindtraining.org/?page_id=6

Best Practices in Small Wind Tower Climbing Safety
smallwindtraining.org/?pageid=64

Small Wind Toolbox fact sheets
renewwisconsin.org/wind/windtoolbox.htm

Distributed Wind Energy Association additional small wind fact sheets
distributedwind.org/zoning-resource-center/

Small Wind Certification Council's list of *fully* certified turbines
smallwindcertification.org/certified-small-turbines/

Intertek's Small Wind Program Directory of turbines and pending certifications
intertek.com/wind/small/directory/

Interstate Turbine Advisory Council unified list of wind turbines
cesa.org/projects/ITAC/itac-unified-list-of-wind-turbines/

Books
Dan Chiras, *Power from the Wind* (New Society, 2009).

Paul Gipe, *Wind Energy Basics* (Chelsea Green, 2009).

Paul Gipe, *Wind Power-Renewable Energy for Home, Farm, and Business* (Chelsea Green, 2004).

Ian Woofenden, *Wind Power for Dummies* (Wiley, 2009).

Small Wind Turbines, Cuisinart for Birds or Red Herring?

Avian mortality is an issue that always seems to come up with wind turbines. Do birds die from running into wind turbine blades, towers, and guy wires? Yes, they do. In large numbers? Hardly. Let's look at the facts. Birds die all the time, in great numbers, due to human endeavors. Muffy the house cat and all her friends are the number 1 bird killer, accounting for 2.4 billion deaths annually.[1] Glass-faced office buildings are number 2. These account for 100 million to 900 million bird deaths per year.[2] Some skyscrapers have been monitored and shown to kill as many as 200 birds per day. Cars and trucks kill another 50 to 100 million birds per year.[3] Cell phone, TV, radio, and other communications towers kill 4 million to 10 million birds per year.[4]

In comparison, the vast majority of large wind plants report small numbers of bird kills,[5] and even the famous Altamont Pass area, in which wind turbine bird kills were first reported, has about 1,000 bird deaths per year. This sounds like a lot, until we note that there are 5,400 turbines on this site. That gives an average of 0.19 bird kills per turbine per year.[6] The problem seems to be the abundance of prey in the tall grasses around the towers, and early tower designs provided a wealth of good perches. It's also located very near a large nesting golden eagle population. In addition, older turbines are very small, sited much closer together than modern machines, and have very high rpm.

That's the true story for large power-production turbines. Bird kills are a non-issue. In comparison, residential-size turbines and towers simply do not pose any significant risk to birds. The towers are too short, and the blade swept area is too small. The chance of your residential turbine ever hitting a bird are vanishingly small. If you or your neighbors are truly concerned about bird deaths, getting rid of Muffy would have much more real effect.

For more information about bird mortality studies specific to small wind, see the Small Wind Toolbox, under Fact Sheets for Permits and Zoning Hearings: renewwisconsin.org/wind/wind toolbox.htm.

—Doug Pratt, revised by Tom Gray

Notes

1. smithsonianscience.org/2013/01/cats-kill-2-4-billion-birds-annually/.
2. Curry & Kerlinger, Dr. Daniel Klem of Muhlenberg College has done studies over a period of 20 years, looking at bird collisions with windows. https://web.archive.org/web/20120905200358/http://www.currykerlinger.com/birds.htm. See also cleantechnica.com/2013/10/31/canada-ranks-top-bird-killers-wind-turbines-even-close-top/.
3. Curry & Kerlinger, "Those statistics were cited in reports published by the National Institute for Urban Wildlife and US Fish and Wildlife Service." https://web.archive.org/web/20120905200358/http://www.currykerlinger.com/birds.htm.
4. Ibid., "US Fish and Wildlife Service estimates that bird collisions with tall, lighted communications towers, and their guy wires result in 4 to 10 million bird deaths a year."
5. https://web.archive.org/web/20120905203746/http://www.currykerlinger.com/windpower.htm
6. nrel.gov/docs/fy04osti/33829.pdf

The Gipe Family Do-It-Yourself Wind Generator Slide Show

All photos: © Paul Gipe

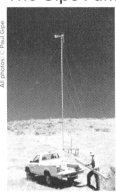

1. Taking delivery of a new air turbine and 45-foot tower.

2. Checking the packing list against parts delivered.

3. Securing the tower's base plate.

4. Aligning the guy anchors.

5. Driving the screw anchors.

6. Unspooling the guy cable.

7. Assembling the mast.

8. Clevis and gated fitting hook.

9. Gin pole and lifting cables.

10. Strain relief for supporting power cables.

11. Final assembly of the turbine.

12. Disconnect switch and junction box.

13. Slowly raising the turbine with a Griphoist-brand hand winch.

14. Air turbine safely installed with a Griphoist-brand hand winch.

REAL GOODS

Hydrogen Fuel Cells

Power Source of Our Bright and Shining Future? Not in the Foreseeable Future

Fuel cells for a while became a darling of the energy world. Intended to bring clean, quiet, reliable, cheap energy to the masses, they would allow us to continue an energy-intensive lifestyle with no penalties or roadblocks. However, this promise has not yet come to fruition on any grand scale since fuel cells first appeared on the market in 2003. Unfortunately, those hoping for the price to go down are still waiting. This is still a very young, developing technology that has yet to take hold in any residential markets. The first small commercial units started appearing in 2003 for relatively outrageous prices. Residential units that will run off natural gas or propane to provide backup power have recently come on the market, and they're fairly expensive. But those fuels are not themselves clean, so even an expensive fuel cell running on conventional fossil fuels is no energy panacea.

Traditional energy production, for either electricity or heat, depends on burning a fuel source like gasoline, fuel oil, natural gas, or coal to either spin an internal combustion engine or heat water, and then piping the resulting steam or hot water to warm buildings or run turbines to make electricity. As we know, combustion produces carbon dioxide and other pollutants, and it is the primary driver of climate change.

Fuel cells are chemical devices that go from a source, like hydrogen or natural gas, straight to heat and electrical output, without the combustion step in the middle. Fuel cells increase efficiency by two- or three-fold and dramatically reduce the undesirable by-products such as carbon. A fuel cell is an electrochemical device, similar to a battery, except fuel cells operate like a continuous battery that never needs recharging. So long as fuel is supplied to the negative electrode (anode) and oxygen, or free air, is supplied to the positive electrode (cathode), they will provide continuous electrical power and heat. Fuel cells can reach nearly 80% efficiency when both the heat and electric power outputs are utilized.

How Fuel Cells Operate

A fuel cell is composed of two electrodes sandwiched around an electrolyte material. Hydrogen fuel is fed into the anode. Oxygen (or free air) is fed into the cathode. Encouraged by a catalyst, the hydrogen atom splits into a proton and an electron. The proton passes through the electrolyte to reach the cathode. The electron takes a separate outside path to reach the cathode. Since electrons flowing through a wire is commonly known as "electricity," we'll make those free electrons do some work for us on the way. At the cathode, the electrons, protons, and oxygen all meet, react, and form water. Fuel cells are actually built up in "stacks" with multiple layers of electrodes and electrolyte. Depending on the cell type, electrolyte material may be either liquid or solid. Like other electrochemical devices such as batteries, fuel cells eventually wear out and the stacks have to be replaced.

Fuel cells are chemical devices that go from a source, like hydrogen or natural gas, straight to heat and electrical output, without the combustion step in the middle. Fuel cells increase efficiency by two- or three-fold and dramatically reduce the undesirable by-products such as carbon.

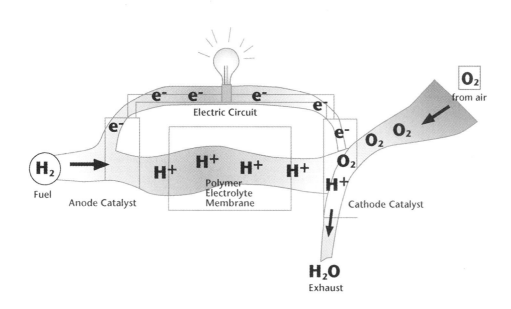

H₂
Fuel

e⁻ e⁻ e⁻
Electric Circuit

e⁻

e⁻

e⁻

H+ H+ H+ H+

Anode Catalyst

Polymer Electrolyte Membrane

O₂

O₂ O₂

O₂

H+

O₂ from air

Cathode Catalyst

H₂O
Exhaust

As an electrical supplier, think of a fuel cell as a vastly improved generator. It burns less fuel, it makes less pollution, there are no moving parts, and it makes hardly any noise. But unless you have a free supply of hydrogen, it will still use non-renewable fossil fuels for power, which will cost something and produce some carbon emissions.

What's "Fuel" to a Fuel Cell?

In purest form, a fuel cell takes hydrogen, the most abundant element in the universe, and combines it with oxygen. The output is electricity, pure water (H_2O), and a bit of heat. Period. Very clean technology! No nasty by-products and no waste products left over. There are some real obvious advantages to using the most abundant element in the universe. We've got plenty of it on hand, and a hydrogen-based economy would be far more clean and sustainable than a petroleum-based economy.

On the other hand, hydrogen doesn't usually exist in pure form on Earth. It's bound up with oxygen to make water, or with other fuels like natural gas or petroleum. If you run down to your friendly local gas supply to buy a tank of hydrogen, what you get will be a by-product of petroleum refining.

Since hydrogen probably isn't going to be supplied in pure form, most commercial fuel cells have a fuel-processing component as part of the package. Fuel processors, or "reformers," do a bit of chemical reformulation to boost the hydrogen content of the fuel. This makes a fuel cell that can run on any hydrocarbon fuel. Hydrocarbon fuels include natural gas, propane, gasoline, fuel oils, diesel oil, methane, ethanol, methanol, and a number of others. But if you feed the fuel cell something other than pure hydrogen, you're going to get something more than pure water in the output, with carbon dioxide usually the biggest component. Importantly, fuel cells emit 40%–60% less carbon dioxide than conventional power generation systems using the same hydrocarbon fuel. Other air pollutants such as sulfur oxides, nitrogen oxides, carbon monoxide, and unburned hydrocarbons are nearly absent, although you'll still get some unsavory trace by-products.

The dream is to build a fuel cell that accepts water input. But it takes more energy to split the water into hydrogen and oxygen than we'll get back from the fuel cell. One potential future scenario has large banks of PV cells splitting water. The hydrogen would be collected and used to run cars and light trucks. Efficiency isn't great, but nasty by-products are zero.

Cruisin' the City on Fuel Cells?

Automobiles, of course, account for one of civilization's biggest uses of fossil fuel. Fuel cells might offer some hope for carbon-free driving, but that promise is a long way off. Auto manufacturers have had fuel cell development programs for fifteen years, but the obstacles to a clean affordable fuel cell vehicle remain high. Currently the only economical way to produce hydrogen involves fossil fuel feedstocks and the burning of additional fossil fuels, while the net energy return on investment for extracting hydrogen from water remains negative. Onboard storage of hydrogen and the deployment of a hydrogen fueling infrastructure are also thorny and expensive problems.

All in all, we're very skeptical about the near-term potential of hydrogen fuel cells to reduce greenhouse gas emissions in the transportation sector. For the foreseeable future, hybrid electric and biofuel-powered vehicles are a much better, more realistic choice. You'll find a fuller explanation of this subject in Chapter 13, "Sustainable Transportation, pages 333–45."

How Do Fuel Cells Compare with Solar or Wind Energy?

As an electrical supplier, a fuel cell is closer kin to an internal combustion engine than to any renewable energy source. Think of a fuel cell as a vastly improved generator. It burns less fuel, it makes less pollution, there are no moving parts, and it makes hardly any noise. But unless you have a free supply of hydrogen, it will still use non-renewable fossil fuels for power, which will cost something and produce some carbon emissions.

Photovoltaic modules, wind generators, hydro generators, and solar hot water panels all use free, renewable energy sources. No matter how much you extract today, it doesn't impact how much you can extract tomorrow, the next day, and forever. This is the major difference between technologies that harvest renewable energy and fuel cells, which are primarily going to continue using non-renewable energy sources.

CHAPTER 4

From Panel to Plug

How to Get Your Renewable Energy Source Powering Your Appliances

A Renewable Energy System Primer: Putting It All Together

IN OTHER CHAPTERS AND SECTIONS of the *Sourcebook*, we provide details about how individual components such as PV modules or inverters work, and why they're needed. Here we explain how and why all the bits and pieces fit together. Read this Renewable Energy System Primer first to get an overall idea of how everything works from a complete system perspective, and then move on to details as needed.

Off-grid and Battery Backup

This section will familiarize you with battery-based systems that are capable of stand-alone operation. For information on simple grid-tied systems with no battery backup, please see the separate section dealing with grid-tied systems, which immediately follows on pages 131–135.

Batteries: The Heart of Your System

The battery, battery pack, or battery bank is the center of every stand-alone renewable energy system. Everything comes and goes from the battery, and much of the safety and overcurrent protection devices (OCPDs) are designed to protect current-carrying conductors going either to or from the battery. The battery is your energy reservoir; it stores up electrons until they're called upon to run any load or appliance.

The size of your reservoir determines how much power you can store but has minimal effect on how fast electrons can be used to charge or discharge the battery, at least in the short term. A smaller battery will limit how long your system will run before recharging is necessary. More battery capacity will allow more days of run time before recharging. However, there are practical limits in each direction. Batteries are like the muscles of your body: They need some exercise to stay healthy but not too much or they get worn out at an early age. We've found it's best to aim for three to five days' worth of battery storage. Less than three and your system will be cycling the battery deeply too frequently, which will shorten the batteries' life. More than five and your battery bank starts getting more expensive than a backup generator or other recharging source.

Once you start discharging power from the batteries, you'd better have a generation source lined up to recharge them or you'll soon have a dead battery bank—which brings us to…

The Charging Source

Batteries aren't the least bit worried about their charging source so long as the voltage from the charging source is a bit higher than the battery voltage, creating the "pressure" the electrons need to flow. Now, you certainly could haul your battery into town to get it charged, or you can connect it to your vehicle charging system and charge it on the way to town and back, which is how our founder, John Schaeffer, came up with the idea to start Real Goods in the first place, but batteries are heavy and this is not a practical

> The battery is your energy reservoir; it stores up electrons until they're called upon to run any load or appliance.

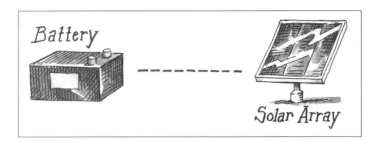

solution. Not to mention that the sulfuric acid in them just loves to eat your clothes, so hauling batteries around isn't recommended. You'll want to keep your batteries where they are and use an on-site charging source to keep them happy. This can be a fossil-fueled or bio-fueled generator, a wind turbine, a hydroelectric generator, solar modules (PVs), a stationary bicycle generator, or combinations of these.

As long as incoming energy keeps up with outgoing energy, your battery will be in great shape. Up to about 80% of full charge, a battery will accept very large current inputs with no problem. As it gets closer to full, we have to start applying some charge control to prevent overcharging. Batteries will be damaged if they are severely or regularly overcharged, voiding the manufacturer's warranty, and leaving you with far less available storage capacity than you need, or having to replace them prematurely.

Batteries Need Charge Control

Someplace in your system, there needs to be a device called a charge controller. Charge controllers use various strategies, but they all have the same purpose: to keep the battery from overcharging and becoming toasted. A wet cell battery that is regularly overcharged will use excessive amounts of water, may run dry, and sheds flakes of active lead from the battery plates, reducing future capacity and eventually rendering it useless. A charge controller monitors the battery voltage, and if it gets too high, the controller will either open the charging circuit or divert excess power into a heater element or other dump load.

We use different strategies depending on the charging source. Solar modules can use simple controls between the solar array and the battery

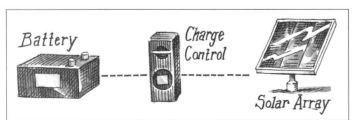

that just open the circuit if voltage gets too high. Rotary generators, such as wind and hydroelectric turbines, use a controller that diverts excess energy into a heater element. (DO NOT open the circuit between a generator and the battery while the generator is spinning! Bad things will happen, and you don't want to learn the hard way.)

Things are looking good, your system is coming along. Now you have a battery, a way to recharge it, a controller to prevent overcharging, but no clue how full or empty it is. You need a system monitor.

Monitoring Your System

System monitors all have a common goal: to keep you informed about your system's state of charge and help you prevent the exceptionally deep discharges that sap the life out of your batteries. Not having a system monitor is like driving a car without any gauges on the dashboard…you'd be much more likely to exceed speed limits, run out of gas, or do accidental damage. Monitors can be as simple as a red or green light, or as complicated as a PC- or web-based system that logs dozens of system readings in real time. At the simplest level, displaying the battery voltage gives an approximate indication of your battery bank's state of charge at any given moment. More robust monitors show how many amps are flowing in or out at

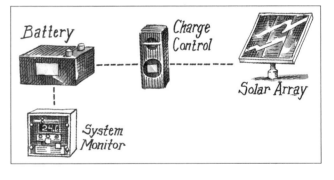

the moment, and the best monitors keep track of all this information, and give it to you as a simple "% of full charge" reading. Many charge controllers offer some simple monitoring abilities, but the best monitors stand alone and get mounted in the kitchen or living room where they're more likely to get some regular attention. You can even view your system monitor's data over the web via your smart phone or any mobile device, giving you some peace of mind when you're away.

Now you have a battery, a way to recharge it, a controller to prevent overcharging, and a system monitor to tell how full or empty the battery is. Next, you need to be able to use the stored energy to run appliances and gadgets.

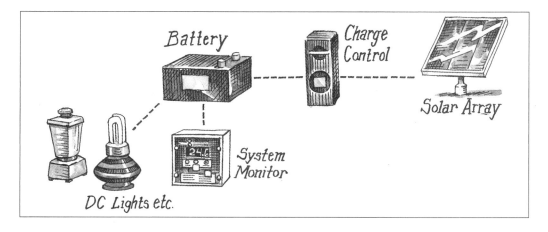

Using Your Power

Batteries deliver direct current, or DC, when you ask them for power. As long as the voltage of your battery and load match, you can tap this energy to run any DC lights or appliances directly. This is fine if you're running the most basic loads such as those found in an RV or a boat; vehicles pretty much run on DC, thanks to their automotive heritage. DC appliances made for RVs and boats are great to use in a tiny cabin as well, as long as you don't need anything too fancy. With a DC breaker or fuse panel and a wiring circuit or two, you're all set to run your DC light or pump straight from the battery (with some overcurrent protection of course, see the next section).

But what if you want to run an entire household, or an AC device? Practically all household appliances run on 120-volt alternating current, or AC. AC power is easier to transmit over long distances, so it has become the world standard. AC appliances won't run on DC from a battery. You could convert all your household lights and appliances to DC, but joining the mainstream as far as appliances are concerned offers some definite advantages. AC appliances are mass-produced, making them less expensive, universally available, and generally of better quality when compared with DC appliances made for the RV and boat industries.

Only DC power can be stored, so all your power storage happens on the DC side of your system in the batteries. AC power has to be generated on demand. To meet this highly variable demand, you need a device called an inverter. The inverter converts low-voltage DC from the battery into high-voltage AC as fast as your household lights and appliances demand it. The appropriate inverter depends on the type of appliances you need to run, and how many you need to run at the same time.

Inverters come in a variety of outputs from 50 watts up through hundreds of thousands of watts, and range in cost from $25 to more than you want to know. You want one that will deliver enough wattage to start and run all the AC appliances you might turn on at the same time, but no bigger than necessary. Bigger inverters cost more and use a bit more power on standby, waiting for something to turn on. However, lower-cost inverters usually leave out some protection equipment, which means a shorter life expectancy. (We've seen really cheap inverters blow up the instant a compact fluorescent lamp was turned on.) If you find yourself tempted to buy a cheap inverter, go ahead and buy a backup too, just to have on hand when you need it.

Many of the larger, household-size inverters come with built-in battery chargers that come on automatically anytime outside AC power becomes available from starting a generator or plugging into an RV park outlet. This provides a simple backup charging source for times when the primary solar or wind charger isn't keeping up. Having a generator on standby is never a bad idea, even if you don't use it much; when you do need it, you'll thank yourself.

Now you have a battery, a way to recharge it, a controller to prevent overcharging, a system monitor to tell how full or empty it is, and a way to use your stored energy. Only one thing is missing, and it's the most important one: Safety!

Overcurrent Protection Devices,
AKA: How to Avoid Burning
Down Your House

Generally speaking, low-voltage DC systems and batteries are a lot safer than higher-voltage AC systems. That being said, if something does short-circuit a battery, look out (and keep your eyes closed)! Because they're close by, batteries can deliver more amps into a short circuit than an AC system can, because the AC is strung out over many miles. So common sense, and the National Electrical Code (NEC), says you *must* put some

The inverter converts low-voltage DC from the battery into high-voltage AC as fast as your household lights and appliances demand it.

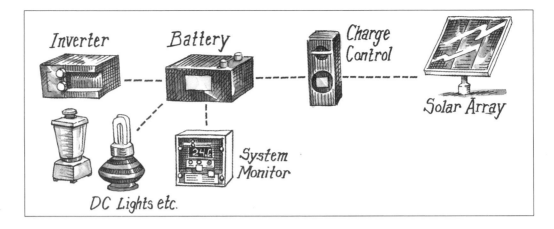

overcurrent protection on any wire attached to a power source. "Overcurrent protection" is code-speak for a fuse or circuit breaker. "Power source" means batteries or any charging source such as PV arrays or generator. These fuses or circuit breakers must be rated for DC use and appropriately sized for the wires they're protecting. Consult the manufacturer's installation manual, code book, or an electrician for appropriately sizing both overcurrent protection devices and conductors. A DC short circuit is harder to interrupt than an AC short, so DC-rated equipment usually contains extra stuff and is therefore more expensive. DC-rated fuses and circuit breakers are built tougher, and you really, really want them that way. You need to see wires glowing red with burning insulation dripping off them only once to appreciate the serious importance of fuses. Most of us think fuses and breakers are to protect the devices in the circuit, which they end up doing by default, but they're really there to protect the conductors from seeing more current than they can handle, by opening the circuit before something bad happens.

And there you have it! Now you have a battery, a way to recharge it, a controller to prevent overcharging, a system monitor to tell how full or empty it is, a way to use your stored energy, and the safety equipment to prevent burning down your house. That's a complete system!

What System Voltage Should You Use?

Renewable energy systems can be put together with the battery bank running at a nominal 12, 24, or 48 volts. The more energy your system processes per day, the higher your system voltage is likely to be. As systems process more and more power, it begins to make sense to use a higher voltage. Remember Ohm's Law: Watts = Amps times Volts. If you double the voltage, you need to push only half the amperage to perform a given job, requiring lower-gauge, easier to work with conductors. None of this is written in stone, but in general, systems processing up to 2,000 watt-hours per day are fine at 12 volts. From 2,000 to about 7,000 watt-hours per day a system will work better at 24 volts, and from 7,000 watt-hours on up, we prefer to run at 48 volts. Remember, these are guidelines, not rules. We can make a system run at any of these three common voltages, but you'll buy less hardware and big wires if system voltage goes up as the watt-hours increase. Almost all of the battery systems used year round run at 48 V, and have practically no DC loads, except maybe for dumping excess power in the summer.

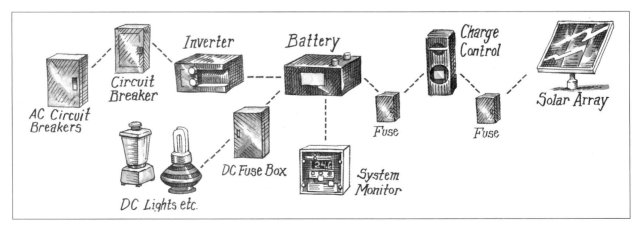

Time to Build!

Are you ready to put together a renewable energy system? Of course you're not. You probably still have a lot of questions about how big, how many, what brand, and more. These questions can be answered by consulting other sections in this chapter: Safety and Fusing, Controllers, Monitors, Batteries, Inverters, Wiring, and the System Sizing Worksheet. Each of these sections explains a particular renewable energy component in more detail and helps you make the best choice for your individual needs. Even if you're not ready to order your dream system, give yourself credit: You've learned a lot already, and you're almost off to the races!

And finally, if you aren't the technical type, and don't get joy from figuring this stuff out yourself—or even if you do but would appreciate a little guidance—you can contact our highly experienced technical staff, which we think is the best in the business. We're available for consultation, system design, and troubleshooting, Monday through Saturday, 10:00 A M to 5:00 P M Pacific time. Call a Real Goods tech at 800-919-2400 or by going to our website: realgoods.com.

The Importance of Conservation

Conservation always comes first, and in the case of off-grid solar, it pays for itself by a factor of $3. That is, every dollar you spend to conserve energy will save you $3 in solar system costs.

The cost to purchase an off-grid solar system is anywhere from $1,000–$3,000 per kilowatt-hour used on a daily basis, depending on how many bells and whistles you want to add to your system. To outfit an average American home (which uses 25 kWh) with solar would cost you between $25,000 and $75,000 with no rebates, incentives, or tax credits.

Now consider the importance of conservation. Look at your light bulbs. If you replace 30 of the 60-watt incandescent bulbs in your home with more efficient compact fluorescent bulbs that run on average five hours per day (national average) and use only 15 watts each, you'll save 6,750 watts per day (45 watts each × 30 bulbs × 5 hours), which equates to as much as $20,250 in savings on your solar system. If you use LEDs, the savings will be even greater. See Chapter 6, pages 209–221.

The point is: Always implement conservation measures first before you build or install your renewable energy system. The money you save will more than pay for those more expensive, longer-lasting, energy-efficient appliances!

System Sizing Worksheets

Off-grid Systems

The following System Sizing Worksheet provides a simple and convenient way to determine approximate total household electrical needs for off-the-grid systems. Take your time with this, because a realistic estimate is important. You do not want to fall short on power, and yet you want to buy the right amount of equipment to get the job done, and not much more. You'll be able to find a better use for those dollars. The average daily watt-hours you come up with will be the basis for all future design considerations, so you want this number to be as accurate as possible. Once the worksheet is completed, our Real Goods technicians can size the photovoltaic and battery system to meet your needs. Do you want to go with grid-tied? That's easy—see the next paragraph. Over ninety percent of the renewable energy systems we design are primarily PV-based, so these worksheets deal primarily with PV. If you are fortunate enough to have a viable wind or hydro power source, you'll find output information for these sources in their respective sections of Chapter 3 of this *Sourcebook*. Our technical staff has considerable experience with these alternate sources and will be glad to help you size a system. Give us a call: 800-919-2400 or realgoods.com.

Grid-tied Systems

These are easy. Whatever your renewable energy system doesn't cover, your existing utility company will. So we don't need to account for every watt-hour beforehand, and if you've already been in the house a while, your utility bills can tell us your energy needs. Tell us either your budget, or how many kilowatt-hours of utility power per month you currently use. For a grid-tied system without batteries, you'll invest about $1,000 for every kilowatt-hour per day your solar system delivers. For a grid-tied system with battery backup that can provide limited power during utility outages, you invest about $3,500–$5,000 for every kilowatt-hour per day generated, and the added cost of battery backup varies greatly depending on how many loads you want running and for how long. These are very general ballpark figures for initial system costs. Remember, state and other rebates can reduce these costs by up to 50% in some states; visit dsiresua.org for information in your state.

General Information

The primary principle for off-the-grid systems is: Conserve, conserve, conserve! As a rule of thumb, it will cost about $1–$3 worth of equipment for every watt-hour per day, or put another way, $1,000 to $3,000 for every kWh you must supply. Trim your wattage to the bone! Don't use incandescent light bulbs or older standard refrigerators. Make sure all your appliances are energy efficient—use ENERGY STAR appliances. Dry your clothes the old-fashioned way, on a line or rack.

If you're intimidated by this whole process, don't worry. Our tech staff is here to hold your hand and help you through the tough parts. We do need you to come up with an estimate of total household watt-hours per day, which this worksheet will help you do. It will get you thinking deeply about what energy habits are absolutely necessary and what you can do without. We can pick up the design process from there. You are in the best position to make lifestyle decisions: How late do you stay up at night? Are you religious about always turning the light off when leaving a room? Are you running a home business, or is the house empty five days a week? Do you hammer on your computer for 12 hours a day? Does your pet iguana absolutely require his rock heater 24 hours a day? These are questions we can't answer for you. So figure out your watt-hours to let us know what we're shooting for. We can always work backwards from your budget, but you'll still want to know if your system's production will be in the ballpark of what you need, or not.

What's It Going to Cost Me to Go Solar?

It takes three easy steps to get a ballpark calculation for a grid-tied system.
1. Find your daily utility usage by dividing the kilowatt-hours (kWh) used on an average month's bill by 30.
2. Divide that number by 5 (the average number of peak sun hours in the US), and multiply by 1.43 to account for system losses. This is the size of the solar system, in kilowatts, that you will need for taking care of 100% of your electrical needs.
3. Multiply that number by $3,500 ($3.50/watt installed) for a good ballpark idea of the gross installed cost.

Can state rebate incentives take a chunk out of that price? Go to dsireusa.org to find out what grants or incentives are available in your state. For instance, in New York you can multiply your gross installed cost by 0.4 to account for rebates and tax credits. Many states offer rebates and incentives either through the state or through utilities, including CO, CT, VT, OR, WA, AZ, NY, NJ, RI, MA, and NH. The Database of State Incentives for Renewables & Efficiency also contains handy maps showing where various kinds of incentives are available. Rebates and incentives are often in flux, and the database does its best to stay up to date, but to be sure, use the links provided in the database to go directly to the official state or utility website to see the current level of rebate or incentive.

What ongoing savings can you expect? Whatever you're paying the utility for electricity now will change to $0, if your system is correctly sized to produce as much power as you use on an annual basis (service charges of $5 to $30 per month will still apply, depending on the utility).

Call our techs at 800-919-2400 for more information on how solar can work for your home.

Determine the Total Electrical Load in Watt-hours per Day

The following form allows you to list every appliance, how much wattage it draws, how many hours per day, and how many days per week it runs. This gives us a daily average for the week, as some appliances, like a washing machine, may be used only occasionally.

Some appliances may give only the amperage and voltage on the nameplate. Practically all appliances will have their electrical specs on them. We need to know their wattage. Use Ohm's Law: Multiply the amperage by the voltage to get wattage. Example: A blender nameplate says, "2.5 A 120 V 60 Hz." This tells us the appliance is rated for a maximum of 2.5 amps at 120 volts/60 cycles per second. 2.5 amps times 120 volts equals 300 watts. Be cautious of using nameplate amperage, however. For safety reasons, this must be listed as the highest amperage the appliance is capable of drawing. Actual running amperage can be much less. If you suspect this could be the case, it may be worth looking in the owner's manual or calling the manufacturer. This is particularly true for refrigerators and entertainment equipment. The Power Consumption Table may help give you a more "real" wattage use. An excellent tool for determining loads is the "Kill-a-Watt" meter. For 120-volt AC appliances up to 1,500 watts, just plug the appliance into the meter and the meter into your household outlet. The meter gives instant power usage readings (watts) as well as the accumulated energy used over a period of time (watt-hours). There's even a power-strip version of the "Kill-a-Watt" than can measure usage of multiple appliances at once! (Get one at realgoods.com.)

SYSTEM SIZING WORKSHEET

AC device	Device watts	×	Hours of daily use	×	Days of use per week	÷	7	=	Average watt-hours per day
		×		×		÷	7	=	
		×		×		÷	7	=	
		×		×		÷	7	=	
		×		×		÷	7	=	
		×		×		÷	7	=	
		×		×		÷	7	=	
		×		×		÷	7	=	
		×		×		÷	7	=	
		×		×		÷	7	=	
		×		×		÷	7	=	
		×		×		÷	7	=	
		×		×		÷	7	=	
		×		×		÷	7	=	
		×		×		÷	7	=	
		×		×		÷	7	=	
		×		×		÷	7	=	
		×		×		÷	7	=	
		×		×		÷	7	=	
		×		×		÷	7	=	
		×		×		÷	7	=	
		×		×		÷	7	=	

1   Total AC watt-hours/day

2   × 1.1 = Total corrected DC watt-hours/day

DC device	Device watts	×	Hours of daily use	×	Days of use per week	÷	7	=	Average watt-hours per day
		×		×		÷	7	=	
		×		×		÷	7	=	
		×		×		÷	7	=	
		×		×		÷	7	=	
		×		×		÷	7	=	
		×		×		÷	7	=	
		×		×		÷	7	=	
		×		×		÷	7	=	
		×		×		÷	7	=	

3   Total DC watt-hours/day

3 (from previous page)	Total DC watt-hours/day	
4	Total corrected DC watt-hours/day from line 2 +	
5	Total household DC watt-hours/day =	
6	System nominal voltage (usually 12 or 24) ÷	
7	Total DC amp-hours/day =	
8	Battery losses, wiring losses, safety factor × 1.2	
9	Total daily amp-hour requirement =	
10	Estimated design insolation (hours per day of sun, see map on page 582) ÷	
11	Total PV array current in amps =	
12	Select a photovoltaic module for your system	
13	Module rated power amps ÷	
14	Number of modules required in parallel =	
15	System nominal voltage (from line 6 above)	
16	Module nominal voltage (usually 12) ÷	
17	Number of modules required in series =	
18	Number of modules required in parallel (from line 14 above) ×	
19	Total modules required =	

BATTERY SIZING

20	Total daily amp-hour requirement (from line 9)	
21	Reserve time in days ×	
22	Percent of useable battery capacity ÷	
23	Minimum battery capacity in amp-hours =	
24	Select a battery for your system, enter amp-hour capacity ÷	
25	Number of batteries in parallel =	
26	System nominal voltage (from line 6)	
27	Voltage of your chosen battery (usually 6 or 12) ÷	
28	Number of batteries in series =	
29	Number of batteries in parallel (from line 25 above) ×	
30	Total number of batteries required	

Line-by-Line Instructions

Line 1: Total all average watt-hours/day in the column above.

Line 2: For AC appliances, multiply the watt-hours total by 1.05 to account for inverter inefficiency (typical by 95%). This gives the actual DC watt-hours that will be drawn from the battery.

Line 3: DC appliances are totaled directly, no correction necessary.

Line 4: Insert the total from line 2 above.

Line 5: Add the AC and DC watt-hour totals to get the total DC watt-hours/day. At this point, you can send the design forms to us, and after a phone consultation, we'll put a system together for you. If you prefer total self-reliance, forge on.

REAL GOODS

Line 6: Insert the voltage of the battery system; 12-volt, 24-volt, or 48-volt.

Line 7: Divide the total on line 5 by the voltage on line 6.

Line 8: This is our fudge factor that accounts for losses in wiring and batteries, and allows a small safety margin. Multiply line 7 by 1.2.

Line 9: This is the total amount of energy that needs to be supplied to the battery every day on average.

Line 10: This is where guesswork rears its ugly head. How many hours of sun per day will you see? Our Solar Insolation Maps in the Appendix (page 424) give the average daily sun hours for the worst month of the year. You probably don't want to design your system for worst possible conditions. Energy conservation during stormy weather, or a backup power source, can allow use of a higher hours-per-day figure on this line and reduce the initial system cost. The National Renewable Energy Laboratories (NREL) provides peak sun hour date for each state; visit pvwatts.nrel.gov for peak sun hour info and more.

Line 11: Divide line 9 by line 10; this gives the total PV current needed.

Line 12: Decide what PV module you want to use for your system, or at least pick a wattage if you can't settle on a manufacturer just yet. You may want to try the calculations with several different modules. It all depends on whether you need to round up or down to meet your needs. Consider your method of mounting the panels and who will be doing the work. If you're designing a system you'll be installing on your own, make it easy on yourself and install a larger quantity of smaller, and easier to work with, modules.

Line 13: Insert the amps of output at rated power for your chosen module.

Line 14: Divide line 11 by line 13 to get the number of modules required in parallel. You will almost certainly get a fraction left over; round up or down to a whole number. We conservatively recommend that any fraction from 0.3 and up be rounded upward.

If yours is a 12-volt nominal system, you can stop here and transfer your line 14 answer to line 19. If your nominal system voltage is something higher than 12 volts, then forge on.

Line 15: Enter the system battery voltage. Usually this will be either 12 or 24.

Line 16: Enter the module nominal voltage. This will generally be 12 for lower-wattage modules or 24 for larger ones.

Line 17: Divide line 15 by line 16. This will be how many modules you must wire in series to charge your batteries.

Line 18: Insert the figure from line 14 and multiply by line 17.

Line 19: This is the total number of PV modules needed to satisfy your electrical needs. Too many modules for you? Reduce your electrical consumption, or add a secondary charging source such as wind or hydro if possible, or a stinking, noisy, troublesome, fossil fuel-gobbling generator (no bias of course).

Battery Sizing Worksheet

Line 20: Enter your total daily amp-hours from line 9.

Line 21: Reserve battery capacity in days. We usually recommend about three to five days of backup capacity. Less reserve will mean that the battery cycles excessively on a daily basis, which results in lower life expectancy. More than five days' capacity starts getting so expensive that a backup power source should be considered.

Line 22: You can't use 100% of the battery capacity (unless you like buying new batteries). The maximum is 80%, and we usually recommend to size at 50% or 60% to be safe. This makes your batteries last longer and leaves a little emergency reserve. Enter a figure from 0.5 to 0.8 on this line.

Line 23: Multiply line 20 by line 21, and divide by line 22. This is the minimum battery capacity you need.

Line 24: Select a battery type. See the Battery section of our website for amp-hour capacity and more details (realgoods.com/catalog search/result/?cat=0&q=batteries). Enter the amp-hour capacity of your chosen battery on this line.

Line 25: Divide line 23 by line 24; this is how many batteries you need in parallel.

Line 26: Enter your system nominal voltage from line 6.

Line 27: Enter the voltage of your chosen battery type.

Line 28: Divide line 26 by line 27; this gives you how many batteries you must wire in series for the desired system voltage.

Line 29: Enter the number of batteries in parallel from line 25.

Line 30: Multiply line 28 by line 29. This is the total number of batteries required for your system.

Power Consumption Table

Appliance	Watts	Appliance	Watts	Appliance	Watts
Coffeepot	200	Electric blanket	2,000	Compact fluorescent	
Coffee maker	800	Blow dryer	1,000–1,500	Incandescent equivalents	
Toaster	800–1,500	Shaver	15	40 watt equiv.	11
Popcorn popper	250	WaterPik	100	60 watt equiv.	16
Blender	300	Computer		75 watt equiv.	20
Microwave	600–1,700	Laptop	50–75	100 watt equiv.	30
Waffle iron	1,200	PC	200–600	Ceiling fan	10–50
Hot plate	1,200	Printer	100–500	Table fan	10–25
Frying pan	1,200	System (CPU, monitor, laser printer)	up to 1,500	Electric mower	1,500
Dishwasher	1,200–1,500	Fax	35	Hedge trimmer	450
Sink disposal	450	Typewriter	80–200	Weed eater	450
Washing machine		DVD Player	25	¼" drill	250
Automatic	500	TV 25" color	150+	½" drill	750
Manual	300	19" color	70	1" drill	1,000
Vacuum cleaner		12" b&w	20	9" disc sander	1,200
Upright	200–700	VCR	40–100	3" belt sander	1,000
Hand	100	CD player	35–100	12" chain saw	1,100
Sewing machine	100	Stereo	10–100	14" band saw	1,100
Iron	1,000	Clock radio	1	7¼" circular saw	900
Clothes dryer		AM/FM car tape	8	8¼" circular saw	1,400
Electric	4,000	Satellite dish/Internet	30–65	Refrigerator/freezer— Conventional ENERGY STAR	
Gas heated	300–400				
		CB radio	5	23 cu. ft.	540 kWh/yr.
Heater		Electric clock	3	20 cu. ft.	390 kWh/yr.
Engine block	150–1,000	Radiotelephone		16 cu. ft.	370 kWh/yr.
Portable	1,500	Receive	5	Sun Frost	
Waterbed	400	Transmit	40–150	16 cu. ft. DC	112
Stock tank	100	Lights		12 cu. ft. DC	70
Furnace blower	300–1,000	100 W incandescent	100	Freezer—Conventional	
Air conditioner		25 W compact fluorescent	28	14 cu. ft.	440
Room	1,500	50 W DC incandescent	50	14 cu. ft.	350
Central	2,000–5,000	40 W DC halogen	40	Sun Frost freezer	
Garage door opener	350	20 W DC compact fluorescent	22	19 cu. ft.	112

Grid-tied Systems

How to Put Your Electric Utility Company to Work for You!

Utility electric costs are rising, storms are becoming more extreme, and the power grid we once took for granted is becoming increasingly unreliable. Adding this concern about reliability to a desire to help mitigate the impacts of climate change, many of our customers are turning to the reliability and consistency of solar power and grid-tied systems. The cost of solar has declined dramatically in recent years (in fact over 60% lower since our last *Sourcebook* was published!), and state and federal rebate programs remain in place. The time for solar is now.

A grid-tied system makes it possible to generate your own solar power and send any energy you don't use on the spot back to the utility company, where it is "stored" for you to use later. Your utility electric meter will run backward anytime your renewable energy system is making more power than your house needs at the moment. If the meter runs backward, your utility company is accounting for the power coming from your system, and if you live in a state with net metering, they will credit your account at the same rate that you pay them. It feels great! Even if your renewable energy system isn't making more energy than your house needs at the moment, any watt-hours your system delivers will displace an equal number of utility watt-hours, directly reducing your utility bill, lowering your carbon footprint, and reducing your exposure to volatile utility rates. If your utility has a tiered rate schedule, meaning the cost per kilowatt-hour gets more and more expensive as your usage increases, supplanting even 50% of your usage can have a dramatic impact on your bill. Don't be discouraged if you can't offset 100% of your usage—every kilowatt-hour counts!

The Legalities of Electricity and Net Metering

The 1978 federal *Public Utilities Regulatory Policies Act* (PURPA) states that any private renewable energy producer in the United States has a right to sell power to their local utility company. The law doesn't state that utilities have to make this easy or profitable, however. Under PURPA, the utility usually will pay its "avoided cost," otherwise known as wholesale rates. This law was an outgrowth of the 1973 Arab oil embargo and was intended to encourage renewable energy producers. But 2–4 cents per kilowatt-hour was not sufficiently encouraging to have much of an effect on home utility intertie installations.

The next wave of encouragement started arriving in the mid-1980s with net metering laws. There is no blanket federal net metering law and we are still waiting to see federal legislation on the matter. Currently, 43 states and the District of Columbia have net metering laws, and most utilities support net metering. These policies allow small-scale renewable energy producers to export excess power to their utility, through the existing electric meter, for standard retail prices. In recent years, we've seen net metering come under scrutiny by utilities, and efforts have been made to make it more difficult to go solar, ending up in court or on the docket of the public utilities commission. We're happy to report that, so far, the solar industry and net metering has prevailed, though not without cost to jobs and businesses while waiting for the rulings to come down.

> If the meter runs backward, your utility company is accounting for the power coming from your system, and if you live in a state with net metering, they will credit your account at the same rate that you pay them.

GRID-CONNECTED SOLAR ELECTRIC SYSTEM

electrical meter

fuse box

inverter

Inverter conditions electricity for use.

How Grid-tied Systems Work

A grid-tied system uses an inverter that takes any available renewable power generated onsite, turns it into conventional AC, and feeds it into your electrical circuit breaker panel. The inverter is essentially connected like any other appliance, it just happens to supply power instead. Electricity flow is much like water flow; it runs from higher to lower voltage levels. When you turn on an electrical appliance, the voltage level in your house slightly decreases, letting current flow in from the utility. The intertie inverter carefully monitors the utility voltage level and feeds its power into the grid at just slightly higher voltage. As long as the voltage level in your house is being maintained by the grid-tied inverter, and your usage and production happen to be in balance, no current flows in from the utility. You are displacing utility power with renewable power. If your house demands more power than the system is delivering at the moment, utility power compensates seamlessly. If your house is using less power than the system is delivering, the excess pushes out through your net meter, turning it backward, back into the grid. This treats the utility like a big 100%-efficient battery. You export power during the day; it sends power back to you at night. If you live in one of the majority of states that allow net metering, then everything goes in and out through a single residential meter. Your utility will swap out your existing meter for one that counts kilowatt-hours going in both directions, usually free of charge. Many utilities have upgraded over the last few years to newer digital or smart meters, so your existing meter might be solar ready without any changes. Grid-tied systems are very simple, and extremely beneficial to renewable energy producers who have utility power available.

Keeping Grid-tied PV Safe!

Utility companies are naturally concerned about any source that might be feeding power into their network. Is this power "clean" enough to sell to your neighbors? Are the frequency, voltage, and waveform within acceptable specifications? And most important, what happens if the utility power goes off? Utility companies take a very dim view of any power source that might send power back into the grid in the event of a utility power failure, as this could be a serious threat to power line repair workers.

All intertie inverters have precise output specifications, voltage limits, and multilayered automatic shut-off protection features as required by UL specification 1741. This specification details all the necessary output and safety requirements as agreed upon by utility and industry representatives and is a specific UL listing for this new family of intertie inverters. If the utility fails, the inverter will disconnect in no more than 30 milliseconds, faster than you can blink! Most utility companies agree this is a faster response than their repair crews usually will manage. Even an "island" situation will be detected and shut off within two minutes. An island happens (theoretically) when your little neighborhood is cut off, isolated, and just happens to require precisely the amount of energy that your intertie inverter is delivering at the moment. Safety has been the top concern during development of intertie technology. These inverters just can't cause any problems for the utility. The worst problem they could cause the homeowner is to shut off the intertie system temporarily because utility power drifts outside specifications. As an increasing number of PV systems come online, it's more important than ever to ensure safety during an outage. Even one can be a safety hazard, not to mention an entire neighborhood.

Do You Need Batteries or Not?

Grid-tied inverters can be divided into two categories: those that support backup batteries, and those that don't. While the vast majority of installations today do not have battery backup, this is still a major fork in the road of system design. The two system types operate very differently, can require different or additional equipment, and have big price differences depending on system design

A typical grid-tied system, with backup power.

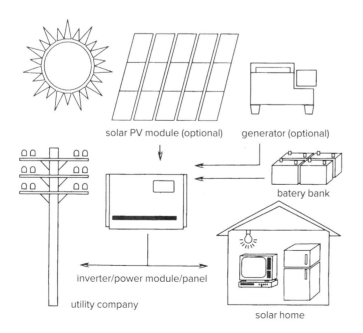

solar PV module (optional) generator (optional)

batery bank

inverter/power module/panel

utility company

solar home

and battery bank capacity. To help you choose wisely, read on.

Grid-tied systems with battery backup are essentially large stand-alone renewable energy systems that use either a sophisticated inverter that also can intertie with the utility, or a second inverter that pulls power and charges batteries. When your system looks and behaves like an off-grid system except that the inverter accepts utility grid input, this is called a DC-coupled system, because the battery bank on the DC side of the system is where the PV and inverter come together. More recently, AC-coupled systems have come on the scene. AC-coupled systems look and behave like a grid-tied system, except that a second inverter and battery bank are added on. They are called AC-coupled because a load center on the AC side of the system is where the on-grid and off-grid components come together.

Several manufacturers offer battery-based intertie inverters. The new Radian series by Outback Power Systems is the DC-coupled system of choice. The Radian series inverters range in output depending on needs, and go as large as 8,000 W of continuous power with 48-volt battery packs. For AC-coupled systems, SMA has the Sunny Island battery-based inverter that can work along with their existing series of Sunny Boy grid-tied inverters. All these inverters will allow selected circuits in your house to continue running on battery and solar power if utility power fails. They probably won't run your entire house, but they can keep the fridge, furnace, and a few lights going through an extended outage, and they'll do it automatically, without noise or fuss. With inverter, safety, charge controller, batteries, and a minimal PV array, these systems start out at approximately $10,000. Average-size systems that will comfortably (partially) support a typical suburban home run about $20,000–$30,000. As always, the price tag will depend on how much battery capacity you're looking for.

AC-coupled systems are becoming the leading choice for those wanting to add battery backup, mainly because of the added efficiency of not sending all your power through a charge controller and battery bank before getting to the inverter. In an AC-coupled system, the battery bank and off-grid inverter patiently await a grid outage to kick in, usually energizing a sub panel, or "backup loads" panel, when the grid goes down. But when the grid is present, they stay out of the equation. Many if not all of the inverters used in a DC-coupled system can function as the second inverter in an AC-coupled system.

Straight grid-tied systems are much simpler.

They consist of PV modules, a mounting structure that holds the modules, the grid-tied inverter(s), a couple of safety switches, and a run of wiring that connects everything. Without batteries, less safety and control equipment is necessary and less physical space is required onsite. Compared to DC-coupled battery backup systems, about 5% to 15% more of your PV power ends up as useful AC power, depending on equipment selection, and more of your investment tends to end up buying PV power. Without batteries, if utility power goes off, so does your grid-tied system, regardless of whether the sun is shining. This is one of the more common misconceptions about grid-tied PV: Remember, without a backup battery bank or generator, your lights will go out too!

Grid-tied inverters without battery backup come in modest to large sizes from 2,000 watts up to 10,000 watts, and are produced by an increasing number of companies, including Power-One, SMA, and Fronius. These inverters start at just under $4,000 for a system that delivers approximately 10 kilowatt-hours per day. The average-size system we've sold over the past few years is around 6 kW, delivers approximately 24 kilowatt-hours per day, and sells for about $20,000 (before any rebates).

Which type of intertie system is right for you? That primarily depends on the reliability of your utility provider. If you live in an area where storms, poor maintenance, or mangled deregulation schemes tend to knock out power regularly, then a battery-based system is going to greatly increase your security and comfort level. If you experience frequent and long-lasting outages, batteries might make sense. Looking out over the darkened neighborhood from inside your comfortable, well-lit home does wonders for your sense of well-being. On the downside, adding batteries will raise your system cost by $6,000–$12,000 and adds components that will need some ongoing maintenance and eventual replacement. If utility power is reliable and well maintained in your area, then you have little incentive for a battery-based system because the additional cost is hard to justify. Those dollars could be better spent on more PV wattage. You'll see more benefits and get to enjoy almost zero maintenance. Just be aware that if utility power does fail, your solar system will automatically shut off too for safety reasons. One nice thing about PV systems is that they are modular in nature, so don't fret over battery backup too much. If the grid becomes less reliable down the road, you can always add batteries later in an AC-coupled configuration.

With inverter, safety, charge controller, batteries, and a minimal PV array, these systems start out at approximately $10,000.

Rebates Can Seriously Reduce Your Total System Cost

Money is available for doing the right thing. A number of states have programs that will pay you an incentive in the form of a rebate based on expenditure or system performance, a tax credit, or a grant to defray the costs of your renewable energy system. It's difficult to keep tabs on all the rapidly changing programs in all the states. The Database of State Incentives for Renewables & Efficiency does a great job on its website (dsireusa.org), but even they have trouble keeping up with all the activity across the nation, so be sure to use the provided links to the state or utility websites for the most up-to-date information. We also pro- vide a list of State Energy Offices in the Appendix. Give yours a call to ask about net metering and money available for renewable energy projects.

Currently the state of New York has some of the most attractive incentives in the country. Between a state rebate based on the capacity (size) of the system, a state tax credit amounting to 25% of the system cost (capped at $5,000), and the federal tax credit for 30% of the system cost, home-owners in New York are only spending about 40% of the total system cost out of pocket. As is the case with most incentives and rebates, they will not be around forever and will decrease as funds are allocated. California has had a great program through the California Solar Initiative, but after

Grid-tied Is a Good Investment

An "enthused but discouraged" customer considering a solar electric system asked us the following questions.

Q: Capital invested in a solar system must come from somewhere. If you take it out of savings, you will give up the interest that the money can earn from savings. If you take it from an investment account, you give up the projected annual return on your investment. If you borrow the money, you give up the interest cost for the investment.

Depreciation is a real cost issue, particularly for batteries. Batteries deteriorate over time; golf cart batteries should be replaced after 3–5 years. Other components such as PV arrays may have a long life, but depreciation over a 30-year period is not immaterial.

Electronic components will age. Controllers and other electronics may be subject to failure and replacement after 10 years.

With a fair, complete consideration of all the financial factors, does solar electric really make economic sense?

A: There is almost no reason to be off-grid if utility power is available. Being off-grid is usually a choice made out of necessity (no grid available), and return on investment is simply not an issue. On the other hand, solar grid-tied systems can be very cost-effective where utility rates are high and rebates are available, such as in California, Hawaii, and much of New England. In these places, going solar is a good investment—and batteries are not required. This investment can be a better use of one's capital than a savings account. The return is better and the risk very low, as long as utility rates keep going up. And with Power Purchase Agreement (PPA, in which the system is owned by a third party and you buy the power from them) and lease options (see the "Converting Sunshine" sidebar, or page 85 in Chapter 3), you can go solar, use clean energy, and save money with $0 out of pocket. The question isn't "How much will this cost me?"—the questions is "How much will this save me?"

For example, the simple pretax first-year return on a 5.5 kW grid-tied PV system in California can be about 17% over a 25-year period. Another way of stating this fact is that the savings generated over 25 years will pay for the investment more than 4 times over. (This figure is based on the following information: System cost after tax credit = $21,580; avoided electric rate = $0.21/kWh; annual electric rate increase = 5%; inverter is replaced after 15 years.) This calculation doesn't even take into account the increase in value of the home due to the home improvement. (For more on that subject, see page 140 for the accompanying sidebar on solar payback, or visit sandia.gov.)

huge success, it ran out of money and is now only available through a few utilities that have not yet used up all of their funds.

Capacity-based vs. Performance-based Incentives (PBI)

The way in which states and utilities incentivize the purchase of PV systems can vary. The two most common types are capacity-based, where the size of the system determines the rebate amount, and performance-based, where the actual kilowatt-hours produced by the system determines the incentive amount. For example, New York currently offers a capacity-based incentive of $0.85/watt up to 25 kW, meaning a 5 kW system would receive a $4,250 rebate—nothing to sneeze at! This rebate drops by about $0.15 cents every few months, so it might already be gone by the time you read this. Massachusetts and New Jersey have performance-based incentives, known as PBIs, where for each kilowatt-hour produced, a monetary credit is earned, for example, around $0.20. Solar renewable energy credits, or SRECs, are earned based on your system's production, and can be sold via a broker, like any other commodity, on an energy credit market. Many states offer a combination of capacity-based and performance-based incentives. Find out what's available in your area or give us a call, we've got rebate and incentive specialists who can give you a quote over the phone: 888-567-6527, realgoodssolar.com.

Good for Your Pocketbook, Good for The Planet

The equity value of a PV system pays for itself the day it's installed. A general rule of thumb is that for every $1 reduction in annual operating expenses, you can expect $20 in increased home value. So if you're saving $1,000 a year in utility bills, that's a $20,000 increase in the value of your home. Sandia National Laboratories provides a calculation based on where you live to see what increased appraisal value you can expect from your system (sandia.gov). And in states with solar incentives, like Colorado, Massachusetts, New Jersey, New York, and others, your system will normally pay for itself with energy savings in 2–10 years, providing a 10%–50% return on investment (ROI). This ROI is way better than the best bonds and CDs or other safe financial investments today.

Most state incentive regulations won't let residential customers deliver more energy than their average electric use. The details vary from state to state, but if you make more power than you use, you'll either give it away or sell it at wholesale rates. Besides the investment payback value, grid-tied is appealing because you are independent of the utility and covering your own electricity needs directly with a clean, nonpolluting, renewable power source. It also feels good to increase the value of your home and make a healthy return on your investment. Solar modules last for decades, require almost no maintenance, and don't borrow from our kids' future, which is probably the best possible reason to invest in renewable energy.

Converting Sunshine into Electricity

*All the Details on How You Can Solarize Your Home,
Slash Your Electric Bill, and Spin Your Meter Backward*

WE'VE TALKED A LOT ABOUT how to solarize your home using clean, safe, reliable, and cost-effective power from the sun. It's easier than you think, and Real Goods residential installation division, RGS Energy, is here to take care of all the details. (See rgsenergy.com, or you can contact RGS at 888-567-6527, solar@rgsenergy.com.) Rebates and tax incentives currently available, combined with today's lower cost for solar panels, have made solar power affordable and cost-effective, providing a double-digit return on investment. There has never been a better time than now to go solar. In most cases, we can provide you with a system that will cost the same or less than what you currently pay for utility power—and in many cases, solar will pay for itself almost instantly. We offer PPAs (power purchase agreements), leases, and financing, so the only thing you really have to lose is more money to your utility company.

Between Real Goods and RGS Energy, we have provided solar energy to over 22,500 homes and businesses in America since 1978, for a total of over 235 megawatts. We are the world's oldest and most experienced supplier of renewable energy. Our customers are our first priority, and we will take care of every detail, from the planning stages to the installation and deployment of your new solar energy system. We've grown the business over the years primarily through customer referrals rather than marketing and advertisements. If there's been a solar system up and running in your neighborhood, whether it's brand new or been there a while, we may well have installed it!

Depending upon the circumstances of individual installation, system payback times are as brief as 2–10 years, with annual rates of return as high as 50%. With PPAs (in which a third party owns the system and you buy the power from them), leases, or financing, you're likely to be cash-flow-positive from year 1 forward. The simple fact is that installing a solar electric system on your home has become cost-effective. With your personal RGS energy consultant walking you through every step of the process, it couldn't be easier. And you'll have peace of mind knowing that you've become part of the energy solution for future generations as we see the twin problems of climate change and dwindling fossil fuel supplies make our world vulnerable to significant or catastrophic disruption. We've still got a long way to go, and a LOT more roofs to solarize. Believe it or not, the US still has less than 1% solar penetration.

Your solar consultant can either come to your home or meet with you virtually via phone/web (if you're in one of our service territories) to review your electrical needs and your last 12 months of electric bills. He or she will then evaluate your home for solar potential, provide you with a financial analysis and payback period, and give you both a price quotation and an installation time, should you decide to proceed. We'll even take care of the paperwork needed for rebates and incentives, utility interconnection, and building permits. We offer a turnkey solar solution.

Today's Solar Economics— Your System Pays for Itself the Day It's Installed!

At Real Goods, we've been selling solar for 36 years. When we started in 1978, the price was well over $100 per watt, or more than $6 per kilowatt-hour. Today, depending on state incentives, that $6 per kWh cost has come down to around $0.07/kWh, almost a 99% drop! This means that you'll be guaranteeing yourself an

Fetzer Vineyards, Hopland, California.

> "The installation of photovoltaic panels on the Fetzer Vineyards administration building marks another milestone in our continuing quest toward sustainability of our winery. We are especially pleased to have our neighbors and friends at Real Goods oversee the project."
>
> PAUL DOLAN,
> PRESIDENT, FETZER VINEYARDS

electric rate of $0.07/kWh for the next 30 years—and we all know that, over that time, utility rates will be sharply increasing; in most places, utility rates have gone up by 4 to 7% annually.

What do these lower prices and governmental incentives mean for the average homeowner? We frequently see over 20% annual returns on investment (ROI)—far better than the stock market, bond market, money markets, and long-term CDs.

According to the National Appraisal Institute (*Appraisal Journal*, October 1999), your home's value increases $20 for every $1 reduction in annual utility bills. This means whether you purchase, finance, or go with a PPA/lease, a monthly utility bill savings of $100 per month would increase the value of your home by $24,000!

State and Federal Incentives
In 1996, the state of California, through its California Energy Commission (CEC), paved the way for the solar industry and created a model for other states by providing cash rebates directly to customers who installed grid-tied solar systems on their homes.

Many states offer capacity-based or performance-based incentives directly, or through utilities, cutting the total cost of a system dramatically. These rebates can usually be assigned to the contractor installing the system, reducing the up-front out-of-pocket expense to the homeowner. Don't get too comfortable, however, because these rebates usually have limited funding, and decrease incrementally as certain installed capacity milestones are reached. Yesterday's $2.00/watt state rebate is today's $0.85/watt rebate. There's usually a lengthy application process, and sometimes rebates drop unexpectedly, so there's always a bit of a sense of urgency in securing them. As the solar industry matures, so does the administration of these incentive programs, to the benefit of all stakeholders. In addition to rebates and incentives, you can take a federal investment tax credit (ITC) of 30% of the cost of your renewable energy system (solar or wind) after rebates. This comes directly off the amount of federal income tax that you owe. The tax credit is set to decrease from 30% to 10% on January 1, 2017, and we're hopeful it will be extended by Congress beyond that date.

Other states, for example Colorado, New Jersey, and New York, have similar rebate and tax credit programs for renewable energy systems. Much of the information we provide here is for California, but it serves as an example of how similar incentives in your state may make the choice to go solar financially attractive. Check with your state energy office or a local renewable energy contractor to learn the details of your state's program (or check out dsireusa.org). Then call us for further consultation.

Businesses can also get a 10% federal income tax credit with no cap. The MACRS five-year accelerated depreciation schedule is also applicable. With these incentives, we have often seen paybacks as low as 1 year and internal rates of return in excess of 50%.

Time-of-use Metering (TOU)
California's net metering laws (over 40 states have them now) mandate that your utility company cannot charge you more for your electricity than they pay you for the solar power you generate on a daily or monthly basis. If you've generated more energy than you used over a 12-month period, you'll probably be reimbursed, but only at the wholesale rate. For this reason, it's generally not a good idea to size your system to produce more than your current or expected energy usage.

The physics of solar power make it the perfect energy source—it puts out the most power in mid-afternoon, which is exactly when California utilities have the greatest demand due to summer air conditioning. To decrease the summer mid-afternoon power load, the utilities have instituted "time-of-use" metering to encourage homeowners to conserve in the afternoon. This is great news for solar system owners!

Time-of-use metering is an excellent choice for PV systems that meet a minimum of 50% of your electrical needs. This means you can sell your solar power to the utility between noon and 6 PM for as much as $0.46/kWh, and then you can buy that same power back at a rate as low as $0.16/kWh. This gives you a huge financial benefit in the long run and also reduces the initial size of your solar system and consequently your ultimate payback time. The benefit from taking advantage of TOU metering is so big, that having additional panels on your west roof can pencil out better financially vs. only using your south-facing roof. Do proceed with caution however, since when you use the most electricity plays a role in how much TOU metering makes sense for you. Consult with your utility and request a TOU analysis, if you're not already on a TOU rate schedule, which will show you how many kilowatt-hours you're using "on peak" and "off peak." Our Solar Experts can use this information to conduct a complimentary rate analysis, to show what the impact will be of switching to TOU when you go solar.

Your Property Tax Won't Go Up
California law prohibits county property tax assessors from increasing your tax assessment because of value increases from your installation of a solar system. (38 states now offer some form of property tax exemption.) This is great news for your cash flow, as you'll be saving hundreds or even thousands of dollars every year from your new solar system, increasing your property's resale value by thousands due to your utility bill savings and not adding even one penny in property tax!

Financing Your Solar System: Own, PPA, or Lease
Real Goods has researched numerous quality home equity lenders and other home improvement loan sources in California. If you choose not to self-fund your solar system, there are many cost-effective alternatives. Many lenders are open to financing solar systems for less than prime rate. Combining a 12-month "same as cash" loan, where there are no payments or interest for

12 months, with a roughly 12-year loan at around 2.9% interest, is one way to enjoy the benefits of system ownership without a large initial outlay. The 12-month, no-interest loan is equal to the amount you'll receive in state and federal tax credits, or any rebates not assigned to the installer, which within 12 months of installation you'll have recouped, and then you can pay back the loan. The remaining balance of the system cost is paid for in monthly payments that usually are less than your utility bills, leaving you cash flow-positive from day one! Call Real Goods at 888-567-6527 to get our latest recommendations on the best financing options. If you can't take advantage of tax credits, or you don't want to make any investment or borrow money, you can still take advantage of a PPA (Power Purchase Agreement; for more information, see Chapter 3, page 85) or lease, for little or no money down. For many homeowners, this means instant positive cash flow. For a more detailed explanation of the payback opportunities, read the sidebar "What's the Payback?"

How Much Power Can Your Solar System Produce?

If you've checked with any of our competitors, you may see somewhat divergent opinions on how much electricity a proposed solar system can produce. While the industry is a lot more mature than it was when Real Goods started, and there's a more consistent standard of calculating system output, we still sometimes see large swings in the estimated output and ROI from installer to installer. Pay close attention to the assumptions being plugged into their formulas—specifically solar access percentage and annual utility inflation rates—as these can have the biggest impact in painting what might be an overly optimistic picture of what a system will do for you. With satellite imagery and in-house design tools able to accurately estimate the production of a system, it's possible for

solar installers to provide a reasonable quote without setting foot on the roof. Remember that these are only estimates and that without taking solar access measurements and getting on the roof to verify that a proposed layout will fit, annual kilowatt production numbers are not guaranteed. At RGS Energy, we're literally decades ahead of our competitors, and have become quite handy at being able to conduct a virtual site survey and rate analysis. After sending out a site technician, we're usually right on the money (or kilowatt-hours as the case may be), but until we've gathered this data, nothing is written in stone.

STC or "Nameplate" Ratings: STC stands for Standard Test Conditions. It is the rated output in watts that the manufacturer puts on its photovoltaic (PV) modules under laboratory-perfect conditions. STC ratings are generally used by the solar industry, installers, our competitors, and the general public.

PTC Ratings: PTC stands for Practical Test Conditions, or the ratings under the PVUSA Test Conditions. This is the standard used by the California Energy Commission (CEC) and in general runs about 6%–12% less than STC. PTC ratings generally are used to calculate the California rebate and have been adopted by numerous other states that look to the CEC for guidance.

Real-life Expectations: Many industry professionals have studied the issue of photovoltaic ratings and are uncomfortable with both STC and PTC ratings because they seem overrated to real-world conditions.

Recently a consensus has emerged that to be conservative, you should expect your solar system to yield in AC output to your electrical panel about 77% of STC (manufacturer's nameplate) ratings (that is, multiply STC by 0.77).

In Summary: Throughout our literature, like all in the solar industry, we will use STC ratings to describe our systems by size. However, to calculate performance for actual real-life expectations, we use the STC rating times 77%. What this means is well summarized by the example at the right featuring a solar panel we commonly use for 3 kW systems.

How RGS Energy Professionals Work to Solarize Your Home

1. Feasibility and Cost/Benefit Analysis: After you contact one of our solar experts, we go to work immediately to determine if solar is right for you. We ask you to provide us with:

- Your utility history for the last 12 months or as long as you've been in your home if it's new
- Your address and confirmation if online maps like Google Earth correctly identify your home
- A description of your roofing material, condition, and age
- An indication of your desire to be 100% energy independent, or what fraction thereof
- An indication if you're already leaning toward purchase, finance, PPA, or lease

Next, our solar specialists will discuss with you if you would prefer we make a personal visit to your home or a virtual appointment via phone/web.

This is how a 3 kW system at the outset becomes a 2 kW system in actual electricity yield.

STC rating	208 W STC×14 = 2,912 watts STC
PTC rating	183.3 W PTC×14 = 2,566 watts PTC
Real-life rating	208 W STC×14×75% = 2,184 watts actual

We'll get back to you with:

- A comprehensive rate analysis, including a recommendation of which solar system is right for your situation
- A price quotation for either purchase, finance, lease, or PPA that includes all equipment, delivery to your home, building permit fees, installation, building inspections, and amount of rebate available from the state or your utility
- A recommendation on whether time-of-use metering will work for you
- A financial analysis of return on investment for your system, years to payback, and internal rate of return, including any state or federal tax credits or incentives available to you

2. Contract Signing, Procurement, and Installation Scheduling: When you have made a decision to proceed with a solar installation, after consultation with us, the next steps are:

- Sign our installation contract or service agreement and, if necessary, provide us with a $250 to $1,000 refundable deposit
- Submit your application (this is handled by us) to your state, utility, or relevant agency, qualifying you for the maximum rebate and an interconnection agreement
- Procure a building permit; you will have anywhere from 3 months to 1 year to install the system, depending on how it is financed, what rebates and incentives are applied for, and how long the interconnection application with your utility allows
- Provide any additional information needed by our solar technician if you are building a new house

3. Installation of Your Solar System: Our professionals will typically install your complete system in one to three days.

PROJECTED RATES OF RETURN ON YOUR 3.5 KW RESIDENTIAL SOLAR SYSTEM

The following figures are calculated based on a PG&E electric bill, which has a tiered structure with a baseline allowance of 351 kWh per 30 days. Tiered structure means that once you've used your baseline allowance, the kWh rate you pay for each additional kWh increases, by quite a bit. Tier 1 is only $0.13/kWh, but Tier 3 power is going to cost you $0.32/kWh. Savings as shown below will vary from these numbers depending upon your utility company. These are averages for California with PG&E as the utility.

If your monthly electric bill is usually:	A 3.5 kW solar electric system will reduce it to:
$100	$20
$150	$35
$200	$60
$250	$95
$300	$140

If you choose to pay cash for your system, the payback period will range from 2 to 10 years, depending upon how you value the discounted pretax cash flow and the anticipated inflation rate. Your return on investment will typically be 10%–30%, far better than any safe investment available today…or you can choose to finance your solar system. The loan payments will probably be low enough compared to your utility bill savings that you'll see positive cash flow from year 1.

For a free rate analysis of your electric bills and to see what your post-solar utility bill will look like, along with payback and ROI, contact RGS Energy. Solar systems for homes with large utility bills are always cost-effective and have shorter payback times and higher returns on investment due to the higher average electricity rates from tapping those expensive upper tiers.

THE BOTTOM LINE

Total installed cost of 3.5 kW system (incl. sales tax) before rebate and tax credit	$14,090
Federal tax credit (30% of post-rebate installed cost)	($4,227)
Total installed cost (incl. sales tax) after rebate and tax credit	$9,863

Estimated monthly cost based on projected production ratings:	
Electricity cost per month before solar @ E1 rate using 720 kWh/mo.	$150
Less solar electricity cost savings per month (avoided electric cost of $.25/kWh, across Tiers 1 to 3) (2,695 kW×$0.25/kWh×5.5 hr./day× 30.4 days/mo.)	($35)
Plus monthly payment on financing $9,863 @ 2.99%, 12-year loan	$84
Total monthly cost for electricity after installing solar system	$119

The bottom line is that, in certain situations, your cash flow can actually be positive from year 1 on by going solar!

3.5 KW RESIDENTIAL SOLAR SYSTEM COMPONENTS

Component	Quantity
3,500 watts of PV modules (14, Sharp 208 W)	3.5 kW
Power-One PVI-3.0 3000 DC-to-AC inverter with metering and integrated combiner and DC disconnect	1
Solar panel mounting kits and roof standoffs	2
Utility-required safety AC disconnect switch	1
Weatherproof PV output cable (50 ft., male-female)	2

- 90% of the payment is due the day we begin installation. The balance is due upon completion of the installation. We will collect the rebate on your behalf so you won't be responsible for any income tax. You pay us only the after-rebate price, without waiting for the rebate
- Upon completion and payment, we'll schedule an inspection with the local building inspector and utility for final approval
- If you don't already have a meter that's solar ready, your utility will install your net meter, usually at no cost
- You'll be making your own clean solar power and contributing to the energy solution!

What's the Payback?

How to Calculate the Return on Your Solar Electric System Investment Before You Buy

FOR YEARS, QUESTIONS ABOUT returns on the expensive investment in a solar electric system were dismissed with the analogy, "What's the payback on your swimming pool?" That sentiment might speak to the converted, but for most people considering a solar energy system, finances are a major deciding factor. Residential PV systems are being looked at as an investment, and commercial systems even more so.

Fortunately, photovoltaic technology has matured enough that the payback question can now be given a serious answer, backed by solid math and accounting. The answers vary significantly by local climate, utility rates, and incentives. In the best cases—California and New York— the compound annual rate of return is well over 15%, the cash flow is positive, and the increase in property resale value more than covers the cost of the PV system (visit sandia.gov). In other parts of the country where electric rates are low and incentives may be less, the return on a grid-tied system may barely cover its potential inverter replacement costs.

This section focuses on residential analyses. Similar calculations can be done for commercial situations, but significant differences in the tax and accounting rules exist. See the resources section in the appendix to learn more about commercial analyses.

What Factors Improve Payback?
The most important factors that make solar an attractive investment are high electric rates, financial incentives, net metering policies, and good sunlight (available in most of the continental US).

Electric rates vary by region, state, and utility. California, Hawaii, and New York have the highest average rates, well above $0.20 per kilowatt-hour. California's tiered pricing structure also penalizes large residential users with prices as high as $0.42 per kilowatt-hour (see Figure 1).

Under most net-metering laws (these also vary by state or utility), solar energy offsets the retail cost of the electricity generated. Even better, in California, solar systems are allowed to operate on a time-of-use rate schedule, which enables users to sell back electricity to the utility at peak rates, which can be even more valuable. The higher the price your PV power can fetch, the better payback scenario you have.

Direct incentives can include tax benefits such as credits or depreciation. The most celebrated incentive is the federal tax credit for solar systems

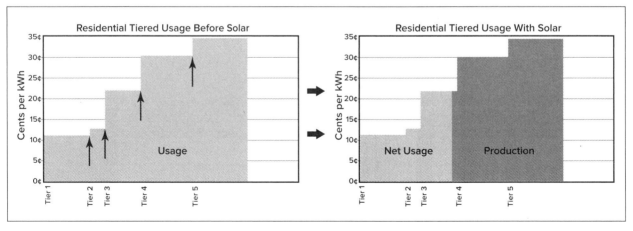

Figure 1. Effect of Tiered Electric Rates on Large Users, and How Solar Helps. Tiered-rate pricing penalizes large users most with high marginal electricity rates. Solar energy offsets highest-tier usage first, making the solar customer look like a small user, with a lower marginal cost.

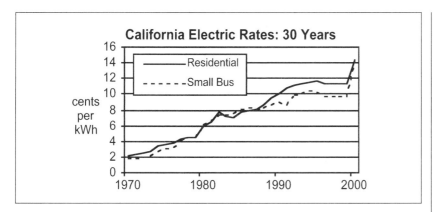

California Electric Rates: 30 Years

cents per kWh

Residential
Small Bus

1970 1980 1990 2000

Fig. 2. California Electric Rates: 30 Years. Inflation is a major factor in photovoltaic system returns. In California, rates increased an average 6.7% per year from 1970 to 2000. This article assumes 5% electric rate inflation going forward. Source: California Public Utility Commission, "Electric Rate Compendium," November 2001.

that went into effect January 1, 2006. This credit is for 30% of the system cost, after any up-front rebates. For PV systems, that means a significant credit on the purchaser's tax return for the year the system was installed. The federal credit can be coupled with state incentives such as rebates, which can discount up to 60% of the system cost. Some states also have a state tax credit, which can further reduce the up-front cost of a system. For example, New York offers a tax credit of 25% of the system cost, capped at $5,000. Consult a certified tax advisor to check the applicability of such incentives to your situation.

There are also forms of direct incentives called PBIs (performance-based incentives) and SRECs (solar renewable energy credits, or "green tags"). Both are paid on a per-kilowatt-hour-produced basis. They don't reduce the up-front cost, but they do increase the cash payments received after commissioning the system. Payments can be as much as $0.22 per kilowatt-hour for a 10-year or longer contract on SRECs. Because these payments often can be combined with net metering value, the PV system is capable of garnering substantial revenue per kilowatt-hour generated.

Inflation in electric rates is another factor that can improve payback (see Figure 2). Solar is an inflation-protected investment because it

offsets electricity costs at the current prevailing retail rate—that is to say, once you've purchased your PV system, the effective kWh rate on your electricity is fixed for the life of the system. As utility rates rise, you save even more money. In the case of some PPAs (power purchase agreements), there can be an annual increase to the solar kWh rate, but even so, these tend to be very low, 1.5% to 2.9%, and also remain fixed, so your solar bill will never be a surprise at year 1 or year 20.

Determining the Payback
The economic value of a solar system can be measured in several ways: *rate of return, cash flow*, and *increase in property resale value*. In strong cases, the returns will be over 20%, the cash

flow will be positive (as early as year 1 with financing, a lease, or a PPA), and the increase in resale value will exceed the system cost. These scenarios are common in certain markets.

The cash flow will be positive, either immediately or within the first few years, especially for homeowners who finance their solar systems, but with increasing utility costs, combined with incentives, payback can occur very soon even when purchasing a system outright. Cash-flow calculation compares the estimated savings on the electric bill with the cost of the loan. Monthly cost is the principal plus interest payment required to pay off the loan, less any tax savings. In the case of "deductible" loans, such as home equity-based loans, the interest is usually tax deductible and thus the loan effectively costs less. Home equity loans are also excellent sources of funds because interest rates on real estate-secured loans are relatively low and payment terms can be long.

Inflation plays an important part in the cash-flow equation because as electric rates rise, savings from a solar system will increase. But inflation doesn't affect loan rates, particularly for fixed-rate loans, so your savings grows while the cost of the loan stays relatively constant (it rises a little over time as the interest portion and thus the tax deductibility of the payment declines). See Figure 3 for an

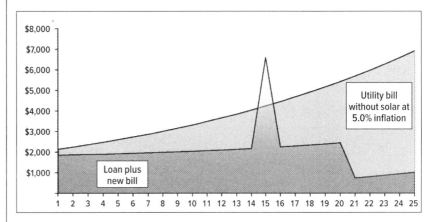

$8,000
$7,000
$6,000
$5,000
$4,000
$3,000
$2,000
$1,000

Utility bill without solar at 5.0% inflation

Loan plus new bill

1 2 3 4 5 6 7 8 9 10 11 12 13 14 15 16 17 18 19 20 21 22 23 24 25

Figure 3. Inflation's Effect on Loan Costs vs. Electric Costs Without Solar. In this case, the photovoltaic system loan term is 20 years, with inverter replacement occurring at year 15 (the large spike).

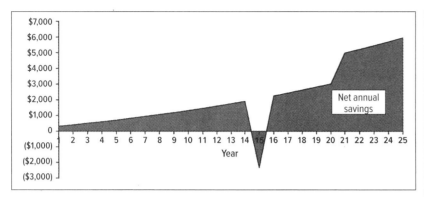

Figure 4. Net Annual Savings with a Photovoltaic System. Net annual savings consists of the old electric bill minus the new bill, loan principal payments, loan net interest (after taxes), maintenance, and inverter-replacement costs. This example system is cash-positive from the first day at no out-of-pocket costs to the purchaser.

example. For example, in California PG&E rates have increased by 6.7% when averaged over the last 30 years.

Figure 4 highlights the difference in the curves in Figure 3. It shows the net annual savings—old bill minus new bill, loan, maintenance, and inverter-replacement costs—and the effects of inflation over time.

Figure 5 shows the accumulation of net annual savings. This accumulation is free and clear with no initial outlay of cash, because that was covered by the loan. In the

example, the system will save the owner about $90,000 over 25 years, with no up-front cost. The savings are small though significant in the first years but really jump when the loan payments stop. You can select any loan term that suits your needs. Note that these are examples of ideal cases in the states with the best economics.

An increase in property resale value occurs in homes with solar electric systems because these systems decrease utility operating costs. The generally accepted rule of thumb in the industry is that a home's value increases $20,000 for every $1,000 reduction in annual operating costs from energy efficiency. A more detailed analysis can incorporate factors

specific to your market (visit sandia .gov for a detailed calculation).

The rationale is that the money from the reduction in operating costs can be spent on a larger mortgage with no net change in monthly cost of ownership. Historically, mortgage costs have an after-tax effective interest rate of about 5%. If $1,000 of reduced operating costs is put toward debt service at 5%, it can support an additional $20,000 of debt. To the borrower, total monthly cost of homeownership is identical. Instead of paying the utility, the homeowner pays the bank, but the total cost is unchanged and you have free electric power.

The column labeled "Appraisal Equity/Resale Increase" in the Sample Scenarios table shows the increase in home value that you can expect. This increase can effectively reduce the payback period to zero years if you chose to sell the property immediately, and it removes the purchase risk. It could even lead to a profit on resale in some cases. An independent study from the Lawrence Berkeley Laboratory found that homes with solar not only sold for more money, but they also sold more quickly (eetd .lbl.gov/ea/emp/reports/lbnl-4476e .pdf). What's more, the housing market at the time of this study was

Payback Calculators and Resources

Database of State Incentives for Renewable Energy: **dsireusa.org**

The Clean Power Estimator: **pge.cleanpowerestimator.com /sites/pge/pge.aspx**

FindSolar.com: **findsolar.com**

PV Watts: **nrel.gov/rredc/pv watts/**

RETScreen: **retscreen.net/ang /home.php**

Solar Energy Industries Association Guide to Federal Tax Incentives: **seia.org/research -resources/solar-tax-manual**

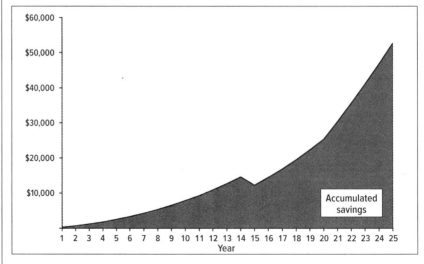

Figure 5. Accumulation of Net Annual Savings. The accumulation of the net annual savings shows total lifetime electric bill savings experienced with a photovoltaic system, net of all costs. All initial costs are included in the chart.

in pretty bad shape, so as that market improves, we expect to see an even greater benefit to homeowners with solar systems looking to sell. PV systems will appreciate over time, rather than depreciate as they age. The appreciation comes from the increasing annual savings the system will yield as electric rates and bill savings rise. Such appreciation cannot continue forever, being limited by the system's remaining life. Here the system is assumed to be worthless at the end of 25 years—a conservative estimate since the panels are warranteed at 25 years to work at 80% of their original capability. Figure 6 shows both the increasing property resale value due to increasing annual savings and the remaining-value limitation that takes over at approximately year 11. As the NREL resale study suggests, however, actual resale could be much higher depending on the market mood for solar.

Creating Markets That Reward Investment

Solar has finally come into its own in many markets. To encourage widespread adoption of solar energy, we need to empower everyone with knowledge of the financial benefits of renewable electricity and expand the programs that make it possible— tiered rates, time-of-use net metering, Renewable Portfolio Standards with solar requirements, and RECs. Market forces will take it from there.

Thanks to Andy Black, who provides solar financial analysis tools and consultation through OnGrid Solar, for providing the original data and analysis upon which this sidebar is based.

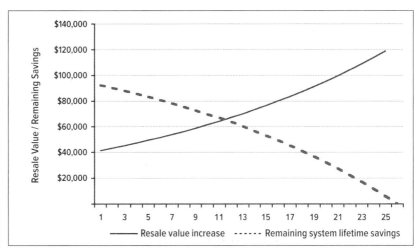

Figure 6. House Resale Value Over Time. The rising curve represents the increasing property value based on 20 times annual savings. The declining curve is the remaining savings of the photovoltaic system before the end of 25 years of conservatively estimated life. The resale value estimate is based on the lower of these two curves. This estimate can be compared with the accumulated savings in Figure 5. In later years, resale value declines, but greater accumulated savings have been enjoyed.

How Does Solar Compare with Remodeling Investments?

Why should a homeowner pay more for a house with a solar system when s/he could buy a nearby non-solar house and install a solar system for less money?

Nationally, decks return an average of more than 90% of their cost on resale (Remodeling Online, remodeling.hw.net). Kitchen and bathroom remodeling have similar results, depending on geography. So it makes sense that in certain regions where the sun shines brightly and electric rates are high, solar would return more than its installed cost, with a national average comparable to other types of house upgrade. The table below lists projected resale value of various solar systems, compared with nationwide averages for other home improvements.

Resale Value Comparisons of Various Home Improvements

Home Improvement Type	Investment Amount/ Net System Cost After Tax Credits	Resale Value Increase	% Return
4.4-kilowatt Photovoltaic System, California, 75% solar offset	$12,000	$17,000	141%
9.9-kilowatt Photovoltaic System, California, 83% solar offset	$23,000	$38,000	165%
Deck Addition	$6,300	$6,700	104%
Bathroom Remodel	$10,100	$9,100	89%
Window Replacement	$9,600	$8,200	85%
Kitchen Remodel	$44,000	$33,000	75%

Analysis based on data from Remodeling Online's 2003 survey, updated 2014 by RGS Energy.

Sample Scenarios for Residential System Payback

Example	Electric Bill Before Solar	Usage Before Solar per Month (kWh)	Solar System Size (Standard Test Conditions)	Final Net Cost with Tax Credits and Rebate	Pretax Annual Return	Appraisal Equity/ Resale Increase in First Year	Net Monthly Cash Flow Compared with 5% 20-Year Loan in First Year
California: Medium System	$180	800	4.4 kW	$12,000	22.6%	$17,000	$35/mo.
California: Large System	$470	1,600	9.9 kW	$23,000	32%	$38,000	+$206/mo.
New Jersey	$240	1,200	9.0 kW	$23,000	23.6%	$28,000	+$132/mo.
Colorado	$100	750	4.4 kW	$11,500	12%	$16,000	$(24)/mo.

Safety and Fusing

Any time we have a power source and wiring, we need to ensure that the number of amps, aka current flow, will be kept within the capability of the wire. Too much current through too small a wire leads to melted insulation or even fires. Renewable energy systems have at least two power sources. The solar, wind, or hydro generator is one (or more), and the battery pack is the other. The battery requires particular concern because it stores a lot of energy and can deliver it very quickly. Even small battery packs store up what can be an awesome amount of energy for future use, and if short-circuited, there's nothing holding back all those electrons. Let's start our conversation about safety with batteries, since they're arguably the most dangerous part of a renewable energy system.

In many ways, batteries are safer than traditional AC power. It's not impossible, but it's pretty difficult to shock yourself with the low-voltage direct current (dc) in the 12- or 24-volt battery systems we commonly use. On the other hand, because batteries are right there in your house, they can deliver many more amperes into an accidental short circuit than a long-distance AC transmission system can. These high-amperage flows from battery systems can turn wires red-hot almost instantly, burning off the insulation and easily starting a fire. (The National Electrical Code says the same thing, although it spends a lot more words saying it.)

And if red-hot wires and melting insulation aren't enough to teach caution, it's difficult to stop a DC short circuit compared to AC power. In a really severe short, popping open a circuit breaker or fuse may not stop the electrical flow. The current will simply arc across the gap and keep on cooking! Once a DC arc is struck, it has much less tendency to self-snuff than an AC arc. Arc-welders greatly prefer using DC for this very reason, but impromptu welding is not the sort of thing you want to introduce into your house. Fuses that are rated for DC use a special snuffing powder inside to prevent an arc after the fuse blows. AC-rated fuses rarely use this extra level of protection. Always double-check the ratings and classifications on your overcurrent protection devices. DC-rated equipment, because it is less common, is more expensive, but don't be tempted to use its less expensive AC cousin…it's not worth the risk.

Many DC circuit breakers have limited current interruption ability, and require an upstream main fuse that's designed to stop very large (20,000-amp) current flow. All the Class T

Real Goods' #1 safety rule is: **Any wire that attaches to a battery must be fused!**

An Outback DC Enclosure with large breaker for the inverter, ground fault protection for the PV array, smaller breakers for the charge controller, and lots of space left over.

Lightning Protection and Grounding

Let's preface this subject with a warning: *Lightning is capricious.* That means it does whatever it wants to do, whenever it wants to do it. This does not mean you're defenseless, however. You can do plenty to avoid lightning damage 99% of the time.

The most important thing to keep in mind about lightning is that it *wants* to go to ground. Your job is to give it a clear, easy path, and then stay out of its way. Good, thorough grounding is your first, and most effective, line of defense. Solar arrays have a nice conductive aluminum frame around each module. But frames are usually anodized, which makes them act as an insulator. There will not be a good electrical path between the frames and the mounting structure. We used to have to run bare copper wire, no smaller gauge than your PV interconnect wiring, across the backs of all the modules, forming a good mechanical connection to each module frame. Stainless steel self-tapping screws with stainless star washers were the best choice. Today we're using more and more grounding clips, made by Wiley Electronics, which bond the modules and rails without the need to run copper ground wire between each module, saving time and money on each installation without compromising safety. You'll still want to ground the racking the old-fashioned way, however. Run this ground wire to your household ground rod, if it's within about 50 feet. If your array is more remote, install a ground rod at the array and ground to it. It's fine to have multiple ground rods in a single system…however…*all ground points must be connected to each other with #6-gauge copper wire.* This is probably the most important point about lightning safety. Connect all the grounds together!

Why? Let's say your array just took a hit, but the lightning took the easy path to ground you provided. Now the earth around your array is at several thousand volts potential, and that energy is looking for a way to spread out, the quicker and easier the better. It sees a nice fat ground rod over at your house, and if it just jumps into this highly conductive copper PV-negative wire that someone nicely provided, it can get over to the house ground rod real easy…so it does. Goodbye charge controller, and whatever else gets in the way. If your ground rods are all connected together with bare copper wire, however, the lightning is going to take the easy route, and pretty much stay out of your system wiring. If your bare ground wire is buried in direct contact with the ground, not inside a conduit, the lightning will like it even better, because it has more ground contact to disperse into.

For wind turbine towers, which obviously are just begging for lightning strikes, you need a ground rod at the base of the tower, and a ground rod at each guy wire anchor. Connect them all together, and then connect back to the house ground.

Good grounding like this will save your system from damage through more than 90% of lightning strikes. Unfortunately, your equipment can't withstand a near or worst-case direct hit, and if this happens, do not fear, it won't happen twice. For that occasional capricious strike that decides to take an excursion through your wiring, you need to have lightning protectors installed. These $45 lightning arrestors short out internally if voltage goes above 300, and route as much energy as possible to ground. You should have at least one on each set of input wires. If you live in a high-lightning zone, a couple of protectors on each input are cheap insurance. Some installers don't bother using lightning arrestors, as there has been some debate about their effectiveness, but at $45 a pop, it's cheap insurance, especially for nearby strikes (see realgoods.com).

Good grounding and lightning protectors are sufficient protection for residential systems. If you're installing in a really severe lightning climate, like a mountaintop (or Florida, according to some folks), seek professional lightning protection assistance! Much more needs to be done for extreme sites, where a simple lightning rod is only the beginning.

fuses we offer have this ability and rating (see real goods.com). When large household-sized inverters started becoming widely available, large DC-rated circuit breakers, DC disconnects, and DC power centers started showing up in the industry. For household systems, these are the safest and most aesthetically pleasing solution.

The National Electrical Code requires "overcurrent protection and a disconnect means" between any power source and any appliance. A DC-rated circuit breaker covers both those requirements, and has become the first choice. For small systems, a breaker box and circuit breakers may get expensive, and folks are tempted to cut corners. Do so at your own peril.

Safety with a large inverter requires some specialized equipment. Full-size 3,000- to 5,000-watt inverters can safely handle up to 500 amps input at times. Systems of this size will use either an OutBack or MidNite Solar DC power center that provides all the fusing, safety disconnect, ground fault protection, and interconnection space for everything DC in one compact, prewired, engineered, UL-approved package. These power centers are the best choice for full-time off-grid living, and they make installation a snap.

For midsize, 400- to 1,000-watt inverters, a 110-amp Class T fuse assembly with an appropriate MidNite Solar or equivalent circuit breaker makes a cost-effective and code-compliant safety disconnect. The small inverters of 100 to 300 watts simply plug or hardwire into a fused outlet. Combiner Boxes, Safety Fuseholders, and Midget Fuses are available for high-voltage PV and advanced battery series connections; 600-volt and 1,000-volt DC UL-rated overcurrent protection equipment is readily available off the shelf. USE 2 wire is now UL rated to 1,000 volts dc.

Fuses are sized to protect the wires in a circuit. See the chart on page 172 for standard sizing/ampacity ratings. These ratings are for the most common wire types. Some types of wire can handle slightly more current, and all types of wires can safely handle much more current for a few seconds. If a particular appliance, rather than the wire, wants protection at a lower amperage than the rest of the circuit, then usually a separate or in-line fuse is used just for that appliance.

Charge Controllers

Almost any time batteries are used in a renewable energy system, a charge controller is needed to prolong battery life and regulate charging. The most basic function of a controller is to prevent battery overcharging. If batteries routinely are allowed to overcharge, their life expectancy will be reduced dramatically. A controller will sense the battery voltage, and reduce or stop the charging current when the voltage gets high enough. This is especially critical with sealed batteries that do not allow replacement of the water that is lost during overcharging. Your batteries will have specific charging set points for different stages of charging, which are programmable in most good charge controllers, so you can be sure to get the most out of your batteries.

The only exception to the need for a charge controller is when the charging source is very small and the battery is very large in comparison. If a PV module produces 1.5% of the battery's ampacity or less, then no charge control will be needed. For instance, a PV module that produces 1.5 amps charging into a battery of 100 amp-hours capacity won't require a controller, as the module will never have enough power to push the battery into overcharge.

PV systems generally use a different type of controller than a wind or hydro system requires.

If batteries routinely are allowed to overcharge, their life expectancy will be reduced dramatically.

PV controllers can simply open the circuit when the batteries are full without any harm to the modules. Do this with a rapidly spinning wind or hydro generator and you will quickly have a toasted controller, and possibly a damaged generator. Rotating generators make electricity whenever they are turning. With no place to go, the voltage will escalate rapidly until it can jump the gap to some lower-voltage point. These mini lightning bolts can do damage. With rotating generators, we generally use diversion controllers that take some power and divert it to other uses. Both controller types are explained below.

PV Controllers

Most PV controllers simply open or restrict the circuit between the battery and PV array when the voltage rises to a set point. Then, as the battery absorbs the excess electrons and voltage begins to drop, the controller will turn back on. With some controllers, these voltage points are factory preset and nonadjustable, while others can be adjusted in the field. Early PV controllers used a relay, a mechanically controlled set of contacts, to accomplish this objective, making audible clicks as they opened and closed. Newer solid-state controllers use power transistors and PWM (Pulse Width Modulation) technology to turn the circuit rapidly on and off, effectively floating the battery at a set voltage. PWM controllers have the advantage of no mechanical contacts to burn or corrode, but the disadvantage of greater electronic complexity. Both types are in common use on smaller systems, with most designs favoring PWM, as electronics have been gaining greater reliability with experience over the years.

The latest generation of PV controls, and now the most common, employs an electronic trick called Maximum Power Point Tracking, or MPPT. This allows the controller to run the PV array at whatever voltage delivers the highest wattage. This is often at a higher voltage level than the batteries would tolerate and is well above their nominal rating. The excess voltage is converted to amperage that the batteries can digest happily before it leaves the controller. So MPPT controllers usually push more amps out than they're taking in. On average, an MPPT controller will deliver about 15% more energy per year than a standard controller. And they do their best work in winter, when most off-grid homes need all the charging help they can get. Cold temperatures tend to elevate PV voltages, and long hours of darkness tend to lower battery voltages. MPPT controls make up most of the PV controller market.

Controllers are rated by how much amperage they can handle. National Electrical Code regulations require controllers to be capable of withstanding 25% over-amperage for a limited time. This allows your controller to survive the occasional edge-of-cloud effect, when for a brief period, sunlight availability can increase dramatically. Regularly or intentionally exceeding the amperage ratings of your controller is the surest way to prematurely destroy it. It's perfectly okay to use a controller with more amperage capacity than you are generating. In fact, with this piece of hardware, buying larger to allow future expansion is often smart planning, and usually doesn't cost much. Because MPPT controllers will output much higher amperages than are coming from the array, it's necessary to size your conductors to handle the maximum output of the controller, even if your array amperage doesn't seem nearly that high.

A PV controller usually has the additional job of preventing reverse-current flow at night. Reverse-current flow is the tiny amount of electricity that can flow backward through PV modules at night, discharging the battery. With smaller one- or two-module systems, the amount of power lost to reverse current is negligible. A dirty battery top will cost you far more power loss. Only with larger systems does reverse-current flow become anything to be concerned about. Much has been made of this problem in the past, and almost all charge controllers now deal with it automatically. Most of them do this by sensing that voltage is no longer available from the modules, when the sun has set, and then opening the relay or power transistor. A few older or simpler designs still use diodes—a one-way valve for electricity—but relays or power transistors have become the preferred methods (see sidebar on Dinosaur Diodes).

Hydroelectric and Wind Controllers

Hydroelectric and wind controllers have to use a different strategy to control battery voltage. A PV controller can simply open the circuit to stop the charging and no harm will come to the modules. If a hydro or wind turbine is disconnected from the battery while still spinning, it will continue to generate power; but with no place to go, the voltage rises dramatically until something gives. With these types of rotary generators, we typically use a diverting charge controller. Examples of this technology are the MorningStar Tristar series and the Outback FX80 (when used as a diversion load controller rather than a PV controller). (You can get these controllers at realgoods.com.) A diverter-type control will monitor battery voltage and, when it reaches the adjustable set point, will turn on some kind of electrical load. A heater element is the most common diversion load. Both air and water heater elements are commonly used, with water heating being the most popular. It's nice to know that any power beyond what is needed to keep your batteries fully charged is going to heat your household water or hot tub. Diversion controllers can be used for PV regulation as well, in addition to hydro or wind duties in a hybrid system with multiple charging sources. This controller type is rarely used for PV regulation alone, due to its higher cost.

Controllers for PV-direct Systems

A PV-direct system connects the PV module directly to the appliance we wish to run. They are used most commonly for water pumping and ventilation. By taking the battery out of the system, initial costs are lower, control is simpler, and maintenance is virtually eliminated. PV-direct systems are elegant, as the appliance operates most strongly when the sun is at its peak, meaning you pump water and cool your attic when those functions are most needed.

Although they don't use batteries, and therefore don't need a controller to prevent overcharging, PV-direct systems often do use a device to

On average, a Maximum Power Point Tracking (MPPT) controller will deliver about 15% more energy per year than a standard controller. And they do their best work in winter, when most off-grid homes need all the charging help they can get.

If a hydro or wind turbine is disconnected from the battery while still spinning, it will continue to generate power; but with no place to go, the voltage rises dramatically until something gives.

Dinosaur Diodes

When PV modules first came on the market a number of years ago, it was common practice to use a diode, preferably a special ultralow forward-resistance Schottky diode, to prevent the dreaded reverse-current flow at night. Early primitive charge controllers didn't deal with the problem, so installers did, and gradually diodes achieved a mystical must-have status. Over the years, the equipment has improved and so has our understanding of PV operation. Even the best Schottky diodes have 0.5- to 0.75-volt forward voltage drop. This means the module is operating at a slightly higher voltage than the battery. The higher the module voltage, the more electrons that can leak through the boundary layer between the positive and negative silicon layers in the module. These are electrons that are lost to us; they'll never come down the wire to charge the battery.

The module's 0.5-volt higher operating voltage usually results in more power being lost during the day to leakage than the minuscule reverse-current flow at night we're trying to cure. Modern charge controllers use a relay or power transistor, which has virtually zero voltage drop, to connect the module and battery. The relay opens at night to prevent reverse flow. Diodes have thus become dinosaurs in the PV industry. Larger, multi-module systems may still need blocking diodes if partial shading is possible. Call our tech staff for help.

conditions. The LCB accomplishes this by taking advantage of some PV module and DC motor operating characteristics. When a PV module is exposed to light, even very low levels, the voltage jumps way up immediately, though the amperage produced at low light is very low. A DC motor, on the other hand, wants just the opposite conditions to start. It wants lots of amps but doesn't care much if the voltage is low. The LCB provides what the motor wants by down-converting some of the high voltage into amperage. Once the motor starts, the LCB will automatically raise the voltage back up as much as power production conditions allow without stalling the motor. Meanwhile, on the PV module side of the LCB, the module is being allowed to operate at its maximum power point, which is usually a higher voltage point than the motor is operating at. It's a constant balancing act until the PV module gets up closer to full output, when, if everything has been sized correctly, the LCB will check out of the circuit, and the module will be connected directly to the motor.

LCBs cost something in efficiency, which is why we like to get them out of the circuit as the modules approach full power. All that fancy conversion comes at a price in power loss—usually about 10%–15%—but the gains in system performance more than make up for this. An LCB can boost pump output as much as 40% by allowing the pump to start earlier in the day and run later. Under partly cloudy conditions, an LCB can mean the difference between running or not. We usually recommend them strongly with most PV-direct pumping systems. Ranchers probably use PV-direct water pumping systems more than anyone else we know, and they swear by the LCBs, because with the added delivery more livestock can thrive from one water source than would be possible otherwise. With PV-direct fan systems, where start-up isn't such a bear, LCBs are less important.

boost pump output in low-light conditions. While not actually a controller, these booster devices are closer kin to controllers than anything else, so we'll cover them here. The common name for these booster devices is LCB, short for linear current booster.

An LCB is a solid-state device that helps motors start and keep running under low-light

Monitors

If you're going to own and operate a renewable energy system, you might as well get good at it. Your reward will be increased system reliability, longer component life, and lower operating costs. The system monitor is the dashboard that allows you to peer into the electrical workings of your system and keep track of what's going on.

The most basic, indispensable, minimal piece of monitoring gear is the voltmeter. A voltmeter measures the battery voltage, which in a lead-acid

battery system can be used as a rough indicator of system activity and battery state of charge. By monitoring battery voltage, you can avoid the battery-killing over- and undervoltages that come naturally with ignorance of system voltage. No battery, not even the special deep-discharge types we use in renewable energy systems, likes to be discharged below 80% of its capacity. This drastically reduces its life expectancy. Since a decent voltmeter is a tiny percentage of the battery

Tracking Mounts or Maximum Power Point Tracking?

Which delivers more power?

Until MPPT devices came on the scene, tracking mounts were the only way to increase the daily output from your PV array. By following the sun from east to west, trackers can increase output as much as 30%. Tracking works best in the summer, when the sun makes its highest arc across the sky. Tracking is the best choice for power needs that peak in the summer, such as water pumping and cooling equipment. In the winter, tracking gains are much less, usually about 10%–15%.

What if you're looking for more winter output? MPPT charge controls can increase your cold-weather output up to 30%. MPPT controls are a refinement of linear current boosting, which allows the PV module to run at its maximum power point, while converting the excess voltage into amps. MPPT controls perform best with cold modules and hungry batteries, a wintertime natural! Some MPPT controllers also allow wiring and transmitting at higher voltage. A 72-volt array can charge a 24-volt battery, for instance. This saves additional costs in wire and reduces transmission losses. When using an MPPT charge controller, we recommend running your array at the highest possible voltage. Your batteries will thank you.

Bottom line: For most residential power systems, an MPPT control will do more for your power well-being than a tracking mount.

cost, there is no excuse for not equipping your system with this minimal monitoring capability. With the growth and development happening in the electric vehicle sector driving battery innovation, mostly with lithium-ion batteries, we're excited about the benefits we'll see in the near future. But for now, hold on to that voltmeter, and keep your batteries happy by not letting their voltage get too low!

Beyond the voltmeter, the next most common monitoring device we use is the ammeter. Ammeters measure current flow (amps) in a circuit. They can tell us how much energy is flowing in or out of a system. Small ammeters are installed in the wire. Larger ammeters use a remote sensing device called a shunt. A shunt is a carefully calibrated resistance that will show a certain number of millivolts drop at a specific number of amps current flow. For instance, our 100-amp shunt is rated at 100 millivolts. For every millivolt difference the meter sees between one side of the shunt and the other, it knows that one amp is flowing. By noticing which side is higher, it knows if the flow is into or out of the battery, and will display a + or − sign. A single shunt gives us the net amperage in or out for the entire system, but won't tell us, for example, that 20 amps are coming down from the PV array, but 15 amps are going straight into the inverter, and only 5 amps are passing through the shunt.

The most popular type of monitoring device is the system-integrated genre. These use a series of shunts and ammeters to keep tabs on the current flows in and out of your system and report them to an LCD or LED display, or directly to a computer IP address, enabling you to check your system remotely from your computer or smart phone and monitor easily and accurately how much energy has been taken from the battery bank. Our most popular monitoring system is the Flex Net DC system. MidNite Solar also makes a fine monitoring system. (These products are available at realgoods.com.) Not every system or every person needs this kind of sophisticated monitor/controller, but for those with an in-depth interest or an aversion to technology, it's available at modest cost.

Install Your Monitor
Where It Will Get Noticed

The best, most feature-laden monitor won't do you a bit of good if nobody ever looks at it. We usually recommend installing your monitor in the kitchen or living room so it's easy for all family members to notice and learn from. Or if your monitoring is web-based, make it your homepage when you get online. Sometimes a dedicated tablet or other mobile device serves as the main monitoring interface. It's surprising how fast kids can learn given the proper incentive. A "No gaming or movies unless the batteries are charged" rule works wonders to teach the basics of battery management. Sometimes this even works on adults, too.

Elastic Voltmeters!

A voltmeter is a basic battery monitoring tool. It can give you a rough indication of how full or empty a lead-acid battery is, because the at-rest voltage will rise or fall slightly with state of charge. HOWEVER… the most important thing to know about a voltmeter is…this is a very e l a s t i c sort of indicator. Voltage stretches up or down when charging or discharging. When charging, a 12-volt battery can read over 14 volts. Almost the instant you stop charging, it will drop under 13 volts. Did you lose power when the voltage dropped? Not at all. It was just the voltage snapping back down to normal. Now put that battery under a heavy load like a toaster or microwave. The voltage will drop well under 12 volts, maybe even under

11 volts, depending on how big the battery is, how well charged it is, and how heavy the load. But it'll bounce right back up as soon as the microwave goes off. The lesson here is that you need to know what else is happening when you read your voltmeter. A reading of 11.8 volts isn't a problem while the microwave is running, but if nothing is on, and the batteries have been sitting awhile to stabilize, then 11.8 volts is a very nearly dead battery. To get a true reading of a battery's state of charge, no charging or discharging should have occurred for at least a few hours. That means early in the morning, when loads have been turned off for the night and before the PV modules have begun charging, is the best time to check your battery voltage.

Large Storage Batteries

Batteries provide energy storage, and are required for any remote, stand-alone, or backup renewable energy system. Batteries accumulate energy as it is generated by various renewable energy devices such as PV modules, wind, or hydro plants. This stored energy runs the household at night or during periods when energy output exceeds energy input. Batteries can be discharged rapidly to yield more power than the charging source can produce by itself, so pumps or motors can be run intermittently. For personal safety and good battery life expectancy, batteries need to be treated with some care, and get properly recycled at the end of their life. If common sense caution is not used, batteries can easily provide enough power for impromptu welding and even explosions.

Batteries 101

A wide variety of differing chemicals can be combined to make a functioning battery. Some combinations are very low-cost, but they have very low power potential; others, like the lithium-ion batteries used in laptops and electric vehicles for their light weight, can store astounding amounts of power but also cost a lot more. Lead-acid batteries offer the best balance of capacity per dollar, and are far and away the most common type of battery storage used in stand-alone power systems.

This battery type is also well known as the common automotive battery. Although storage batteries and starting batteries have very different internal construction and appropriate appli-

cations, they're both lead-acid types, and that's what we're going to cover in this section.

Simplified Lead-acid Battery Operation

What exactly is a lead battery and how does it work, you ask? The lead-acid battery cell consists of positive and negative lead plates made of slightly different alloys, suspended in a diluted sulfuric acid solution called an electrolyte. This is all contained in a chemically and electrically inert case. If we lower the voltage on the battery terminals by turning on a load such as a light bulb, the cell will discharge. Sulfur molecules from the electrolyte bond with the lead plates, releasing electrons, which flow out the negative battery terminal, through the light, and back into the positive terminal. This flow of electrons from one voltage potential to another is what we call electricity. If we raise the voltage on the battery terminals by applying a charging source, the cell recharges. Electrons push back into the battery, bond with the sulfur compounds, and force the sulfur molecules back into the electrolyte solution. Electrons will always flow from higher voltage to lower voltage as long as a path is available.

This back-and-forth energy flow isn't perfectly efficient. Lead-acid batteries average about 20% loss. For every 100 watt-hours you put into the battery, you can pull about 80 watt-hours back out. Efficiency is better with new batteries, and drops gradually as batteries age. But even with

> Lead-acid batteries offer the best balance of capacity per dollar, and are far and away the most common type of battery storage used in stand-alone power systems.

a fair loss in efficiency and the need for eventual replacement, this is the best energy storage medium for the price.

A single lead-acid cell produces approximately 2 volts, regardless of size. Each individual cell has its own cap. A battery is simply a collection of individual cells. A 12-volt automotive starting battery consists of six cells, each producing 2 volts, connected in series. Larger cells provide more storage capacity; we can run more electrons in and out, but the voltage output stays in the 2.0- to 2.5-volt potential of the chemical reaction that drives the cell. Cells are connected in series and parallel to achieve the needed voltage and storage capacity. You can have the same power in a low-voltage battery with many amp-hours as a higher-voltage battery with fewer amp-hours—it all boils down to your application.

Battery Capacity

Think of a battery as a bucket of energy. It will hold a specific amount, and no amount of shoving, compressing, or wishing is going to make it hold any more power. For more capacity, you need bigger buckets or more buckets.

A storage battery's capacity is rated in amp-hours: the number of amps it will deliver, times how many hours. How fast or slow we pull the amps out will affect how much energy we get. A slower discharge will yield more total amps. So battery capacity figures need to specify how many hours the test was run. The 20-hour rate is the usual standard for storage batteries, and is the standard we use for all the storage batteries in our publications. It's also the closest to a 24-hour day, and our energy usage routine. A 220-amp-hour golf cart battery, for example, will deliver 11 amps for 20 hours. This rating is designed only as a means to compare different batteries to the same standard and is not to be taken as a performance guarantee. Remember, length of charge/discharge time, and age/number of charge cycles, will affect the battery's performance. Batteries are electrochemical devices, and are also sensitive to temperature, charge/discharge cycle history, and age. The performance you get from your batteries will vary with location, climate, and usage patterns. But in the end, a battery rated at 200 amp-hours will provide you with twice the storage capability of one rated at 100 amp-hours.

What if the battery you're looking at only has a rating for cold cranking amps? That's a battery designed for engine-starting service, not storage. Stay away from it. Starting batteries suffer short, ugly lives when put into storage service. If you see cold cranking amps, often shown as CCA,

Why Batteries?

It's a reasonable question. Why don't renewable energy systems simply produce standard 120-volt AC power like the utility companies? It's simple. No technology exists to store AC power; it has to be produced as needed. For a utility company, with a power grid spread over half a state or more, considerable averaging of power consumption takes place. So it can (usually) deliver AC power as demanded. A single household or remote homestead doesn't have the advantage of power averaging. Batteries give us the ability to store energy when an excess is coming in, and dole it out when there's a deficit. Batteries are essential for remote or backup systems. Utility intertie systems make do without batteries, by using the utility grid like a battery to make up shortfalls or overages. But this requires the presence of grid power and doesn't provide any backup ability in case of a power outage caused by grid failure.

followed by a number, that's the way to tell it's not what you're looking for. They're much less expensive, and so a few folks out there are tempted to defy convention, but the results are always disappointing.

Batteries are less-than-perfect storage containers. For every 1.0 amp-hour you remove from a battery, it is necessary to pump about 1.2 amp-hours back in, to bring the battery back to the same state of charge. This figure varies with temperature, battery type, and age, but is a good rule of thumb for approximate battery efficiency.

Advantages and Disadvantages of Lead-acid Batteries

Lead-acid batteries are the most common battery type. Thanks to the automotive industry, they are well understood, and suppliers for purchasing, servicing, and recycling them are practically everywhere. Lead-acid batteries are 100% recyclable and one of the most recycled items in US society, averaging better than a 95% return rate in most states. Of all the energy-storage mediums available, lead-acid batteries offer the most bang for the buck by a very wide margin. As such, all the renewable energy equipment on the market is designed to work within the typical voltage range of lead-acid batteries. That's the good part.

Now the bad part: The active ingredients, lead and sulfuric acid, are toxins in the environment, and need to be handled with great respect and care. (See "Real Goods' Battery Care Class," pages 157–59 for more info.) Sulfuric acid can cause burns, and just loves to eat holes in your clothes, so you'll probably want a dedicated outfit you're not attached to for doing battery maintenance.

> Batteries are less-than-perfect storage containers. For every 1.0 amp-hour you remove from a battery, it is necessary to pump about 1.2 amp-hours back in, to bring the battery back to the same state of charge.

Lead is a danger if it enters the water cycle, is a strategic metal, and has salvage value, so it should be recycled. Lead is also incredibly heavy. More lead equals more storage capacity, but also more weight. Work carefully with batteries to avoid strains and accidents. Consider two batteries instead of one, simply to transport and put in place more easily. Heavier batteries have heavy-duty handles built into them for two people to carry.

Lead-acid batteries produce hydrogen gas during charging, which poses a fire or explosion risk if allowed to accumulate. The hydrogen must be vented to the outside. See our drawings of ideal battery enclosures for both indoor and outdoor installations at the end of this section. Lead-acid batteries will sustain considerable damage if they are allowed to freeze. This is harder to do than you may imagine. A fully charged battery can survive temperatures as low as −40°F without freezing, but as the battery is discharged, the liquid electrolyte becomes closer to plain water. Electrolyte also tends to stratify, with a lower concentration of sulfur molecules near the top. If a battery gets cold enough and/or discharged enough, it will freeze. At 50% charge level, a battery will freeze at approximately 15°F. This is the lowest level you should intentionally let your batteries reach. If freezing is a possibility, and the house is heated more or less full-time, the batteries should be kept indoors, with a proper enclosure and vent. For a cabin used occasionally, the batteries may be buried in the ground within an insulated box.

> Lead-acid batteries produce hydrogen gas during charging, which poses a fire or explosion risk if allowed to accumulate.

Unit of Electrical Measurement

Most electrical appliances are rated by the watts they use. One watt consumed over one hour equals one watt-hour of energy. Wattage is the product of current (amps) times voltage (Ohm's Law). This means that one amp delivered at 120 volts is the same amount of wattage as 10 amps delivered at 12 volts. Wattage is independent of voltage. A watt at 120 volts is the same amount of power as a watt at 12 volts. To convert a battery's amp-hour capacity to watt-hours, simply multiply the amp-hours times the voltage. The product is watt-hours. To figure how much battery capacity is required to run an appliance for a given time, multiply the appliance wattage times the number of hours it will run to yield the total watt-hours. Then divide by the battery voltage to get the amp-hours. For example, running a 100-watt light bulb for one hour uses 100 watt-hours. If a 12-volt battery is running the light, it will need 8.33 amp-hours (100 watt-hours divided by 12 volts equals 8.33 amp-hours). It's a pretty easy and straightforward way to determine the necessary battery bank size, but if you're feeling unsure, give us a call, we've done it thousands of times.

Batteries will slowly self-discharge. Usually the rate is less than 5% a month, but with dirty battery tops to help leak a bit of current, it can be 5% a week. Luckily this is easily avoidable: just remember to clean the terminals regularly. Batteries don't fare well sitting around in a discharged state. Given some time, the sulfur molecules on the surface of the discharged battery plates tend to crystallize into a form that resists recharging, and will even block off large areas of the plates from doing active service. This is called sulfation, and it kills batteries far before their time. Over the years we've seen "de-sulfators" marketed to reverse the effects of sulfation, but we've yet to see one that really can get the batteries to bounce back. The cure is regular charge and discharge exercise, or lacking that, trickle charging that will keep the battery fully charged. Special pulse chargers that deliver high-voltage spikes of charging power have proven very effective at reducing and preventing sulfation in batteries that don't get much exercise.

Lead-acid batteries age in service. Once a bank of batteries has been in service for six months to a year, it generally is not a good idea to add more batteries to the bank. A battery bank performs like a chain, pulling only as well as the weakest link. New batteries will perform no better than the oldest cell in the bank. All lead-acid batteries in a bank should be of the same capacity, age, and manufacturer as much as possible. Sometimes a "bad battery" is due to a single bad cell, but unfortunately it's not easy and sometimes impossible to replace a battery cell, so the entire battery bank would need replacement.

Why Can't I Use Car Batteries?

We get this question A LOT. Batteries are built and rated for the type of "cycle" service they are likely to encounter. Cycles can be "shallow," reaching 10%–15% of the battery's total capacity, or "deep," reaching 50%–80% of total capacity. No battery can withstand 100% cycling without damage, often severe.

Automotive starting batteries and deep-cycle storage batteries are both lead-acid types, but they have important construction and materials differences. Starting batteries are designed for many, many shallow cycles of 15% discharge or less. They deliver several hundred amperes for a few seconds, and then the alternator takes over and the battery is quickly recharged. Deep discharges cause the lead plates to shed flakes and sag, reducing life expectancy. Deep-cycle batteries are designed to deliver a few amperes for hundreds of hours between recharges. Their lead plates are thicker, and use different alloys that don't soften with deep discharges. Neither battery type is well suited to doing the other's job, and will deteriorate quickly and not last very long if forced to do the wrong service.

Automotive engine starting batteries are rated for how many amps they can deliver at a low temperature, hence the term cold cranking amps (CCA). This rating is not relevant for storage batteries. Beware of any battery that claims to be a deep-cycle storage battery and has a CCA rating.

Sealed or Wet?

Lead-acid batteries come in two basic flavors, traditional "wet cells" or "sealed cells." Wet cells have caps that can be removed. In fact, you have to remove them every month or two to add distilled water. Wet-cell batteries cost less, their problems are more easily diagnosed, and they're usually the right choice for remote homesteads or frequently used systems where upkeep and maintenance is happening regularly. On the other hand, they do require routine maintenance, they produce hydrogen gas during charging that needs to be vented outside, and they have to ship as hazardous freight, which makes them expensive to move. Much like the muscles of your body, wet-cell batteries need regular exercise to maintain good performance. Without regular cycling, they become "stiff" chemically, and won't perform well until they're limbered up. This makes them less than ideal for emergency and backup power systems. Wet-cell batteries should not be used for any battery backup situation.

This is where sealed batteries come in. Sealed cells do better at sitting around for long periods waiting for activity. Plus, nobody has to remember to water them periodically, and they can be installed in places where hydrogen gassing couldn't be tolerated. Sealed batteries come in two basic styles. Gel batteries have the liquid electrolyte in a jellied form. AGM (absorbed glass mat) batteries use a fiberglass sponge-like mat to hold liquid electrolyte between the plates. There is intense debate and rivalry as to which technology is better. AGM is less expensive and fussy to build, and the gels tend to be a bit more robust and longer lived. We offer small sealed batteries as AGM types, and larger sealed batteries as gel types (see realgoods.com).

Sealed batteries require special charge controllers. To prevent gassing and the irreplaceable loss of water, charging voltage on sealed cells needs to be held to a maximum of 14.1 volts (or 2.35 volts per cell). Higher charging voltages, as commonly used for wet cells, will seriously reduce the life expectancy of sealed batteries. Almost all the charge controllers we offer are either voltage adjustable, or have a selection switch for sealed or wet-cell batteries. Be sure to consult both the charge controller and battery manuals to make sure they're in sync with each other.

How Big a Battery Bank Do I Need?

We usually size remote household battery banks to provide stored power for three to five days of autonomy during cloudy weather. For most folks, this is a comfortable compromise between cost and function. If your battery bank is sized to provide a typical three to five days of backup power, then it will also be large enough to handle any surge loads that the inverter is called upon to start. A battery bank smaller than three days' capacity will get cycled fairly deeply on a regular basis. This isn't good for battery life and they won't last as long. A larger battery bank cycled less deeply will cost less in the long run. Banks larger than five days' worth start getting more expensive than a backup power source (such as a modest-sized generator). This highlights the importance of two topics we've already covered: taking some time and care with the load analysis, so you don't over- or undersize your batter bank; and monitoring, so you can keep a close eye on the batteries' state of affairs.

As a general rule of thumb, you'll be better off building your battery bank with the fewest number of cells. Larger cells with more ampacity means fewer interconnects, fewer cells to water, less maintenance time, and a smaller chance that one cell will fail early, causing the entire bank to be replaced before its time. Bigger batteries tend to be higher quality, and also have longer lives.

Occasionally we run into situations with ¾-horsepower or larger submersible well pumps or stationary power tools requiring a larger battery bank simply to meet the surge load when starting. Call the Real Goods technical staff for help if you are anticipating large loads of this type. Our System Sizing Worksheet at the beginning of Chapter 4 or the Appendix has a quick battery sizing section to help you out.

New Technologies?

Compared to the electronic marvels in the typical renewable energy package, the battery is a very simple, proven technology. Tremendous amounts of research have been directed recently into energy-storage technology. The auto industry, specifically electric vehicle manufacturers, are the pioneers right now in battery innovation and development. Years ago, a lightweight battery

If your battery bank is sized to provide a typical three to five days of backup power, then it will also be large enough to handle any surge loads that the inverter is called upon to start.

How Long Will Batteries Last?

Life expectancy depends mostly on what battery type you purchase initially, and then how much love you give your batteries. Any fool can destroy even the best battery pack within six months, while a few saints manage to make their batteries last twice the average expectancy. We suggest starting out with a "training wheel" battery pack, since practically everyone ruins their first battery bank for one reason or another, learns from their mistakes, and then gets a nicer bank that lasts at least as long as it's rated. We've learned from experience approximately what to expect on average from different batteries.

From low to high:
- RV Marine types: 1.5–2.5 years
- Golf Cart types: 3–5 years
- L-16 types: 6–8 years
- Solar Gel (sealed) types: 5–10 years
- Industrial-quality traction cells (IBE and Hawker [formerly Yuasa] brands): 15–25 years

with high energy density didn't exist, but we're starting to see them now, and although they're more expensive, we're starting to see costs come

down. Several dozen new battery technologies have been under intense development in the laboratory. Several of these earlier technologies have begun to mature, after bearing fruit for cell phones, laptops, mobile devices, and most recently hybrid and electric vehicles. Unfortunately, they are still too expensive for the amount of energy storage required in a renewable energy system. The nickel-metal hydride battery bank used in some EVs costs around $30,000, and would only make a very modest-sized remote home battery bank.

Lead-acid batteries are also undergoing research. The possibilities of lead-acid technology are far from tapped out. This old dog is still capable of learning some new tricks. For now we must coexist with traditional battery technology—a technology that is nearly one hundred years old, but is tried and true and requires surprisingly little maintenance. The care, feeding, cautions, and dangers of lead-acid batteries are all well understood. Safe manufacturing, distribution, and recycling systems for this technology are in place and work well. Could we say the same for a lithium-ion or sulfur-bromine battery?

Real Goods' Battery Troubleshooting Guide

What Could Go Wrong?

Famous last words.... Judging by the phone calls we get, plenty can and often does go wrong. Folks have more trouble with batteries than with any other single component of their systems. Here's a troubleshooting guide.

Got Water?

Check the fluid level in all the cells. It should be above the plates, but not above the split ring about ½ inch below the top of the cell. It's bad for the plates to be exposed to air. Add distilled water as needed. In an emergency, clean tap water is

Triple the Life of Your Batteries

With regular cleanings, water top-off, and equalizations, you can triple the life of your batteries. Here's how:

Lesson 1: Equalize every time the batteries drop below 50% state of charge, sometimes two to four times a month.

Lesson 2: Every chance you get, fully recharge your batteries. "Mostly" charging most of the time doesn't cut it.

Lesson 3: Check your batteries often. Look for crud on top—it can build up and actually short between the posts, causing a constant discharge. And be sure to give your batteries

a good cleaning every two to three months. Get on your grubby clothes, eye protection, and gloves, and ensure you have plenty of baking soda next to your batteries. Then mix up a mild soda and water solution and start cleaning. (Use an old paintbrush and rinse all connections with clean water while checking for tightness and corrosion.) Treat any posts and connections as required. Equalize the battery bank, top off the water, and check each cell's voltage and log it.

Lesson 4: Don't overcharge a gel cell battery. There's no way to replace the expelled electrolyte, and you'll shorten the battery's lifespan.

better than no water, but not by much. Distilled water is really the right stuff. Always have some around just in case.

Auto parts stores sell special battery-filler bottles with a nifty spring-loaded automatic shut-off nozzle. Just stick it in the cell, push down, and wait until it stops gurgling. The fluid level will be perfect, and there will be no drips or mess. It's a tool worth having. Or consider automatic battery-watering systems, which make refills on large industrial batteries a snap. (Also, see the PRO-FILL 6-Volt Battery Watering System on our website: realgoods.com.)

Connections Tight and Clean?

Check battery connections by wiggling all the cables, snug up the hardware, and make sure no corrosion is growing.

Got corrosion? This is like cancer—you've got to clean it all out and quickly, or it will return. Disassemble the connection, wire brush and scrape off all the crud you can, then attack what's left with a baking soda and water solution until all traces of corrosion are gone. Clean the bare metal with sandpaper or a wire brush, assemble, and then coat all exposed metal with grease (such as NO-OX-ID, available at realgoods.com) or petroleum jelly to prevent future corrosion.

Is It Charged?

Every system needs some way to monitor the battery state of charge. A voltmeter is the absolute minimum. Voltage can be tricky, however. This is a very elastic sort of indicator. Voltage will stretch upward when a battery is being charged, and it will contract downward when a battery is being discharged. Only when the battery has been sitting for a couple hours with no activity can you get a really accurate sense of the battery's state of charge with a voltage reading. First thing in the morning is the best time.

A fully charged 12-volt battery will read 12.6 volts, or a bit higher. (Those with 24- or 48-volt packs multiply accordingly.) Lower voltage readings mean a lower state of charge, or that something is turned on. At 12.0 volts, you're in the caution zone, and at 11.6 volts any further electrical use is doing damage that will reduce the battery life expectancy. When charging, the voltage needs to get up to 14.0 volts or higher periodically. If your battery doesn't climb above 13.5 volts, it isn't getting close to being fully charged. You don't need to achieve a full charge every day—the more the better—but at least once a week is good. All the voltage numbers in the above paragraph are for 12-volt batteries. For 24-volt battery banks, simply double those numbers, and for 48-volt battery banks, quadruple the numbers.

Is It Healthy?

The very best tool for checking your battery's state of health is a hydrometer. This looks like an oversized turkey baster with a graduated float inside it. You must wait at least 48 hours after adding water before you run a hydrometer test. A sample of electrolyte is drawn up from a cell until the float rises. Note at what level it's floating. It measures the specific gravity of the electrolyte to three decimal places. A typical reading would be 1.220. The decimal point is universally ignored, so our example reads as "twelve twenty." Under normal conditions, all the cells in a battery pack should read within 10 points of each other. That would be 1.215–1.225 in our example. As a general rule, it doesn't matter how high or low the readings are, so long as they're all about the same. We're looking only for differences between cells.

> Voltage can be tricky, however. This is a very elastic sort of indicator. Voltage will stretch upward when a battery is being charged, and it will contract downward when a battery is being discharged.

Volt Readings for your Deep-Cycle Battery

12 DC Volts 13

Get the longest life from deep-cycle batteries by keeping the state of charge at 60% or more for lead-acid and 70% for gel cells. For accurate volt readings, don't charge or discharge batteries for 2 hours.

Voltage Reading	State-of-Charge
12.75	100
12.70	95
12.65	90
12.60	85
12.55	80
12.50	75
12.45	70
12.40	65
12.30	55
12.25	50
12.25	45

When a cell starts to go bad, it will read much higher or lower—usually lower—than its neighbor cells. At 20–30 points difference, you may only need a good equalizing charge to bring all the cells into line. At 40–50 points difference, you probably have a failing cell. Plan for replacement within two or three months. At 60 points or more difference, you have a failed cell that's sucking the life out of all the surrounding cells, and needs to be replaced as soon as possible.

Common Battery Questions and Answers

What voltage should my system run? I notice that components for 12-, 24-, and 48-volt battery banks are available.

Voltage is selected for ease of transmission on the collection and storage side of the system. Low voltage doesn't transmit well, so as home power systems get bigger and are producing greater amounts of energy, it's easier to do at higher voltages. Generally, systems under 2,000 watt-hours per day are fine at 12 volts, systems up to 7,000 watt-hours per day are good at 24 volts, and larger systems should be running at 48 volts. These aren't ironclad guidelines by any means. If you need to run hundreds of feet from a wind or hydro turbine, then higher voltage will help keep wire costs within reason. Or if you're heavily invested in 12-volt with a large inverter and 12-volt lights, you can still grow your system without upgrading to a higher voltage. 12 V and 24 V systems are nice to have for smaller homes, weekend cabins, and RVs and boats, as there are plenty of marine and RV appliances made for these voltages, so you can run more directly off the battery and use a smaller inverter.

> Generally, systems under 2,000 watt-hours per day are fine at 12 volts, systems up to 7,000 watt-hours per day are good at 24 volts, and larger systems should be running at 48 volts.

I'm just getting started on my power system. Should I go big on the battery bank assuming I'll grow into it?

The answer is yes and no. Yes, you should start with a somewhat larger battery bank than you absolutely need, perhaps sizing for four to five days of autonomy instead of the bare minimum three days. Over time, most folks find more and more things to use power for once it's available. But if this is your first venture into remote power systems and battery banks, then we usually recommend that you start slowly. You're bound to make some mistakes and do some regrettable and embarrassing things with your first set of batteries. You might as well make mistakes with inexpensive ones. So no, don't invest too heavily in batteries your first couple of years. The golf cart-type deep-cycle batteries make excellent trainers. They are modestly priced, will accept moderate abuse without harm, and are commonly available. In three to five years, when the golf cart-type trainers wear out, you'll be much more knowledgeable about what your energy needs are and what quality you're willing to pay for.

The battery bank I started with two years ago just doesn't have enough capacity for us anymore. Is it okay to add some more batteries to the bank?

Lead-acid batteries age in service. The new batteries will be dragged down to the performance level and life expectancy of the old batteries. Your new batteries will be giving up two years of life expectancy right off the bat. Different battery types have different life expectancies, so we really need to consider how long the bank should last. For instance, it would be acceptable to add more cells to a large set of forklift batteries at two years of age because this set is only at 10% of life expectancy. But an RV/marine battery at two years of age is at 100% of life expectancy.

I keep hearing rumors about some great new battery technology like "flywheels" that will make lead-acid batteries obsolete in the near future. Is there any truth to this, and should I wait to invest in batteries?

The truth is that several dozen battery technologies are currently under intense development and have been for some time. Some of them, like nickel-metal hydride, look very promising, but none of them are going to give lead-acid a run for your money within the foreseeable future. Lead-acid is also in the laboratory. Lead-acid technology is going to be around, and is going to continue to give the best performance per dollar for a long time to come. We're not yet seeing any of the new technologies impact the renewable energy market.

Real Goods' Battery Care Class

Basic Battery Safety

1. Protect your eyes with goggles and hands with rubber gloves. Battery acid is a slightly diluted sulfuric acid. It will burn your skin after a few minutes of exposure, and your eyes almost immediately. Keep a box or two of baking soda and at least a quart of clean water in the battery area. Flush any battery acid contact with plenty of water. If you get acid in your eyes, flush with clear water for 15 minutes and then seek medical attention.

2. Tape the handles of your battery tools or treat them with Plastic-Dip so they can't possibly short out between battery terminals. Even small batteries are capable of awesome energy discharges when short-circuited. The larger batteries we commonly use in renewable energy systems can easily turn a 10-inch crescent wrench red hot in seconds while melting the battery terminal into a useless puddle, and for the grand finale possibly explode and start a fire at the same time. This is more excitement than most of us need in our lives.

3. Wear old clothes! No matter how careful you are around batteries, you'll probably still end up with holes in your jeans. Wear something you can afford to lose, or at least have holes in—at least you're making a renewable energy fashion statement!

4. Now stop thinking, "Oh, none of that will happen to me!" Even the most experienced professionals can make mistakes if they aren't careful. Melted terminals, fires, explosions, and personal injury are not worth the risk of cutting corners. Don't take chances: Safety measures are easy, and the potential harm is permanent. We've heard horror stories from many customers with years of experience, so if you're ever getting too comfortable around batteries, it might be time to review our safety rules.

Simple Lead-acid Battery Care

1. Take frequent voltage readings. The voltmeter is the simplest way to monitor battery activity and approximate state of charge.
2. Baking soda neutralizes battery acid. Keep at least a couple of boxes on hand, inside the battery enclosure, in case of spills or accidents.
3. Batteries should be enclosed and covered to prevent casual access to terminals. Wet-cell batteries must be vented to prevent accumulation of hydrogen and other harmful gasses. A 2" PVC vent at the highest point in the battery box is sufficient.
4. Do not locate any electrical equipment inside a battery compartment. It will corrode and fail.
5. Check the water level of your batteries once a month until you know your typical usage pattern. Use distilled water only. The trace minerals in tap water kill battery capacity. Get a battery-filler bottle from the auto parts store to make this job easy.
6. Be extra careful with any metal tools around batteries. Tape handles or treat with Plastic-Dip so tools can't possibly short between terminals if dropped.
7. Protect battery terminals and any exposed metal from corrosion. After assembly, coat them with grease or Vaseline, or use one of the professional sprays to cover all exposed metal.
8. Never smoke or carry an exposed flame around batteries, particularly when charging.
9. Wear old clothes when working with batteries. Electrolyte loves to eat holes.
10. Chemical processes slow down at colder temperatures. Your battery will act as if it's smaller. At 0°F, batteries lose about 50% of their capacity. In cold climates, install your batteries in the warmest practical location.
11. All batteries self-discharge slowly, and will sulfate if left unattended for long periods with no trickle charging. Clean, dry battery tops reduce self-discharging.
12. Batteries tend to gain a memory for typical use and will resist wider discharge cycles initially. Like the muscles of your body, periodic stretching exercises increase strength and flexibility. It's good to do an occasional deep discharge and equalizing charge to maintain full battery capacity.
13. A few large cells are better than many small cells when building a battery pack. Larger cells tend to have thicker plates and longer life expectancies. Fewer interconnections and cells mean less maintenance time, and less chance of one cell failing early and requiring the entire battery pack to be replaced early.

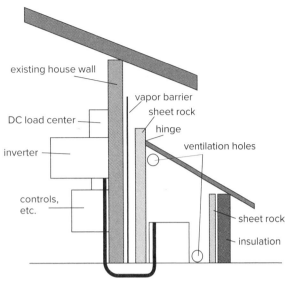

existing house wall

vapor barrier

sheet rock

DC load center

hinge

inverter

ventilation holes

controls, etc.

sheet rock

insulation

Wires and cables pass through wall at bottom of battery box to prevent hydrogen from entering house (hydrogen rises.)

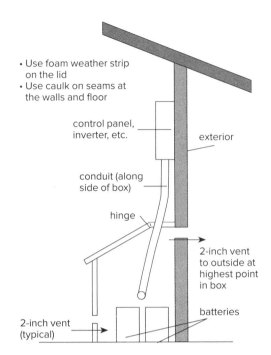

• Use foam weather strip on the lid
• Use caulk on seams at the walls and floor

control panel, inverter, etc.

exterior

conduit (along side of box)

hinge

2-inch vent to outside at highest point in box

batteries

2-inch vent (typical)

Left: Ideal exterior battery enclosure.
Right: Ideal interior battery enclosure.

Regular Maintenance for Wet-Cell Batteries

1. Check the water level after charging and fill with distilled water. Batteries use more water when they're being fully charged every day, and they'll also use more water as they get older. Check the water level every month. Fluid level should be above the plates, but not above the split ring about ½" below the top of the cell. It's bad for the plates to be exposed to air. Add distilled water as needed. In an emergency, clean tap water is better than no water, but not by much. Distilled water is really the right stuff. (Note: we hate supporting WalMart, but they seem to have distilled water much cheaper than anywhere else…)

Don't fill before charging; the little gas bubbles will cause the level to rise and spill electrolyte. Only pure water goes back into the battery. The acid doesn't leave the cell, so you never need to replenish it.

Auto parts stores sell special battery-filler bottles with a nifty spring-loaded automatic shut-off nozzle. Just stick it in the cell, push down, and wait until it stops gurgling. The fluid level will be perfect and there will be no drips or mess. It's a tool worth having. Consider an automatic battery-watering device if you have a large industrial battery bank.

2. Clean the battery tops. The condensed fumes and dust on the tops of the batteries start to make a pretty fair conductor after a few months. Batteries will discharge significantly across the dirt between the terminals. That power is lost to you

forever! Sponge off the tops with a baking soda and water solution, or use the battery cleaner sprays you can get at the auto parts store. Follow the baking soda cleaner with a clear water rinse. Make sure that cell caps are tight and that none of the cleaning solution gets into the battery cell! This stuff is deadly poison to the battery chemistry. Clean your battery tops once or twice a year.

3. Clean and/or tighten the battery terminals. Lead is a soft metal and will gradually "creep" away from the bolts. If you were smart and coated all the exposed metal parts around your battery terminals with grease or Vaseline when you installed them, they'll still be corrosion-free. If not, then take them apart, scrub or brush as much of the corrosion off as possible, then dip or brush with a baking soda and water solution until all fizzing stops and then scrape some more. Keep this up until you get all the blue/green crud off. Then carefully cover all the exposed metal parts with grease when you put it back together. If your terminals are already clean, then just gently snug up the bolts.

4. Run a hydrometer test on all the cells. Wait at least 48 hours after adding water before you run this test. Your voltmeter is great for gauging battery state of charge; the hydrometer is for gauging state of health. Use the good kind of hydrometer with a graduated float (not the cheapo floating-ball type). We sell one for $9 (realgoods .com). A sample of electrolyte is drawn up from a cell until the float rises. Note at what level it's floating. It measures the specific gravity of the electrolyte to three decimal places. A typical reading would be 1.220. The decimal point is universally ignored, so our example reads as "twelve twenty." Under normal conditions, all the cells in a battery pack will read within 10 points of each other. That would be 1.215–1.225 in our example. What we're looking for in this test is not the state of charge, but the difference in points between cells. In a healthy battery bank, all cells will read within 10 points of each other. Any cell that reads 20–25 points lower is probably starting to fail, although you should run an equalizing charge before casting judgment. You may have three to six months to round up a replacement set. At 60 points difference or more, the bad cell is sucking the life out of your battery bank and needs to get out now! Don't pay any attention to the color-coded good-fair-recharge markings on the float. These only pertain to automotive batteries, which use a slightly hotter acid.

5. Run an equalizing charge (wet-cell batteries only). An equalizing charge is a controlled slight overcharge. Your charge controller will either have an equalize button, or if it's a nicer model, a programmable equalizing regime that will occur automatically. This will even out any small differences among cells that have developed over time, and help push any sulfation back into the electrolyte. It's like a minor tune-up for your batteries. It's a good idea to do an equalization every one to six months. Equalizing is more important in the winter when batteries tend to run at lower charge levels than in the summer. *Do NOT equalize sealed batteries! Permanent damage will occur!*

Run your batteries up to about 15.0 volts in a 12-volt system (or 2.50 volts per cell) and hold them at between 15.0 and 15.5 volts (2.5–2.6 volts per cell) for two to four hours. You'll get fairly vigorous gassing and bubbling on all cells. Do not take the caps off or loosen them during the equalizing charge, you'll just lose more water and spatter acid over the top of the battery. Do NOT top up the water just before equalizing. All that gas production can raise the electrolyte level above the cell top and get messy. Top up the water a day or two later.

Battery state-of-charge			
Voltage Reading			Percent of Full Charge
12 V	24 V	48 V	
12.6	25.2	50.4	100%
12.4–12.6	24.8–25.2	49.6–50.4	75%
12.2–12.4	24.4–24.8	48.8–49.6	50–75%
12.0–12.2	24.0–24.4	48.0–48.8	25–50%
11.7–12.0	23.4–24.0	46.8–48.0	0–25%

Annual Maintenance for Sealed Batteries

1. Clean the battery tops. The crud and dust on the tops of the batteries start to make a pretty fair conductor eventually, even on sealed batteries. Batteries can discharge significantly across the dirt between the terminals. That power is lost to you forever! Sponge off the tops with a baking soda/water solution, or use the battery cleaner sprays you can get at the auto parts store. Follow the baking soda cleaner with a clear water rinse. Cleaning the tops of your sealed batteries once a year is sufficient.

2. Clean and/or tighten the battery terminals. Lead is a soft metal and will gradually "creep" away from the bolts. If your terminals are already clean, then just gently snug up the bolts. Terminal corrosion is rare, but not impossible on sealed batteries. If needed, take them apart, scrub or brush as much of the corrosion off as possible, then dip or brush with a baking soda and water solution until all fizzing stops and then scrape some more. Keep this up until you get all the blue/green crud off. (It's like cancer; if you don't get it all, it will return.) Then carefully cover all the exposed metal parts with grease when you put it back together.

For both sealed and gel batteries, keep a maintenance log, as you would with a car. You don't keep a a maintenance log for your car? At least do it with your batteries! And remember, if you keep that maintenance log in the battery room, it will decay because of the hydrogen gas emissions!

Rechargeable Batteries and Chargers

If you've waded this deeply into the *Sourcebook*, we probably don't need to bore you with why you ought to be using rechargeable batteries. They're far less expensive, they don't add toxins to the landfill, and they don't perpetuate a throwaway society. You know all that already, right? Well, just in case you want a review, we'll provide a brief one. Are you already feeling secure in your

Americans toss over 3 billion small consumer batteries into the landfill every year. The rest of the world adds another few billion to the total.

knowledge and just want to choose the best rechargeable battery or charger for your purposes? Then skip ahead to "Choosing the Right Battery" on page 161.

Why We Need to Use Rechargeable Batteries

Americans currently toss over three billion small consumer batteries into the landfill every year. The rest of the world adds another few billion to the total. The vast majority of these are alkaline batteries. Alkaline batteries are the common Duracell, Energizer, and Eveready brands you find in the grocery store checkout lane. While domestic battery manufacturers have refined their formulas in the past few years to eliminate small amounts of mercury, alkaline batteries are still low-level toxic waste. Casual disposal in the landfill may be acceptable to your local health officials, but what a waste of good resources! Our grandchildren, maybe even our children, will be part of the next gold rush, when we start mining our 20th-century landfills for all the refined metals and other depleted resources that are waiting there.

So throwaway batteries are wasteful and a legitimate health hazard. They're also surprisingly expensive. Alkaline cells cost around $0.90–$2.50 per battery, depending on size and brand. Use it once, and then throw it away—so 20th century! In comparison, rechargeable nickel-metal hydride

cells—including the initial cost of the battery, a charger, and one or two cents worth of electricity—cost $0.04 to $0.10 per cycle, assuming a very conservative lifetime of 400 cycles. Chargers outlive batteries, so these rechargeable cost figures err on the high side. Typically, throwaway batteries cost $0.10/hour to operate, while rechargeable batteries cost only 0.1¢/(1/10 of one cent) per hour. Not to mention the pleasure of never having to guess if a battery is good or not, especially if it came from the junk drawer. If you're unsure, just throw it on the charger.

When *Not* to Use Rechargeable Batteries

Rechargeable batteries cost far less, they reduce the waste stream, and they usually present fewer disposal health hazards. What's not to love about them? It isn't a perfect world; rechargeable technology does have deficiencies. Specifically, lower capacity, lower operating voltage, self-discharge, and, in the case of NiCad cells only, toxic waste. Let's look at each issue individually.

Lower capacity. This disadvantage is rapidly disappearing. The latest generation of NiMH (nickel-metal hydride) AA-size batteries actually has *more* storage capacity than the premium name-brand alkalines! This trend also extends to the larger C- and D-size cells, so capacity simply isn't as big an issue as it once was. NiMH batteries

will support high-drain appliances such as digital cameras or remote-control toys longer than any other battery chemistry, including throwaway batteries. And they will do it hundreds of times before hitting the trash.

Lower operating voltage. NiCad and NiMH batteries operate at 1.25 volts per cell. The alkaline cells that many battery-powered appliances are expecting to use operate at 1.50 volts per cell. This quarter of a volt is sometimes a problem for voltage-sensitive appliances, or devices such as large boom boxes that use many cells in series. The cumulative voltage difference over eight cells in series can add up to a problem if the device isn't designed for rechargeable battery use.

Self-discharge. All battery chemistries discharge slowly when left sitting. Some chemistries, like the alkaline and lithium cells used for throwaway batteries, have shelf lives of five to ten years. Rechargeable chemistry will self-discharge much faster, with a shelf life of only two to four months. These batteries aren't the ones you want in your glove box emergency flashlight, and probably aren't the best possible choice for battery-powered clocks.

Toxic NiCad cells. NiCad cells contain the element cadmium. Human beings do not fare well when ingesting even tiny amounts of cadmium. Real Goods stopped selling NiCad batteries several years ago as soon as a better alternative was available. These cells absolutely must be properly recycled. Do not dispose of NiCad cells casually! They are toxic waste! Now, that said, let us point out that the newer, higher-capacity nickel-metal hydride batteries are not toxic at all, and will recharge just fine in any NiCad battery charger you might have already.

Choosing the Right Battery

There are three basic rechargeable battery chemistries to choose from: rechargeable alkaline, nickel-metal hydride (NiMH), and nickel cadmium (NiCad). Each has strengths and weaknesses, although NiCads compare so poorly now that we aren't even going to dignify them by charting. We've summarized everything and made the comparisons with standard throwaway alkaline batteries in the accompanying chart.

Regardless of your chemistry choice, buy several sets of batteries so you can always have one set in the charger while another set is working for you.

> Rechargeable batteries aren't the ones you want in your glove box emergency flashlight, and probably aren't the best possible choice for battery-powered clocks.

Rechargeable Battery Comparisons			
	Standard Alkaline	**Rechargeable Alkaline**	**NiMH**
Volts	1.5	1.5	1.25
Capacity (compared to alkaline)	100%	90% initially, diminishes each cycle	70%–110%, no loss with cycles
Capacity in mAh			
AAA	750 mAh	750 mAh	750 mAh
AA	2,000 mAh	1,800 mAh	2,200 mAh
C	5,000 mAh	3,000 mAh	3,500 mAh
D	10,000 mAh	7,200 mAh	7,000 mAh
Avg. Recharge Cycles	1	12–25	400–600
Self-Discharge Rate	Negligible; 5-year shelf life	Negligible, 5-year shelf life	Modest, 15% loss @ 60 days
Strengths	1.5-volt standard, long shelf life, high capacity	1.5-volt standard, long shelf life, good capacity	Sustains high draw, no voltage fade, many recharges, nontoxic
Weaknesses	Highest cost, throwaway	Loses capacity each cycle, special charger required	1.2-volt, self-discharges
Best Uses	Clocks, emergency stuff	Clocks, emergency stuff, toys, games, radios, flashlights	Digital cameras, remote-control toys, flashlights, high-drain appliance

Choosing the Right Battery Charger(s)

Now that you've seen the choices in rechargeable batteries, you'll also need to decide which charger(s) to use. The three types are AC charger (plugs into a wall socket), solar charger (charges directly from the sun), and 12-volt charger (plugs into your car's cigarette lighter or your home's renewable energy system).

Solar Chargers

The great benefit of solar battery chargers is that you don't need an outlet to plug them into, you just need the sun! They can be used anywhere the sun shines. It used to be that solar chargers worked much more slowly than plug-in types, and it would take several days to charge up a set of batteries. With the advent of companies like Joos and Goal Zero in recent years, this is no longer the case (see their products on our website, real goods.com). We recommend you buy an extra set of batteries with your solar charger, so you can charge one set while using the other. Given time and patience (prerequisites for a prudent, sustainable lifestyle in any case), solar chargers are the best way to go.

DC Chargers

The 12-volt charger will recharge your batteries from a car, boat, or from your home's 12-volt renewable energy system (if you're so blessed). It provides a variable charge rate depending on battery size. In contrast, most AC chargers put the same amount of power into every battery, regardless of size. Even though the 12-volt charger is a fast charger, it will not drain your car battery unless you forget and leave it charging for several days. Some 12-volt chargers can overcharge batteries, however, so pay close attention to how long it takes to charge, and don't leave them in there much longer. Leaving them in the 12-volt charger for one-and-a-half times the recommended duration will not harm the batteries.

AC Chargers

The AC (plug in the wall) chargers are by far the most convenient for the conventional utility-powered house. Most of us are surrounded by AC outlets all day long, and all we have to do is plug in the charger and let it go to work. All these chargers turn off when charging is complete, and go into a trickle charge maintenance mode. Your batteries will always be fully topped up and waiting for you. Most AC chargers automatically figure out what chemistry and size each battery is, then charges it accordingly.

Frequently Asked Questions About Rechargeable Batteries

Is it true that rechargeable batteries develop a "memory"?

This was a NiCad problem many years ago. If you use only a little of the battery's energy every time, and then charge it up, the battery will develop a "memory" after several dozen of these short cycles. It won't be able to store as much energy as when new. This was true only with NiCads, not with NiMHs or alkalines. Improved chemistry has practically eliminated any memory problems. Short cycles have never been a problem for NiMH or alkaline batteries. We've been selling NiMh batteries for quite a while now and have never heard of any "memory" issues, at least that we can remember.

How do I tell when a NiMH (or NiCad) battery is charged?

Because these batteries always show the same voltage from 10%–90% of capacity, it's tricky to know when they're charged. A voltmeter tells little or nothing about the battery's state of charge. Either use one of the smart chargers like the AccuManager (available at realgoods.com), or time the charge cycle. (Calculate the recommended charging time by dividing the battery capacity by the charger capacity and add 25% for charging inefficiency. For instance: A battery of 1,100 mAh capacity in a charger that puts out 100 milliamps will need 11 hours, plus another 3 hours for charging inefficiency, for a total of 14 hours to recharge completely.) A cruder method to use is the touch method. When the battery gets warm, it's finished charging. (This method is only applicable with the plug-in chargers.) When in doubt, refer to the battery and/or charger manufacturer for instructions.

My new NiMH (or NiCad) batteries don't seem to take a charge in the solar-powered charger.
All batteries are chemically "stiff" when new. NiMH batteries are shipped uncharged from the factory, and are difficult to charge initially. A small solar charger may not have enough power to overcome the battery's internal resistance; hence, no charge. The solution is to use a plug-in charger for the first few cycles to break in the battery. Another solution is to install only one battery at a time in the solar charger. Once the batteries have been broken in, the problem disappears. These batteries will give you many, many years of use for no additional cost with solar charging. We've often received phone calls reporting D.O.A. batteries, but after a couple rounds in an AC charger, they came back from the dead and worked like a charm.

My light saber (or other device) doesn't work when I put NiMHs in. What's wrong?
Some battery-powered equipment will not function on the lower-voltage NiMH batteries, which produce 1.25 volts (versus 1.5 volts produced by fresh alkalines). Some manufacturers use a low-voltage cutoff that won't allow using rechargeables. If the cutoff won't let you use NiMHs, it's also forcing you to buy new alkalines when they're only 50% depleted! Sometimes with devices that require a large number of batteries (six or more), the .25-volt difference between rechargeables and alkalines gets magnified to a serious problem. Even a fully charged set of NiMHs will be seen as a depleted battery pack by the appliance. You need a new appliance that's designed for rechargeable batteries.

Do I have to run the rechargeable battery completely dead every time?
No. For greatest life expectancy, it's best to recharge NiMH batteries when you first sense the voltage is dropping. You'll be down to 10% or 15% capacity at this point. Further draining is not needed or recommended.

Alkaline rechargeable batteries really hate deep cycles. They'll do best when they're run only down about 50%.

Inverters

What is an inverter, and why would you want one in your renewable energy system? This is a reasonable question. Some small renewable energy systems don't need an inverter.

The inverter is an electronic device that converts direct current (DC) into alternating current (AC). Renewable energy sources such as PV modules and wind turbines make DC power, and batteries will store DC power. Batteries are the least expensive and most universally applicable energy storage method available. Batteries store energy as low-voltage DC, which is acceptable, in fact preferable, for some applications. A remote sign-lighting system, or a small cabin that only needs three or four lights, will get by just fine running everything on 12-volt DC. But most of the world operates on higher-voltage AC. AC transmits more efficiently than DC and so has become the world standard. No technology exists to store AC, however. It must be produced as needed. If you want to run conventional household appliances with your renewable energy system, you need a device to produce AC house current on demand. That device is an inverter.

Inverters are a relatively new technology. Until the early 1990s, the highly efficient, long-lived, relatively inexpensive inverters that we have now were still a pipe dream. The world of solid-state equipment has advanced by extraordinary leaps and bounds. More than 95% of the household power systems we put together now include an inverter, and 100% of the grid-tied systems we install have one.

Benefits of Inverter Use

The world runs on AC power. In North America, it's 120-volt, 60-cycle AC. Other countries may have slightly different standards, but it's all AC. Joining the mainstream allows the use of mass-produced components, wiring hardware, and appliances. Appliances may be chosen from a wider, cheaper, and more readily available selection. Electricity transmits more efficiently at higher voltages, so power distribution through the house can be done with conventional 12- and 14-gauge Romex wiring using standard hardware, which electricians appreciate and inspectors understand. Anyone who's wrestled with

If you want to run conventional household appliances with your renewable energy system, you need a device to produce AC house current on demand. That device is an inverter.

the heavy 10-gauge wire that a low-voltage DC system requires will see the immediate benefit. A home could be wired just as if it were going to have utility power available, and all you'd need to connect to the main panel to distribute power would be an inverter.

Modern brand-name inverters are extremely reliable. The household-size models we sell have failure rates well under 1%. Efficiency averages about 85%–95% for most models, with peaks at up to 97%. In short, inverters make life simpler, do a better job of running your household, and ultimately save you money on appliances and lights.

An inverter/battery system is the ultimate clean, uninterruptible power supply for your computer, too. In fact, an expensive UPS (uninterruptible power supply) system is simply a battery and a small inverter with an expensive enclosure and a few bells and whistles. Just don't run the power saw or washing machine off the same inverter that runs your computer. We often recommend a small secondary inverter just for the computer system. This ensures that the water pump or some other large, unexpected load can't possibly cause problems for other sensitive equipment.

Description of Inverter Operation

If we took a pair of switching transistors and set them to reversing the DC polarity (direction of current flow) 60 times per second, we would have low-voltage alternating power of 60 cycles or Hertz (Hz). If this power was then passed through a transformer, which can transform AC power to higher or lower voltages depending on design, we could end up with crude 120-volt/60 Hz power. In fact, this was about all that early inverter designs of the 1950s and 1960s did. As you might expect, the waveform was square and very crude (more about that in a moment). If the battery voltage went up or down, so did the output AC voltage, only ten times as much. Inverter design has come a long, long way from the noisy, 50% efficient, crude inverters of the 1950s. Modern inverters hold a steady voltage output regardless of battery voltage fluctuations, and efficiency is typically in the 90%–95% range. The waveform of the power delivered has also been improved dramatically. The power put out by inverters today is so clean that it is hard to tell the difference from grid power.

Waveforms

The AC electricity supplied by your local utility is created by spinning a bundle of wires through a series of magnetic fields. As the wire moves into, through, and out of the magnetic field, the voltage gradually builds to a peak and then gradually diminishes. The next magnetic field the wire encounters has the opposite polarity, so current flow is induced in the opposite direction. If this alternating electrical action is plotted against time, which is what an oscilloscope does, we get a picture of a sine wave, as shown in our waveform gallery. Notice how smooth the curves are. Transistors, as used in all inverters, turn on or off abruptly. They have a hard time reproducing curvy sine waves.

Early inverters produced a square-wave alternating current (see the wave-form gallery again). While a square wave does alternate, it is considerably different in shape and peak voltage. This causes problems for many appliances, especially sensitive electronics and audio/video equipment. Heaters or incandescent lights are fine, motors usually get by with just a bit of heating and noise, but solid-state equipment has a really hard time dealing with it, resulting in loud humming, overheating, or failure. It's hard to find

Electrical Terminology and Mechanics

Electricity can be supplied in a variety of voltages and waveforms. As delivered to and stored by our renewable energy battery system, we've got low-voltage DC. As supplied by the utility network, we've got house current: nominally 120 volts AC. What's the difference? DC electricity flows in one direction only, hence the name direct current. It flows directly from one battery terminal to the other battery terminal. AC current alternates, switching the direction of flow periodically. The US standard is 60 cycles per second. Other countries have settled on other standards, but it's usually either 50 or 60 cycles per second. The electrical term for the number of cycles per second is Hertz, named after an early electricity pioneer. So AC power is defined as 50 Hz or 60 Hz.

Our countrywide standard also defines the voltage. Voltage is similar to pressure in a water line. The greater the voltage, the higher the pressure. When voltage is high, it's easier to transmit a given amount of energy, but it's more difficult to contain and potentially more dangerous. House current in this country is delivered at about 120 volts for most of our household appliances, with the occasional high-consumption appliance like a washer/dryer using 240 volts. So our power is defined as 120/240 volts/60 Hz. We usually use the short form, 120 Vac, to denote this particular voltage/cycle combination. This voltage is by no means the world standard, but is the most common one we deal with. Inverters are available in international voltages by special order, but not all models from all manufacturers.

new square-wave inverters anymore, and nobody misses them.

Most modern inverters produce a hybrid waveform called a quasi-, synthesized, or modified sine wave. In truth, this could just as well be called a modified square wave, but manufacturers are optimists. Modified sine wave output cures many of the problems associated with the square wave. Most appliances will accept it and hardly know the difference. There are some notable exceptions to this rosy picture, however, which we'll cover in detail below in the Inverter Problems section.

Full sine wave inverters have been available since the mid-1990s, but because of their higher initial cost and lower efficiency, they were only used for running very specific loads. This is changing rapidly, and today we're mainly seeing large pure sine wave inverters being installed. A variety of high-efficiency, moderate-cost pure sine wave inverters is available now. Magnum, OutBack, and other manufacturers have unveiled a whole series of sine wave units, and it's obviously the standard today. (You can find these products on our website: realgoods.com.) With very rare exceptions, sine wave inverters will happily run any appliance that can plug into utility power. Motors run cooler and more quietly on sine wave, and solid-state equipment has no trouble. True sine wave inverters deliver top-quality AC power and are almost always a better choice for household use.

Inverter Output Ratings

Inverters are sized according to their output in wattage. More wattage capability will cost more money initially. Asking a brand-name inverter for more power than it can deliver will result in the inverter shutting down. Asking the same of a cheapo inverter may result in the unit going up in smoke. All modern inverters are capable of briefly sustaining much higher loads than they can run continuously, because some electric loads, like motors, require a surge to get started. This momentary overcapacity has led some manufacturers to fudge on their output ratings. A manufacturer might, for instance, call its unit a 200-watt inverter based on the instantaneous rating, when the continuous output is only 140 watts. Happily, this practice has faded into the past by now. Most manufacturers today take an honest approach as they introduce new models and label inverters with their continuous wattage output rather than some fanciful number. In this age of the Internet and social media, untrustworthy manufacturers are weeded out quickly. In any case, we've always

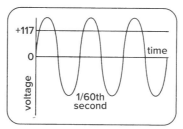

AC sine wave power.

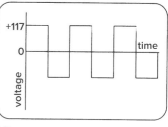

AC square wave power.

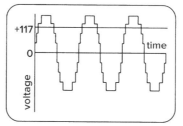

AC modified sine wave power.

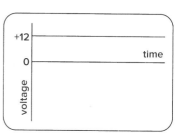

DC 12-volt power.

been careful to list the continuous power output of all the inverters we carry. Just be aware that the manufacturer calling the inverter a 200-watt unit does not necessarily mean it will do 200 watts continuously. Read the spec sheet and manual.

How Big an Inverter Do You Need?

The bigger the inverter, the more expensive the initial cost. So you don't want to buy a bigger one than you need. On the other hand, an inverter that's too small is going to frustrate you because you'll need to limit your power usage.

A small cabin with an appliance or two to run doesn't need much of an inverter. For example, if you want to run a big screen TV, a boom box, and a light all at once, total up all the wattages (about 100 for the TV, 25 for the boom box, and 20 for a compact fluorescent light, a total of 145 watts), pick an inverter that can supply at least 145 watts continuously, and you're all set. This is obviously a simplified example, but that's why we provide the System Sizing Worksheet (see pages 127–8).

To power a whole house full of appliances and lights might take more planning. Obviously not every appliance and light will be on at the same time. Midsize inverters of 600–1,000 watts do a good job of running lights, entertainment equipment, and small kitchen appliances, in other words, most common household loads. What a 1,000-watt inverter will not do is run a mid- to full-size microwave, a washing machine, or larger handheld power tools. For those loads, you need a full-size 3,000-plus-watt inverter. In truth, most households end up with one of the full-size inverters because household loads tend to grow, and larger inverters are often equipped with very

True sine wave inverters deliver top-quality AC power and are almost always a better choice for household use.

powerful battery chargers. This is the most convenient way to add battery-charging capability to a system (and the cheapest, too, if you're already buying the inverter).

Chargers, Lights, Bells, and Whistles

As you go from small plug-in-the-lighter-socket inverters to larger household-size units, you'll find increasing numbers of bells and whistles, most of which are actually pretty handy. At the very least, all inverters have an LED light showing that it's turned on. Midsize units may feature graphic volt and ammeters; better units may have an LCD display and be able to signal generators and other remote devices to turn on or off. Many brands feature monitoring that integrates data from your array, charger, batteries, and inverter all in one LCD display or web/computer-based platform.

Battery chargers are the most useful inverter extra and, for most folks, the cheapest and easiest

> Most households end up with one of the full-size inverters because household loads tend to grow, and larger inverters are often equipped with very powerful battery chargers.

way to add a powerful, automatic battery charger to their system. Chargers are pretty much standard equipment on all larger household-size inverters. With a built-in charger, the inverter will have a pair of "AC input" terminals. If the inverter detects AC voltage at these terminals, because you just started the backup generator, for instance, it will automatically connect the AC input terminals to the AC output terminals, and then go to work charging the batteries. Inverter-based chargers tend to be extremely robust and adjustable, and will treat your batteries nicely.

Many financially challenged off-the-grid systems start out with just a generator, an inverter/charger, and some batteries. You run the generator every few days to recharge the batteries. Add PV charging as money allows, and the generator gradually has to run less and less.

How to Damage or Destroy Your Inverter

Do you want to reduce your inverter's ability to provide maximum output? Here's how: Keep your battery in a low state of charge; use long, undersized cables to connect the battery and inverter; and keep your inverter in a small airtight enclosure at high temperatures. You'll succeed in damaging, if not outright destroying, your inverter. Low battery voltage will severely limit any inverter's ability to meet a surge load. Low voltage occurs when the battery is undercharged, or undersized. If the battery bank isn't large enough to supply the energy demand, then voltage will drop, and the inverter probably won't be able to start the load. For instance, an Outback 3,000 W inverter DR2412 will easily start most any washing machine, but not if you've only got a couple of golf cart batteries. The starting surge will demand more electrons than the batteries can supply, voltage will plummet, and the inverter will shut off to protect itself (and your washing machine motor). Now suppose you have eight golf cart batteries on the same Outback inverter, which is plenty to start the washer with ease, but you decided to save some money on the hookup cables, and used a set of $16.95 automotive jumper cables. The battery bank is capable, the inverter is capable, but not enough electrons can get through the undersized jumper cables. The result is the same: low voltage at the inverter, which shuts off without starting the washer. Do not skimp on inverter cables or on battery interconnect cables. These items are inexpensive compared to the cost of a high-quality inverter and good batteries. Don't cripple your system by scrimping on the electron-delivery parts.

Reading Modified Sine Wave Output with a Conventional Voltmeter

Most voltmeters that sell for under $100 will give you weird voltage readings if you use them to check the output of a modified sine wave inverter. Readings of 80 to 105 volts are the norm. This is because when you switch to "AC Volts" the meter is expecting to see conventional utility sine wave power. What we commonly call 110- to 120-volt power actually varies from 0 to about 175 volts through the sine wave curve. 120 volts is an average called the Root Mean Square, or RMS for short, that's arrived at mathematically. More expensive meters are RMS corrected; that is, they can measure the average voltage for a complex waveform that isn't a sine wave. Less expensive meters simply assume if you switch to "AC Volts" it's going to be a sine wave. So don't panic when your new expensive inverter checks out at 85 volts: The inverter is fine, but your meter is being fooled. Modern inverters will hold their specified voltage output, usually about 117 volts, to plus or minus about 2%. Most utilities figure they're doing well by holding variation to 5%.

Your inverter manual will tell you the proper size to use, or you can always give us a call.

All inverters produce a small amount of waste heat. The harder they work, the more heat they produce. If they get too hot, they will shut off or limit their output to protect themselves. Give the inverter plenty of ventilation. Treat it like a piece of stereo gear: dry, protected, and well ventilated.

How Big a Battery Do You Need?

We usually size household battery banks to provide stored power for three to five days of autonomy during cloudy weather. For most folks, this is a comfortable compromise between cost and function. With three- to five-day sizing, the battery will be large enough to handle any surge loads that the inverter is called upon to start. A battery bank smaller than three days' capacity will cycle fairly deeply on a regular basis. This isn't good for battery life. A larger battery bank cycled less deeply will cost less in the long run, because it lasts longer. Our System Sizing Worksheet covers sizing the battery bank based on your estimated needs (see pages 127–29). Banks larger than five days' capacity are more expensive than a backup power source (such as a modest-sized generator). However, we occasionally run into situations with one-horsepower or larger submersible well pumps or stationary power tools requiring a larger battery bank simply to meet the surge load when starting. Call the Real Goods technical staff for help if you are anticipating large loads of this type (800-919-2400) or realgoods.com.

Inverter Safety Equipment and Power Supply

In some ways, batteries are safer than conventional AC power. It's fairly difficult to shock yourself at low voltage. But in other ways, batteries are more dangerous: They can supply many more amps into a short circuit, and once a DC arc is struck, it has little tendency to self-extinguish. One of the very sensible things the National Electrical Code requires is fusing and a safety disconnect for any appliance connected to a battery bank. Fusing is extremely important for any circuit connected to a battery! Without fusing, you risk burning down your house.

Several products exist to cover DC fusing and disconnect needs for inverter-based systems. For full-size inverter fusing, you can use a properly sized OutBack, Magnum, or MidNite Solar DC

Happy Inverters

1. Keep the inverter as close to the battery as possible, but not in the battery compartment. Five to 10 feet and separated by a fireproof wall is optimal. The high-voltage output of the inverter is easy to transmit. The low-voltage input transmits poorly.
2. Keep the inverter dry and as cool as possible. They don't mind living outdoors if protected.
3. Don't strangle the inverter with undersized supply cables. Most manufacturers have recommendations in the owner's manual. If in doubt, give us a call.
4. Fuse your inverter cables and any other circuits that connect to a battery!

power center, which offer simple installation as well. Our technical staff can answer your questions about which one is better for you. Midsize inverters can use a Class T fuse with a small surface-mounted circuit breaker to provide safe and compliant connection. (Power centers and fuses are available at realgoods.com.) All this safety and connection equipment is covered in the Safety section of this chapter (see page 144). We've recommended fusing or breaker sizing, and required cable sizing with all larger inverters. This isn't the place to get creative—instead, go by the book.

The size of the cables providing power to the inverter is as important as the fusing, as we noted earlier. Do not restrict the inverter's ability to meet surge loads by choking it down with undersized or lengthy cables. Ten feet is the longest practical run between the battery and inverter. This is true even for small 100- to 200-watt inverters. Put the extension cord on the AC side of the inverter! Larger models will include cable and circuit breaker requirement charts. Batteries should either be the sealed type, or live in their own enclosure. Don't put your inverter in with the batteries! Some of the nicer off-grid setups we've seen have the battery enclosure on one side of a wall, and the inverter and other equipment on the other side, making for a very short inverter cable run.

Treat your inverter like a piece of stereo gear: dry, protected, and well ventilated.

Ten feet is the longest practical run between the battery and inverter. This is true even for small 100- to 200-watt inverters.

Potential Problems with Modified Sine Wave Inverter Use—*and How to Correct Them*

We have been painting a rosy picture up to this point, but now it is time for a little brutal honesty to balance things out. If you know about these problems beforehand, it is usually possible to work around them when selecting appliances.

Waveform Problems

Waveform problems are probably the biggest category of inverter problems we encounter. We talked about sine wave versus modified sine wave in the technical description above. The problems discussed here are caused only by modified sine wave inverters. You usually don't want to use one for a home-size system or any sensitive electronics. If you are planning to use a pure sine wave unit, you can skip this section.

Audio Equipment

Some audio gear will pick up a 60-cycle buzz through the speakers. It doesn't hurt the equipment, but it's annoying to the listener. There are too many models and brands to say specifically which have a problem and which do not. We've had better luck with new equipment recently. Manufacturers are starting to put better power supplies into their gear. We can only recommend that you try it and see.

Some top-of-the-line audio gear is protected by SCRs or Triacs. These devices are installed to guard against powerline spikes, surges, and trash (nasties that don't happen on inverter systems). However, they see the sharp corners on modified sine wave as trash and will sometimes commit electrical hara-kiri to prevent that nasty power from reaching the delicate innards. Some are even smart enough to refuse to eat any of that ill-shaped power, and will not power up. The only sure cure for this (other than more tolerant equipment) is a sine wave inverter. If you can afford top-of-the-line audio gear, chances are you've already decided that sine wave power is the way to go.

> Computers run happily on a modified sine wave.

Computers

Computers run happily on a modified sine wave. In fact, most of the uninterruptible power supplies on the market have modified sine wave or even square wave output. The first thing the computer does with the incoming AC power is to run it through an internal power supply. We've had a few reports of the power supply being just a bit noisier on a modified sine wave, but no real problems. Running your family computer off an inverter will not be a problem. What can be a problem is large start-up power surges. If your computer is running off the same household inverter as the water pump, power tools, and microwave, you're going to have trouble. When a large motor, such as a skilsaw, is starting, it will pull the AC system voltage way down momentarily. This can cause computer crashes. The fix is a small, separate inverter that only runs your computer system. It can be connected to the same household battery pack, and have a dedicated outlet or two.

Laser Printers

Many laser printers are equipped with SCRs, which cause the problems detailed above. Laser printers are a very poor choice for renewable energy systems anyway, due to their high standby power use keeping the heater warm. Lower-cost inkjet printers can do almost anything a laser printer can do while using only 25–30 watts instead of 400–900 watts.

Ceiling Fans

Many variable-speed ceiling fans will buzz on modified sine wave current. We've also had some reports of fan remote controls not working on modified sine wave. The fans spin up fine, but the noise is annoying, and you can't change speeds.

Potential Problems with ALL Inverters— *and How to Correct Them*

You sine wave buyers can stop smirking here, the following problems apply to all inverters.

Radio and TV Interference

All inverters broadcast radio noise when operating. The FCC put out new guidelines concerning interference, and as manufacturers come out with new models like the OutBack FX series or the SMA Sunny Island series, this is much less of a problem (see realgoods.com). Most of this interference is on the AM radio band. Do not plug your radio into the inverter and expect to listen to the ball game; you'll have to use a battery-powered radio and be some distance away from the inverter. This is occasionally a problem with TV interference when inexpensive TVs and smaller inexpensive inverters are used together. Distance helps. Put the TV (and the antenna) at least 15 feet from the inverter. Twisting the inverter input cables may also limit their broadcast power (strange as it sounds, it works).

Phantom Loads and Vampires

A phantom load isn't something that lurks in your basement with a half-mask, but it's closely related. Many modern appliances remain partially on when they appear to be turned off. That's a phantom load. Any appliance that can be powered up with a button on a remote control must remain partially on and listening to receive the "on" signal. Most TVs and audio gear these days are phantom loads. Anything with a clock—VCRs, coffee makers, microwave ovens, or bedside radio clocks—uses a small amount of power all the time.

Vampires suck the juice out of your system. These are the black power cubes that plug into an AC socket to deliver lower-voltage power for your answering machine, electric toothbrush, power-tool charging stand, or any of the other huge variety of appliances that use a power cube on the AC socket. These villainous wastrels usually run a horrible 60%–80% inefficiency (which means that for every dime's worth of electricity consumed, they throw away 6 or 8 cents' worth). Most of these nasties always draw power, even if no battery/toothbrush/razor/cordless phone is present and charging. It would cost their manufacturers less than 25 cents per unit to build a power-saving standby mode into the power cube, but since you, the consumer, are paying for the inefficiency, what do they care? The appliance might be turned off, but the vampire keeps sucking a few watts. Have you ever noticed that power cubes are usually warm? That's wasted power being converted to heat. By the way, cute and appropriate as it is, we can't take credit for the vampire name. That's official electric industry terminology (bless their honest souls). Keep those cubes unplugged when not in use, or have them on dedicated circuits that can easily be turned off.

So what's the big deal? It's only a few watts, isn't it? The problem isn't the power consumed by vampires and phantom loads, it's what is required to deliver those few watts 24 hours per day. When there's no demand for AC power, a full-size inverter will drop into a sleep mode. Sleep mode keeps checking to see if anything is asking for power, but it takes only a tiny amount of energy, usually 2–3 watts. If the cumulative phantom and vampire loads are enough to awaken the inverter, it consumes its own load plus the inverter's overhead, which is typically 15–20 watts. The inverter overhead is the real problem. An extra 15–20 watts might sound miniscule, but over 24 hours every day, this much power lost to inefficiency can easily add a couple of $250 modules to the size of the PV array and maybe another battery or two to the system requirements. You want to make sure your inverter can drop into sleep mode whenever there's no real demand for AC power. This is easy to test, and well worth getting a handle on and checking periodically.

Keeping the Vampires at Bay

With some minimal attention to detail, you can keep the vampires and phantoms under tight control and save yourself hundreds or thousands of dollars on system costs.

The solution for clocks is battery power. A wall-mounted clock runs for nearly a year on a single AA rechargeable battery. We have found a wide selection of good-quality battery-powered alarm clocks, and most of the other timekeepers anyone could possibly require, just by looking around. Clocks on house current are ridiculously wasteful.

The solution for phantoms and vampires is outlets or plug strips that are switched off when not in use (such as our Smart Kill A Watt Power Strip, available at realgoods.com). The outlets only get switched on when the appliance is actually being used or recharged. As a side benefit, this cures PMS—perpetual midnight syndrome—on your VCR, too!

If you're aware of the major energy-wasting gizmos we all take for granted, it's easy to avoid or work around them. When you consider the minor inconvenience of having to flip a wall switch before turning on the TV against the $200–$300 cost of another PV module to make up the energy waste, it all comes into perspective. A few extra switched outlets during construction looks like a very good investment. If your house is already built, then use switched plug strips or smart power strips.

> Do not plug your radio into the inverter and expect to listen to the ball game; you'll have to use a battery-powered radio and be some distance away from the inverter.

> Many modern appliances remain partially on when they appear to be turned off. That's a phantom load.

Inverter Conclusions

Inverters allow the use of conventional, mass-produced AC appliances in renewable energy-powered homes, aka living comfortably. They have greatly simplified and improved life off the grid. Modern solid-state electronics have made inverters efficient, inexpensive, and long lived. Batteries and inverters need to live within 10 feet of each other, but not in the same enclosure. Modified sine wave inverters will not be perfectly problem-free with every appliance. Expect a few rough edges. The best inverters produce pure sine wave AC power, which is acceptable to all appliances. Sine wave units tend to cost a bit more, but are strongly recommended for all full-size household systems. Keep the power-wasting phantom loads and power cube vampires under tight control, or be prepared to spend enough to overcome them. Sound good? We thought so!

Wire, Adapters, and Outlets

The solution for phantoms and vampires is outlets or plug strips that are switched off when not in use.

It is important to use common sense when dealing with electricity, and this might be best done by acknowledging ignorance on a particular subject and requesting advice and help from experts when you are unsure of something.

Wires are freeways for electrical power. If we do a poor job designing and installing our wires, we get the same results as with poorly designed roads: traffic jams, accidents, and frustration. Big, wide roads can handle more traffic with ease, but costs go up with increased size, so we're looking for a reasonable compromise. Properly choosing wire and wiring methods can be confusing, but with sufficient planning and thought, we can create safe and durable paths for energy flow.

The National Electrical Code (NEC) provides broad guidelines for safe electrical practices. Local codes may expand upon or supersede this code. It is important to use common sense when dealing with electricity, and this might be best done by acknowledging ignorance on a particular subject and requesting advice and help from experts when you are unsure of something. The Real Goods technical staff is eager to help with particular wiring issues, although you should keep in mind that local inspectors will have the final say on what they consider the most appropriate means to the end of a properly installed electrical system. Advice from your local inspector or local electrician, who will be familiar with local code requirements and renewable energy systems, is probably your best advice.

Wire

Wire comes in a tremendous variety of styles that differ in size, number, material, and type of conductors, as well as the type and temperature rating of insulation protecting the conductor. One of the basic ideas behind all wiring codes is that current-carrying conductors must have at least two layers of protection. Permanent wiring should always be run within electrical enclosures, conduit, or inside walls. Wires are prone to all sorts of threats, including but not limited to abrasion, falling objects, gnawing critters, and children using them as a jungle gym. The plastic insulation around the metal conductor offers minimal protection, based on the assumption that the conductors will be otherwise protected. Poor installation practices that forsake the use of conduit and strain-relief fittings can lead to a breach of the insulation, which could cause a short circuit and fire or electrocution. Electricity, even in its most apparently benign manifestations, is not a force to be managed carelessly.

A particular type and gauge (thickness) of wire is rated to carry a maximum electrical current. The NEC requires that we not exceed 80% of this current rating for continuous-duty applications. With the low-voltage conditions that we run across in independent energy systems, we also need to pay careful attention to voltage drop, or the loss of power over the distance of the wire. As voltage decreases, amperage must increase in order to perform any particular job. So lower-voltage DC systems are likely to be pushing higher amperages. This problem becomes more acute with smaller conductors over longer runs. Some voltage drop is unavoidable when moving electrical energy from one point to another, but we can limit this to reasonable levels if we choose the proper wire size. Acceptable voltage drop for 120-volt circuits can be up to 5%, but for 12- or 24-volt circuits, 2–3% is the most we want to see. We provide a good all-purpose chart and formula on page 172 for figuring wire size at any voltage drop, any distance, any voltage, and any current flow. Use it, or give us a call, and we'll help you figure it out.

DC Adapters and Outlets

With the rising popularity and reliability of inverters that let folks run conventional AC appliances, demand for DC fixtures and outlets continues to diminish, which is mostly a good thing. The roots of the independent energy movement grew from the automobile and recreational vehicle industries, so plugs and outlets for low-voltage applications are usually based on that somewhat flimsy creation, the cigarette lighter socket. This is an unfortunate choice. The limited metal-to-metal contact in this plug configuration seriously limits the current-carrying capacity. Cigarette lighter plugs and outlets are usually rated for 15 amps of surge, but 7–8 amps continuously is absolutely the maximum you can expect without meltdown. Lighter sockets are not approved by the National Electrical Code. There is no plug-and-socket combo that is approved officially for 12-volt use. A large, open receptacle accessible to small children presents an obvious safety hazard. If you must use lighter socket outlets, be sure they're fused!

The conventional rules for DC outlets and plugs say the center contact of the receptacle (female outlet) and the tip of the plug (male adapter) should be positive, and the outer shell of the receptacle and side contact of the plug should be negative. However, the polarity can be reversed in a variety of ways during installation. For some appliances, such as incandescent or halogen lights, it doesn't make a bit of difference. But for those expensive 12-volt compact fluorescent lamps, or that costly DC television you just bought, it matters greatly. It may even be a matter of life or death. Don't assume. Check the polarity with a handheld voltmeter before plugging in.

Another convention is that the power source is presented by the female outlet. This way, it is more difficult to accidentally short out against something or electrocute yourself. Some small solar modules have DC male adapters on them, even though they are a power source, so they can plug into the dashboard lighter socket and trickle charge the battery. These are such a small power source that they present little to no hazard. It will be difficult to get a visible spark from these trickle charger panels.

Some people would rather not use lighter sockets and plugs in their houses because they look dangerous…and they can be. Little fingers and tools fit conveniently within the supposedly "protected" live socket. Probably the safest option is to use an oddball, but still locally available, AC plug-and-socket configuration that has the prongs perpendicular, slanted, or in some other configuration that makes it impossible to plug in to a standard 120-volt AC plug. These have lots of metal-to-metal contact, and will be rated for at least 15 amps continuously. Higher-amperage sockets are readily available if you need them. Although it is tempting to use standard inexpensive 120-volt AC outlets, we've seen this lead to disaster too often, and strongly advise against it. Murphy's Law finds enough ways to trip us up without leaving the welcome mat out. We know from personal experience that any savings are quickly lost when a wrong-voltage appliance is plugged in and destroyed.

Do NOT mix AC and DC wiring within a single electrical box! That's a big-time Code (and common sense) violation. It's fine to put your AC and DC outlets right next to each other—but they need to be in separate boxes and should be labeled.

Wire Sizing Chart/Formula

This chart is useful for finding the correct wire size for any voltage, length, or amperage flow in any AC or DC circuit. For most DC circuits, particularly between the PV modules and the batteries, we try to keep the voltage drop to 3% or less. There's no sense using your expensive PV wattage to heat wires. You want that power in your batteries!

Note that this formula doesn't directly yield a wire gauge size, but rather a VDI number, which is then compared to the nearest number in the VDI column, and then read across to the wire gauge size column.

1. Calculate the Voltage Drop Index (VDI) using the following formula:
   VDI = Amps × Feet ÷ (% Volt Drop × Voltage)
   Amps = Watts divided by Volts

Feet = One-way wire distance
% Volt Drop = Percentage of voltage drop acceptable for this circuit (typically 2%–5%)

2. Determine the appropriate wire size from the chart below.
   A. Take the VDI number you just calculated and find the nearest number in the VDI column, then read to the left for AWG wire gauge size.
   B. Be sure that your circuit amperage does not exceed the figure in the Ampacity column for that wire size. (This is not usually a problem in low-voltage circuits.)

Example: Your PV array consisting of four Sharp 80-watt modules is 60 feet from your 12-volt battery. This is actual wiring distance, up pole mounts, around obstacles, etc. These modules are rated at 4.63 amps × 4 modules = 18.5 amps maximum. We'll shoot for a 3% voltage drop. So our formula looks like:

$$VDI = (18.5 \text{ A} \times 60 \text{ ft.}) \div (3\% \times 12 \text{ V}) = 30.8$$

Looking at our chart, a VDI of 31 means we'd better use #2 wire in copper, or #0 wire in aluminum. Hmmm. Pretty big wire.

What if this system was 24-volt? The modules would be wired in series, so each pair of modules would produce 4.4 amps. Two pairs × 4.63 amps = 9.3 amps max.

$$VDI = (9.3 \text{ A} \times 60 \text{ ft.}) \div (3\% \times 24 \text{ V}) = 7.8$$

Wow! What a difference! At 24-volt input, you could wire your array with little ol' #8 copper wire.

Wire Size (AWG)	Copper Wire		Aluminum Wire	
	VDI	Ampacity	VDI	Ampacity
0000	99	260	62	205
000	78	225	49	175
00	62	195	39	150
0	49	170	31	135
2	31	130	20a	100
4	20	95	12	75
6	12	75	—	—
8	8	55	—	—
10	5	30	—	—
12	3	20	—	—
14	2	15	—	—
16	1	—	—	—

Chart developed by John Davey and Windy Dankoff. Used with permission.

RGS Energy Solar Project Case Study: Ironhouse Sanitary District

THIS 1.1 MW PROJECT FEEDS two meters at the Ironhouse Sanitary District and comprises a ground-mounted single-axis tracker and a carport. The following are some of the factors taken into consideration when designing and installing this system:

Environmental Impact. The Sacramento Delta is located a half-mile north of the parcel selected for the ground-mounted system; a seasonal wetland is located immediately south. RGS Energy worked with a local environmental consultant and biologist to devise an environmental mitigation plan that was approved for a Negative Mitigated Declaration under the *California Environmental Quality Act.* This mitigation plan included:

- Implementation of a Storm Water Pollution Prevention Plan (SWPPP)
- Performing work after the nesting season of birds
- Installing temporary barriers around the seasonal wetland and canal to restrict animal movement into the construction zone, protecting sensitive habitat.

Engineering. Proximity to the Sacramento Delta posed several engineering obstacles, including a low water table, saline air, and caustic soil conditions prone to liquefaction. RGS Energy worked with the client to mitigate these issues:

- Corrosion engineers were consulted to recommend the appropriate coatings for the post-driven piers used in the tracker racking
- The thickness of the driven pier foundations was increased to ensure racking strength throughout the life of the system despite the highly corrosive soil
- Stainless steel hardware and PVC-coated conduit was used where necessary for long-term reliability
- Modules and other electronics were raised to sit above the 100-year Federal Emergency Management Agency (FEMA) flood plain
- Engineered soil was brought in for the inverter foundation pad.

Performance. While the environmental conditions of this site dictated the design, we were able to implement a few performance-enhancing design choices:

- A Power-One ULTRA-1100 inverter with 97.5% CEC efficiency was chosen to maximize production
- The PV Hardware Single-Axis Tracker provides daily sun tracking with only four AC motors driving the entire site. Redundant controllers ensure that any malfunction will only impact a fraction of the total project
- Levelized Cost of Energy optimization calculations were performed to arrive at optimal 36% Ground Coverage Ratio, maximizing production while minimizing required space and equipment
- 1,000-volt system design enables massive strings of 20 modules in series, reducing losses due to voltage drop, and a DC balance-of-systems reduction of nearly 40% compared to an equivalent 600-volt system design.

RGS Energy Solar Project Case Study:
Little Quittacas Solar Farm

THIS 4 MW SOLAR FARM WILL provide approximately 5,000 megawatt-hours of clean electricity each year for the City of New Bedford's (MA) second-largest energy consumer, the Municipal Water Department. The installation of 28 acres of solar panels required the removal of some dense forest and substantial site grading. RGS Energy designed the system and managed all aspects of the project. About 25 electricians, civil engineers, and other construction workers were onsite each day, all coordinated and managed by the RGS Energy Project Manager and Site Supervisor.

Aesthetics. RGS Energy needed to create a site plan that accommodated the needs of all stakeholders, including the system owner, the town permitting authority, and nearby residents.

Residents were concerned about how well the array would be screened from the road. Maintaining the rural character of the neighborhood was important to both the Town and residents, but they had differing opinions regarding the best strategy. RGS Energy proposed a solution that satisfied all stakeholders: a 900-foot-long, 10-foot-high tiered berm with landscape plantings.

Environmental Impact. Little Quittacas Pond, the source of drinking water for the surrounding area, is located just north of the site, and the site is also directly adjacent to wetlands. To prevent any construction runoff from contaminating either of these sensitive areas, RGS Energy worked with a local civil engineer to create a detailed Storm Water Pollution Prevention Plan (SWPPP) that included:

- Creation of a Spill Prevention and Control Plan, including mandatory onsite spill kits
- Surrounding the site with grass berms to direct runoff into infiltration basins
- Installation of a silt fence around the site perimeter that extended 6 inches below ground.

The system was completed with zero spills or discharges.

Completed Ahead of Schedule Despite Challenges. This project needed to be completed by the end of 2013 in order for the system owner to be able to achieve specific tax benefits. Despite several challenges, we completed this project 12 days ahead of schedule.

- Several other projects across the state were subject to the same deadline, putting a strain on utility and subcontractor resources.
- Throughout the seven months of site work/equipment installation, the site was subject to multiple extreme rainfall events. By bringing in additional personnel, as well as overlapping onsite activities, we were able to manage these obstacles without schedule delay, and without sacrificing safety or quality.
- Installing the mounting system became difficult when we encountered densely packed stones that were not discovered during geotechnical borings and pre-construction investigations. By implementing a combination of strategies, including larger and bigger pile drivers, partial rock excavation/removal, and alternative foundations, we avoided delays to the overall project schedule.

CHAPTER 5

Emergency Preparedness and Solar Mobility

Be Empowered When the Grid Goes Down

YOU DON'T HAVE TO LOOK VERY HARD to find plenty of news out there that demonstrates dramatically that the world can be a hazardous place even in the best of times. Power grid failures, disastrous storms and fires, fuel price spikes, terrorist attacks: We all know in our deepest fears that our society has become vulnerable to disruption of the various infrastructures that most of us depend on in our daily lives. Who can doubt that factors as diverse as growing population pressures, grinding poverty, economic globalization, political conflict, climate change, and other natural disasters will continue to place enormous strains on the networks that supply us with food, water, power, and other necessities?

Short of living completely off the grid, most of us are likely to experience several of these kinds of disruptions and breakdowns in the foreseeable future. You will rest easier knowing that you've taken steps to protect yourself and your family from events beyond your control that threaten to undermine your ability to meet your basic needs. Real Goods offers a variety of products and resources to get you up to speed on emergency preparedness (real goods.com). We thank our friends at Goal Zero for contributing a new perspective on how to empower yourself for emergencies or how to manage your portable power needs when you're away from the grid.

> Most of us are likely to experience several kinds of infrastructure disruptions and breakdowns in the foreseeable future. You will rest easier knowing that you've taken steps to protect yourself and your family from events beyond your control that threaten to undermine your ability to meet your basic needs.

Want Some Protection, or at Least a Little Backup? How About Total Freedom?

The place to start, of course, is with electric power. The key point is that, while not always apparent, many domestic systems, appliances, and operations that are not fueled by electricity are nevertheless dependent on it to function. When the grid goes down, you're going to lose a lot more than your lights and your computer.

The renewable energy industry in general and Real Goods in particular have spent many years learning how to provide highly reliable home energy systems under adverse and remote conditions. No utility power? No problem! No generator fuel? No problem! No easy access to the site? No problem! Real Goods can provide off-the-shelf solutions for short-term, long-term, or lifetime energy problems. Like individual lifestyles, energy needs and the necessary equipment to meet them are individual matters. Give our experienced technical staff a call for additional assistance. We're masters at putting together

Whether you choose a portable solar panel and battery that's meant for backpacking or travel, you have the freedom to keep your smart phone, GPS, or camera charged while you're on the move. Freedom from the grid, freedom to do what you love—that's an incredible value add to backup power in your kit.

custom systems to meet individual needs, climates, and sites. Reliable power for your peace of mind is our business.

In addition to reliability, renewable energy provides FREEDOM. Whether you choose a portable solar panel and battery that's meant for backpacking or travel, you have the freedom to keep your smart phone, GPS, or camera charged while you're on the move. Freedom from the grid, freedom to do what you love—that's an incredible

value add to backup power in your kit. This same kit can keep your phone charged in an emergency, and is so easy and reliable that you can use it on all of your weekend adventures. Larger solar backup systems can provide portable power for camping, mobile worksites, or off-grid adventures. Meanwhile, you'll have your electronic gear charged up and ready to go so you're prepared in case of a power outage.

Steps to a Practical Emergency Power System

Following are some practical solutions to be ready for utility disruption. We've arranged them in order of cost with the least expensive options first. Be aware that lowest-cost options will involve some sacrifices and will be acceptable solutions only for brief time periods—emergencies, in other words. As backup systems get more robust, they impose fewer restrictions and are more pleasant to live with.

1. Install a High-quality Backup Generator

A generator will be useful for short power outages, or for longer ones if you have fuel and if you, and the neighbors, are willing to endure the noise. Higher-quality generators tend to be much quieter. Propane fuel is a good choice. It's delivered in bulk to your site (in all but the most remote locations), it doesn't deteriorate or evaporate with age, and it's piped to the generator, so you never need to handle it. Generators live longer on propane because it produces less carbon buildup. Your electrician will need to install a transfer switch between the utility line and the generator so your generator doesn't try to run the neighborhood and doesn't threaten any utility workers.

Simple backup generators use manual starting and have a manual transfer switch. When power fails, somebody has to start the generator, then throw the switch. Fancier systems will do all this automatically. With any generator-based backup system, even automated ones, there will be some dead time between when the power goes out and when the backup systems come online. Many people prefer non-automated generator systems, as they are more fail-safe and keep you

abreast of what's going on, rather than relying on remote automatic systems.

Most generator-based systems will require you to practice some conservation. Don't plan on running big watt-sucker appliances like air conditioning, electric room or water heaters, or electric clothes dryers unless you're willing to support a huge generator of 12,000 watts or larger. Generators run most efficiently at 50%–80% of full load. A big generator is going to suck almost as much fuel to run a few lights as it is to run all the lights and the air conditioning. Generators are not a case of "bigger is better."

2. Add Batteries and an Inverter/Charger

The generator supplies power only when it's running, and it isn't practical or economical to run it all the time. Running a load of just 200 watts is particularly expensive with a generator. They're happiest running at about 80% of full-rated wattage, which keeps carbon buildup to a minimum and is the most efficient in terms of watts delivered per quantity of fuel consumed. Adding batteries and an inverter/charger allows the best use of your generator. Whenever the generator runs, it will automatically charge the batteries. This means the generator is working harder, building up less carbon, and delivering more energy for the fuel consumed. Energy stored in the batteries can be used later to run lights, entertainment equipment, a phone or laptop computer, and smaller loads without starting the generator.

Many medium- to larger-size inverters come with built-in automatic battery chargers. So long as the inverter senses that utility or generator power is available, that line power is passed right through. If outside AC power fails, the inverter will start using stored battery power automatically to supply any AC demands. When outside AC power returns, the batteries will be recharged automatically. For a power failure that lasts more

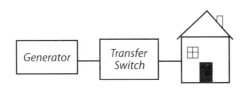

Generator — Transfer Switch — [house]

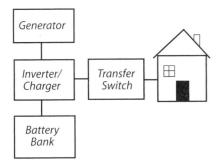

than a couple of days, this means you'll need to run the generator only for a few hours per day, yet you'll have AC power full-time. Putting an inverter and battery pack into the system means that any AC power interruptions will be measured in seconds, or less, depending on the equipment chosen.

There are a variety of battery and inverter solutions on the market. Goal Zero prides themselves not just in their technology, but also on innovating the user experience. By seamlessly building in the inverter, designing the user experience as "plug and play," and providing an expandable solution where every battery and solar panel can be daisy-chained to other panels and batteries for more power, the system is enjoyable and easy to use. In an emergency situation, simplicity and readiness will be greatly appreciated.

Another advantage to adding a battery to your system is portability. Because these batteries don't produce any toxic fumes, they are safe for use indoors. Simply charge your battery from your generator, then bring the battery inside your home. The clean, simple design of the batteries makes them easy to use—anyone in your family can safely and efficiently charge their essential devices. Simple LCD screens tell you how much power you are consuming and indicate how much power you have left before you need a recharge. Goal Zero battery solutions all have built-in inverters and multiple AC, USB, or DC ports for a variety of devices.

3. Add Solar Power for Battery Charging Freedom

A traditional gas generator has the lowest initial cost for power backup; however, it's cheap only if you don't run it. Generators are expensive and noisy to operate, and fuel supplies have to be replenished—we saw the disastrous results of fuel rationing during Hurricane Sandy, when people were waiting in gas lines for 6+ hours. If you anticipate power outages lasting longer than a few weeks, then solar power will save you money

in the long run and deliver a reliable long-term energy supply—and there are no fossil fuels to contend with.

Once you have the battery pack, it can recharge from any source. Electrons all look the same to your battery. Whether they come from a belching cheapo gas generator or a clean silent PV module, the effect is the same on the battery. Renewable energy sources have a higher initial cost but very low costs over time for both your wallet and the environment. Not to mention that solar generators provide power without noise, gasoline, or caustic fumes.

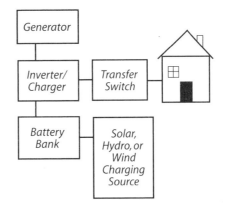

Solar solutions can sometimes require working knowledge of electrical systems, wiring, charge controllers, and inverters. Some companies on the market create simple plug-and-play solutions, making them ideal for anyone's backup kit. Goal Zero's array of solar recharging kits is flexible and versatile enough to fit whatever level of backup you desire. Their innovative solar generator line is plug-and-play, and chainable—meaning you can use their systems with your current battery packs for long-lasting, off-grid power.

Emergency and Backup Power System Design

The average American home uses around 20 kWh of electricity per day. While it's certainly possible to put together an emergency power system to supply this level of use, most folks would find it prohibitively expensive. Most of us are willing to modify our lifestyle in the face of an emergency. Energy conservation is the cornerstone of reduced consumption, which will make getting by in case of disruptions that much more manageable. For example, opening a refrigerator multiple times a day consumes more power, but by simply limiting the amount of times the refrigerator door is opened, you can significantly reduce the amount of power it consumes.

What Do You Really Need in an Emergency?

Obviously, any heating solution that doesn't involve electricity will keep you warm more effectively following a natural disaster.

If you have your own well or other private water supply that requires a pump for water delivery, you'll need some kind of power system that can run your water pump.

Refrigeration, heat, drinking water, lights, cooking equipment, and communications/Internet are on the typical short list of what you really need in an emergency. Let's take a quick look at each of these necessities and see how they can best be supplied when disaster strikes and the grids go down.

Refrigerators and Freezers

Refrigeration is a major concern for most emergency or backup systems. The fridge will probably be your major power consumer, typically using 3–5 kWh of energy per day. It's going to require a 1,500-watt generator or a robust 1,000-watt inverter, at the very least, to start and run a home refrigerator or freezer. If your refrigerator is 1993 vintage or older, strongly consider getting a new one. Refrigerators have made enormous advances in energy efficiency over the past decade. The average new 22-cubic-foot fridge will use half the power of one produced before 1993. If you're shopping, look for the EnergyStar label and those yellow EnergyGuide tags. They really level the playing field. All models are compared to the same standard, which happens to be a pessimistic 90°F ambient temperature. Most of us will see slightly lower operating costs most of the time. Any fridge model sporting a tag for under 500 kWh per year will serve you well and drop energy use to near 1.5 kWh per day. The very best mass-produced fridges will have ratings for 400 kWh/year or less. Top-and-bottom units have a slight energy advantage over side-by-side units. Check out the efficient models sold by Real Goods: realgoods.com.

If you expect power to be out for more than a few weeks, then a propane-powered fridge might be a better choice. Although small by usual American standards (about 7–8 cubic feet), they'll handle the real necessities for about 1.5 gallons of propane per week. Propane freezers are also available.

Heat

If a natural disaster happens in winter, keeping warm suddenly may become overwhelmingly important, while keeping the food from spoiling is of no concern at all. Anyone with a passive solar house, a wood stove, and a supply of firewood will be sitting pretty when the next ice storm hits. Obviously, any heating solution that doesn't involve electricity will keep you warm more effectively following a natural disaster. Beyond wood stoves, other nonelectric solutions may include gas-fired wall heaters or portable gas or kerosene heaters. Make sure you get one with a standing pilot instead of electric ignition, or you'll be out of luck. Also be aware that if your thermostat is electric, it won't work in an extended power outage and you'll need to use manual controls.

Wall heaters are preferable, because they're vented to the outside. Use extreme caution with portable heaters or any other "ventless" heater that uses room oxygen and puts its waste products into the air you're breathing! Combustion waste products combined with depleted oxygen levels are a dangerous and potentially deadly combination—neither brain damage nor freezing are acceptable choices. If you live in earthquake country, don't count on a natural gas supply to continue following a large earthquake.

If a central furnace or boiler is your only heat choice, you'll need some electricity to run it. If your generator- or inverter/battery-based emergency power system is already sized to run a fridge, it probably will be able to pick up the heating chores instead of the fridge during cold weather. (Put the food outside; Mother Nature will keep it cold.) Furnaces, which use an air blower to distribute heat, will generally use a bit more power than the average refrigerator. Boilers, which use a water pump or pumps to distribute heat, will generally use a bit less power than a fridge. Depending on the capabilities of your emergency power system, you might need to limit other uses while the heater is running.

Pellet stoves are a very poor choice for emergency heating. Although the fuel is compact and stores nicely, pellet stoves require a steady supply of power to operate. These machines have blowers for combustion air, other blowers for circulation of room air, and pellet-feed motors.

Water

If you're hooked up to a city-supplied water system, you may have a problem that an emergency power system can't solve. Most cities use a water tower or other elevated storage to provide water pressure. They can continue to supply water for only a few days, or maybe just a few hours, when the power fails. Other than filling the bathtub or draining the water heater tank, which can get you by for a few days, you may want to provide yourself with some emergency water storage. Water storage tanks are available from any farm supply store and some larger building and home supply stores. There are "utility" and "drinking water" grade tanks. Make sure you buy the drinking water grade with an FDA-approved non-leaching coating.

If you have your own well or other private water supply that requires a pump for water delivery, you'll need some kind of power system that can run your water pump. AC water pumps require a large surge current to start, the size of the surge depending on pump type and horsepower. Submersible pumps require more starting power than surface-mounted pumps of similar horsepower.

For short-term backup, the easiest and cheapest solution is to use a generator to run the pump once or twice a day to fill a storage or large pressure tank. While the generator is running, you can also catch a little battery charging for later use to run more modest power needs that can be supplied by an inverter/battery system.

Longer-term solutions will require either a bigger inverter and battery bank to run the water pump on demand, or a smaller pump that can run more easily on solar or battery power. Our knowledgeable technical staff will by happy to discuss your situation and make the best recommendations for your needs and budget.

Water Quality (Is It Safe to Drink?)

Natural disasters often leave municipal water sources polluted and unsafe to drink, sometimes for weeks. So even though you may have water, you'd better think twice before drinking it. Disease carried by polluted drinking water does far more human harm than natural disasters themselves.

With just a little preparation, and at surprisingly low cost, you can be ready to treat your drinking water. And considering how truly miserable, long-lasting, and even-life threatening many waterborne diseases are, an ounce of prevention is worth many tons of cure. You can boil, which is time- and energy-intensive; you can treat with iodine, which is cheap and effective but distasteful; or you can filter, which is quick and effective with the right equipment. The "right equipment" usually means either ultrafine ceramic filtration or a reverse osmosis system. We offer several models of both types at realgoods.com.

Lighting

If you use compact fluorescent lamps (CFLs) instead of common incandescent lights, you'll get four to five times as much light for your power expenditure; you'll do even better with LED lights, which put out more lumens per watt and last three times as long as CFLs. Because power is valuable and expensive in an emergency or

The Goal Zero Lighthouse 250, an LED light that can be charged by solar.

backup situation, these efficient lights are a must. If CFLs or LEDs happen to be part of your normal household lighting anyway—and they should be, if you're concerned about global warming and saving money—you'll enjoy some energy savings while waiting for The Big One. Since CFLs will outlast 10–13 normal incandescents and will use 75% less energy (and LEDs will outlast more than 30 incandescents), they'll end up paying for themselves by putting hundreds of dollars back in your pocket over their lifetime. CFLs and LEDs are a great investment, both for everyday living and for energy-efficient emergency backup. For a more in-depth discussion of efficient lighting, see Chapter 6, pages 209–14. Efficient light bulbs are available from Real Goods at realgoods.com.

Smaller battery-powered, solar-recharged lanterns and flashlights can be extremely helpful during short-term outages and can easily be carried to wherever light is needed. We offer several quality models that can be charged with their on-board solar cells, or simply have fresh batteries slipped in when time and conditions won't allow solar recharging. Reliable flashlights are important for emergency preparation. Goal Zero's Lighthouse 250 is a great example of an LED light that is energy-efficient, can be recharged from solar, and also has multiple lighting levels in order to preserve battery life in an emergency (see real goods.com).

Cooking

Cooking is one of the easier problems to tackle in a power failure. Let's start by looking at what you're using to cook and bake with now.

Electric Stoves and Ovens. These use massive amounts of electricity. It's not going to be practical to run your electric burners or oven with a backup system, or even with a generator. Break out the camping gear! Camp stoves running on white gas or propane are a good alternative and widely available at modest cost. Charcoal

Disease carried by polluted drinking water does far more human harm than natural disasters themselves.

If you use compact fluorescent lamps (CFLs) instead of common incandescent lights, you'll get four to five times as much light for your power expenditure; you'll do even better with LED lights, which put out more lumens per watt and last three times as long as CFLs.

It's not going to be practical to run your electric burners or oven with a backup system, or even with a generator. Break out the camping gear!

If the weather allows, solar ovens do a great job at zero operational cost.

barbecues can be used as well, but *only outdoors!* They produce carbon monoxide, which is odorless and deadly if allowed to accumulate indoors. Want a longer-term solution? Consider a gas stove.

Gas Stoves and Ovens. These can be fired by either propane or natural gas. Propane, which depends only on a small locally installed tank, may be more dependable in a major emergency, particularly an earthquake. Major earthquakes break buried natural gas lines, though these systems won't be affected by electrical outages. In either case, a power failure will put your modern spark-ignition burner lighters out of commission. Just use a match or camping-type stove igniter for the burners. Older stoves use a pilot light for ignition, which will work as normal with or without electricity.

Your gas oven probably won't work without electricity. Older stoves, using a standing pilot light, will be unaffected; all others use either a "glow bulb" or a "spark igniter." Spark igniters, like those used on stovetop burners, will allow you to light the oven with a match. This is good. On the other hand, glow bulbs require 200–300 watts of electrical power all the time the oven is on and will not allow match lighting. This is a problem. You can tell which type of igniter you have by opening the bottom broiler door and then turning on the oven. You'll hear or see the spark igniter, and the unit will light up quickly. Glow bulbs take 30 seconds or longer to light up, and you'll probably see the glowing orange-red bulb once it gets warmed up.

Solar Ovens. If the weather allows, solar ovens do a great job at zero operational cost. They do require full sunlight and are excellent for anything, including a cup of rice, a batch of muffins, or a pot of stew. We offer several ovens, kits, and plans at realgoods.com.

Communications
Simple communication could well be your biggest concern in the aftermath of a disaster—being able to call your friends and family can provide a little relief and normality in tough situations. And even in countries where food is scarce 365 days a year, let alone after a disaster, communication and power are a priority. Goal Zero knows this first-hand from their humanitarian aid work in Haiti after the devastating earthquake in 2010, where villagers took solar recharging kits over 50 lb. bags of rice because they understood the importance of having power for communication.

A Sun Oven solar cooker.

Ensuring you have a means to charge a smart phone in the wake of a power outage can be as simple as having a backup battery. Typical smartphones require about 7–10 watts for a full charge, so pick a battery with the appropriate power capacity. For added convenience, you can purchase a solar panel to charge up your backup battery or your phone directly from the sun. We have several chargers at realgoods.com, including smaller solar recharging kits, perfect for keeping phones charged during any situation. For simple communication needs, we offer several models of radios with wind-up and/or solar power sources.

Keeping larger electronics like laptops and tablets charged means increasing the size of your battery backup source. Tablets can take anywhere from 25–40 watts for a full recharge, average laptops around 50 watts. Keeping the power requirements of your devices in mind, it'll be easier to pick a suitable backup battery.

Conclusion

An ancient Chinese curse says, "May you live in interesting times." Like it or not, we have been thoroughly cursed with interesting times. We can make the best of it with some modest preparation, or we can dally, do nothing, and wait to see what gets dealt to us and how much it hurts. When you're shopping for emergency preparedness, choose the level of protection that feels comfortable to you. After you're prepared, you can sit back feeling secure, knowing that you're independent from the grid—at least for a while. If you're really ambitious, talk to our technicians about a full-time off-the-grid system. And go ahead and make your own interesting times and use your gear before an emergency comes up. With Goal Zero's smart design and portable batteries, you'll find all sorts of opportunities to use backup batteries, lights, or solar panels in your everyday life as well.

CHAPTER 6

Energy Conservation

Superefficient Lighting, Heating, and Cooling Are the First Steps

WE LIVE IN A DYNAMIC SOCIETY with a constantly growing economy. All economic activities require energy inputs, so, realistically, economic growth demands increased energy consumption. To fuel this process, according to the conventional wisdom, we need to pump more oil out of the ground and build more power plants to generate more electricity. Right?

Fortunately, no, this is wrong. It is a well-documented fact that conservation is the cheapest, most cost-effective way to "produce" energy: Renewable energy guru Amory Lovins, years ago, introduced us to the concept of "negawatts, or watts we never need to use." You know the drill: Raised on a steady diet of cheap fossil fuel energy during the 20th century, especially since World War II, our society is enormously wasteful of energy. Not only are those fossil fuel supplies finite and dwindling, but we now also recognize the dire ecological impact of fossil fuel energy on Earth's climate. Energy conservation is the fundamental starting point for a sustainable future, coming before renewable energy and before solar.

Did we forget to mention that energy conservation saves us all money, too? A lot of money.

Lighting is a ubiquitous example of energy that is consumed day and night. In fact, lighting accounts for approximately 28% of the electricity used by businesses and for 5%–10% of total energy use in the average American home. Relatively new LED technology makes it possible for this energy drain to be drastically reduced, and it is fairly inexpensive when you look at the true costs of what it takes to turn on a light. LEDs have evolved very quickly in the past five years, although they've been around since the 1960s in every electronic device that has a little indicator light. In short order, it seemed like every flashlight and many streetlight signals were equipped with LEDs, and now they're being used in our cell phones, flat screen TVs, car headlights, and many more applications. LEDs have significant advantages over compact fluorescent lights, not to mention incandescent lights. They offer better light quality, use less energy, last longer, emit less heat, and use no mercury.

Here are a couple of quick examples to illustrate the glories of energy conservation. The California Energy Commission states that when every household in the state replaces four 100-watt incandescent light bulbs with four 27-watt compact fluorescent bulbs (the CF equivalent of a 100-watt light), burning on average for five hours a day, the state will save 22 gigawatt-hours per day (a gigawatt is 1,000 megawatts)—enough energy to *shut down* 17 power plants! The better news is that LEDs offer even more efficiency and lifetime hours: Replacing those incandescent bulbs with LEDs could reduce the demand for electricity enough to shut down 50 power plants in California! According to the *Clean Energy Act* of 2007, light bulbs have to be 30% more efficient than previous standards. This law resulted in the phase out of 100 W, 60 W, and 40 W incandescent light bulbs. These bulbs, which have been in our lives since the late 19th century, can no longer be manufactured in or imported into the US, though stores are allowed to sell through their old inventories. Soon the high-pressure sodium and inefficient metal halide lights, responsible for illuminating our roadways, shopping malls, and warehouses, will meet the same fate as the incandescent bulbs. It is high time for us to move beyond outdated 19th- and 20th-century thinking.

> It is a well-documented fact that conservation is the cheapest, most cost-effective way to "produce" energy.

> For every dollar you spend on energy conservation, you save $3 or more on the cost of your solar electric system.

The Benefits of LED Lights

LEDs:
- Offer better light quality
- Use less energy
- Last longer
- Emit less heat
- Use no mercury

In a similar vein, if each of those California households replaced one average-flow shower-head with a low-flow, energy-saving showerhead, California would save an additional 19.2 GWh per day—enough to shut down another 15 power plants. Conservation is indeed a very powerful tool. Or try this: For every dollar you spend on energy conservation, you save $3 or more on the cost of your solar electric system.

> The lower your household energy consumption, particularly for heating and electricity, the more easily renewable resources can contribute to meeting your needs with clean, reliable, sustainable energy.

Conserving energy is truly a no-brainer. Conserving energy reduces greenhouse gas emissions, slows down the depletion of natural resources, decreases environmental pollution, takes strain off the planet's organic life-support systems, and saves a significant amount of money! None of us can afford *not* to conserve energy. The further beauty of conservation is that regardless of what not-so-enlightened pundits and critics say, conservation does *not* mean sacrifice. Many European and Scandinavian societies enjoy a comparable standard of living to those of us in the United States, but they do it on a much tighter energy budget. As the authors of *Limits to Growth: The 30-Year Update* note, "It seems certain that the US economy could do everything it now does, with currently available technologies and at current or lower costs, using half as much energy" (p. 96), which would bring it to the efficiency levels of Western Europe which uses ⅓ of the energy on average as the US. It's all about awareness, habits, and using energy efficiently. The tools, techniques, and technologies are readily available and well tested. We just have to start implementing them on a much wider basis.

Energy conservation is especially important if you want to achieve some degree of self-sufficiency, independence from conventional utility networks, or off-the-grid living—even if you want to net meter your grid-tied solar power to near zero. The lower your household energy consumption, particularly for heating and electricity, the more easily renewable resources can contribute to meeting your needs with clean, reliable, sustainable energy. But regardless of your mode of living, the same few conservation principles apply to everyone: Build smaller rather than larger. Make sure your building envelope is tight. Use passive solar strategies to minimize your heating and cooling loads. Install low-flow showerheads, low-flush or composting toilets, and compact fluorescent lights. Buy the most efficient appliances possible for your needs and budget. Strive to use renewable energy resources.

Real Goods supplies many types of products, appliances, systems, and informational resources to help families conserve energy and consume it efficiently. The material in this chapter focuses on domestic space heating and cooling, household appliances, and lighting. Energy conservation as applied to water heating and transportation is addressed in later chapters. Read this chapter carefully before even considering the size of your future solar electric system, as you will save lots of money by needing a much smaller system after you've first taken steps to conserve energy.

We'd like to mention one other idea. If you want to take the concept of energy conservation to its furthest reaches, consider the notion of *embodied energy*. Every commodity or product that you consume "embodies" all the energy it took to produce it and get it to you. From the big-picture perspective, then, you can also conserve by assessing the comparative energy required to produce and transport various goods that you choose to buy. If you live on the East Coast, for example, a locally grown organic tomato in season embodies much less energy than a California or Central America organic tomato that's calling out to your taste buds in January. The scenarios and possible comparisons are endless, of course, and it's complicated to figure out how much energy is embodied in any given thing, or to measure the actual impact of one choice over another. Thinking about this issue will probably take you places you may not want to go, and drive you crazy in the process. Nonetheless, the concept of embodied energy has real environmental import. We suggest that cultivating the habit of thinking about embodied energy is a conscientious act of global citizenship in service to sustainability.

REAL GOODS

Living Well but Inexpensively

Imagine a pair of similar suburban homes on a quiet residential street. Both house a family of four living the American suburban lifestyle. The homes appear to be identical, yet one spends under $50 per month on utilities, and the other over $400. How can that be? This huge cost difference demonstrates the dramatic savings that careful building design, landscaping, and selection of energy-efficient appliances make possible without affecting a family's basic lifestyle.

Our example isn't based on some bizarre, untested construction technique, or appliances that can be operated only by rocket scientists, but on simple, common sense building enhancements and off-the-shelf appliances, all of which you can learn about in this *Sourcebook*. Studies have shown that no investment pays as well as conservation. Banks, mutual funds, real estate investments…none of these options will bring the 100%–300% returns on investment that are achievable through simple, inexpensive conservation measures. Not even our own dearly beloved renewable energy systems will repay your investment as quickly as conservation.

At Real Goods, we are experts in energy conservation. We have to be! After more than 36 years designing renewable electrical systems for remote locations, we have learned to squeeze the maximum work out of every precious watt. Although our focus is on energy generation through solar modules or wind and hydroelectric generators, we often have to backtrack a bit to basic building design, or retrofitting of existing buildings, before we start selecting appliances. We'll tackle the broad, multifaceted subject of energy conservation by first looking at those topics. Then we'll delve more into space heating and cooling, energy-efficient appliances, and superefficient lighting.

It All Starts with Good Design

The single most important factor that affects energy consumption in your house is design. All the intelligent appliance selection or retrofitting in the world won't keep an Atlanta, Georgia, family room with a 4 × 6-foot skylight and a west-facing 8-foot sliding glass door from overheating every summer afternoon and consuming massive amounts of air conditioning energy. Intelligent passive solar design is the best place to start. Passive solar design works compatibly with your local climate conditions, using the seasonal sun angles at your latitude to create an interior environment that is warm in winter and cool in summer, without the addition of large amounts of energy for heating and cooling. Read about some real-life examples of this with the Solar Living Center design (page 408) and SunHawk (page 70).

Passive solar buildings cost no more to design or build than energy-hog buildings, yet they cost only a fraction as much to live in. You don't have

House orientation affects heating and cooling costs.

to build a totally solar-powered house to take advantage of some passive solar savings. Just a few simple measures, mostly invisible to your neighbors, can substantially improve your home's energy performance!

For an in-depth examination of passive solar design strategies, from the simple to the complex, see Chapter 2, "Land and Shelter."

Retrofitting: Making the Best of What You've Got

If you're like many of us, you already have a house, have no intention of building a new one, and just want to make it as comfortable and economical as possible. By retrofitting your home with energy savings in mind, you can lower operating costs, boost efficiency, improve comfort levels, save money, and feel good about what you've accomplished.

Whether you want to do this kind of work yourself or have it done by professionals, the

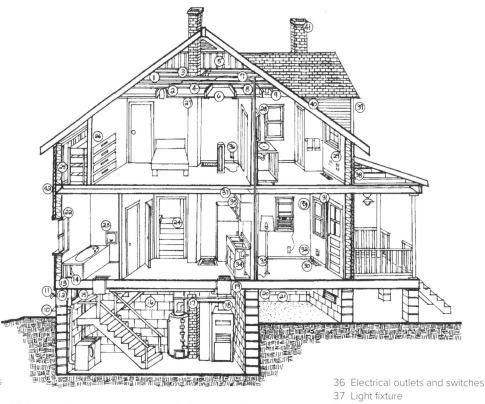

Attic:
1 Dropped ceiling
2 Recessed light
3 Chimney chase
4 Electric wires and box
5 Balloon wall
6 Attic entrance
7 Partition wall top plate
8 Plumbing vent chase
9 Exhaust fan

Basement and crawlspace:
10 Dryer vent
11 Plumbing/ utility penetrations
12 Sill plate
13 Rim joist
14 Bathtub drain penetration
15 Basement windows and doors
16 Block wall cavities
17 Water heater and furnace flues
18 Warm-air ducts
19 Plumbing chase
20 Basement/crawlspace framing
21 Floorboards

Living area:
22 Window sashes and doors
23 Laundry chute

24 Stairwell
25 Kneewall/framing intersection
26 Built-in dresser
27 Chimney penetration
28 Built-in cabinet
29 Cracks in drywall

30 Warm-air register
31 Window and door frames
32 Baseboards, coves, interior trim
33 Plumbing access panel
34 Sink drain penetration
35 Dropped soffit

36 Electrical outlets and switches
37 Light fixture

Exterior:
38 Porch framing intersection
39 Missing siding and trim
40 Additions, dormers, overhangs
41 Unused chimney
42 Floor joist

Each year in the US, about $13 billion worth of energy, in the form of heated or cooled air—or about $150 per household—escapes through the holes and cracks in residential buildings.

AMERICAN COUNCIL FOR AN ENERGY-EFFICIENT ECONOMY

place to start is with a full home energy audit. A trained auditor will investigate your house top to bottom, identify all the weatherization work that needs to be done, and give you an itemized list in order of cost-effectiveness. That's important because, for example, while many people think first about replacing leaky windows, that's usually an expensive proposition and many other actions provide more conservation and efficiency bang for the buck.

Weatherization and Insulation

Increasing insulation levels and plugging air leaks are favorite retrofitting pastimes, which produce well-documented, rapid paybacks in comfort and savings on your space-heating bills. Short of printing your own money, weatherization and insulation are the best bets for putting immediate and easy cash in your wallet—and they're a lot safer in the long run than counterfeiting!

Weatherization

Weatherization—the plugging and sealing of air leaks—can save 25%–40% of your heating and cooling bills. The average unweatherized house in the US leaks air at a rate equivalent to a 4-foot-square hole in the wall. Weatherization is the first place for the average homeowner to concentrate your efforts. You'll get the most benefit for the least effort and expense.

Where to Weatherize

(The following suggestions are adapted from *Homemade Money*, by Richard Heede and the Rocky Mountain Institute. Used with permission.)

Here's a basic checklist to help you get started. Weatherization points are keyed in parentheses to the illustration above.

1. In the attic
- Weather-strip and insulate the attic access door (6).
- Seal around the outside of the chimney with metal flashing and high-temperature sealant such as flue caulk or muffler cement (3).
- Seal around plumbing vents, both in the attic floor and in the roof. Check roof flashings (where the plumbing vent pipes pass through the roof) for signs of water leakage while you're peering at the underside of the roof (8).
- Seal the top of interior walls in pre-1950s

houses anywhere you can peer down into the wall cavity. Use strips of rigid insulation, and seal the edges with silicone caulk (7).

- Stuff fiberglass insulation around electrical wire penetrations at the top of interior walls and where wires enter ceiling fixtures (but not around recessed light fixtures unless the fixtures are rated IC [for insulation contact]). Fluorescent fixtures usually are safe to insulate around; they don't produce a lot of waste heat. Incandescent fixtures should be upgraded to compact fluorescent lamps, but that's another story; see below, pages 216–17 (2, 4).
- Seal all other holes between the heated space and the attic (9).

2. In the basement or crawl space

- Seal and insulate around any accessible heating or A/C ducts. This applies to both the basement and the attic (18).
- Seal any holes that allow air to rise from the basement or crawl space directly into the living space above. Check around plumbing, chimney, and electrical penetrations (11, 14, 19, 21).
- Caulk around basement window frames (15).
- Seal holes in the foundation wall as well as gaps between the concrete foundation and the wood structure (at the sill plate and rim joist). Use caulk or foam sealant (12, 13).

3. Around windows and doors

- Replace broken glass and reputty loose panes. See the Window section, page 188, about retrofitting or upgrading to better windows (22).
- Install new sash locks, or adjust existing ones on double-hung and slider windows (22).
- Caulk on the inside around window and door trim, sealing where the frame meets the wall and all other window woodwork joints (31).
- Weather-strip exterior doors, including those to garages and porches (31).
- For windows that will be opened, use weather stripping or temporary flexible rope caulk.

4. In living areas

- Install foam rubber gaskets behind electrical outlet and switch trim plates on exterior walls (36).
- Use paintable or colored caulk around

bath and kitchen cabinets on exterior walls (26, 35).
- Caulk any cracks where the floor meets exterior walls. Such cracks are often hidden behind the edge of the carpet (32).
- Got a fireplace? If you don't use it, plug the flue with an inflatable plug, or install a rigid insulation plug. If you do use it, make sure the damper closes tightly when a fire isn't burning. See the section on Heating and Cooling (pages 192–200) for more tips on fireplace efficiency (41).

5. On the exterior

- Caulk around all penetrations where electrical, telephone, cable, gas, dryer vents, and water lines enter the house. You may want to stuff some fiberglass insulation into the larger gaps first (9, 10, 11).
- Caulk around all sides of window and door frames to keep out the rain and reduce air infiltration.
- Check your dryer exhaust vent hood. If it's missing the flapper, or it doesn't close by itself, replace it with a tight-fitting model (10).
- Remove window air conditioners in winter; or at least cover them tightly, and make rigid insulation covers for the flimsy side panels.
- Caulk cracks in overhangs of cantilevered bays and chimney chases (27, 40).

Insulation

Many existing homes are woefully under-insulated. Exterior wall cavities or underfloors are often completely ignored, and attic insulation levels are sometimes more in tune with the 1950s, when heating oil was 12¢ per gallon, than they are with the 21st century, when oil prices are 30 times higher and promising to continue to escalate. Adequate insulation rewards you with a building that is warmer in the winter, cooler in the summer, and much less expensive to operate year-round.

Recommended insulation levels vary according to climate and geography. Your friendly local building department can give you the locally mandated standards, but for most of North America, you want a minimum of R-11 in floors, R-19 in walls, and R-30 in ceilings. These levels will be higher in more northern climes and lower in more southern climes. Or take a big step like we've done with our Solar Living Center and go for the R-65 provided by straw bale construction, or the R-35 with the Rastra construction at Sun-Hawk.

R-Values of Loose-Fill Insulation		
MATERIAL	R-VALUE PER INCH	USE
Fiberglass (low density)	2.2	Walls and ceilings
Fiberglass (medium density)	2.6	Walls and ceilings
Fiberglass (high density)	3.0	Walls and ceilings
Cellulose (dry)	3.2	Walls and ceilings
Wet-spray cellulose	3.5	Walls and ceilings
Rock wool	3.1	Walls and ceilings
Cotton	3.2	Walls and ceilings

R-Values of Rigid Foam and Liquid Foam Insulation		
MATERIAL	R-VALUE PER INCH	USE
Expanded polystyrene	3.8–4.4	Foundations, walls, ceilings, and roofs
Extruded polystyrene	5.0	Same
Polyisocyanurate	6.5–8.0	Same
Roxul (mineral wool)*	4.3	Foundation exteriors
Icynene	3.6	Walls and ceilings
Air Krete	3.9	Walls and ceilings

*Rigid board insulation made from mineral wool.

Source: Dan Chiras, *The Solar House: Passive Heating and Cooling*

Insulate It Yourself?

Some retrofit insulation jobs, like hard-to-reach wall cavities, are best done by professionals who have the appropriate tools and experience to do the job quickly, correctly, and with minimal disruption. We advise you to find an insulation and weatherproofing company in your area that will inspect your home and give you an estimate to bring it up to an acceptable standard.

Easier-to-reach spaces, like attics, are improvable by the do-it-yourselfer with either fiberglass batt insulation or blown-in cellulose insulation, which can be easier and quicker to install. Many lumber and building supply centers will rent or loan the blowing equipment (and instructions!) when you buy the insulation. Some folks may choose to leave this potentially messy job to the pros.

Underfloor areas are usually easy to access, but working overhead with fiberglass insulation is no picnic. You might consider a radiant barrier product for underfloor use because of the ease of installation. A radiant barrier stapled on the bottom of floor joists performs equally to R-14 insulation in the winter. Performance will be less in the summer when keeping heat out (R-8 equivalent), but this isn't a big consideration in most climates.

Radiant Barriers—
The Newest Wrinkle in Insulation

Over the past decade, a new type of insulation called "radiant barrier" has become popular. Radiant barriers work differently from traditional "dead-air space" insulation. See the sidebar for a detailed explanation.

A radiant barrier will stop about 97% of radiant heat transfer. Multiple layers of barrier provide no cumulative advantage; the heat practically all stops at the first layer. Radiant barriers do not affect conducted or convected heat transfers, so they are usually best employed in conjunction with conventional dead-air space types of insulation, such as fiberglass.

R-values can't be assigned to radiant barriers because they're only effective with the most common form of heat transfer, radiant transfer. But for the sake of having a common understanding, the performance of a radiant barrier is often expressed in R-value equivalents. For instance,

Adequate insulation rewards you with a building that is warmer in the winter, cooler in the summer, and much less expensive to operate year-round.

Insulation

Since warm air rises, the best place to add insulation is in the attic. This will help keep your upper floors warm. Recommended insulation levels depend on where you live and the type of heating system you use. For most climates, a minimum of R-30 will do, but colder areas, such as the northern tier and mountain states, may need as much as R-49. You can find useful information about insulating materials and insulation levels needed in various locations at energy.gov/energysaver/articles/insulation-materials.

There are four types of insulating materials: batts, rolls, loose fill, and rigid foam boards. Each is suitable for various parts of your house. Batts, designed to fit between the studs in the walls or the joists in the ceilings, are usually made of fiberglass or rock wool. Rolls are also made of fiberglass and can be laid on the attic floor. Loose-fill insulation, which is made of cellulose, fiberglass, or rock wool, is blown into attics and walls. Rigid board insulation, designed for confined spaces such as basements, foundations, and exterior walls, provides additional structural support. It is often used on the exterior of exposed cathedral ceilings. Several types of insulation can also be spray applied, including foam as well as cellulose, rock wool, and fiberglass.

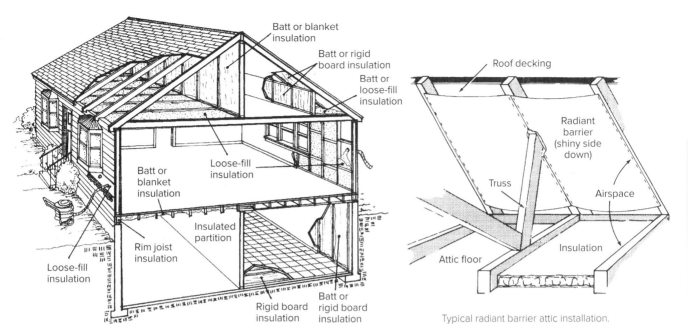

Batt or blanket insulation

Batt or rigid board insulation

Batt or loose-fill insulation

Loose-fill insulation

Batt or blanket insulation

Insulated partition

Rim joist insulation

Loose-fill insulation

Rigid board insulation

Batt or rigid board insulation

Roof decking

Radiant barrier (shiny side down)

Truss

Airspace

Attic floor

Insulation

Typical radiant barrier attic installation.

Where to insulate. Illustration adapted from Reader's Digest (1982), *Home Improvements Manual*, page 360, Pleasantville, NY.

when heat is trying to transfer downward, like from the underside of your roof into your attic in the summer, a radiant barrier tacked to the underside of the rafters will perform equivalently to R-14 insulation. When that same heat is trying to transfer back out of your attic at night, however, the radiant barrier only performs equivalently to R-8, since it has no effect on the convection transfer that is in the upward direction. Therefore, conventional insulation is best for keeping warm in the winter when all the heat is trying to go up through convection. Radiant barrier material is most effective when it's reflecting heat back upward (the direction of convection). So, two of the best applications of this product are under rafters to keep out unwanted summer heat, and under floors to keep desirable winter heat inside your house.

The independent nonprofit Florida Solar Energy Center (FSEC) has tested radiant barrier products extensively and has concluded that installation on the underside of the roof deck will

Two of the best applications of radiant barriers are under rafters to keep out unwanted summer heat, and under floors to keep desirable winter heat inside your house.

How Radiant Barriers Work

Heat is transmitted three ways:

Conduction is heat flow through a solid object. A frying pan uses heat conduction.

Convection is heat flow in a liquid or gas. The warmer liquid or gas rises and the cooler liquid or gas comes down. Floor furnaces heat the air, which then rises in the room.

Radiation is the transmission of heat without the use of matter. It follows line of sight, and you can feel it from the sun and from a wood stove. Radiant heat flow is the most common method of heat transmission.

Radiant barriers use the reflective property of a thin aluminum film to stop (i.e., reflect) radiant heat flow. Gold works even better, but its use is limited mainly to spacecraft, for obvious reasons. A thin

aluminum foil will reflect 97% of radiant heat. If the foil is heated, it will emit only 5% of the heat a dull black object of the same temperature would emit. For the foil to work as an effective radiant barrier, there must be an air space on at least one side of the foil. A vacuum is even better and is used in glass thermos bottles, where the double glass walls are aluminized and the space between them is evacuated. Without an air space, ordinary conduction takes place and there is no blocking effect.

A radiant barrier blocks only heat flow. It has no effect on conduction or convection. But then, so long as there's a small air space, the only practical way that heat can be transmitted across it is by radiant transfer.

The R-value of a single-pane window is a miserable R-1. Nearly half of all residential windows in the United States provide only this negligible insulation value. An "energy-saving" thermal pane (or double-pane) window is worth only about R-2.

reduce summer air conditioning costs by 10%–20%. And that's in a climate where humidity is more of a problem than temperature. This means rapid payback for a product as inexpensive, clean, and easy to install as a radiant barrier. FSEC also paid close attention to roof temperatures, as there's a logical concern about where all that reflected heat is going. They found that shingle temperatures increased by 2°–10°F. To a roof shingle, that's beneath notice and will have no effect on roof lifespan.

Air-to-air Heat Exchangers (aka Heat-recovery Ventilators)

Building and appliance technologies have made tremendous gains in the past 30 years. Improvements in insulation, infiltration rates, and weather stripping do much to keep our expensive warmed or cooled air inside the building. In fact, we've gotten so good at making tight buildings and reducing infiltration of outside air that indoor air pollution has become a problem. (For more on this subject, see Chapter 9.) A whole new class of household appliances has been developed to deal with the by-products of normal living. Air-to-air heat exchangers, also known as heat-recovery ventilators, take the moisture, radon, and chemically saturated indoor air and exchange it for outdoor air. They strip up to 75% of the heat from the outbound air in the winter or the inbound air in the summer, thereby increasing your home's energy efficiency. These heat exchangers use minimal power, and controls can be triggered by time, humidity, or usage. If you use gas for cooking, live in a moist climate, have radon or allergy problems, or simply live in a good, tight house, then air-to-air heat exchangers are worth considering.

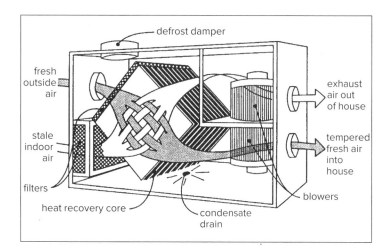

defrost damper

fresh outside air

stale indoor air

filters

heat recovery core

condensate drain

exhaust air out of house

tempered fresh air into house

blowers

The air-to-air heat exchanger. A heat recovery ventilator's heat exchanger transfers 50%–70% of the heat from the exhaust air to the intake air. Illustration courtesy of Montana Department of Natural Resources and Conservation.

Windows

Window technology is an ever-evolving field of building technology. And it's about time! Windows are the weakest link in any building's thermal barrier. Until recently, a window was basically a hole in the wall that let light in and heat out. The R-value of a single-pane window is a miserable R-1. Nearly half of all residential windows in the United States provide only this negligible insulation value. An "energy-saving" thermal pane (or double-pane) window is worth only about R-2. When you consider that the lower-cost wall around the window is insulated to at least R-12, and probably to over R-16, a window needs a solid justification for being there. In cold climates, windows are responsible for up to 25% of a home's winter heat loss. In warmer climates, solar radiation, entering through improperly placed or shaded windows, can boost air conditioning bills similarly.

Worth Retrofitting?

Major innovations in window construction have occurred in the past decade or so. The most advanced superwindows now boast R-values up to R-10. While superwindows cost only 20%–50% more than conventional double-pane windows, the payback period for such an upgrade in an existing house is 15–20 years due to installation costs. That's too long for most folks. But if you're planning to replace windows anyway due to remodeling or new construction, the new high-tech units are well worth the slightly higher initial cost and will pay for themselves in just a few years.

Sneaky R-values

When shopping for windows, be forewarned that until recently, most windows were rated on their "center of glass" R-value. Such ratings ignore the significant heat loss through the edge of the glass and the frame. Inexpensive aluminum-framed windows, which transmit large amounts of heat through the frame, benefit disproportionately from this dubious standard. A more meaningful measure is the newer "whole-unit" value. Make sure you're comparing apples to apples. Wood- or vinyl-framed units outperform similar metal-framed units, unless the metal frames use "thermal break" construction, which insulates the outer frame from the inner frame. The *Consumer Guide to Home Energy Savings*, published by the American Council for an Energy-Efficient Economy, provides an in-depth overview of how to understand various window ratings (get a copy at realgoods.com).

Start with Low-cost Options

With windows, as with all household energy-saving measures, start with jobs that cost the least and yield the most. A common sense combination of three or four of the ideas in the next section will result in substantial savings on your heating/cooling bill, minimized drafts, more constant temperatures, and enhanced comfort, especially in areas near windows.

Window Solutions for Efficient Heating

(The following suggestions (cold- and warm-weather window solutions) are adapted from *Homemade Money*, by Richard Heede and the Rocky Mountain Institute. Used with permission.)

First, stop the wind from blowing in and around your windows and frames by caulking and weather stripping. After you've cut infiltration around the windows, the main challenge is to increase the insulating value of the window itself while continuing to admit solar radiation. Here are some suggestions for beefing up your existing windows in winter.

Install Clear Plastic Barriers on the Inside of Windows

Such barriers work by creating an insulating dead-air space inside the window. After caulking, this is the least expensive temporary option for cutting window heat loss. Such barriers can cut heat loss by 25%–40%.

Repair and Weatherize Exterior Storm Windows

If you already own storm windows, just replace any broken glass, reputty loose panes, install them each fall, and seal around the edges with rope caulk.

Add New Exterior or Interior Storm Windows

Storm windows are more expensive than temporary plastic options but have the advantages of permanence, reusability, and better performance. Storm windows cost about $7.50–$12.50 per square foot and can reduce heat loss by 25%–50%, depending on how well they seal around the edges. Exterior storm windows will increase the temperature of the inside window by as much as 30°F on a cold day, keeping you more comfortable.

Apply Low-e Films

Low-e films substantially reduce the amount of heat that passes through a window, with minimal effect on the amount of visible light passing through. The accompanying sidebar explains how these films work. Don't use low-e films on, or built into, south-facing windows if you want solar gain! They can't tell the difference between winter and summer. Note that low-e films will retard either heat loss or heat gain, depending on where they are applied.

Exotic Infills

The other advanced technology often found in new windows is exotic infills. Instead of filling the space between panes with air, many windows are now available with argon or krypton gas infills that have lower conductivity than air and boost R-values. Krypton has a higher R-value but costs more. These inert gases occur naturally in the atmosphere and are harmless even if the window breaks. (Krypton-filled windows are also safe from Superman breakage.) All this technology adds up to windows with R-values in the 6–10 range.

Install Tight-fitting Insulating Shades

These shades incorporate layers of insulating material, a radiant barrier, and a moisture-resistant layer to help prevent condensation. Several designs are available. One popular favorite is Window Quilts. This material consists of several layers of spun polyester and radiant barriers with a cloth outer cover. Depending on style, they fold or roll down over your windows at night, providing a tight seal on all four sides, high R-value insulation, privacy, and soft quilted good looks.

> With windows, as with all household energy-saving measures, start with jobs that cost the least and yield the most.

Low-e Films

The biggest news in window technology is "low-e" films, for low-emissivity. These thin metal coatings allow the shortwave radiation of solar energy to pass in but block most of the longwave thermal energy trying to get back out. A low-e treated window has less heat leakage in either direction. A low-e coating is virtually invisible from the inside, but most brands tend to give windows a semi-mirror appearance from the outside.

Low-e windows are available ready-made from the factory, where the thin plastic film with the metal coating is suspended between the glass panes; or low-e films can be applied to existing windows.

We offer a do-it-yourself product that is applied permanently with soap and a squeegee, or it can be applied professionally (see realgoods.com). These films offer the same heat-reflecting performance as the factory-applied coatings and are relatively modest in cost (especially compared with new windows). They do require a bit of care in cleaning, as the plastic film can be scratched. Some utility companies offer rebates for after-market low-e films.

Pop-in panels aren't ideal, as they require storage whenever you want to look out the window, but they are cheap, simple, and highly effective.

Appropriate landscaping on the west and south sides of your house provides valuable summertime shading that will reduce unwanted heating by as much as 50%. Those are better results than we get from more expensive projects like window and insulation upgrades!

This allows your windows to have all the daytime advantages of daylighting and passive heat gain, while still enjoying the nighttime comfort of high R-values and no cold drafts. Because there is little standardization of window sizing, Window Quilts are generally custom cut to size. You can order directly from the manufacturer, or they can direct you to a local retailer who can supply installation services. Contact them at 802-246-4500 or 1windowquilts.com.

Construct Insulated Pop-in Panels or Shutters

Rigid insulation can be cut to fit snugly into window openings, and a lightweight, decorative fabric can be glued to the inside. Pop-in panels aren't ideal, as they require storage whenever you want to look out the window, but they are cheap, simple, and highly effective. They are especially good for windows you wouldn't mind covering for the duration of the winter. Make sure they fit tightly so moisture doesn't enter the dead-air space and condense on the window.

Close Your Curtains or Shades at Night

The extra layers increase R-value, and you'll feel more comfortable not being exposed to the cold glass.

Open Your Curtains During the Day

South-facing windows let in heat and light when the sun is shining. Removing outside screens for the winter on south windows can increase solar gain by 40%.

Clean Solar Gain Windows

Keep those south-facing windows clean for better light and a lot more free heat. Be sure to keep those same windows dirty in the summer. (Just kidding!)

Window Solutions for Efficient Cooling

The main source of heat gain through windows is solar gain—sunlight streaming in through single or dual (or these days, even triple) glazing. Here are some tips for staying cool:

Install White Window Shades or Miniblinds

Using shades or blinds is a simple old-fashioned practice. Since our grandparents didn't have air conditioners, they knew how to keep the heat out. Miniblinds can reduce solar heat gain by 40%–50%.

Close South- and West-facing Curtains

Do this during the day for any window that lets in direct sunlight. Keep these windows closed too.

Install Awnings

Awnings are another good old-fashioned solution. Awnings work best on south-facing windows where there's insufficient roof overhang to provide shade. Canvas awnings are more expensive than shades, but they're more pleasing to the eye, they stop the heat on the outside of your building, and they don't obstruct the view.

Hang Shades Outside the Window

Hang tightly woven screens or bamboo shades outside the window during the summer. Such shades will reduce your view, but they are inexpensive and stop 60%–80% of the sun's heat from getting to the window.

Plant Trees or Build a Trellis

Deciduous (leaf-bearing) trees planted to the south or, particularly, to the west of your building provide valuable shade. One mature shade tree can provide as much cooling as five air conditioners (although they're a bit difficult to transplant at that stage, so the sooner you plant the better). Deciduous trees block summer sun but drop their leaves to allow half or more of the winter sun's energy into your home to warm you on clear winter days.

Low-e Films and Exotic Gas Infills

Both low-e films and inert-gas infill windows will improve cooling efficiency as well as heating efficiency. See above, page 189, and the sidebar on page 189 for further details.

Landscaping

When retrofitting existing buildings, most folks don't think about the impact of landscaping on energy use. Appropriate landscaping on the west and south sides of your house provides valuable summertime shading that will reduce unwanted heating by as much as 50%. Those are better results than we get from more expensive projects like window and insulation upgrades! If your landscaping is deciduous, with trees losing their leaves in winter to let the warming winter sun through, you have the best of both worlds. Landscaping that makes outdoor patio and yard spaces cooler, more livable, and inviting in the summer can also block cold winter winds that push through the little cracks and crevices of a typical house. Check out the south-facing landscaping at our Solar Living Center. For more information, see Home Cooling on page 200.

Conserving Electricity

Appliance selection is one area we generally have control over. If you're renting, you may not have control over your house design, window selection and orientation, or heating plant. But you can select the light bulbs in your lamps, the showerhead in your bathroom, and ultraefficient ENERGY STAR appliances. And you can determine whether appliances get turned off (really off!) when not in use. Electricity generation is one of the largest contributors of greenhouse gas emissions, along with the internal combustion engine. So in addition to saving you money, conserving electricity contributes meaningfully to saving the planet.

Many household appliances are important enough to justify entire chapters in the *Sourcebook*. Lighting, heating and cooling (with refrigeration), water pumping, water heating, composting toilets, and water purification are big topics that have significant impact on total home energy use. Please see the individual chapters or sections on the above products. Here we provide general information that applies to all appliances.

Don't Use Electricity to Make Heat

Avoid products that use electricity to produce heat when you have a choice. Making heat from electricity is like using bottled water for your lawn. It gets the job done just fine, but it's terribly expensive and wasteful. Electric space heaters are not the only electrical appliances that produce heat; in fact, most of them do. This includes electric water heaters, electric ranges and ovens, hot plates and skillets, waffle irons, waterbed heaters, and the most common household electric heater, the incandescent light bulb. Most of these appliances can be replaced by other appliances that cost far less to operate. Incandescent bulbs are one of the most dramatic examples, although they are being rapidly phased out, in a clear public victory on behalf of a sane national energy policy that will help to mitigate climate change. Standard light bulbs return only 10% of the energy you feed them as visible light. The other 90% disappears as heat. Compact fluorescent lamps return better than 80% of their energy as visible light, and they last more than 10 times longer per lamp.

If you have a choice about using gas or electricity for residential heating, water heating, or cooking, use gas, even if this means buying new appliances. If you live in an area that has natural gas service, your monthly bills will decline by 25%–30%. As of this writing, propane and oil are more expensive heating fuels then electricity, but costs vary by region. The same goes for clothes drying. Better yet, use the zero-energy-cost option: the clothesline, which has the side benefit of making your clothes last longer and smell wonderfully like fresh air instead of perfumed detergent or dryer sheets. Dryer lint is evidence of your clothes wearing out by tumbling.

Phantoms and Vampires!

Now what's that odd quote in the margin about televisions using power when "off"? Appliances that use power even when they're off create what we call phantom loads. Any device that uses a remote control is a phantom load, because part of the circuitry must remain on in order to receive the "on" signal from the remote. For most TVs, this power use is 15–25 watts. DVDs players are typically 10–50 watts. Together that'll cost you over $20–$25 per year. Either plug these little watt-burners into a switched outlet, or use a switched plug strip so they really can be turned completely off when not in use.

The other increasingly common villains to watch out for are the little transformer cubes that live on the end of many small-appliance power cords. In the electric industry, these are officially known as "vampires," because they constantly suck juice out of your system, even when there's no electrical demand at the appliance. Ever noticed how those little cubes are usually warm to the touch? That's wasted wattage being converted to heat. Most vampire cubes draw a few watts continuously. Vampires need to be unplugged when not in use, or maybe easier just to connect to switched outlet strips.

Go Low-flow, Too

Showers typically account for 32% of home water use. A standard showerhead uses about 2–5 gallons of water per minute, so even a five-minute shower can consume 25 gallons. According to the US Department of Energy, heating water is the second-largest residential energy user. With a low-flow showerhead, energy use and costs for heating water for showers may drop as much as 50%. This is particularly important if you heat your water with electricity, which is the most expensive and energy-inefficient way. A study a few years ago showed that changing to a low-flow showerhead saved 27¢ worth of water and 51¢ of electricity per day for a family of four. So, besides being good for the Earth, a low-flow showerhead will pay for itself in about two months! When we conducted our "elimination of the production of three billion pounds of carbon dioxide from the

> All the remote control televisions in the US, when turned to the "off" position, still use as much energy as the output of one Chernobyl-sized plant.
>
> AMORY LOVINS,
> THE ROCKY MOUNTAIN INSTITUTE

atmosphere" campaign at Real Goods, we found that around one-third of our successful elimination came from the conversion to low-flow showerheads—and it's painless.

Water-saving Showerheads
As noted above, showers account for one-third of the average family's home water use, and heating water is the second-largest residential energy user. Our low-flow showerheads can easily cut shower water consumption by 50% (see realgoods.com). Add an instantaneous water heater for even greater savings—see Chapter 8 for details.

House Heating and Cooling

Heating and cooling a home typically accounts for more than 40% of a family's energy bill (as much as two-thirds in colder regions) and costs on average well over $800 a year—and clearly much more in regions with higher than average heating or cooling demands. That's a good bit of money, and any of us could find more rewarding ways to spend it. Fortunately, it isn't difficult to cut our heating and cooling bills dramatically with a judicious mixture of weatherization, additional insulation, window upgrades, landscaping, improved heating and cooling systems, and careful appliance selection.

Making improvements to your building envelope with weatherization, insulation, window treatments, and landscaping for energy conservation is covered in the preceding section. In this section, we'll cover home heating and cooling systems, looking at sustainability issues, selecting the best options for your home, and improvements and fine-tuning to keep those systems working efficiently.

Sustainability and Renewables
Whether you're concerned about costs, environmental impact, or both, it makes sense to consider home heating and cooling from a sustainability perspective. We know that the fossil fuels are going to run out eventually; we know that fossil fuel prices are very likely to increase, probably dramatically, as supplies become used up during the 21st century; we know that our continuing reliance on oil from the Middle East carries various risks that portend the real possibility of price hikes; and we know that our enormous consumption of fossil fuels produces tremendous, potentially catastrophic, environmental damage. The solutions to all these problems depend upon greater use of renewable energy resources: sun, wind, water, biomass, and geothermal heat.

Home heating and cooling options can be more or less sustainable, depending on your choice of strategies, systems, and fuels. As discussed above and in Chapter 2, maintaining a tight building envelope and incorporating passive solar strategies to whatever extent possible are the obvious places to start.

Passive Solar Heat
Solar energy is the most environmentally friendly form of heating you can find. The amount of solar energy we take today in no way diminishes the amount we can take tomorrow, or in the next decade, or even in the next millennium. Free fuel is always the cheapest fuel. Passive solar heating is nonpolluting and does not produce greenhouse gases. It uses no moving parts, just sunshine through south-facing insulated windows and thermal mass in the building structure to store the heat. (Those south of the equator, please make the usual direction adjustments.)

A carefully designed passive solar building can rely on the sun for half or more of its heating needs in virtually any climate in the United States. Many successful designs have cut conventional heating loads by 80% or better. Some buildings, such as the Real Goods Solar Living Center in Hopland, California (page 408), SunHawk (page 70) or the Rocky Mountain Institute's headquarters at high elevation in Snowmass, Colorado, have no need whatsoever for a central heating

Sun path diagrams. Passive solar heating is practicable in every climate. This simplified illustration shows the importance of a calculated roof overhang to allow solar heating in the winter but prevent unwanted solar heat gain in the summer. Adapted from an illustration by E SOURCE (1993), *Space Heating Technology Atlas*.

> If a house isn't resource-efficient, it isn't beautiful.
>
> AMORY LOVINS, THE ROCKY MOUNTAIN INSTITUTE

system. RMI even grows banana trees indoors! By the same token, in hot climates, good design, shading, and passive cooling strategies can eliminate the need for mechanical cooling or air conditioning. Our Solar Living Center can survive over four weeks of daily high temperatures exceeding 100°F—and a steady flow of overheated visitors—while the interior temperature stays below 78°F without using air conditioning or mechanical cooling in any way.

Adding south-facing windows or a greenhouse is a way to retrofit your house to take advantage of free solar gain. See Chapter 2 for more detailed information about passive solar design.

Active Solar Heating: Best for Hot Water

In many climates around the US, flat-plate collector systems can provide enough solar heat to make a central heating system unnecessary, but in most regions, it is prudent to add a small backup system for cloudy periods. Active solar systems can be added at any time to supplement space heating and water heating needs. However, active solar space heating has a high initial cost, is complex, and has the potential for high maintenance. We don't recommend it when it's so very easy to employ passive solar! The beauty of passive systems is in their simplicity and lack of moving parts. Remember the KISS rule: Keep It Simple, Stupid.

The best and most common use of active systems is for domestic water heating. See Chapter 8 for a detailed discussion of solar hot water systems.

Buying a New Heating System

If your old heating system is about to die, you're probably in the market for a new one. Or, if you're currently spending over $1,000 per year on heating, it's likely that the $800–$4,500 you'll spend on a more efficient heating system will reduce your bills enough to pay for your investment in several years. Your new system also will be more reliable, and it will increase the value of your house. As you shop, keep in mind the following factors that will save you money and increase your comfort:

If you've weatherized and insulated your home—and we'll remind you again that these are by far the most cost-effective things you can do—then you can downsize the furnace or boiler without compromising the capability of your heating system to keep you warm. An oversized heater will cost more to buy up front and more to run every year, and the frequent short-cycle on and off of an oversized system reduces efficiency.

Ask your heating contractor to explain any sizing calculations and make sure he or she understands that you have a tight, well-insulated house, to verify that you don't get stuck with an oversized model (which the contractor gets to charge more money for).

Wood Stoves, Masonry Stoves, and Pellet Stoves

Wood heat is a mixed blessing. If harvested and used in a responsible manner, firewood can be a sustainable resource. As long as the wood you burn is part of a managed cycle that includes replanting trees, heating with wood is a carbon-neutral process that does not make a net contribution to climate change. That's because the CO_2 released when the wood burns is equal to the CO_2 the tree absorbed while it was growing. Burning it does emit CO_2, but a tree newly planted to replace it will absorb an equivalent amount over its lifetime. The carbon impact of burning wood is generally less significant than that of burning long-buried fossil fuels like oil.

But, to be honest, most wood gathering is not done sustainably. If you have your own woodlot, you can create a sustainable forest management plan with the help of a professional forester, if necessary. If you buy wood, you can try to ascertain if your wood supplier cuts in a responsible manner or buys logs from someone who does—and you can encourage him or her to do so. You can also participate in reforestation programs run by organizations such as American

A carefully designed passive solar building can rely on the sun for half or more of its heating needs in virtually any climate in the United States.

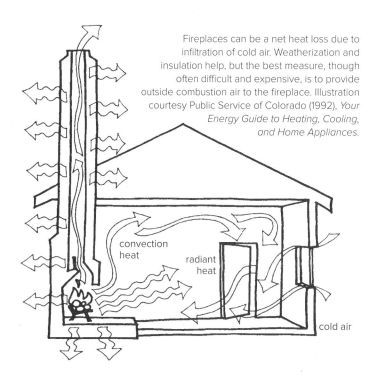

Fireplaces can be a net heat loss due to infiltration of cold air. Weatherization and insulation help, but the best measure, though often difficult and expensive, is to provide outside combustion air to the fireplace. Illustration courtesy Public Service of Colorado (1992), *Your Energy Guide to Heating, Cooling, and Home Appliances.*

convection heat

radiant heat

cold air

Forests; it costs very little to offset your annual fuel wood consumption by funding the planting of new trees.

Pollution is still a problem, though progress is being made. There are approximately 12 million wood-burning stoves and fireplaces in the United States, the majority of them older pre-EPA designs, that contribute millions of tons of pollutants to the air we breathe. In Washington State, for example, wood heating contributes the majority of the particulates in two of the counties with the worst air pollution problems. Modern wood stoves have catalytic combustors and/or other features that boost their efficiencies into the 60%–75% range while lowering emissions by two-thirds. Older wood stoves without air controls have efficiencies of only 20%–30%. Wood burns at a higher temperature, and the gases go through a secondary combustion in a catalytic—or a new technology, noncatalytic—wood stove, thereby minimizing pollution and creosote buildup.

If you're buying a new wood stove, make sure it's not too large for your heating needs. This will allow you to keep your stove stoked with the damper open, resulting in a cleaner and more efficient burn. Some stoves incorporate more cast iron or soapstone as thermal mass to moderate temperature swings and prolong fire life. If you live in a climate that requires heat continuously for several months, you definitely want the heaviest stove you can afford. Masonry heaters take the thermal mass idea one very comfortable step further.

Masonry Stoves

Masonry stoves, also called Russian stoves or fireplaces, developed by a people who know heating, have long been popular in Europe and Scandinavia. These large freestanding masonry fireplaces work by circulating the heat and smoke from the combustion process through a long twisting labyrinth masonry chimney. The mass of the heater slowly warms up and radiates heat into the room for hours. Masonry stoves are best suited for very cold climates that need heat most of the day. They are much more efficient than fireplaces. They have very low emissions because airflow is restricted only minimally during burning, giving a clean high-temperature burn. How many wood stoves have seats built into them so that you can snuggle in? Russian fireplaces commonly do. As Mark Twain observed, "One firing is enough for the day…the heat produced is the same all day, instead of too hot and too cold by turns."

Masonry stoves are expensive initially but will last for the life of the home and are one of the cleanest, most satisfying ways to heat with wood. They're easiest to install during initial construction and are best suited to well-insulated homes with open floor plans that allow the heat to radiate freely. The Tulikivi masonry stove at SunHawk stays warm for 48 hours after firing one load of wood (see page 73).

Pellet Stoves

If you can't burn wood but are looking for a similar way to wean yourself from oil, a pellet stove may be right for you. As renewable energy expert Greg Pahl observes, pellets have clear environmental benefits: "Pellets are a renewable resource, so burning pellets does not add any net carbon dioxide to the atmosphere. Because of the extremely hot combustion in most pellet-fired appliances, pellets burn cleanly with extremely low emissions and produce virtually no creosote. Compared with other fuels, burning 1 ton of pellets instead of heating with electricity will save 3,323 pounds of carbon emissions. You'll save 943 pounds of emissions per ton by replacing oil with pellets and 549 pounds per ton if you're replacing natural gas."

Pellet stoves have the advantage that the fuel pellets of compressed sawdust, cardboard, grass, or agricultural waste are fed automatically into the stove by an electric auger. The feed rate is dialed up or down with a rheostat, so they'll happily run all night if you can afford the pellets. You fill the reservoir on the stove once every day or so. Pellet stoves are cleaner burning than wood stoves because combustion air is force-fed into the burning chamber. Unlimited access to oxygen and drier fuel than firewood mean a more efficient combustion. But the pellets may be slightly more expensive than firewood in most parts of the country and simply unavailable in some areas. Pellets must be kept bone dry in order to auger and burn.

Pellet stoves have disadvantages, too. They have more mechanical parts than wood stoves do, so they're subject to maintenance and breakdown, and they require electricity. No power, no fire—the fire will go out without the combustion blower running. Pellet stoves usually cannot be used with restricted renewable electricity systems, but many renewable energy systems have considerable extra power in the winter, like from a hydro system.

Displace Oil with Biodiesel

If you're stuck with an oil-burning furnace or boiler, don't just sit there and grimace about all

Masonry stoves are best suited for very cold climates that need heat most of the day.

the greenhouse gases you're putting into the atmosphere—you *can* do something about it (besides making sure your house is tight and your system efficiency is high): biodiesel. The original diesel engine exhibited at the Paris World's Fair in 1900 ran on vegetable oil, and the concept remains more than viable. Biodiesel is a renewable fuel that can be made easily from a variety of biomass feedstocks, even from recycled cooking oil. Biodiesel has been promoted primarily as a vehicle fuel, but it also works as a home heating fuel additive. Increasing numbers of people are burning B100, or pure biodiesel, while others burn a blend of #2 heating oil mixed with 10% or 20% biodiesel (B20); no conversion is required, and the extra maintenance needed is minimal. B20 blends tend to be slightly more expensive than heating oil, but the price is bound to become more competitive over time. Biodiesel manufacturing facilities are springing up everywhere, and fuel distributors are beginning to supply it to consumers. Seek and ye shall find.

Any biodiesel you use to displace oil from your heating system lowers carbon dioxide emissions and reduces particulate pollution. Call a local fuel distributor or your state energy office for more information about obtaining biodiesel for heating purposes. See Chapter 12, "Sustainable Transportation," for details on biodiesel as an alternative motor vehicle fuel.

Fuel Switching: Natural Gas Furnaces and Boilers

When available, natural gas is the lowest-cost heating fuel for most homes. If you have a choice, switching fuels may be cost-effective but only if this does not also require substantial changes to your existing heat distribution system. For example, replacing the old hot water boiler with a high-efficiency forced-air furnace would require installation of all new forced-air ducts.

It is always cost-effective to pay a little more up front in return for more efficiency. But it may not be cost-effective to pay a lot more for the most efficient gas unit, since your reduced heating needs mean a longer payback. The differences

> Biodiesel is a renewable fuel that can be made easily from a variety of biomass feedstocks, even from recycled cooking oil. Biodiesel has been promoted primarily as a vehicle fuel, but it also works as a home heating fuel additive.

To Burn or Not to Burn

I'm totally confused. The subject is wood, a material I have used to heat my home for the past 30 years. I know it's more work to burn wood, but I like the exercise. It makes more sense to burn calories splitting and hauling hardwood than running in front of a television set on a treadmill at the health spa. I enjoy the fresh air. I like the warmth. I like the flicker and glow of embers.

Sometimes, when I am stacking wood so that a summer's worth of sunshine can make it a cleaner, hotter fuel, I think of those supertankers carrying immense cargos of black gold from the ancient forests of the Middle East. I think of the wells that Saddam Hussein set afire in Kuwait, and the noxious roar of the jet engines of the warplanes we used to ensure our country's access to cheap oil.

My wood comes from hills that I can see. It's delivered by a guy named Paul, who cuts it with his chain saw, then loads it in his one-ton pickup. We talk about the weather, the conditions in the woods, and the burning characteristics of different species. We finish by bantering about whether he's delivered a large cord, a medium, or a small. I can't imagine having the same conversation with the man who delivers oil or propane.

Unfortunately (and this is where I begin to get confused), there is another side to the wood-burning issue. Along comes evidence that the emissions from airtight stoves contain insidious carcinogens, and further indications that a home's internal environment can be more adversely affected by burning wood than by passive cigarette smoke. So now I burn my wood in a "clean" stove. But my environmental activist friends say that the only good smoke is no smoke and that even my supposedly high-tech stove is smogging up their skies.

Theoretically, I should be able to go back to feeling good about wood, but life is never that simple. On the positive side of the ledger comes the information that wood burning, in combination with responsible reforestation, actually helps the environment by reversing the greenhouse effect. The oxidation of biomass, whether on the forest floor or in your stove, releases the same amount of carbon into the atmosphere. It makes more sense for wood to be heating my home than contributing to the brown skies over Yellowstone.

This is obviously a simplistic analysis, but wood should be a simple subject. In the meantime, I have to keep warm, so I'm choosing to burn wood. I keep coming back to the fact that humans have been burning oil for more than 100 years and wood for 5 million.

—Stephen Morris

between various models of gas furnaces and boilers can be significant, and it's worth your while to examine efficiency ratings before making a decision.

If you are currently using electricity for heating, switching to any other fuel will be cost-effective. See the next subject.

Electric Resistance Heating

Electric baseboard heating is by far the most expensive and least efficient way to warm your home; it's cheap to buy and install, but it costs two to three times more to heat with electricity than it does to heat with gas. The life-cycle cost of electric heating is, without exception, far higher than that of gas-fired furnaces and boilers, even taking into account the higher installation cost of the latter, and even in regions with exceptionally low electricity prices. Electric resistance is an energy-inefficient way to produce heat, and because grid-generated electric power is a key greenhouse gas culprit, this is a terrible environmental choice. (Electric heating may be excused if you have a plentiful, seasonal supply of renewably generated power, for example from a year-round micro-hydro system.)

If you must heat with electricity, making sure that your home is well insulated and weatherized, along with installing a programmable thermostat, are no-brainer investments that you should do immediately. If you also use air conditioning and live in a mild winter climate, consider switching to a heat pump.

> The life-cycle cost of electric heating is, without exception, far higher than that of gas-fired furnaces and boilers.

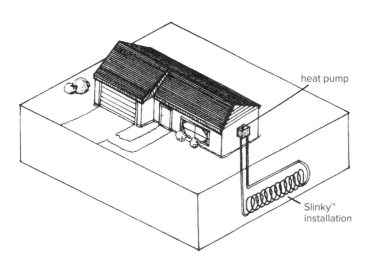

Ground-source Heat Pump. Ground-source heat pumps can be highly cost-effective in new construction over the life cycle of heating and cooling equipment compared to more conventional options. Several designs are on the market; this illustration shows the Slinky™ design. Adapted from an illustration by E SOURCE (1993), *Space Heating Technology Atlas.*

Geothermal Energy: Heat Pumps

Geothermal heating is a simple and elegant concept. Earth has been absorbing and storing solar energy for millions of years, and it's a massive object: Voilà, the epitome of the phrase "thermal mass." A heat pump is a mechanical device that makes use of this fact, and the fact that the temperature just below the surface remains constant year-round.

Heat pumps are considered a renewable energy system because they consume no fuel to generate heat. In fact, they don't generate heat at all; they operate by moving heat from a warmer place to a cooler place. However, a heat pump uses electricity to perform this sleight of hand. Heat pumps, therefore, can be considered the most efficient form of electric heat. And they are very efficient.

There are three types of heat pumps: air-to-air, ground-source, and water-source. They collect heat from the air, water, or ground depending on type, concentrate the warmth, and distribute it through the home. Heat pumps typically deliver three times more energy in heat than they consume in electric power. Heat pumps also can be used to cool homes by reversing the process—collecting indoor heat and transferring it outside the building. Some newer models are designed to provide an inexpensive source of household hot water as well.

Not all heat pumps are created equal, but fortunately the Air Conditioning and Refrigeration Institute rates all the ones on the market. Look at the efficiency ratings and purchase a system designed for a colder climate. Some electric utilities offer rebates and other incentives to help finance the higher capital costs of these more efficient systems.

Air-to-air heat pumps are the most popular, the least expensive initially, and the most expensive to run in colder weather. Air-to-air units must rely on inefficient backup heating mechanisms (usually electric resistance heat) when outside temperatures drop below a certain point. That point varies from model to model, so make sure you buy one that's designed for your climate— some models can cope with much colder weather before resorting to backup heat. A heat pump running on backup is no better than standard electric resistance heating. Some newer pumps have more efficient gas-fired backup mechanisms, but if gas is available, why not use it directly?

Ground- or water-source heat pumps are among the most efficient heating systems around and are substantially more cost-effective in colder climates. Because of the extensive burial of plastic pipe required, these systems are best installed during new construction. Tens of thousands of ground-source heat pumps have been installed in Canada, New England, and other frost-belt areas. See the annualized heating/cooling system cost chart below for a comparison.

Fireplace Inserts

As noted by the authors of the *Consumer Guide to Home Energy Savings* (available at realgoods .com), a fireplace is essentially part of a room's décor, not a source of heat. A customary fireplace installation generally will lose more heat than it provides. However, if outfitted with a well-sealed insert, a fireplace can provide some useful heat.

A fireplace insert fits inside the opening of a fireplace, operates like a wood stove, and offers improved heat performance. An insert is difficult to install because it must be lifted into the fireplace opening and the space around it covered with sheet metal and sealed to reduce air leaks. Such inserts are only 30%–50% efficient, which is a good improvement over a fireplace but less efficient than a good freestanding wood stove, which is free to radiate heat off all sides.

A fireplace can be retrofitted with an airtight wood stove installed in, or better yet in front of, the fireplace. This is the most efficient choice, since all six heat-emitting sides are inside the living space, on the hearth. The stove's chimney is linked into the existing flue, and good installations line the chimney all the way to the top with stainless pipe to limit creosote buildup and provide easy flue cleaning. Simply shoving a few feet of wood stove pipe into the existing chimney is begging for a chimney fire of epic proportions. Any creosote in the smoke stream will condense on the large, cool chimney walls, providing a spectacular amount of high-temperature fuel in a few years. This has been well proven to be an excellent way to burn a house down—further demonstrations and experiments aren't necessary. Line your chimney when retrofitting a wood stove.

Gas Fireplaces

Gas fireplaces have combustion efficiencies up to 80%, whereas the initially less expensive gas logs are only 20%–30% efficient. However, as a precaution against gas leaks, many local codes require dampers to be welded open. This reduces the overall efficiency of these units unless they have tight-fitting doors to cut down the amount of warmed interior air allowed to float out freely.

People Heaters

It often makes sense to use a gas or electric radiant or convection heater to warm only certain areas of the home or, as in the case of radiant heaters, to keep individual people warm and comfortable. Radiant systems keep you and objects around you warm, just like the sun warms your skin. This allows you to set the thermostat for the main heating system lower by 6–8°F.

Even if you heat with gas, electric convective or radiant spot heaters can save you money, depending on how the system is used. Radiant systems are designed and used more like task lights; you turn them on only when and where heat is needed, rather than heating the whole house. Most central heating systems give you little control in this respect, since they are designed to heat the entire home.

Heating Costs Comparison

While operating costs will vary by climate and region, and fuel and electricity prices differ across the country, the following chart represents estimated installation and operating costs for selected heating and cooling systems in a typical single-family house. It assumes that heat pumps will provide air conditioning in the summer as well as space heating in the winter.

Fine-tuning Your Heating System

Let's assume that your home is already well insulated and weather-tight. You have reduced the

Comparing Heating Fuel Costs*	
IN DOLLARS PER MILLION BTU	
ELECTRICITY	
Ground-source heat pump (COP** = 3.0)	$11
Air-source heat pump (COP = 2.0)	$15
Baseboard resistance heater (COP = 1.0)	$36
NATURAL GAS	
Central furnace (AFUE*** = 85%)	$13
PROPANE	
Central furnace (AFUE = 85%)	$38
FUEL OIL	
Central furnace (AFUE = 80%)	$28
CORDWOOD	
Airtight stove (65% efficient)	$13
WOOD PELLETS	
Pellet stove (80% efficient)	$15

*Fuel costs are 2013 national averages from eia.gov /neic/experts/heatcalc.xls: electricity = $0.12/kWh; natural gas = $1.10/therm.; propane = $2.68/gal.; fuel oil = $3.93/gal.; cordwood @ $200/cord; wood pellets @ $0.10/lb.
**COP = Coefficient of Performance
***AFUE = Annual Fuel Utilization Efficiency

amount of heat escaping in the most cost-effective way. It now makes sense to look at how efficiently your heating system produces and delivers heat to your living space. This section gives tips for making your existing heating system run more efficiently, whether it is powered by gas, electric, oil, wood, or solar energy. The following suggestions are either relatively inexpensive or free. All are highly cost-effective.

If you have an older oil burner, installing a flame-retention burner head (which vaporizes the fuel and allows more complete combustion) typically will pay back your investment in two to five years.

Remember, if your existing system appears close to a natural death, or its maintenance costs are high, these fine-tuning tips may not be particularly cost-effective. Modern heating systems have seen dramatic efficiency gains over the past few years, so replacing an old one could well save you money.

For safety reasons, adjustments, tune-ups, and modifications to your heating system are best done by a heating system professional. Homeowners and renters can improve system performance by insulating ducts and pipes, cleaning registers, replacing filters, and installing programmable thermostats.

Furnaces and Boilers
Heating System Tune-ups
Gas furnaces and boilers should be tuned every two years, oil units once a year. Your fuel supplier can recommend or provide a qualified technician. Expect to pay $60–$150—money well spent. Also have the technician do a safety test to make sure the vent does not leak combustion products into the home. You can do a simple test by extinguishing a match a couple of inches from the spillover vent: The smoke should be drawn up the chimney.

During a furnace tune-up, the technician should clean the furnace fan and its blades, correct the drive-belt tension, oil the fan and the motor bearings, clean or replace the filter (make sure you know how to perform this monthly routine maintenance), and help you seal ducts if necessary.

Efficiency Modifications
While the heating contractor is there to tune up your system, he or she may be able to recommend some modifications, such as reducing the nozzle (oil) or orifice (gas) size, installing a new burner and motorized flue damper, or replacing the pilot light with an electronic spark ignition. If you have an older oil burner, installing a flame-retention burner head (which vaporizes the fuel and allows more complete combustion) typically will pay back your investment in two to five years.

Turn off the Pilot Light During the Summer
This simple act will save you about $2–$4 per month. Do this only if you can safely light it again yourself, so you don't have to pay someone to do it. Federal regulations now require that new natural gas-fired boilers and furnaces be equipped with electronic ignition, saving $30-$40 per year in gas bills. Propane-fired units cannot be sold without a pilot light and cannot be retrofitted safely with a spark igniter unless you install expensive propane-sniffing equipment. Propane is heavier than air, and any leaking propane can pool around the unit, creating an explosion potential when the unit tries to start. Natural gas is lighter than air.

Insulate the Supply and Return Pipes on Steam and Hot Water Boilers
Use a high-temperature pipe insulation such as fiberglass wrap for steam pipes. The lower temperatures of hydronic or hot water systems may allow you to use a foam insulation (make sure it's rated for at least 220°F).

Clean or Change Your Filter Monthly
A clogged furnace filter impedes airflow, makes the fan work harder, makes the furnace run longer, and cuts overall efficiency. Filters are designed for easy service, so the hardest part of this maneuver is turning into the hardware store parking lot with a note in hand of the correct size filter. While you're there, pick up several, since you'll want to replace the filter every month during the heating season. Most full-house air conditioning systems use the same fan, ducts, and filter. So monthly changes during A/C season are a must, too. They'll set you back a couple of bucks apiece. For five bucks, you can buy a reusable filter that will need washing or vacuuming every month but will last for a year or two.

Seal and Insulate Air Ducts
Losses from leaky, uninsulated ducts—especially those in unheated attics and basements—can reduce the efficiency of your heating system by as much as 30%. Don't blow your expensive heated air into unheated spaces! Seal ducts thoroughly with mastic, caulking, or duct tape, and then insulate with fiberglass wrap.

A clogged furnace filter impedes airflow, makes the fan work harder, makes the furnace run longer, and cuts overall efficiency. Filters are designed for easy service, so the hardest part of this maneuver is turning into the hardware store parking lot with a note in hand of the correct size filter.

Radiators and Heat Registers

Vacuum the cobwebs out of your registers. Anything that impedes airflow makes the fan work harder and the furnace run longer.

Reflect the Heat from Behind Your Radiator

You can make foil reflectors by taping aluminum foil to cardboard, or use radiant barrier material (left over from your attic or basement). Place them behind any radiator on an external wall with the shiny side facing the room. Foil-faced rigid insulation also works well.

Vacuum the Fins on Baseboard Heaters

Vacuuming up dust improves airflow and efficiency. Keeping furniture and drapes out of the way also improves airflow from the radiator.

Bleed the Air out of Hot Water Radiators

Trapped air in radiators keeps them from filling with hot water and thus reduces their heating capacity. Bleeding out the air should also quiet down clanging radiators. You can buy a radiator key at the hardware store. Hold a cup or a pan under the valve as you slowly open it with the key. Close the valve when all the air has escaped and only water comes out. If it turns out that you have to do this more than once a month, have your system inspected by a heating contractor.

Thermostats

Install a Programmable Thermostat

One-half of homeowners already turn down their heat at night, saving themselves 6%–16% of heating energy. A programmable "setback," or clock thermostat can do this for you automatically. In addition to lowering temperatures while you sleep, it also will raise temperatures again before you get out from under the covers. It not only sounds good, it's cost-effective too. Such thermostats can be used to drop the house temperature during the day as well if the house is unoccupied. *Home Energy* magazine reports savings in excess of 20% with two eight-hour setbacks of 10°F each.

Electronic setback thermostats are readily available for $25–$150. If you have a heat pump or central air conditioning, make sure the thermostat is designed for heat and A/C. Installation is simple, but make sure to turn off the power to the heating plant before you begin. If connecting wires makes you jittery, have an electrician or heating technician do it for you. It's well worth the investment.

Real Goods now sells a smart thermostat that you can control from anywhere via the Internet: find it at realgoods.com.

Heat Pumps

Change or Clean Filters Once a Month

Dirty filters are the most common reason for heat pump failures.

Keep Leaves and Debris Away from the Outside Unit

Good airflow is essential to heat pump efficiency. Clear a 2-foot radius. This includes shrubbery and landscaping.

Listen Periodically to the Compressor

If the compressor cycles on for less than a minute, something is wrong. Have a technician adjust it immediately.

Dust off the Indoor Coils of the Heat Exchanger at Least Once a Year

This dust-off should be part of the yearly tune-up if you pay a technician for this service.

Have the Compressor Tuned Up

Every year or two, a technician should check refrigerant charge, controls, and filters, and oil the blower motors.

Wood Stoves

Use a Wood Stove Only if You're Willing to Make Sure It Burns Cleanly

Wood stoves aren't like furnaces or refrigerators; you can't just turn them on and forget them. "Banking" a wood stove for an overnight burn only makes the wood smolder and emit lots of pollution. Sure, we've all done it, but we didn't know any better. Now we do. A well-insulated house will hold enough heat that you shouldn't

Thermostat Myths

Myth: Turning down your thermostat at night or when you're gone means you'll use more energy than you saved when you have to warm up the house again.
Fact: You always save by turning down your thermostat no matter how long you're gone. (The only exception is older electric heat pumps with electric resistance heaters for backup.)
Myth: The house warms up faster if you turn the thermostat up to 90°F initially.
Fact: Heating systems have only two "speeds": ON or OFF. The house will warm up at the same rate no matter what temperature you set the thermostat at, as long as it is high enough to be "ON." Setting it higher only wastes energy by overshooting the desired temperature.

Fireplaces offer a good light show and friendly ambience, but are not an effective source of heat. In fact, fireplaces are usually a net heat loser because of the huge volume of warm interior air sucked up the chimney, which has to be replaced with cold air leaking into the house.

need to keep the stove going all night. Heavier masonry stoves are a good idea in colder climates. Once the large mass is warmed up, it emits slow, steady heat overnight without a smoky fire.

Burn Well-seasoned Wood and No Trash
Wood should be split at least six months before you burn it. Never burn garbage, plastics, plywood, or treated lumber.

Hotter Fires Are Cleaner
Another good case for masonry stoves, which burn hot, clean, short fires with no air restriction. The heat is soaked up by the masonry and emitted slowly over many hours.

If a Wood Stove Is Your Main Source of Heat, Use an EPA-approved Model
EPA regulations now require wood stoves to emit less than 10% of the pollution that older models produced.

Check Your Chimney Regularly
Blue or gray smoke means your wood is not burning completely.

Fireplaces
Fireplaces offer a good light show and friendly ambience, but are not an effective source of heat. In fact, fireplaces are usually a net heat loser because of the huge volume of warm interior air sucked up the chimney, which has to be replaced with cold air leaking into the house. Fireplaces average minus 10% to plus 10% efficiency. If you've done a good job of weatherizing your house and sealing up air leaks, you may lack adequate fresh air for combustion, or the chimney won't draw properly, filling the house with smoke. The two best things to do with a fireplace are (1) don't use it and plug the flue, or (2) install a modern wood stove and line the chimney.

Improve the Seal of the Flue Damper
To test the damper seal, close the flue, light a small piece of paper, and watch the smoke. If the smoke goes up the flue, there's an air leak. Seal around the damper assembly with refractory cement. (Don't seal the damper closed.) If the damper has warped from heat over the years, get a sheet-metal shop to make a new one, or consider a wood stove conversion.

Install Tight-fitting Glass Doors
Controlling airflow improves combustion efficiency by 10%–20% and may help reduce air leakage up a poorly sealed damper.

Use a C-shaped Tubular or Blower-equipped Grate
Tube grates draw cooler air into the bottom and expel heated air out the top. Blower-equipped grates do the same job with a bit more strength.

Caulk Around the Fireplace and Hearth
Do this where they meet the structure of the house, using a butyl rubber caulk.

Use a Cast-iron Fireback
Firebacks are available in a variety of patterns and sizes. Fireplace efficiency is improved by reflecting more radiant heat into the room.

Locate the Fire Screen Slightly Away from the Opening
Moving the screen forward into the room a bit allows more heated air to flow over the top of the screen. While fire screens do prevent sparks from flying into the room, they also prevent as much as 30% of the heat from entering the room.

If You Never Use Your Fireplace, Put a Plug in the Flue to Stop Heat Loss
Seal the plug to the chimney walls with good-quality caulk and tell this to anyone who might build a fire (or leave a sign). Temporary plugs can be made with rigid board insulation, plywood with pipe insulation around the edges, or an inflatable plug.

Home Cooling Systems

Cooling your home, like heating it, can be accomplished in sensible, inexpensive, energy-efficient ways, or in the wasteful way that Americans have favored for the past several decades in the era of cheap fossil fuels: Install a big air conditioner, crank it up, and pay the bills. It's much better for your pocketbook, and for the planet, to pursue the common-sense approach: Reduce your cooling load, take advantage of passive solar cooling strategies, utilize low-energy methods like fans, and increase the efficiency of your existing air conditioner. Finally, if you determine that it's cost-effective, you should explore the prospect of buying new, efficient cooling equipment.

Reduce the Cooling Load Through Smart Design and Passive Solar Strategies

The best strategy for keeping a dwelling cool is to keep it from getting hot in the first place. This means preventing outside heat from getting inside, and reducing the amount of heat generated inside by inefficient appliances such as incandescent lights, unwrapped water heaters, or older refrigerators.

In a hot, humid climate, 25% of the cooling load is the result of infiltration of moisture; 25% is from outdoor heat penetrating windows, walls, and roof; 20% is from unwanted solar gain through windows; and the remaining 30% is from heat and moisture generated within the home. It makes a lot of economic sense to cut these loads before getting that new monster air conditioner. You'll be able to get by with a smaller, less-expensive unit that costs less to operate for the rest of its life. And by cutting your power consumption, you'll significantly reduce your personal emissions of greenhouse gases and other environmental pollutants.

The same kinds of smart design choices that reduce your heating load and make your home energy-efficient also serve to reduce the cooling load. Weatherizing, insulating, radiant barriers, good-quality windows, environmental landscaping, effective ventilation, and roof whitening each has the capacity to reduce your cooling costs substantially, especially in combination. Funny how we keep hammering those points. These tricks for keeping outside heat from getting inside should be considered, and implemented, before you move on to even low-energy methods that consume electricity—especially if you rely on renewable power systems.

Each of these methods, plus others, is recapped briefly here. For more-detailed discussions of these aspects of energy-efficient building design, see Chapter 2.

Weatherization

Weatherization, usually done in northern climates to keep heat in, can also effectively keep heat and humidity out. Cutting air infiltration by half, which is easily done, can cut air conditioning bills by 15% and save $50–$100 per year in the average house in the southern United States. Much of the work done by an air conditioner in humid climates goes into removing moisture, which increases comfort. Reducing air infiltration is more important in humid climates than in dry ones.

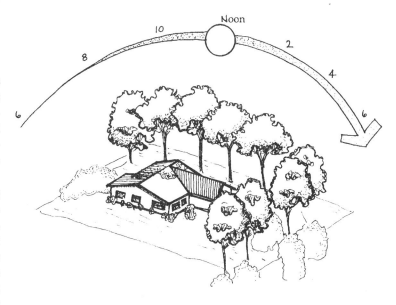

Mature trees shading your house will keep it much cooler by lowering roof and attic temperatures, blocking unwanted direct solar heat radiation from entering the windows, and providing a cooler microclimate outdoors. Adapted from an illustration by Saturn Resource Management, Helena, Montana.

Insulation

Insulation slows heat transfer from outside to inside your home, or vice versa in the winter. Attic insulation levels should be R-19 or higher. If you also need to heat your house in the winter, insulation is even more important because of the greater temperature differential between inside and outside. In that case, you'll want to have insulation levels of R-30 or higher. In hot climates, if you already have some significant amount of insulation, it will probably be more cost-effective to install a radiant barrier instead of adding more insulation.

Radiant Barriers

Radiant barriers are thin metal films on a plastic or paper sheet to give them strength. They will

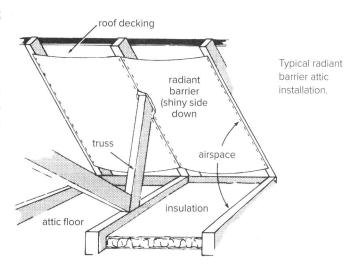

Typical radiant barrier attic installation.

Shading that blocks summer sun on the east, south, and west sides of your house, but not summer breezes, is one of the most effective ways to keep your home cooler.

reflect or stop approximately 97% of long-wave infrared heat radiation. Such barriers typically are stapled to the underside of attic rafters to lower summertime attic temperatures. Laying these barriers on the attic floor is also an option, but research indicates that the effectiveness of the barrier is reduced by dust buildup. Lowering attic temperatures—which easily can reach 160°F on a sunny day—reduces heat penetration into the living space below and is frequently a cost-effective measure to lower air conditioning bills. A study by the Florida Solar Energy Center found that properly installed radiant barriers reduced cooling loads by 7%–21%. The study also found that temperatures of the roofing materials were increased by only 2–10°F, which will have little or no impact on the life expectancy of a shingle. A radiant barrier will reduce, but not eliminate, high attic temperatures, so insulating to at least R-19 is still advisable to slow down heat penetration. The cost for radiant barrier material is very moderate at approximately 20¢–40¢ per square foot. We sell Reflectix at realgoods.com.

Windows

For too long, a window has simply been a hole in the wall that lets light and heat through. Windows, even so-called thermal pane (dual-glazed) windows, have very low R-values. Single-pane units have an R-value of 1; thermal pane windows have an R-value of 2. Both are pretty miserable.

Conventional windows conduct heat through the glass and frame, permit warm, moist air to leak in around the edges, and let in lots of unwanted heat in the form of solar radiation. South-facing windows with properly sized roof overhangs or awnings won't take in the high summer sun but will accept the lower winter sun when it usually is wanted. East- and west-facing windows cannot be protected so simply. See Chapter 2 for a complete discussion of window treatments, including newer high-tech windows with R-values up to 8 or 10.

Landscaping for Shade and Passive Cooling

You can take advantage of so-called environmental landscaping to provide shading and/or to channel cooling breezes through your home.

Shading that blocks summer sun on the east, south, and west sides of your house, but not summer breezes, is one of the most effective ways to keep your home cooler. Planting shade trees, particularly on the west and south sides of your house, can greatly increase comfort and

coolness. (Remember that deciduous trees, which lose their leaves in winter but grow a thick canopy of leaves in the summer, should be used on the south side in colder climates to maximize solar gain in winter and minimize it in summer.) Awnings, porches, or trellises on those same sides of a building will reduce solar gain through the walls as well as through the windows. A home's inside temperature can rise as much as 20°F or more if the east and west windows and walls are not shaded.

Trees and shrubs planted in the proper configurations also can funnel and concentrate breezes to increase airflow through and around a house. Increased airflow translates into increased cooling action, all thanks to Mother Nature.

Ponds are an effective method of passive cooling as well. The pond at SunHawk (page 72) provides much summer cooling by the convective effect of air passing over the cool water of the pond and blowing toward the house.

Mechanical Shading

In addition to trellises and awnings, various types of mechanical window-shading devices can be cost-effective. Some are internal (such as drapes, curtains, and shades); some are external (such as louvers, shutters, and exterior roller shades). In the overall scheme of things, these devices are relatively inexpensive, and they operate effectively to block sun and heat from entering your home. Explore the options and decide which ones will work best in various applications.

Roof Whitening

A reflective white roof is an obvious solution. If you're replacing a roof, using a white or reflective roofing surface will reduce your heat load on the house. If you are not reroofing, coating the existing roof with a white elastomeric paint is a cost-effective measure in some climates. Be sure to get a specialty paint specifically designed for roof whitening: Normal exterior latex won't do.

The Florida Solar Energy Center reports savings from white roofs of 25%–43% on air conditioning bills in two poorly insulated test homes, and 10% savings in a Florida home with good R-25 roof insulation already in place. Costs are relatively high—the coatings cost 30¢–70¢ per square foot of roof surface area—and adding insulation or a radiant barrier may be a more cost-effective measure. Light-colored asphalt or aluminum shingles are not nearly as effective as white paint in lowering roof and attic temperatures.

Passive Nighttime Cooling

Opening windows at night to let cooler air inside is an effective cooling strategy in many regions. This method is far more effective in hot, dry areas than in humid areas, however, because dry air tends to cool down more at night and because less moisture is drawn into the house. The Florida Solar Energy Center found that when apartments in humid areas (and Florida certainly qualifies) opened their windows at night to let the cooler air in, the air conditioners had to work much harder the next day to remove the extra moisture that came along with the night air. Storing "coolth" works better if your house has thermal mass such as concrete, brick, tile, or adobe that will cool down at night and help prevent overheating during the day. These are the same things that make passive solar houses work better at heating in the winter.

Active Solar Ventilation

Ventilating your attic is important for moisture control and keeping cool. Roofs can reach temperatures of 180°F, and attics can easily exceed 160°F. Unless you get rid of this heat, it will eventually soak through even the best insulation. Solar-powered direct-vent fans are a very cost-effective way to move this unwanted heat out, exactly when you need it, when the sun is at its hottest trajectory. By using solar power, they'll

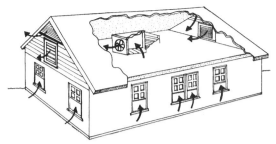

A whole-house fan can cool the house by bringing in cooler air—especially at night—and lowering attic temperatures. Adapted from an illustration by Illinois Department of Energy and Natural Resources (1987), *More for Your Money…Home Energy Savings.*

run only when the sun shines, and the brighter the sun, the faster they'll work. We offer complete kits in two different sizes; they include fan, temperature switch—so it will run only when the attic temperature is above 80°F—photovoltaic module(s), and mounting. (See realgoods.com.)

Whole-house, Ceiling, and Portable Fans

A whole-house fan is an effective means of cooling and is far less expensive to run than an air conditioner. It can reduce indoor temperatures by 3–8°F, depending on the outside temperature. Through prudent use of a whole-house fan, you can cut air conditioner use by 15%–55%. This is true particularly if you live in a climate that reliably cools off in the evening. A high R-value house can be buttoned up during the heat of the

> Storing "coolth" works better if your house has thermal mass such as concrete, brick, tile, or adobe that will cool down at night and help prevent overheating during the day. These are the same things that make passive solar houses work better at heating in the winter.

Renewable Energy Systems and Home Cooling

Traditional air conditioners, even small room-size units, and renewable energy systems have generally been incompatible. The equipment and knowledge have been readily available—the problem is the sheer quantity of electricity required for air conditioning, and the long hours of continuous operation, augmented by the high initial cost of renewable energy systems. It used to be a rule of thumb that it would take about $3.50 worth of renewable energy components for every watt-hour needed. Thus, a 1,500-watt room-size A/C unit running for six hours and consuming 9,000 watt-hours would have required a $30,000 system.

No longer. At today's prices, the PV, mount, and controls required to power that air conditioner would cost approximately $2,500. One reason this is feasible is that, in most climates, the A/C load is only active when the sun is shining. Once the sun goes down, mechanical cooling isn't necessary. That means you don't need to increase your battery capacity to support the air conditioner—and batteries are expensive. According to renewable energy pioneer Jeff Oldham of Regenerative SOLutions, an off-the-grid A/C system including photovoltaics is currently cheaper than an earth tube cooling system (large pipes buried 6–9 feet in the ground pulling earth-cooled air into the home via a fan). Jeff also

points out that because the PVs are so cheap, efficiency may not be as high a priority as it used to be, even recently. For example, a super, high-efficiency air conditioner at about a ton of cooling can cost $1,600, while the least efficient unit available in the US costs $600 for that same ton. It takes an extra $400 of PV to power the low-efficiency unit, which translates to a $600 saving on the overall cost of the cooling system.

The market for efficient air conditioners is evolving rapidly. Look for models by LG, Panasonic, Mitsubishi, and perhaps Fujitsu, which tend to be the most efficient brands.

Cool drinks, the Real Goods Solar Hat Fan, a minimum of lightweight clothing, siestas, going for a swim, or sitting on a shady porch—all of these personal cooling methods work.

day, heating up inside slowly like an insulated ice cube. In the evening, when it's cooler outside than inside, a whole-house fan blows the hot air out of your house while drawing cooler air in through the windows. It should be centrally located so that it draws air from all rooms. Be sure your attic has sufficient ventilation to get rid of the hot air. To ensure a safe installation, the fan must have an automatic shut-off in case of fire.

Ceiling, paddle, and portable fans produce air motion across your skin that increases evaporative cooling. A moderate breeze of one to two miles per hour can extend your comfort range by several degrees and will save energy by allowing you to set your air conditioner's thermostat higher or eliminate the need for air conditioning altogether. Less frequent use of air conditioning by setting the thermostat higher will cut cooling bills greatly.

Evaporative Cooling

If you live in a dry climate, an evaporative, or "swamp" cooler is an excellent cooling choice, saving 50% of the initial cost and up to 80% of the operating cost of an air conditioner. The lower the relative humidity, the better they work. Evaporative coolers are not effective when your relative humidity is higher than about 40%.

Personal Cooling

Cool drinks, the Real Goods Solar Hat Fan, a minimum of lightweight clothing, siestas, going for a swim, or sitting on a shady porch—all of these personal cooling methods work (realgoods .com).

Improving Air Conditioner Efficiency

Once you've reduced heat penetration and taken advantage of passive solar and low-energy cool-

ing methods, if you still have a need for additional cooling, it makes sense to improve the efficiency of your air conditioner.

Check Ducts for Leaks and Insulate

With central A/C systems, inspect the duct system for leaks and seal with duct tape or special duct mastic where necessary. The average home loses more than 20% of its expensive cooled or heated air before it gets to the register. It's well worth your time to find and seal these leaks. Just turn the fan on and feel along the ducts. Distribution joints are the most common leaky points.

If your ducts aren't already insulated in the attic or basement, insulate them with foil-faced fiberglass duct insulation. Batts can be fastened with plastic tie wraps, wire, metal tape, or glue-on pins. Your A/C technician can do this job or recommend someone who can, if you're not up for it.

Have Your Heat Pump or Central A/C Serviced Regularly

Regular maintenance will include having the refrigerant charge level checked—both under- and overcharging compromise performance—oiling motors and blowers, removing dirt buildup, cleaning filters and coils, and checking for duct leakage. Proper maintenance will ensure maximum efficiency as well as the equipment's longevity.

- Set your A/C to the recirculate option if your system has this choice—drawing hot and humid air from outside takes a lot more energy. Many newer systems only recirculate, so if you can't find this option, it's probably already built in.
- Set your A/C thermostat at 78°F—or higher if you have ceiling fans. For each degree you raise the thermostat, you'll save 3%–5% on cooling costs.
- Don't turn the A/C thermostat lower than the desired setting—cooling only happens at one speed. The house won't cool any faster, and you'll waste energy by overshooting.
- Turn off your A/C when you leave for more than an hour. It saves money.
- Close off unused rooms or, if you have central A/C, close the registers in those rooms and shut the doors.
- Install a programmable thermostat for your central A/C system to regulate cooling automatically. Such a thermostat can be programmed to ensure that your house is cool only when needed.
- Trim bushes and shrubs around outdoor

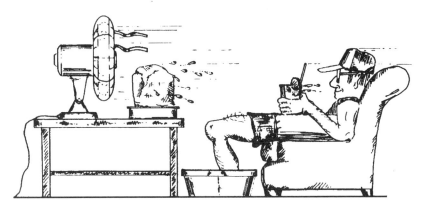

Personal cooling. Buying a block of ice isn't cost-effective, but feeling cool is important. Illustration courtesy of Saturn Resource Management, Helena, Montana.

condenser units so they have unimpeded airflow; a clear radius of 2 feet is adequate. Remove leaves and debris regularly.

- Provide shade for your room A/C or for the outside half of your central A/C if at all possible. This will increase the unit's efficiency by 5%–10%.
- Remove your window-mounted A/C each fall. Their flimsy mounting panels and drafty cabinets offer little protection from winter winds and cold. If you are unable to remove the unit, at least close the vents and tightly cover the outside.
- Clean your A/C's air filter every month during cooling season. Normal dust buildup can reduce airflow by 1% per week.
- Clean the entire unit according to the manufacturer's instructions at least once a year. The coils and fins of the outside condenser units should be inspected regularly for dirt and debris that would reduce airflow. This is part of the yearly service if you're paying a technician for the service.

If You Live in a Humid Climate

An important part of what a conventional air conditioner does is removing moisture from the air. This makes the body's normal cooling—sweating—work better, and you feel cooler. Unfortunately, some of the highest-efficiency models don't dehumidify as well as less-efficient air conditioners. A/C units that are too big for their cooling load also have this problem. High humidity leads to indoor condensation problems and gives us a sweaty mid-August New York City feeling—minus the physical and aural assaults and the great food and nightlife. People tend to set their thermostats lower to compensate for the humidity, using even more energy. Here are some better solutions.

- Reduce the fan speed. This makes the coils run cooler and increases the amount of moisture that will condense in the A/C rather than inside your house.
- Set your A/C to recirculate if this is an option and you haven't already done so.
- Choose an A/C with a "sensible heat fraction" (SHF) less than 0.8. The lower the SHF, the better the dehumidification.
- Size your A/C carefully. Contractors should calculate dehumidification as well as cooling capacity. Many air conditioning systems are oversized by 50% or more. The old contractor's rule of thumb—1 ton of cooling capacity per 400 square feet of living space—results in greatly oversized

A/C units when reasonable weatherization, insulation, and shading measures have been implemented to reduce cooling load. One ton per 800–1,000 square feet may be a more reasonable rule of thumb.

- Investigate using a Dinh-type heat pipe/pump. These devices cool the air before it runs through the A/C evaporator and warm it as it blows out, and this without energy input. They can help your A/C dehumidify several times more effectively.
- Explore "desiccant dehumidifiers," which, coupled with an efficient A/C, may save substantial energy. Desiccant dehumidifiers use a water-absorbing material to remove moisture from the air. This greatly reduces the amount of work the A/C has to do, thereby reducing your cooling bill.

Buying a New Air Conditioner

Air conditioners have become much more efficient over the last 25 years, and top-rated models are 30%–50% better than units made in the 1970s. Federal appliance standards have eliminated the least efficient models from the market, but builders and developers have little incentive to install a model that's more efficient than required, since they won't be paying the higher electric bills.

If you have an older model, it may be cost-effective to replace it with a properly sized, efficient unit. The EPA estimates that if your air conditioner is more than 12 years old, replacing it with a new efficient model could reduce your cooling costs by 30% and provide an excellent return on investment.

Air Conditioner Efficiency

Several studies have found that most central air conditioning systems are oversized by 50% or more. To avoid being sold an oversized model, or one that won't remove enough moisture, ask your contractor to show you the sizing calculations and dehumidification specifications. Many contractors may simply use the "1 ton of cooling capacity per 400 square feet of living space" rule of thumb. This will be far larger than you need, especially if you've effectively weatherized and insulated your home and done some of the other measures we discuss here to reduce heat gain and the need for cooling.

One ton of cooling capacity—an archaic measurement equaling approximately the cooling capacity of one ton of ice—is 12,000 Btus per hour. Room air conditioners range in size from ½ ton to 1½ tons, while typical central A/Cs range from 2 to 4 tons.

Air conditioners have become much more efficient over the last 25 years, and top-rated models are 30%–50% better than units made in the 1970s.

Apply the weatherization, insulation, shading, and alternative cooling tips we have discussed, and your need for energy-intensive mechanical cooling will be eliminated in all but the most severe climates. Real Goods sells a selection of DC-powered cooling appliances that can be run from batteries or directly from photovoltaic modules.

Desuperheaters

It's possible to use the waste heat removed from your house to heat your domestic water. In hot climates, where A/C is used more than five months per year, it makes economic sense to install a desuperheater to use the waste heat from the A/C to heat water. Some utilities give rebates to builders or homeowners who install such equipment.

Evaporative Coolers

If you live in a dry climate, an evaporative, or "swamp," cooler can save 50% of the initial cost and up to 80% of the operating cost of normal A/C. An evaporative cooler works by evaporating water, drawing fresh outside air through wet porous pads. This is an excellent choice for most areas in the southwestern Sunbelt. The lower the relative humidity, the better they work. At relative humidity above 30%, performance will be marginal; above 40% humidity, evaporative coolers will be totally ineffective. "Indirect" or "two-stage" models that yield cool dry air rather than cool moist air are also available, but the same low humidity climate restrictions apply.

Heat Pumps

If you live in a climate that requires heating in the winter and cooling in the summer, consider installing a heat pump in new construction or if you are replacing the existing cooling systems. The ground- or water-coupled models are the most efficient; see the discussion on page 196 and the comparison chart included there. The air conditioning efficiency of a heat pump is equivalent to that of a typical air conditioner, as they're basically the same thing, but the heating mode will be much more efficient than the cooling mode, outperforming everything except natural gas-fired furnaces.

Appliances for Cooling

Since our core business is renewable energy, and the high energy consumption of air conditioning equipment makes renewable energy systems very expensive, we don't offer these appliances. Apply the weatherization, insulation, shading, and alternative cooling tips we have discussed, and your need for energy-intensive mechanical cooling will be eliminated in all but the most severe climates.

Real Goods sells a selection of DC-powered cooling appliances that can be run from batteries or directly from photovoltaic modules. These include small, highly efficient evaporative coolers, a selection of fans for attic venting or personal cooling, and some ceiling-mounted paddle fans. Check out our webstore at realgoods.com.

Energy-efficient Appliances

Why Bother to Shop for Energy Efficiency?

Wasted energy translates into carbon dioxide production, air pollution, acid rain, and lots of money down the drain. The average American household spends more than $1,100 per year on appliances and heating and cooling equipment. You can easily shave off 50%–75% of this expense by making intelligent appliance choices.

For example, simply replacing a 20-year-old refrigerator with a new energy-efficient model will save you about $100 per year in reduced electric bills, while saving 1,000 kWh of electricity and reducing your home's CO_2 contribution by about a ton per year. Highly efficient appliances may be slightly more expensive to buy than comparable models with lower or average efficiencies. However, the extra first cost for a more efficient appliance is paid back through reduced energy bills long before the product wears out.

What Appliances Can I Buy from Real Goods?

Many appliance manufacturers prefer to sell through established appliance stores. Your choices in ordering through the Real Goods website (realgoods.com) are limited to more specialized or high-efficiency appliances. Feel free to call our technical staff at 1-800-919-2400 to inquire about what we carry. At present, we offer several varieties of incredibly efficient clotheslines; gas and DC refrigerators and freezers (Sun Frost, Sundanzer, Dometic, TruckFridge, and Crystal Cold) specifically for off-the-grid folks; high-efficiency tankless water heaters in both gas and electric; and the most problem-free solar water heaters available. We want to enable our customers to make intelligent appliance purchases.

Refrigerators and Freezers

The refrigerator is likely to be the largest single power user in your home aside from air conditioning and water heating. Refrigerator efficiency has made enormous strides in the past 20 years, largely due to insistent prodding from the Feds with tightening energy standards. An average new fridge with top-mounted freezer sold today uses under 500 kilowatt-hours per year, compared with more than 1,800 kilowatt-hours annually for the average 1973 model. The most efficient models available today are under 400 kilowatt-hours per year and still dropping. The typical refrigerator has a lifespan of 15–20 years. The cost of operation over that time period will easily be two to three times the initial purchase price, so paying somewhat more initially for higher efficiency offers a solid payback. It may not be worth scrapping your 15-year-old clunker to buy a new energy-efficient model. But when it does quit, or it's time to upgrade, buy the most efficient model available. A great source for listings of highly efficient appliances (as well as other kinds of products) is the Environmental Protection Agency's ENERGY STAR® Website, energystar.gov/certified-products/certified-products?c=products.pr_find_es_products. Be sure to have the Freon removed professionally from your clunker before disposal. Most communities now have a shop or portable rig to supply this necessary service at reasonable cost.

Refrigerator Shopping Checklist

- Smaller models obviously will use less energy than larger models. Don't buy a fridge that's larger than you need. One large refrigerator is more efficient than two smaller ones, however.
- Models with top- or bottom-mounted freezers average 12% less energy use than side-by-side designs.
- Features like through-the-door ice, chilled water, or automatic icemakers increase the profit margin for the manufacturer, and the purchase price by about $250. So ads and salespeople tend to push them. They also greatly increase energy use and are far more likely to need service and repair. Avoid these costly troublesome options.
- Make sure that any new refrigerator you buy is CFC-free in both the refrigerant and the foam insulation.
- Be willing to pay a bit more for lower operating costs. A fridge that costs $75 more initially but costs $20 less per year to operate due to better construction and insulation will pay for itself in less than four years.

Also note that snazzy features, such as side-by-side refrigerator-freezer units and through-the-door ice or cold water, increase energy consumption. If you can do without, it's cost-effective to avoid these features.

Refrigeration and Renewable Energy Systems

When your power comes from a renewable energy system, which has a high initial cost, the payback for most ultraefficient appliances is immediate and handsome. You will save the price of your appliances immediately in the lower operating costs of your power-generating system. This is particularly true for refrigerators.

Many owners of smaller or intermittent-use renewable energy systems prefer gas-powered fridges and freezers, especially when the fuel is already onsite anyway for water heating and cooking. Gas fridges use an environmentally friendly mixture of ammonia and water, not Freon, as a refrigerant. We offer gas-powered freezers and refrigerator/freezer combinations.

For many years, Sun Frost held the enviable status as manufacturer of the "world's most efficient refrigerators," and that has produced a steady flow of renewable energy customers willing to pay the high initial cost in exchange for the savings in operating cost. Real Goods has been an unabashed supporter of this product for many years. But times are changing. The major

Refrigerators and CFCs

All refrigerators use some kind of heat-transfer medium, or refrigerant. For conventional electric refrigerators, this has always been Freon, a CFC chemical and prime bad guy in the depletion of the ozone layer. Manufacturers all have switched to non-CFC refrigerants, and by now it should be nearly impossible to find a new fridge using CFC-based refrigerants in an appliance showroom.

What most folks probably didn't realize is that, of the 2.5 pounds of CFCs in the typical 1980s fridge, 2 pounds are in the foam insulation. Until very recently, most manufacturers used CFCs as blowing agents for the foamed-in-place insulation. Most have now switched to non-CFC agents, but if you're shopping for new appliances, it's worth asking.

The CFCs in older refrigerators must be recovered before disposal. For the 0.5 pound in the cooling system, this is easily accomplished, and your local recycler, dump, or appliance repair shop should be able either to do it or to recommend a service that will. To recover the 2 pounds in the foam insulation requires shredding and special equipment, which sadly happens only rarely.

The average American household spends more than $1,100 per year on appliances and heating and cooling equipment. You can easily shave off 50%–75% of this expense by making intelligent appliance choices.

A new, more efficient refrigerator can save you $70–$80 a year on electricity and will completely pay for itself in nine years.

appliance manufacturers have managed to implement huge energy savings in their mass-produced designs over the past several years. While few of the mass-produced units challenge Sun Frost for king-of-the-efficiency-hill, many are close enough, and less expensive enough, to deserve consideration. The difference between a $900 purchase and a $2,700 purchase will buy quite a lot of photovoltaic wattage. A number of mass-produced fridges have power consumption in the 500 to 1,500 watt-hours per day range and can be supported on renewable energy systems quite easily. Any reasonably sized fridge with power use under 1.5 kilowatt-hours per day is welcome in a renewable system; that's $45–$50 per year on the yellow EnergyGuide tag that every new mass-produced appliance wears in the showroom.

In Sun Frost's defense, we need to point out that conventional refrigerators run on 120 volts AC, which must be supplied with an inverter and will cost an extra 10% due to inverter efficiency. Also, if your inverter should fail, you could lose refrigeration until it gets repaired. Sun Frosts (if ordered as DC-powered units) will run on direct battery power, making them immune to inverter failures.

Improving the Performance of Your Existing Fridge

Here is a checklist of tips that will help any refrigerator do its job more easily and more efficiently.

- Cover liquids and wrap foods stored in the fridge. Uncovered foods release moisture (and get dried out), which makes the compressor work harder.
- Clean the door gasket and sealing surface on the fridge. Replace the gasket if damaged. You can check to see if you are getting a good seal by closing the refrigerator door on a dollar bill. If you can pull it out without resistance, replace the gasket. On new refrigerators with magnetic seals, put a flashlight inside the fridge some evening, turn off the room lights, and check for light leaking through the seal.
- Unplug the extra refrigerator or freezer in the garage. The electricity the fridge is using—typically $130 a year or more—costs you far more than the six-pack or two you've got stashed there. Take the door off, or disable the latch so kids can't possibly get stuck inside!
- Move your fridge out from the wall and vacuum its condenser coils at least once a year. Some models have the coils under the fridge. With clean coils, the waste heat is carried

off faster and the fridge runs shorter cycles. Leave a couple of inches of space between the coils and the wall for air circulation.
- Check to see if you have a power-saving switch or a summer-winter switch. Many refrigerators have a small heater (yes, a heater!) inside the walls to prevent condensation buildup on the fridge walls. If yours does, switch it to the power-saving (winter) mode.
- Defrost your fridge if significant frost has built up.
- Turn off your automatic icemaker. It's more energy-efficient to make ice in ice trays.
- If you can, move the refrigerator away from any stove, dishwasher, or direct sunlight.
- Set your refrigerator's temperature between 38° and 42°F, and your freezer between 10° and 15°F. Use a real thermometer for this, as the temperature dial on the fridge doesn't tell real temperature.
- Keep cold air in. Open the refrigerator door as infrequently and briefly as possible. Know what you're looking for. Label frozen leftovers.
- Keep the fridge full. An empty refrigerator cycles frequently without any mass to hold the cold. Beer makes excellent mass, and you probably always wanted a good excuse to put more of it in the fridge, but it tends to disappear. In all honesty, plain water in old milk jugs works just as well.

Buying a New Refrigerator

A new more efficient refrigerator can save you $70–$80 a year on electricity and will completely pay for itself in nine years. It also frees you from the guilt of harboring a known ozone killer: Since 1999, all new fridges in the showroom are completely CFC-free.

Shop wisely by carefully reading the yellow EPA EnergyGuide label found on all new appliances. Use it to compare models of similar size.

The American Council for an Energy-Efficient Economy is a nonprofit organization dedicated to advancing energy efficiency as a means of promoting economic development and environmental protection. Their invaluable *Consumer Guide to Home Energy Savings* lists all the most efficient mass-produced appliances by manufacturer, model number, and energy use. We strongly recommend consulting this guide before venturing into any appliance showroom. It has an amazingly calming effect on overzealous sales personnel and allows you to compare energy use against the top-of-the-class models.

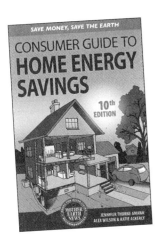

The *Consumer Guide to Home Energy Savings*, now in it's 10th edition, is a fantastic and well-regarded resource about all things energy conservation.

Clothes Washers and Dryers

Laundry is another domestic activity that consumes considerable power. When it comes to appliances for washing and drying clothes, however, the energy-efficiency focus is decidedly lopsided. Much progress has been made recently on resource-efficient washing machines. Clothes dryers, on the other hand, are not yet required to display EnergyGuide labels, so it is difficult to compare their energy usage, and apparently, manufacturers have little incentive to improve efficiency. (In 2012, the EPA announced an ENERGY STAR "specification process" for dryers.) The good news about drying, of course, is that it's easy to dry clothes using solar power—by hanging them up on a line or a rack, indoors or outdoors—thereby eliminating or greatly reducing the need to use a power-hungry machine.

Most of the energy use in washing clothes is for heating water. The new resource-efficient washing machines use less water than conventional models do, which cuts their energy consumption by up to two-thirds. Some models are horizontal-axis front loaders, in which the clothes tumble through the water. Others are redesigned vertical-axis top loaders that spray the clothes or bounce them through the water rather than submerging them. Both types also use higher spin speeds to extract more water from the clean clothes, thereby reducing the energy required for drying (solar or otherwise).

The only energy-saving features available on dryers are sensors that shut off the machine when the clothes are dry. (Some European models are using heat pumps, but these have not hit the US market yet.) Most better-quality models on the market today have this feature. In addition to using a resource-efficient washer that spins more water out of your clean clothes, and taking advantage of free solar energy as much as possible, here are a few more tips for using a clothes dryer efficiently (courtesy of the American Council for an Energy-Efficient Economy).

The EPA and Refrigerator Power Use Ratings

We're all familiar with the yellow energy-use tags on appliances now. These level the playing field and make comparisons between competing models and brands simple. With some appliances, like refrigerators, the EPA takes a "worst-case" approach. Fridges are all tested and rated in 90°F ambient temperatures. The test fridge is installed in a "hot box" that maintains a very stable 90° interior temperature. This results in higher energy use ratings than will be experienced in most settings. At the mid-70° temperatures that most of us prefer inside our houses, most fridges will actually use about 20%–30% less energy than the EPA tag indicates.

- Separate clothes and dry similar types together.
- Dry two or more loads in a row to take advantage of residual heat already in the machine.
- Clean the filter after each use. A clogged filter will reduce airflow.
- Dry full loads when possible. Drying small loads wastes energy because the dryer will heat up to the same extent regardless of the size of the load.
- Make sure the outdoor exhaust vent is clean and that the flapper opens and closes properly. If that flapper stays open, cold air will infiltrate into the house through the dryer, increasing your wintertime heating costs.

For more information about the efficient washers and dryers that Real Goods sells, check out our website: realgoods.com.

Superefficient Lighting

Thanks to Lindsay Wood, aka The LED Lady (the ledlady.com), for a through revision of the material on lighting in this chapter.

A revolution has taken place in commercial and residential lighting. Until recently, we were bound by Thomas Edison's ingenious discoveries. But that was more than a century ago, and Mr. Edison's magic now appears crude, even outdated. Mr. Edison was never concerned with how much

energy his incandescent bulb consumed, only that it produced light. Today, however, economic, environmental, and resource issues force us to look at energy consumption much more critically. Incandescent literally means, "to give off light as a result of being heated." With standard light bulbs, typically only 10% of the energy consumed is converted into light, while 90% is given off as heat. So, in short, incandescent lights are in

In short, incandescent lights are in reality "incandescent heaters"! Fluorescent technology started changing that ratio and brought it more even, and now with LED technology, 90% of the energy is given off as light and less than 10% as heat.

reality "incandescent heaters"! Fluorescent technology started changing that ratio and brought it more even, and now with LED technology, 90% of the energy is given off as light and less than 10% as heat.

New technologies are available today that improve on both the reliability and the energy consumption of Edison's invention. Although the ini-tial price for these new energy-efficient products is higher than that of standard light bulbs, their cost-effectiveness is far superior in the long run. (See the discussion of whole-life cost comparison below.) When people understand just how much money and energy can be saved with compact fluorescents or LED lights, they often replace regular bulbs that haven't even burned out yet.

Lighting Characteristics

To better understand the lighting options available to us, it is important to understand the characteristics of light regardless of the technology used to produce it.

Beam Spread

Before we get into the different types of lights and the various fixtures well suited for them, we need to understand beam spread. Have you ever used a flashlight that can be manipulated so you can twist and change the spread of light from a wider beam to a smaller spot light? In a theater when a soloist is performing a spot light will often illuminate that person, leaving the rest of the stage dark. This beam spread exists in every light bulb we buy. As we describe the different types of lights, we'll mention beam spread for different applications.

Types of Lamps (aka Bulbs)

There are many types of lamps, also known as bulbs. In the lighting industry, the bulb is referred to as a lamp, even though the general public often considers a lamp a fixture. Whenever we use the terms "bulb" or "lamp," please know that they refer to the actual light source. Below are images of the most widely used bulbs. Note the number that follows the "A," "PAR," "BR," or "MR" indicates the width of the bulb in one-eighth-inch units. For example, a PAR 20 lamp would be 20 one-eighth-inch units, otherwise known as 2½ inches. The lighting industry has long been using this terminology, and it's not changing anytime soon.

A Series or A Line Lamp

The A series is the classic type of light bulb that has been the most commonly used type for general-purpose lighting applications since the early 20th century. This A series lamp started out as pear-shaped and is equipped with an "Edison Screw Base." Although A series (or A Line) bulbs started out incandescent, CFL and LED lamps have adopted the shape as well.

Floor and table lamps are an excellent application for LED A bulbs, as they give off a wider beam spread of light, about 320°. CFLs may not fit well into lamps that have a metal harp that supports the lampshade because of the base that holds the ballast. An inexpensive replacement harp can solve this problem. Be aware also that heavier CFLs may change a lightweight but stable lamp into a top-heavy one. However, LEDs do not need as much room for ballast technology and fit well into these lamps.

Desk lamps can be difficult to retrofit if they have smaller odd-shaped halogen lights. However, desk lights that incorporate an A lamp are readily available, and you can easily buy LED or CFL A lamps. Be sure to test the CFL in the lamp, as it may need an extender base.

> Floor and table lamps are an excellent application for LED A bulbs, as they give off a wider beam spread of light, about 320°.

Incandescent 40 W

CFL 13 W, 26 W+

LED 8 W, 13 W, 17 W

Parabolic Aluminum Reflectors, or PAR series lamp: PAR 20, 30, 38, 40

Three-way sockets are most commonly used in table and desk lamps. Manufacturers of LEDs and CFLs make three-way bulbs. While more efficient than conventional three-way bulbs, they are more expensive than one-way LEDs and CFLs. While the lifespan of an LED is not reduced by three-way technology, three-way CFLs have a shorter lifespan than one-way CFLs.

PAR Series Lamp

PAR stands for parabolic aluminum reflector. This lamp, commonly found in PAR 20, 30, and 38 sizes, is widely used in can or track lighting in homes or retail. Depending on the beam spread, it is advantageous for its ability to light a specific area or display. In a track system, the PAR bulb needs to have a track head to attach to the track and be able to move into the position where you want the light. A PAR 30 lamp is commonly used for a 6" can or track lighting. The halogen version of the PAR 30 consumes about 65 W or more; the equivalent LED runs about 15 W. When you want much more light, use an 8" can or track system to accommodate the PAR 38. Halogen PAR 38s can run between 75–150 W and can be replaced by a 17–20 W PAR 38 LED.

BR or R Series Lamp—Bulb Reflector

Bulb reflector, also known as reflector. This bulb has the look of a PAR lamp, except for the bulbous shape of the lens on top. This bulbous shape allows for a wider beam spread. Please note: The BR series lamps may also be listed as R lamps. For example, a BR30 can also be listed as an R30.

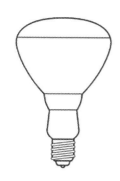

Bulb reflector, or R series lamp: BR 20, 30, 38, 40

Recessed can lights are ideal for either the PAR or BR series of bulbs, and this is where beam spread really makes all the difference. To determine which size PAR lamp to use, first you need to know the diameter of the opening of the can. Typically cans come in 4", 6", and 8" sizes. LEDs fit well in any of them, as their design mirrors that of the halogen PAR lamps originally intended for these cans. If you want to create a spot light effect on your floor, then choosing a PAR 25° or 10° is the way to go. However, if you want to have a wide spread of light, then choosing a PAR 38° would be more appropriate. To get an even larger, softer beam spread, choosing the BR series will give 120° of light, which is ideal for a living room or kitchen equipped with cans. CFLs often have difficulty fitting into the can and making contact with the Edison screw-in base because of their ballast—however, you can purchase an extender base for these CFLs.

MR Series Lamp—Multifaceted Reflector

Multifaceted reflectors are smaller versions of the PAR lamp. These lamps are most commonly found on track lighting systems used in a home or gallery to light specific areas, art, etc. When installing MR16s on a track, the consumer has a choice among track heads from basic to architectural. The typical MR16 runs 50 W and puts out a warm orange/reddish light. The LED equivalent consumes only 7 W and generates the same if not better quality light. However, the LED version does not get as orange when dimmed all the way down. LEDs still run on the cooler color temperature scale, but are perfect for a location that does not need such warm light. LED MR16 systems would be ideal for off-grid solar homes.

Note that MR16s come in a variety of bases:
- MR16 GU 5.3—has two pins that you "push" in; runs on 12 V
- MR16 GU5.10—has two knobs and twists to go in; runs on 120 V
- MR11—has smaller and tighter pins; runs on 12 V

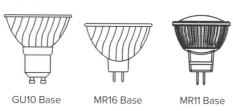

GU10 Base   MR16 Base   MR11 Base

Multifaceted Reflector, or MR series lamp: MR16 GU10, MR16, MR11

Track lighting is very common today. MR16 lamps are often found in Ikea furniture. When choosing MR16 lamps, it's very important to pay attention to the base. You can either purchase an MR16 12 volt, which would be an MR16 GU5.3 base, or an MR16 120 volt, which would be an MR16 GU5.10 base, as seen in the illustration. You either twist in the light (MR16 GU5.10, 120 V, commonly found in IKEA furniture) or push in the light (MR16 GU5.3, 12 V). MR16 track lights

The light output of a lamp is measured in lumens; energy input to a lamp is measured in watts. The 60-watt incandescent puts out 855 lumens and was replaced with the 13-watt CFL that puts out 780 lumens. Enter LEDs. The same 60 W incandescent and 13 W CFL are now being replaced by the 13 W LED that puts out 800 lumens. And what's really striking is that the incandescent lasts for 1,000 hours, the CFL for 12,000 hours, and the LED for 25,000 hours!

are very well suited for DC renewable energy systems because they are already a 12 V system and they only use around 3–10 watts, depending on the amount of light desired.

Lumens

The light output of a lamp is measured in lumens; energy input to a lamp is measured in watts. Sadly, this is where we all lack education about lighting. The 60 W incandescent puts out 855 lumens and was replaced with the 13 W CFL that puts out 780 lumens. Enter LEDs. The same 60 W incandescent and 13 W CFL are now being replaced by the 13 W LED that puts out 800 lumens. And what's really striking is that the incandescent lasts for 1,000 hours, the CFL for 12,000 hours, and the LED for 25,000 hours!

The efficiency of a lamp is expressed as lumens per watt. In the efficiency chart below, this number is listed for various lamps. High-pressure sodium and low-pressure sodium lights are primarily used for security and street lighting.

Light Color

The terms "color temperature" and "color rendering" are technical terms used to describe light. But while they sound simple or innocuous, these can be very complex topics.

Color Temperature, also known as Kelvin Temperature

To best understand Kelvin temperature, we need to know first that Lord Kelvin discovered absolute zero, and the Kelvin temperature scale is named in his honor. While we're accustomed to seeing temperature depicted in degrees, the degree symbol is often left off of lighting information. Physicists began using the Kelvin scale to describe the colors of metal when heated at different temperatures. For example, think of an iron; as the iron is heated, it goes through color changes from red to orange to blue to white. If a light bulb is listed as 2700 K, that means 2,700 degrees Kelvin, and the light is similar to a sunset that is orange in color. If the bulb is listed as 5000 K, it is similar to the bluish color of a midday sun. The Kelvin scale is shown below.

Color Rendering Index (CRI)

The Color Rendering Index (CRI) describes the ability of a light source to replicate colors. It is measured as a number between 0 and 100.

Since the light bulb was invented, developers of artificial light have been striving to match the sun's 100 CRI. Think of a green apple in the sun-light: We would see all the natural colors. Place that same green apple under a high-pressure sodium (HPS "orange") streetlight. The orange light, or 2,200° Kelvin, would turn the apple more yellow and less green.

Incandescent light achieved a high CRI rating because the technology was based on heating a piece of metal protected from oxidization. Fluorescent technology depends on the use of phosphor blends. These phosphors are added to the lamps to give it a specific color (or Kelvin temperature). A common complaint about CFLs is that people don't like the color of the light. Recently, however, there has been notable improvement of fluorescent linear lamps using tristimulus phosphors, and they now have CRIs in the 80s. At the same time, LED lamps are now getting 80–95 CRI using the same phosphors to generate a specific Kelvin temperature. These CFLs and LEDs incorporate relatively expensive phosphors with peak luminance in the blue, green, and red portions of the visible spectrum (those that the human eye is most sensitive to), and produce about 15% more visible light than those with standard phosphors. Wherever people spend much time around lighting with a higher (80+) CRI, it is better on many levels. The accompanying table shows how the CRI stacks up from one type of lighting technology to another.

When choosing a light bulb, color temperature is one of the more important choices, and it is a personal choice. Here are some general guidelines to consider when choosing your lights.

- Warmer whites (2700 K–3500 K) are preferred in living, dining, and reception areas to create a more relaxed environment.
- Natural whites (4000 K–4500 K) are preferable for kitchens and bathrooms, where tasks are performed.

Light, Color Rendering Index, and Color Temp		
Type of Light	CRI	°Kelvin
Incandescent	90–95	2,700
Cool white fluorescent	62	4,100
LED	80–95	2,700
		3,000
		4,100
		5,000
Warm white fluorescent	51	3,000
Compact fluorescent	82	2,700

CRI = Color Rendition Index. A measure of how closely a light source mimics sunlight. 100 is perfect sunlight, 0 is darkness. Most standard CFLs are around 82 CRI; most full-spectrum lights are 90–92 CRI. Most LEDs are around 82 CRI, and some LEDs are at 93.

- Daylight whites (5000 K–6000 K) are favored by retailers and offices, though natural whites could be utilized here.
- Cool whites (6000 K–7000 K) are found mainly in industrial and hospital areas.

Older eyes often react better in cooler color temperatures. Women often prefer warmer colors than men. Task lighting is better if cooler. Cooler whites can raise attention and awareness. Warmer whites can soften environments and make for a more relaxed space. Warmer whites are more likely to mask the true color of objects, adding a yellow tint.

Wattage

The terms "power" and "energy" are frequently confused. Power is the rate at which energy is generated or consumed and is measured in units (watts) that represent "energy per unit time." For example, when an incandescent light bulb with a power rating of 100 W is turned on for one hour, the energy used is 100 watt-hours (W/h). This same amount of energy would light a 40-watt bulb for 2.5 hours, or an LED 13 W bulb for 7.69 hours. Most confusion centers around the 60-watt bulb and the amount of light (lumens)

we perceive it puts out. When a 13 W LED can generate the same amount of light, it becomes less about the actual wattage and more about lumens and lumens per watt.

Lifetime Hours

The claims of 30,000, 50,000, and 100,000 hours that LED technology lasts are simply way beyond the reality of what we're used to with incandescent and fluorescent/CFL lifetime hours. An article released by the company Fuse LED puts it quite well: Instead of asking why LEDs last so long, the question really should be why do the other bulbs burn out so quickly? To summarize, it all boils down to the fact that incandescent lamps use sensitive metals that when heated and cooled eventually experience fatigue, until one unsuspecting moment we go to turn on the lamp and we hear that popping sound of the metal filament breaking. With fluorescent tubes, the filaments at each end wear out and produce that blackened look. CFLs have no filament, but their electrodes that emit electrons also wear out.

Lamp life estimates for incandescent and fluorescent technology are based on a three-hour duty cycle, meaning that the lamps are tested by turning them on and off once every three hours

Most confusion centers around the 60-watt bulb and the amount of light (lumens) we perceive it puts out. When a 13 W LED can generate the same amount of light, it becomes less about the actual wattage and more about lumens and lumens per watt.

Technology	Lamp Type	Hours of Life Expectancy	Years of Life Expectancy*	End of Life-span Result
LED	13 W A Lamp	30,000	27	70% of original brightness
CFL	13 W A lamp	6,000	6	Burn out
Incandescent	60 W A lamp	1,000	2	Burn out

* based on 3 hrs a day

Efficiency and Lifetime of Lamps (Electric Rate: $0.15 per kWh)				
Lamp Type	Power	Lumens	Lumens per Watt	Lifetime in Hours	Lamps per Highest Lifetime Hour
Incandescent A lamp	60 W	855	14	1,000	30
CFL A lamp	13 W	780	60	12,000	2.5
LED A lamp	13 W	800	61.5	30,000	1
PAR 30 halogen lamp	55 W	960	17.6	1,000	40
PAR 30 CFL lamp	15 W	700	47	10,000	4
PAR 30 LED lamp	14 W	850	60	40,000	1
MR16 halogen	50 W	920	18.4	2,000	15
MR16 LED	7 W	525	75	30,000	1
Metal halide, floodlight	250 W	18,750	75	12,000	8.3
LED, floodlight	60 W	6,000	100	100,000	1
T8, 4-ft. fluorescent tube	32 W	2600	86.0	20,000	2.5
T8 4-ft. LED tube	20 W	2000	100	50,000	1

until half of the test batch of lamps burn out. Turning fluorescent and CFL lamps on and off more frequently decreases lamp life.

However, LED technology is entirely different. LEDs use no filament, no electrodes, and involve no oxidation. They actually never burn out—instead, they lose brightness. Both incandescents and CFLs are fragile to handle, but LEDs are solid-state devices that can stand up to vibration, weather (if rated for it), and extreme cold temperatures. The accompanying chart compares the similar CFL and LED to a a 60 W incandescent lamp. The LEDs simply fade rather than burn out.

Heat Output

All light technologies put out heat. Incandescent bulbs put out 90% light and 10% heat. These lights were ideal for incubating chicks, heating a small room, and making cookies in the Holly Hobbie oven. Fluorescent tubes and CFLs give off heat, but much less than incandescents. LEDs utilize a heat sink, which is a path that removes heat from the light emitting diode. The most common types of heat sinks employed by LED lighting manufacturers are conductive and convective; some use thermal radiation. Heat dissipation can also be achieved with both conduction and long-operating fans like those you would find inside a computer housing. As LED technology advances, the heat sinks are becoming integrated with design and aesthetics, so much so that a 90 W halogen PAR 38 lamp looks very similar to a 17 W LED PAR 38 lamp that conducts the heat into a solid material such as metal. The removal of heat is very important to the efficiency and lifetime of the lamp.

Dimming

LED technology is getting better and better with dimming features. Unlike fluorescents and CFLs, LED is a technology that lends itself to being controlled. With their instant-on and dimmable capabilities, LEDs are extremely efficient while at the same time offering the benefits of high-quality light. The packaging should say dimmable if the LED is in fact dimmable.

LEDs dim differently than incandescents. An incandescent dimmed all the way down gives off an orange color because the filament is heating at a much lower temperature. When some LEDs get to the bottom limit of the dimmer, they shut off. That's one reason it is important to test the LED lights you want to dim to see *how* they dim. The good news is when you dim your lights, you use less energy.

Most common "white" LEDs convert blue light from a solid-state device to an approximate white light spectrum using photoluminescence or colored phosphors added to the LEDs, the same principle used in conventional fluorescent tubes.

Light Technology

Many different technologies have provided the world with light since the electric light bulb was invented in the 1800s. Four subsequent technological breakthroughs are notable: the incandescent light and long-lasting filament; the mercury vapor lamp; fluorescent technology; and LED technology. The good news is that as the technology of lighting advances so does the efficiency of delivering the same, if not better, light. But many people are not aware of recent changes in technology, and consumers generally have not been educated about the various aspects of light described above. This makes it difficult for consumers to know what they are buying. Not surprisingly, the early adopters were the first to discover the flaws of the emerging lighting technologies and to recognize the discrepancies between the grandiose claims made by lighting manufacturers and the reality of their products. We saw this first with the bulb lifetime claims made by CFL manufacturers, and then with the claims of LED manufacturers that their products generated as much light as a 60 W incandescent bulb.

Thomas A. Edison, the father of light, made the following statement when asked how he felt about failing 10,000 times to make the electric light bulb. "I have not failed. I've just found 10,000 ways that won't work." This quote puts the first-, second-, and third-generation light bulbs into perspective. Edison's first electric bulb ran for only 48 hours. Now that same type of bulb using LED technology can last 40,000 hours!

Below we describe the different technologies of light still available to us. Each will be reviewed and compared to the most current LED technology. And that's where we'll start.

Light Emitting Diode (LED)

The term "solid state" refers to light emitted by solid-state electroluminescence, as opposed to the light emitted by incandescent bulbs (which use thermal radiation) or fluorescent tubes. Compared to incandescent lighting, solid-state lighting (SSL) creates visible light with reduced heat generation. Most common "white" LEDs convert blue light from a solid-state device to an approximate white light spectrum using photo-luminescence or colored phosphors added to the LEDs, the same principle used in conventional fluorescent tubes.

LED technology is following a similar trajectory as computer advancement. Output doubles about every 18 months while prices slowly drift downward. The light emitting diode was invented in the early 1900s, but the first visible-spectrum (red) LED wasn't developed until 1962, by Nick J. Holonyak while working at General Electric. Holonyak is considered by some as the "father of LEDs." Until 1968, LEDs cost $200 per unit and were thus too costly for most uses. Dr. Jean Hoerni at Fairchild Semiconductor first figured out how to mass-produce these red LEDs using compound semiconductor chips fabricated with his planar process.

Shuji Nakamura of Nichia Corporation demonstrated the first high-brightness blue LED in 1994. The existence of blue and high-efficiency LEDs quickly led to the development of the first white LED, which is the basis for the LED lights available to consumers today.

As LED technology has evolved, efficiency and light output has risen exponentially, with a doubling occurring approximately every 36 months since the 1960s. This trend is generally attributed to the parallel development of other semiconductor technologies and advances in optics and material science. Within one year of this book being published, we will see more efficient LEDs that offer even better light quality (CRI).

Leading Benefits of LEDs
LEDs offer many advantages over incandescent, fluorescent, and compact fluorescent lights.

- **More efficient.** LEDs are 75% more efficient than incandescents and 25%–50% more efficient than fluorescents or CFLs.
- **Longer lasting.** LED bulbs can last 20,000–40,000 hours, while some LED fixtures last 100,000 hours.
- **Better light quality.** The color rendering of LEDs is typically 80–95. Businesses with security camera systems like the ability to depict true colors.
- **Low heat output.** LEDs give off 95% light and 5% heat. You can save on air conditioning costs when using LEDs.
- **Use no mercury.** LEDs are made with metals, electronics, and plastic. No toxic mercury is used, unlike in CFLs.
- **Give off no ultraviolet light.** LED technology does not involve electrifying mercury gas, which is the cause of UV light emissions.
- **Instant-on.** LEDs are very easily controlled and do not have any start-up time; they also do not die earlier than expected due to being turned on and off.
- **Not sensitive to vibration.** LEDs aren't bothered by vibration or modest impacts, unlike CFLs. Your flashlight won't die just because you dropped it.
- **Cold weather.** LEDs love the cold, unlike fluorescents and CFLs that need a specific and consistent temperature to work well.

The Federal Trade Commission (FTC) bulb label.

LEDs offer many advantages over incandescent, fluorescent, and compact fluorescent lights. They: are more efficient; are longer lasting; give better quality light; give off little heat; give off no UV light; are "instant-on"; are not sensitive to vibration; love cold weather; and work in outdoor wet locations.

Lighting Facts labels.

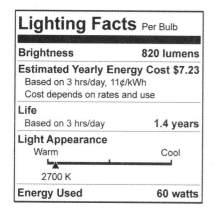

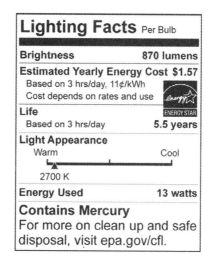

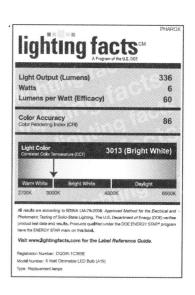

- **Wet applications.** LEDs also work well in outdoor wet locations, as long as the light bulb is "wet" rated.

Lighting Facts Label

Ingredients labeling has revolutionized our food, cleaning supplies, and clothing industries because consumers have a right to know what's inside the product they are buying. Lighting is no different, and in 2011, the Federal Trade Commission (FTC) mandated new "Lighting Facts labeling" on all lighting products. For the first time, the label on the front of the package will emphasize the bulbs' brightness as measured in lumens, rather than giving a measurement of watts. The new front-of-package labels also will include the estimated yearly energy cost for the particular type of bulb.

While the FTC provided a labeling system for all light bulbs, the Department of Energy chose to create a Lighting Facts label for the LED industry alone. (The DOE labeling system is voluntary, while the FTC label is mandatory.) The rapid growth of LED lighting means that an increasing number of new products is coming onto the market. While many of these products showcase the energy-saving potential and performance attributes of LED lighting, underperforming products are also appearing. Since bad news travels fast, such inferior products could discourage consumers from accepting this new technology. This occurred when CFLs were introduced, slowing their acceptance among the American public. DOE further developed a colored LED Lighting Facts label to avoid this problem for LEDs. The FTC label is black and white.

Can you easily tell which label is for an incandescent 60 W bulb and which one is for a CFL? The DOE label for the LED light (on the right) is colored. Can you see the light spectrum and determine the Kelvin temperature of the LED light?

Incandescent

After oil and gas lamps were found to be smelly and smoky, Thomas Edison, a very aggressive inventor, created the first incandescent electric light. Incandescent means white, glowing, and luminous with intense heat. What we know today is that the incandescent light puts off 90% heat and 10% light. A typical 60 W light bulb produces 16 lumens per watt (lm/W) while a CFL or LED produces 62 lm/W for the same light output. Edison experimented with all kinds of vegetable fibers including bamboo, baywood, cedar, flax, and cotton. The first incandescent light to work for two days used a cotton filament. It wasn't until Edison used a carbon filament that he was able to

get the light to last about 1,200 hours. Compared now to an LED that lasts 30,000 hours and puts off 90% light and 10% heat, light technology has come a long way.

Tungsten-halogen

Tungsten-halogen (or quartz) lamps are really just "turbocharged" incandescents. They are typically only 10%–15% more efficient than standard incandescents, a step in the right direction, but nothing to write home about. Compared with standard incandescents, halogen fixtures produce a brighter, whiter light and are more energy-efficient because they operate their tungsten filaments at higher temperatures than do standard incandescent bulbs. In addition, unlike the standard incandescent light bulb, which loses approximately 25% of its light output before it burns out, a halogen light's output depreciates very little over its life, typically less than 10%.

Low- and High-pressure Sodium

A sodium vapor lamp is a gas discharge lamp that uses sodium in an excited state to produce light. There are two types of these sodium vapor lights, low-pressure and high-pressure. Low-pressure lamps are more efficient than high-pressure lamps; however, both give off an orange glow that is predominantly used in outdoor streetlight applications. Another good reason for using sodium vapor lights in cities was to reduce light pollution into the night sky. Cities and businesses with these lights will soon be moving toward LEDs, as their light quality is quite poor and LEDs offer better light quality, are more efficient, and last longer.

Metal Halide

Metal halide is an electric lamp that produces light by an electric arc through a gaseous mixture of vaporized mercury and metal halides. They are often referred to as a high-intensity gas discharge lamp (HID). Developed in the 1960s, they are similar to mercury vapor lamps, but contain additional metal halide compounds in the arc tube, which improve the efficacy and color rendition of the light.

Metal-halide lamps average 75–100 lumens per watt, which is about twice that of mercury vapor lights and 3 to 5 times that of incandescent lights, and produce an intense white light. However, the lamp life only lasts 6,000 to 15,000 hours, and they require a warm-up period of several minutes to reach full light output. Before LEDs were an option, this lifespan was considered long. Like sodium lamps, metal halide lamps are used for wide-area overhead lighting of commercial, industrial, and public spaces.

Fluorescent Lights

Fluorescent lights are still trying to overcome a bad reputation. For many people, the term "fluorescent" connotes a long tube emitting a blue-white light with an annoying flicker, a death-warmed-over color, and headaches. However, these limitations have been overcome completely with technological improvements.

The fluorescent tube, however it is shaped, contains a special gas at low pressure. Fluorescent lamps work by ionizing mercury vapor in a glass tube. This causes electrons in the gas to emit photons at UV frequencies. The UV light is converted into standard visible light using a phosphor coating on the inside of the tube. Depending on the amount of phosphors (seen as yellow or orange), fluorescents will give off either a cool white (5000 K) or warm white (3500 K) light.

The fluorescent tube may be long and straight, as with standard fluorescents, or there may be a series of smaller tubes, compacted in a configuration that can be screwed into a common light fixture. This latter configuration is called a compact fluorescent light, or CFL. They are also known to as "curly Q bulbs."

Consumer awareness of and interest in compact fluorescents exploded over the past decade, and the market for them grew significantly. They have become the bulb of choice among energy-conscious consumers and a symbol of environmental awareness, often seen in the imagery used by organizations promoting conservation and environmental stewardship. In many ways, however, CFLs have been superseded by LEDs, which happen to be coming on strong just as 40 W, 60 W, and 100 W incandescents are being phased out. It would be unfortunate if the lofty status accorded to CFLs gets in the way of consumers' appreciation of and understanding of LEDs, which truly offer many more benefits.

Fluorescent Ballasts

All fluorescent lights require a ballast to operate, in addition to the bulb. In tubes the ballast is separate from the tube, and in CFLs the ballast is in the lamp, just above the screw-in base. Fixtures with constricted necks or deeply recessed sockets may require a socket extender (to extend the lamp beyond the constrictions). These are readily available at most hardware stores.

The ballast regulates the voltage and current delivered to a fluorescent lamp and is essential for proper operation; each type/wattage requires ballast specifically designed to drive it. There are two types of ballasts that operate on AC: magnetic and electronic. The magnetic ballast, the standard since fluorescent lighting was first developed, uses electromagnetic technology. The electronic ballast, only recently developed, uses solid-state technology. All DC ballasts are electronic devices.

Magnetic ballasts can last up to 50,000 hours and often incorporate replaceable bulbs. Magnetic ballasts flicker when starting and take a few seconds to get going. They also run the lamp at 60 cycles per second, and this flicker affects some people adversely.

Electronic ballasts weigh less than magnetic ballasts, operate lamps at a higher frequency (30,000 cycles per second vs. 60 cycles), are silent, generate less heat, and are more energy-efficient. They start almost instantly with no flickering, a big advantage for people who suffer from the "60-cycle blues." However, electronic ballasts cost more, particularly DC units. They last about 10,000 hours, the same as the bulb, and most do not have replaceable bulbs. Electronic ballasts with a much longer life are possible, but costs go up dramatically.

Note: There is no difference between a fluorescent tube for 120 volts and one for 12 volts. Only the ballasts are different.

Fluorescents and Remote Energy Systems

Fluorescent lights are available for both AC and DC applications. Most people using an inverter choose AC lights because of the wider selection, a significant quality advantage, and lower price. Some older inverters may have problems running some magnetic-ballasted lights. If you have an older Heart inverter, buy one light to try it first.

Most inverters operate compact fluorescent lamps satisfactorily. However, because all but a few specialized inverters produce an alternating current having a modified sine wave (versus a pure sinusoidal waveform), they will not drive compact fluorescents that use magnetic ballasts as efficiently or "cleanly" as possible, and some lamps may emit an annoying buzz. Electronic-ballasted compact fluorescents, on the other hand, are tolerant of the modified sine wave input and will provide better performance, silently.

Compact Fluorescent Applications

Due to the need for a ballast, a compact fluorescent light bulb is shaped differently from an incandescent. This is the biggest obstacle in retrofitting light fixtures. Compact fluorescents are longer, heavier, and sometimes wider. The ballast, the widest part, is located at the base, right above

In many ways, compact fluorescent lights have been superseded by LEDs, which happen to be coming on strong just as 40 W, 60 W, and 100 W incandescents are being phased out.

the screw-in adapter. Today, "mini-spiral" CFLs are available that fit into just about any fixture that currently uses a standard incandescent bulb. CFLs quickly became the most popular model in the marketplace, especially as mandates and rebates became the norm.

Additionally, the CFL bulb became the bulb of choice for being eco-friendly. As manufacturers become attuned to this relatively new and mandated market, more light fixtures suited for compact fluorescents became available. Now CFL technology is rapidly being replaced by LED, which offers better light quality, lifetime use, and a few other features that work better with controls, existing fixtures, and even DC systems. Real Goods offers a limited selection of both CFL and LED lamps and fixtures; see realgoods.com for the latest offerings in this rapidly expanding field.

Start-up Time
The start-up time of CFLs has long challenged consumers to really fall in love with them. Now with the instant-on feature of LEDs, you don't have to wait for your light to brighten or feel stuck with technology that simply doesn't perform as well as you'd like. You can also turn them on and off as much as you like and as rapidly as you like. Such is not the case for CFLs.

Full-spectrum Fluorescents
Literally, "full-spectrum" refers to light that contains all the colors of the rainbow. As several manufacturers use the term, "full-spectrum" refers to the similarity of their lamps' light (including ultraviolet light) to the midday sun. While we do believe that the closer an electric light source matches daylight, the healthier it is, we have several practical reservations. The quality of light produced by "full-spectrum" lighting available today is very "cool" in color tone and, in our opinion, not flattering to people's complexions or most interior environments. Also, the light intensity found indoors is roughly 100 times less than that produced by sunlight. At these low levels, we question whether people receive the full benefits offered by full-spectrum lighting.

Ideally, we encourage you to get outdoors every day for a good dose of sunshine. Indoors, we think it is best to install lighting that is complimentary to you and the ambiance you desire to create. For most applications, people prefer a warmer-toned light source. Most of the compact fluorescent lamps we offer mimic the warm rosy color of 60-watt incandescent lamps, but we do offer a small selection of full-spectrum lamps for those who prefer them.

Health Effects
Migraine headaches, loss of concentration, and general irritation have all been blamed on fluorescent lights. These problems are caused not by the lights themselves, but by the way they operate. Common magnetic ballasts run the lamps at the same 60 cycles per second that is delivered by our electrical grid. This causes the lamps to flicker noticeably 120 times per second, every time the alternating current switches direction. Approximately one-third of the human population is sensitive to this flicker on a subliminal level. These factors offer further support for choosing LED technology, which avoids the problem because it is a solid-state technology.

Recycling
Another major benefit to people and planet is that LEDs do not contain mercury. Mercury is a toxic substance and is used to make fluorescent and CFL lamps, which need a special hazardous waste facility to recycle the burned out bulbs. Contractors who work on commercial jobs are required to bring lamps with mercury in them to certified hazardous waste recyclers. However, the millions of residential consumers who aren't even aware that there is mercury inside their "energy-efficient CFLs" may unknowingly toss the light into the garbage, aka landfill. Once inside a landfill that may not be well protected, the mercury can leak into our groundwaters and make its way to our oceans.

If you have any fluorescent tubes or CFLs, please take them to your local hardware store, which typically has a program to take back the bulbs for recycling. Even if the fee is more than the purchase of the bulb, consider the possibility of further hazardous leakage as the instigator for doing the right thing…and tell your friends. One of the downsides to LEDs is that people will be ditching their CFL lighting for the LED option in large numbers. This sadly is not being well managed by the same companies who are now bringing us LED technology, and could create a serious mercury pollution problem.

With the long lifespan that LED technology offers, many LED manufacturers are not considering what will be done with all the LED lamps when they reach 70% brightness and need to be replaced. LED lights are made with a mix of plastic, metals, semiconductor materials, precious metals, and phosphors. Having all these materials mixed together does not bode well for tossing it into our recycling can, for which one should it go into? In reality, these LED lamps are most likely going to be treated as e-waste, and that industry is

Another major benefit to people and planet is that LEDs do not contain mercury. If you have any fluorescent tubes or CFLs, please take them to your local hardware store, which typically has a program to take back the bulbs for recycling.

mostly unregulated. News stories have appeared about children in China scouring through piles of computer waste for precious metals.

The lighting industry has a responsibility to its consumers to do the right thing, and fortunately LED technology gives us some time to figure out how to handle the hazardous wastes appropriately.

Savings

The main justification for buying fluorescent, CFL, or LED lights is to save energy and money over using incandescent, halogen, metal halide, or high-pressure sodium lights. Compact fluorescents provide opportunities for savings over incandescent, however, compared to more efficient LEDs, less so. It is often stated that LEDs are expensive, but most consumers only look at the immediate price of buying the bulb and not the cost of the energy it takes to run the bulb. Additionally the amount of money spent on having to buy the bulbs you need to match the lifespan of the LED is an added fee. In addition, the inconveniences of start-up time and poor dimmability vs. LED having instant-on and being dimmable, make buying fluorescents or CFLs less desirable. In the competition to accumulate negawatts—the ability to provide the same services people want, such as hot showers, cold drinks, and lit rooms, with much less energy—LED technology wins hands down.

The next question, of course, is, "*How much do CFLs or LEDs save when compared to our well-known incandescent?*" The more a light is used, the more energy and money you can save by replacing it with a more efficient bulb. Installing LEDs in the fixtures that are used the most hours and use high wattage saves the most energy.

An excellent candidate, for example, is the typical 60 W A lamp. The calculations in the accompanying Cost Comparison table assume that the light is on for an average of six hours per day

and that power costs 15¢ per kWh (the approximate national average). The total cost of operation is the cost of the bulb(s) plus the cost of the electricity used. The table below shows that LEDs, while costing $14.97 vs. a CFL for $6.67 or an incandescent for $0.97, are still the low-cost option when you look at how long they last and how little energy they use. Which would you rather pay: $241.06 for a 60 W incandescent or $61.94 for an LED?

Below is another comparison showing how LED outperforms its predecessor, metal halide. Once again the price to purchase the LED option is considerably higher, $488.62 vs. $23.00 for metal halide. However, when you bring into the equation that the LED will last 23 years while

Cost Comparison of Incandescent, CFL, and LED			
	60 W Incandescent Light Bulb	13 W CFL Light Bulb	13 W LED Light Bulb
Cost of bulb	$0.97	$6.67	$14.97
Product life hours	1,000	8,000	25,000
Years of life per 6 hrs. a day	.46 yr.	3.65 yrs.	11.41 yrs.
Bulbs used per 25,000 hours	25	3.125	1
Total bulb cost per 25,000 hours	$24.25	$20.84	$14.97
Energy cost annually (.15/kWh)	$19.71	$4.27	$4.27
Total cost bulbs + energy per 11.41/yrs.	$241.06	$69.56	$61.94

Cost Comparison of One Metal Halide Replaced with LED		
	400 W Metal Halide and Ballast	130 W LED Light and Driver
Cost of light and ballast/driver	$23.00	$488.62
Product life hours	20,000	100,000
Years of life per 12 hrs. a day	4.56 yrs.	22.83 yrs.
Bulbs used per 100,000 hours	5	1
Total bulb cost per 100,000 hours	$115.00	$488.62
Energy cost annually (.5/kWh)	$262.80	$85.41
Total cost bulbs + energy per 22.83/yrs.	$6,159.40	$2,443.62

Brand	Cree
Part Number	BA19-080270 MF-12DE26-2U400
Item Weight	2.4 pounds
Product Dimensions	18 × 8.9 × 6 inches
Item Model Number	BA19-080270 MF-12DE26-2U400
Color	White
Shape	A bulb
Lifetime Hours	25,000 hours life: 22.8 years (based on 3 hrs/day)
Lumens	800
Voltage	120 volts
Fixture Features	Damp rated
Certification	Lighting Facts
Type of Bulb	LED A series
Base Type	B15d
Luminous Flux	800 Lumens
Wattage	9.5 watts
Incandescent Equivalent	60 watts
Bulb Features	Dimmable
Color Temperature	2700 Kelvin
Color Rendering Index (CRI)	83
Bulb Diameter	2.4 inches
Bulb Length	4.7 inches
Start-up Time	Immediate
Lumen Maintenance Factor at the End of Life	85

How much do CFLs or LEDs save when compared to our well-known incandescent? The more a light is used, the more energy and money you can save by replacing it with a more efficient bulb. Installing LEDs in the fixtures that are used the most hours and use high wattage saves the most energy.

the metal halide only lasts 4.66 years, the formula changes. On top of longevity, LED offers the energy savings by providing as good if not better light for 130 W for what metal halide provides at 400 W. $177.39 saved per year is not something to sneeze at.

DC Lighting

When designing an independent power system, it is essential to use the most efficient appliances possible. LED lighting is 75% more efficient than incandescent lighting. This means you can produce the same quantity of light with only 10%–15% of the power use. When the power source is relatively expensive, like PV modules, or noisy and expensive, like a generator, this becomes terribly important. LED lighting is quite simply the best and most efficient way to light your house. Until LED technology came along, those with DC-based electric systems only had fluorescent or CFL lights available, which was expensive and sensitive to power surges.

We used to recommend against using DC lights in remote home applications except in very small systems where there is no inverter. However, now that LEDs are in the marketplace and readily available for a low-wattage DC system, you can have the best of both worlds: LEDs use minimal energy, and have been developed long enough that the bugs have been worked out in the electronics used to manage power surges and sign waves.

For your home, think of using a 12 V track lighting system where each MR16 consumes only 7 watts but puts out the equivalent of a 50 W halogen lamp. (Note that for a wide beam spread use 38°, or for spot lights use 25°.) RV and boating websites are great places to find a wide variety of LEDs that run on DC systems. An Internet search on "LEDs for boats" or "LEDs for RVs" will give you a whole host of products to choose from. Real Goods sells LED lights too: realgoods.com.

How to Read a Specification (Spec) Sheet

The most common place you will be reading lighting specifications is online. These pages could be filled with different types of specification layouts that manufacturers provide because there is no standard for what has to be listed other than a Lighting Facts FTC label. The information below is typical of what you may find on a manufacturer's website. You will recognize most of the information from the characteristics of light provided earlier.

Shown above is a list of specifications related to every light bulb. First, it is important to know what type of bulb you have so you can match it to the space you are lighting. Second, the amount

An Energy Tale, or Real Goods Walks Its Talk

Several years ago, Real Goods headquarters moved into a 10,000-square-foot office building. Built in the early 1960s, the building had conventional 4-foot fixtures, each holding four 40-watt cool white tubes: the dreary, glaring "office standard." Light levels were much brighter than recommended when employees were using computers a lot. Plus, a number of our employees were having headache, energy level, and "attitude" problems after moving, probably caused by the 60-cycle flicker from the old magnetic ballasts. Reducing overall light levels for less eyestrain was needed, as was giving employees a healthier working environment. And if we could save some money too, it wouldn't hurt.

Our solution was to retrofit the existing fixtures with electronic ballasts for no-flicker lighting, install specular aluminum reflectors, and convert from four cool-white 40-watt T-12 lamps per fixture to two warm-colored 32-watt T-8 lamps. Power use per fixture was reduced by over 60%, but desktop light levels, because of the reflectors and more efficient lamps and ballasts, were reduced by only 30% or less. The entire retrofit cost about $6,000 (not counting the $2,300 rebate from our local utility), yet our electrical savings alone are close to $5,000 per year. Plus our employees enjoy flicker-free, warm-colored, nonglaring light and greatly reduced EMF levels. The entire project paid for itself in less than a year! Improved working conditions came as a freebie!

Editor's Note: Even though this story is now more than 15 years old, its continuing relevance justifies repeating it again.

lumens the bulb puts out determines how bright it will be. Third, knowing how much energy a lamp uses (in watts) is important, because an A lamp, 60 W incandescent lasts only 1,200 hours while the same LED A Lamp can last more than 25,000 hours. Fourth, the color temperature rating of the bulb is an important personal choice, especially if you want a warm color (2700 K or cool white 5000 K).

CHAPTER 7

Water Development

Reckoning with Our Most Precious and Endangered Resource

WATER IS THE SINGLE MOST IMPORTANT INGREDIENT in any homestead. Without a dependable water source, you can't call any place home for very long. An age-old saying goes, "You buy the water and the land comes for free." Today, unfortunately, many people literally *are* buying water, unnecessarily and at outrageous expense, most of whom can least afford it. Some are swayed by multimillion-dollar ad campaigns by corporations selling bottled water; others are victims of the privatization of previously public water supplies. Around the world, *nearly 800 million* people still lack access to clean water, and as supplies increasingly become polluted, privatized, or gobbled up by for-profit enterprises, that number is projected to steeply rise. See the accompanying sidebars for a shocking picture of the current state of our most precious and endangered resource. We send a shout out to Brad Lancaster, award-winning author of *Rainwater Harvesting for Drylands and Beyond*, for help in writing this chapter. Brad has more than 20 years experience as a permaculture designer and consultant with a specialization in dryland environments.

Real Goods offers a wide variety of water development solutions for remote, and not so remote, homes. We specialize in products that are made specifically for solar-, battery-, wind-, gravity-powered, or water-powered pumping. These pumps are designed for long hours of dependable duty in out-of-the-way places where the utility lines don't reach. For instance, ranchers are finding that small solar-powered submersible pumps are far cheaper and more dependable than the old wind pumps that still dot the Great Plains. Many state and national parks use our renewable-energy-powered pumps for their backcountry campgrounds. The great majority of pumps we sell are solar- or battery-powered electric models, so we'll cover them first.

The initial cost of solar-generated electricity is high. Therefore, most solar pumping equipment is scaled toward modest residential needs, rather than larger commercial or industrial applications. Solar-powered pumps tend to be far more efficient than their conventional AC-powered cousins. By wringing every watt of energy out of a system, we're able to keep the start-up costs reasonable.

> Around the world, *nearly 800 million* people still lack access to clean water, and as supplies increasingly become polluted, privatized, or gobbled up by for-profit enterprises, that number is projected to steeply rise.

The Three Components of Every Water System

Every rural water system has three basic components.

A Source. This can be a well, a spring, a pond, a creek, a roof or other catchment surface from which you collect rainwater runoff, or the big expensive tanker truck that hauls in a load every month. In addition, you can reuse water once used—such as with a greywater-harvesting system.

A Storage Area. This is sometimes the same as the source, and sometimes it is an elevated or pressurized storage tank. (Note: that unlike other

Begin by planting the rain—don't drain it. Simple water-harvesting earthworks and their associated plantings help plant the rain by slowing, spreading, and sinking/infiltrating surface water flow into the soil. This reduces erosion, mitigates drought, lessens downstream flooding, and improves stormwater quality.

cleaner water, it is typically NOT a good idea to store greywater. Direct greywater directly into the soil instead where soil life can filter and utilize the greywater and nutrients.)

A Delivery System. This used to be as simple as the bucket at the end of a rope. But given a choice, most of us would prefer to have our water arrive under pressure from a faucet. Hauling water, because it's so heavy, and because we use such surprising amounts of it, gets old fast.

The first two components, source and storage, you need to produce locally; however, we can offer a few pointers based on experience and some of the better, and worse, stories we've heard.

Sources

Rainwater

Rainwater is the mother of all fresh water. It is naturally distilled through evaporation prior to cloud formation and thus is one of purest sources of water. It has about 100 times less total dissolved solids (TDS) than ground and surface water in Tucson, Arizona. This is why rainwater is known around the world as "sweet water."

Begin by planting the rain—don't drain it. Simple water-harvesting earthworks and their associated plantings help plant the rain by slowing, spreading, and sinking/infiltrating surface water flow into the soil. This reduces erosion, mitigates drought, lessens downstream flooding, and improves stormwater quality. We can then access that planted rain by planting "living pumps" of vegetation through which the water is captured in the produce from a fruit tree, shade, lumber, wind breaks, erosion control, livestock and wildlife fodder, etc. In addition, the act of planting the rain also helps recharge our wells, springs, streams, and other surface water via the more consistent, slow release of the harvested water from the soil.

Rainwater harvesting tanks can be considered for collecting roof runoff (cleaner than ground surface runoff because cows and dogs don't fly) after reducing the needed size (and thus cost) of the tanks by reducing your need for stored water. Planting the rain does just that by storing the rain falling on the land in the land, thereby reducing the need to apply additional irrigation waters. The soil is our largest and least expensive storage media. But as most people drain the rain, it is no wonder that 30% to 70% of the average American household's drinking water consumption is the irrigation of their landscape and garden. Treated drinking water (municipal water) should not be used to irrigate our plantings. Instead, rainwater

should be the primary irrigation source of our plantings, and greywater should be the secondary irrigation source. Drinking water if used at all should only be a supplementary source. See the "Planting Rain and Abundance" sidebar and Brad Lancaster's best-selling, award-winning book *Rainwater Harvesting for Drylands and Beyond* for more information. (Available at Harvesting Rainwater.com or realgoods.com.)

Wells

The most common domestic water source is the well, which can be hand dug, driven, or drilled. Wells are less prone than springs, streams, or ponds to picking up surface contamination from animals, pesticides and herbicides, or other runoff. Hand-dug wells are usually about 3 feet or larger in diameter and rarely more than 40 feet deep. These are difficult to build, they need regular cleaning, and it is dangerous to have them lying around unused on your property. Drilled or driven wells are standard practice now. The most common size for a residential driven or drilled well is either 4 or 6 inches. Beware of the "do-it-yourself" well-drilling rigs, which rarely can insert pipe larger than 2 inches. This restricts your pump choices to only the least efficient jet-type pumps. A 4-inch casing is the smallest you want—that's the minimum for a submersible pump.

Springs

Many folks with country property are lucky enough to have surface springs that can be developed. At the very least, springs need to be securely fenced to keep out wildlife. More commonly, springs are developed with either a backhoe and 2- or 3-foot concrete pipe sections sunk into the ground, or drilled and cased with one of the newer, lightweight horizontal drilling rigs. This helps to ensure that the water supply won't be contaminated by surface runoff or animals. To be respectful to your land and the critters that lived there before you came, prior to developing a spring, it's a good idea to give some consideration to any ecosystems it supports. If there are no other springs nearby, try to ensure that some runoff will still continue after development.

Ponds, Streams, and Other Surface Water

Some water systems are as simple as tossing the pump intake into the existing lake or stream and turning on the switch. However, surface water usually is used only for livestock or agricultural needs—although for many homesteads with gardens and orchards, this can be 90% of your

Planting Rain and Abundance:
Making the Most of Your Free Onsite Water *by Brad Lancaster*

As with most communities the world over, in my community of Tucson, Arizona—where we only get 11 inches of annual precipitation—in an average year of rainfall *more rain falls on the surface area of the community than all its (half million-plus) citizens consume of utility water in that same year*.

This rainwater is the best water for our plants and soil. It is salt-free (salts common in our groundwater and imported surface waters such as those from the Colorado River can build up in irrigated soil and impede plants' ability to photosynthesize and utilize water). Rainwater is a natural fertilizer (containing sulfur, beneficial microorganisms, mineral nutrients, and nitrogen). And it's free.

Nonetheless, we drain the vast majority of that high-quality rainwater out of our communities almost as quickly as it arrives via mound-like landscapes, soil scraped and raked bare, excessive paving, and our streets and storm drains. What are we thinking?

This unnaturally exposed soil and pavement, along with the sun-baked exterior walls of our homes, schools, and other buildings, then drain still more water by absorbing the heat of the sun during the day and reradiating that heat back out at night, increasing temperatures up to 10°F, which leads to greater evaporative loss of water. In dryland environments (about half the Earth's land, and the majority of the land west of the Mississippi River) the potential water loss to evaporation *exceeds* our average annual rainwater gain. The more temperatures increase, the greater the evaporative loss.

Use this information to act, and enhance life with six steps:

1. Improve your water gain by planting the rain before you plant any vegetation. (Or plant the rain beside vegetation if the plants are already in the ground.) Plant the rain within bowl-like, as opposed to mound-like, shapes in your landscape to capture and infiltrate, rather than drain, the rain.

2. Emphasize the placement of these basin-shaped rain gardens next to and below impervious surfaces like roofs, roads, and patios from which water runs off. That way you can double or even triple the available rainfall in the basins by capturing both rainfall and runoff, which becomes run*on*, and that's right on!

3. Decrease potential water loss by planting shading vegetation, like low-water-use, native, food-producing trees that will then grow to shade and cool roads, patios, and the east-, west-, and even north-facing walls of adjoining buildings. This will reduce unwanted sun exposure on our buildings' walls and windows in the morning and afternoon of the hot months. (But leave the winter sun/south-facing wall, beneath an appropriately sized roof overhang or awning, open to the winter sun low in the southern sky, so you can get free heat, light, and solar power when you need it most.) The runoff from the buildings and paved surfaces then freely runs into the rain gardens to irrigate the trees, while the trees passively shade and cool the pavement—reducing water loss to both wasteful runoff and evaporation.

This will also reduce water consumed to generate power. Electricity produced from burning coal consumes just under a half gallon of water per kWh of power produced. The average household consumes about 1,000 kWh of electricity a month,

to sewer

Reproduced with permission from *Rainwater Harvesting for Drylands and Beyond* by Brad Lancaster

and thus about 500 gallons of water per month for the electricity if provided by a coal-burning power plant. Increase that number to 100,000 households, and the monthly water consumption to generate the homes' power jumps to over 51 million gallons of water a month. Using less power, by providing more of your cooling and heating by summer shade and winter sun, will reduce this water consumption/loss.

Note that the City of Tucson just passed an ordinance that all new city streets must be designed and built to harvest at least a half-inch rainstorm's worth of water to freely irrigate streetside vegetation shading the street and walkways.

4. Mulch the surface of the soil to make it more porous to speed up the rate at which water infiltrates, while reducing the loss of soil moisture to evaporation. Compost and woody organic matter are the best mulch as they increase the fertility of the soil and plant growth. Furthermore, this mulch feeds beneficial soil microorganisms, such as mycorrhizal fungi, which tap into and expand the surface area of associated plants' roots. The plants can then more efficiently uptake harvested water, as the fungi give the plants water and minerals, while the plants give the fungi carbohydrates and sugars. At the very least, don't rake up and throw away your fallen leaves. They are called "leaves" because you are supposed to *leave* them as mulch beneath your plantings. Fallen

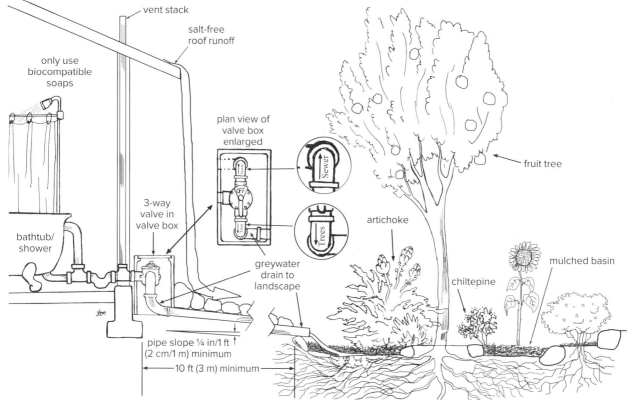

seedpods of mesquite, ironwood, palo verde, and other trees also make great mulch.

5. If you want any higher-water-use plantings such as fruit trees, be sure to plant your greywater before you plant your fruit tree(s). (Or if your fruit tree is already planted, then plant the greywater next to the fruit tree.) Greywater is the drainwater from household bathroom sinks, showers, bathtubs, and washing machines. The volume of greywater running down the drain of the average Arizona family household is enough to meet about half of the average family's landscape irrigation demand. If you use the right non-toxic, salt-free soaps and detergents, your greywater can be directed to, and planted within, the same mulched basins that capture your rainwater. No tanks, pumps, or filters are needed. In times of rain, the basins act as rain gardens. In times of no rain, they act as greywater gardens. As long as you are home, that greywater flow to your plants can be perennial—even in the driest of times.

See the "Greywater Harvesting" page at HarvestingRainwater .com for "Soap and Detergent Info" on what ingredients and products are good or bad to use. Also see Chapter 10 on Greywater and Composting Toilets, pages 279–90.

6. If you have an air conditioner, direct its salt-free condensate water to the rain gardens instead of the sewer. You'll only get about a ¼ gallon per day of condensate from a home air conditioner in the dry season, but it can be as much as 18 gallons a day in the humid season. Condensate from commercial air conditioners equals hundreds of gallons a day.

Taking these steps that harvest, rather than drain free, local waters transform *dehydrating* landscapes into *rehydrating* landscapes that provide myriad additional benefits such as more local food, enhanced flood control, diverse wildlife habitat, beauty, and more life that can also lead to more rain.

This is because clouds are more likely to form from cooled atmospheric moisture evapotranspired through plant leaves than from the warmer moisture evaporated from bare soil. In addition, raindrops are more likely to condense around tiny, richly textured, airborne particles of organic matter generated by the vegetation (such as pollen) than around the less textured and less cool particles of dust from exposed dirt. We can choose to work with these natural systems or against them.

I think you'll find going *with* the flow is always the easiest and most abundant path.

Brad Lancaster is the author of the award-winning books *Rainwater Harvesting for Drylands and Beyond*, Volumes 1 and 2 and Harvesting Rainwater.com, which show you how to make the most of the above six steps and a whole lot more—even if you are renting your home and don't have a yard.

Rainwater Harvesting Resources

Rainwater Harvesting for Drylands and Beyond, Volume 1, 2nd Edition: *How to Welcome Rain Into Your Life and Landscape* (book) by Brad Lancaster, RainSource Press, 2013, 8.5 × 11, 304 pages, 280+ illustrations and photos, ISBN 978-0-9772464-3-4.

Rainwater Harvesting for Drylands and Beyond, Volume 2: *Water-Harvesting Earthworks* (book) by Brad Lancaster, RainSource Press, 2008, 8.5 × 11, 448 pages, 450+ illustrations and photos, ISBN 978-0-977246-41-0.

Water Storage (book) by Art Ludwig, Oasis Design, 2005, 8.5 × 11, 125 pages, ISBN 978-0-964343-36-8.

HarvestingRainwater.com

To be respectful to your land and the critters that lived there before you came, prior to developing a spring, it's a good idea to give some consideration to any ecosystems it supports.

total water use. If you don't have a surface water source on your property, consider putting in a pond. Ponds are one of the least expensive and most delightfully pleasing methods of supplying water, not only for your own needs but also for providing rich habitat for the wildlife population that your pond will soon support. They often are used to hold winter and spring runoff for summertime use. Don't drink or cook with surface water unless it has been treated or purified first (see Chapter 9 for more help on this topic).

Storage

You might need to provide some means of water storage for any number of reasons. The most common are: To get through long dry periods, to provide pressure through elevated storage and gravity feed or pressurized air, to keep drinking water clean and uncontaminated, and to prevent freezing.

Surface Storage

Those with ponds, lakes, streams, or springs may not need any additional storage. Let the livestock find their own way to the water, or pump straight to whatever needs irrigation. However, many systems will need to pump water to an elevated storage site or a large pressure tank in order to develop pressure through a gravity-fed system. Remember, every 2.3 feet of vertical fall translates into 1 psi of pressure. You may need seasonal freeze protection, which means protected or underground storage.

Pond Liners

Custom-made polyvinyl liners are available for non-clay lined ponds or leaking tanks. Liners have revolutionized pond construction. They are designed for installation during pond construction and are then buried with 6 inches of dirt around the edges, practically guaranteeing a leak-free pond even when working with gravel, sand, or other problem soils. When buried, the life expectancy of these liners is 50 years or more. They make reliable pond construction possible in locations that normally could not accommodate such inexpensive water storage methods. Note that all three of the ponds at our Solar Living Center are lined and have been operating flawlessly with no leaks for 20 years now.

Global Water Crisis

NEARLY 800 million people today don't have access to clean drinking water, and millions die every year for lack of it. Water resources around the globe are threatened by climate change, misuse, and pollution. But there are solutions: By using innovative water-efficiency and conservation strategies, community-scale projects, smart economics, and new technology, we can provide for people's basic needs while protecting the environment.

That almost 800 million people lack access to clean water is surely one of the greatest development failures of the modern era. That as many as 5 million people—mainly children—die every year from preventable water-related disease is surely one of the great tragedies of our time.

Unfortunately, despite a growing recognition that more must be done to help those without clean water or adequate sanitation, almost nothing is being done. The problem is not merely a lack of aid (although more money is needed) or a lack of technology. It is a failure of vision and will. According to many international water experts, hundreds of billions of dollars are needed to bring safe water to everyone who needs it. Since international water aid is so paltry, many of these experts claim that privatization of water services is the only way to help the poor.

But many critics of this approach note that community-scale infrastructure and efficiency and conservation can bring basic water services to the millions who need it without breaking the bank. And many critics of the "gold-plated" approach argue that water privatization, although it can play some productive role, will never be able to bring water to the world's poorest people.

There are solutions to the global water crisis that don't involve massive dams, large-scale infrastructure, and tens or hundreds of billions of dollars. First and foremost, we must use what the Pacific Institute calls "soft path" solutions to the global water crisis. Soft path solutions aim to improve the productivity of water rather than seek endless new supply; soft path solutions complement centrally planned infrastructure with community-scale projects; and soft path solutions involve stakeholders in key decisions so that water deals and projects protect the environment and the public interest.

For more information on global water issues, challenges, and solutions, visit pacinst.org and worldwater.org.

—Courtesy of Peter Gleick, Pacific Institute

The Real Cost of Bottled Water

San Franciscans and other Bay Area residents enjoy some of the nation's highest-quality drinking water, with pristine Sierra snowmelt from the Hetch Hetchy reservoir as our primary source. Every year, our water is tested more than 100,000 times to ensure that it meets or exceeds every standard for safe drinking water. And yet we still buy bottled water. Why?

Maybe it's because we think bottled water is cleaner and somehow better, but that's not true. The federal standards for tap water are higher than those for bottled water.

The Environmental Law Foundation has sued eight bottlers for using words such as "pure" to market water that contains bacteria, arsenic, and chlorine. Bottled water is no bargain either: It costs 240 to 10,000 times more than tap water. For the price of one bottle of Evian, a San Franciscan can receive 1,000 gallons of tap water. Fifty percent of bottled water should be labeled bottled *tap* water because that's exactly what it is (47.8% in 2009 according to Food and Water Watch, foodandwaterwatch.org/briefs/bottling-our-cities-tap-water/). Recently Pepsi and Nestlé have been forced to acknowledge that their Aquafina and PureLife brands are sourced from the tap. And while, as a result of pressure from activist corporate accountability groups the market for bottled water has stagnated, consumer demand remains unacceptably high.

Clearly, the popularity of bottled water is the result of huge marketing efforts. The global consumption of bottled water reached 65 billion gallons in 2011, a 50% increase since 2004. Even in areas where tap water is clean and safe to drink, such as in San Francisco, demand for bottled water is increasing—producing unnecessary garbage and consuming vast quantities of energy. So what is the real cost of bottled water?

Most of the price of a bottle of water goes for its bottling, packaging, shipping, marketing, retailing, and profit. Transporting bottled water by boat, truck, and train involves burning massive quantities of fossil fuels. More than 5 trillion gallons of bottled water is shipped internationally each year. Here in San Francisco, we can buy water from Fiji (5,455 miles away) or Norway (5,194 miles away) and many other faraway places to satisfy our demand for the chic and exotic. These are truly the Hummers of our bottled water generation. As further proof that the bottle is worth more than the water in it, since 2007 the state of California has been refunding 5¢ for recycling a small water bottle and 10¢ for a large one. The next time you see someone drinking "Fiji" water, do them a favor and let them know they're blowing it!

Just supplying Americans with plastic water bottles for one year consumes more than 50 million gallons of oil, enough to take 100,000 cars off the road and 1 billion pounds of carbon dioxide out of the atmosphere, according to the Container Recycling Institute. In contrast, San Francisco tap water is distributed through an existing zero-carbon infrastructure: plumbing and gravity. Our water generates clean energy on its way to our tap—powering our streetcars, fire stations, the airport, and schools.

More than 1.9 million tons of plastic water bottles ended up in the nation's trash in 2011, taking up valuable landfill space, leaking toxic additives such as phthalates into the groundwater, and taking 1,000 years to biodegrade. That means bottled water may be harming our future water supply.

The rapid growth in the bottled water industry means that water extraction is concentrated in communities where bottling plants are located. This can have a huge strain on the surrounding ecosystem.

So it is clear that bottled water directly adds to environmental degradation, global warming, and a large amount of unnecessary waste and litter. All this for a product that is often inferior to San Francisco's tap water. Luckily, there are better, less expensive alternatives:

- In the office, use a water dispenser that taps into tap water. The only difference your company will notice is that you're saving a lot of money.
- At home and in your car, switch to a stainless steel water bottle and use it for the rest of your life, knowing that you are drinking some of the nation's best water and making the planet a better place.

© Jared Blumenfeld and Susan Leal, *San Francisco Chronicle*, February 18, 2007, reprinted with permission and revised with updated statistics. Jared Blumenfeld is the EPA Regional Administrator for the Pacific Southwest. Susan Leal was the general manager of the San Francisco Public Utilities Commission and is the author of *Running Out of Water: The Looming Crisis and Solutions to Preserve Our Most Precious Resource* (2010).

John Schaeffer's pond in Hopland, California.

Approximate Daily Water Use for Home and Farm

Usage	Gallons per Day
Home	
National average residential use	75 per person
As above, without flush toilet	55 per person
Drinking and cooking water only	4–6 per person
Lawn—Garden—Pool	
Lawn sprinkler, per 1,000 sq. ft., per sprinkling	600 (approx. 1 inch)
Garden sprinkler, per 1,000 sq. ft., per sprinkling	600 (approx. 1 inch)
Standard garden hose	10 gallons per minute
Swimming pool maintenance, per 100 sq. ft. surface area	30 per day
Farm (maximum needs)	
Dairy cows	20 per head
Dry cows or heifers	15 per head
Calves	7 per head
Beef cattle, yearlings (90°F)	20 per head
Beef, brood cows	12 per head
Sheep or goats	2 per head
Horses or mules	12 per head
Swine, finishing	4 per head
Swine, nursing	6 per head
Chickens, laying hens (90°F)	9 per 100 birds
Chickens, broilers (90°F)	6 per 100 birds
Turkeys, broilers (90°F)	25 per 100 birds
Ducks	22 per 100 birds
Dairy sanitation—milk room and milking parlor	500 per day
Flushing barn floors	10 per 100 sq. ft.
Sanitary hog wallow	100 per day

Tanks

A covered tank of one sort or another is the most common and longest-lasting storage solution. The cover must be screened and tight enough to keep out light and critters such as lizards, mice, and squirrels from drowning in your drinking water (always an exciting discovery). The most common tank materials are polypropylene, fiberglass, and concrete or ferro-cement.

With rainwater tanks, a great option is to use an angled downspout or "rainhead" screen. Placed below the downspout opening of a roof gutter, the 45° or so screen diverts leaves, critters, insects, and other debris out of the water flow, while the water runs through the screen into your pipe and tank. Another advantage of a rainhead downspout screen is that, unlike a gutter screen, you can see from the ground if it's clogged or damaged, and then you can quickly fix it. (These rarely clog.)

Plastic Tanks. Both fiberglass and polypropylene tanks are commonly available. Your local farm supply store is usually a good source. All plastic tanks will suffer slightly from UV degradation in sunlight. Simply painting the outside of the tank will prevent this problem and probably make the tank nicer looking. If this tank is for your drinking water, make sure it's internally coated with an FDA-approved material for drinking water. Most farm supply stores will offer both "drinking water grade" and "utility grade" tanks. Utility grade tanks often use recycled plastic materials and probably will leach some polymers into the water, particularly when new. Some plastic tanks can be partially or completely buried for protection against freezing. Ask before you buy if this is a consideration for you.

Concrete Tanks. Concrete is one of the best and longest-lasting storage solutions, but these tanks are expensive and/or labor-intensive initially. All but the smallest concrete tanks are built on site. They can be concrete block, ferro-cement, or monolithic block pours. Although monolithic pours require hiring a contractor with specialized forming equipment, these tanks are usually the most trouble-free in the long run. Any concrete tank will need to be coated internally with a special sealer to be watertight. Concrete tanks can be buried for freeze protection and to keep the water cool in hot climates. They can cost 40¢–60¢ per gallon, but in the long run, they're well worth it.

Pressure Tanks. These are used in pumped systems to store pressurized water so that the pump

REAL GOODS

doesn't have to start for every glass of water. They work by squeezing a captive volume of air, since water doesn't compress. Pressure tanks are rated by their total volume. Draw-down volume—the amount of water that actually can be loaded into and withdrawn from the tank under ideal conditions—is typically about 40% of total volume. So a "20-gallon" pressure tank can really deliver only about 8 gallons before starting the pump to refill. Pressure tanks are one of the few things in life where bigger really is better. With a larger pressure tank, your pump doesn't have to start as often, it will use less power and live longer, your water pressure will be more stable, and if power fails, you'll have more pressurized water in storage to tide you over. Forty-gallon capacity is the minimum for residential use, and more is better.

Delivery Systems

A few lucky folks are able to collect and store their water high enough above the level of intended use that the delivery system will simply be a pipe or hose, and the weight of the water will supply the pressure free of charge. This is known as a gravity-fed system. (You need 46 vertical feet for 20 pounds per square inch. Said another way, for every foot the height of the water is above its destination, you gain 0.43 pounds per square inch of pressure.) For those using gravity to pressurize your water, avoid unnecessary constrictions in your distribution system that will reduce your pressure. For example, use full port valves that do not constrict to smaller interior diameters. (Search "full port valves" at HarvestingRainwater .com for more.) In addition, 1-inch-diameter pipe will give you better flow than smaller-diameter pipe in low-pressure situations. But most of us are going to need a pump, or two, to get the water up from underground and/or to provide pressure.

Water Conversions	
1 gal. of water	= 8.33 lb.
1 gal. of water	= 231 cu. in.
1 cu. ft. of water	= 62.4 lb.
1 cu. ft. of water	= 7.48 gal.
1 acre ft. of water	= 326,700 gal.
1 psi of water pressure	= 2.31 ft. of head

We'll cover electrically driven solar and battery-powered pumping first, then water-powered and wind-powered pumps.

The standard rural utility-powered water system consists of a submersible pump in the well delivering water into a pressure tank in some location that's safe from freezing. A pressure tank extends the time between pumping cycles by saving up some pressurized water for delivery later. This system usually solves any freezing problems by placing the pump deep inside the well, and the pressure tank indoors. The disadvantage is that the pump must produce enough volume to keep up with any potential demand, or the pressure tank will be depleted and the pressure will drop dramatically. This requires a ⅓ hp pump minimally, and usually ½ hp or larger. Well drillers often will sell a much larger than necessary pump because it increases their profit and guarantees that no matter how many sprinklers you add in the future, you'll have sufficient water-delivery capacity. (And they don't have to pay your electric bill!) This is fine when you have large amounts of utility power available to meet heavy surge loads, but it's very costly to power with a renewable energy system because of the large equipment requirements. We try to work smarter and smaller while using less-expensive resources to get the job done.

Solar-powered Pumping

Where Efficiency Is Everything

PV modules are still not cheap, and water is still surprisingly heavy. These two facts define the solar-pumping industry. At 8.3 pounds per gallon, it takes a lot of energy to move water uphill. Anything we can do to wring a little more work out of every last watt of energy is going to make the system less expensive initially. Because of these economic realities, the solar-pumping industry tends to use the most efficient pumps available. For many applications, that means a positive-displacement type of pump. This class of pumps prevents the possibility of the water slipping from

high-pressure areas to lower-pressure areas inside the pump. Positive-displacement pumps also ensure that even when running very slowly—such as when powered by a PV module under partial light conditions—water still will be pumped. As a general rule, positive-displacement pumps deliver four to five times the efficiency of centrifugal pumps, particularly when lifts over 60 feet are involved. Several varieties of positive-displacement pumps are commonly available. Diaphragm pumps, rotary-vane pumps, piston pumps, and the newest innovation, helical-rotor pumps, are all available from Real Goods.

Beware of Privatizing the Essence of Life

JUST LIKE AIR, WATER IS PRECIOUS and sustains all life on Earth. Increased demand for water by industry and agriculture is draining away the planet's rivers, lakes, and other freshwater sources. Meanwhile, a profit-driven industry increasingly controls our water supply. How we address the impending water crisis will have tremendous implications for people's health and the environment in the 21st century.

Though only a little less than 1% of Earth's water is available for human use, there is still more than enough fresh water to sustain every person on the planet—all 7 billion of us. Because of its essential, even sacred, role in life, many believe that water is a common resource to be shared by all.

In North America, most people receive water from a public utility. But not everyone in the world has access to the water they and their families need. The United Nations estimates that almost 800 million people lack access to safe drinking water. Hundreds of millions must walk miles every day to gather enough water to survive. The need to travel great distances to collect water is a leading reason that young girls are not able to attend school in many African countries. And 1.5 million children die every year of diseases caused by unsafe water. These situations are likely to worsen. According to the UN, by 2025 two-thirds of the world's population—more than 5 billion of us—will lack access to water and "could be under stress conditions." The World Bank has predicted that the wars of tomorrow will no longer be fought over oil but over water.

How is it possible that the water crisis could explode within a single generation?

There are many causes: the escalating use and abuse by water-intensive industries such as mining, paper production, and power generation; a growing population and an increasing need for irrigation; a spread of industrial pollution that fouls lakes and rivers, especially in developing countries; and spreading droughts induced by climate change. The World Bank estimates that rising temperatures and decreasing rainfall associated with climate change will reduce the amount of rain-fed farmland by approximately 10% within the lifetimes of today's children.

There's Huge Profit in Thirst

A limited water supply, coupled with the growing demand for water, is seen by corporations as a huge profit-making opportunity. *Fortune Magazine* maintains that "water will be to the 21st century what oil was to the 20th century," and corporations are racing to stake claims to this "liquid gold."

In fact, corporations have already been meeting behind closed doors for years, vying for control of the world's water resources. They have pushed officials at the World Bank and International Monetary Fund to make industry-friendly water policies a condition of debt assistance to developing countries. Water corporations have pushed trade ministers and officials at the World Trade Organization to craft industry-biased trade agreements. In March 2006, Coca-Cola sponsored the World Water Forum, where giant corporations met with representatives of the United Nations, governments, and the World Bank, to promote profit-oriented water policies around the world.

Nowhere is the corporate water grab more insidious than its escalating control of drinking water. Supplying water is already a $420-billion-a-year business. Throughout the world, powerful corporations are gaining control of public water systems, reducing a shared common resource into simply another opportunity to profit. For example, Suez—the corpo-ration that built the Suez Canal—has recently been snatching up government contracts to take over municipal water systems.

And if controlling our taps were not enough, Coke, Pepsi, and Nestlé are bottling our water and selling it back to us at prices that are hundreds, even thousands, of times greater than what tap water costs. Today, water is one of the world's fastest-growing branded beverages.

Bottled Water: Is It Better?

In countries like the US, most water services are hidden from public view. We catch a glimpse of them when we turn on the shower or flush the toilet. Then they retreat into the background, and we go about our day.

Bottled water is an exception. Bottled water corporations aim to brand the water we drink and turn it into a status symbol. But these companies, led by Coke, Nestlé, and Pepsi, have sold us a bill of goods. Misleading advertising is fueling the explosive growth of the bottled water business. In 2005, bottled water corporations spent over $158 million to portray their products as "pure," "safe," "clean," "healthy," and superior to tap water. (Advertising for bottled water dropped to "only" $61 million in 2012.) Today, three out of four Americans drink bottled water, and one in five drinks *only* bottled water, even though it is much more expensive than tap water and can sometimes be less safe.

Water bottling is one of the least regulated industries in the US. Tap water and bottled water are subject to similar standards, but tap water is tested far more frequently and is more rigorously monitored and enforced by the EPA. In contrast, there are significant gaps in FDA regulation of bottled water, and the agency largely relies on the corporations to police themselves. A comprehensive study conducted by the Environmental Working Group in 2008 found that 10 popular brands of bottled water, purchased from grocery stores and other retailers in 9 states and the District of Columbia, contained 38 chemical pollutants, with an average of 8 contaminants in each brand.

People are paying a high price for this deception, and price gouging is only the beginning. Bottled water corporations use their political and economic clout to secure sweetheart deals, block legislative efforts to protect local water rights, and pursue costly and time-consuming litigation against individuals and governments.

Water bottling is one of the least regulated industries in the US. Tap water and bottled water are subject to similar standards, but tap water is tested far more frequently and is more rigorously monitored and enforced by the EPA. In contrast, there are significant gaps in FDA regulation of bottled water.

Communities Resist Privatization

Imagine that you live in a town where some of the wells are contaminated with elevated levels of naturally occurring radiation. The contamination is known to the managers of the corporate-controlled water system, who shut down the wells when government inspectors arrive to take water samples. Their deception discovered, the managers are fired, indicted, and replaced with new managers. Months later, you turn on the tap while entertaining guests for a Memorial Day picnic, only to find it dry. You call the water company, but your call is disconnected after 40 minutes on hold because no one at the private water company could be reached. Welcome to Toms River, New Jersey, where Suez controls the water system.

Imagine residing in a town where the local water has become unfit to drink. The private corporation that supplies tap water refuses to repair the water system or to connect to another source and instead provides 25 gallons of bottled water a week to each household. Welcome to San Jerardo, California.

Imagine a major US city that spends $1 million on a marketing campaign to boost people's confidence in the public water system, which is widely regarded as one of the highest-quality systems in the country. Then imagine opening the newspaper one day to discover that the city has simultaneously been spending $2 million to provide expensive bottled water for city workers. Welcome to San Francisco.

In Stanwood, Michigan, concerned citizens fought Nestlé for 10 years. Nestlé has drained tens of millions of gallons of water from local water sources and ecosystems, causing significant damage to the local environment—a stream, two lakes, and rich diverse wetlands have been harmed. A local environmental group, Michigan Citizens for Water Conservation (MCWC), won a major court victory in 2003 that shut down the well field where Nestlé bottled water. But Nestlé retaliated. The corporation used its political and financial leverage to appeal the ruling and won temporary permission to continue to extract and bottle 218 gallons of water per minute. The citizens group appealed the case to the Michigan Supreme Court and in 2009 an out-of-court settlement forced Nestlé to reduce its extraction rate by half. This is a victory for the environment, but it came at great cost and the corporation continues to put stress on local water resources.

The reality behind this industry's slick public relations and marketing is that bottled water threatens our health and our ecosystems, costs far more than tap water, and undermines local democratic control of a common resource. Bottled water corporations take water from underground springs and municipal sources without regard to scarcity or human rights. Corporations are trying to make a profit-driven commodity out of a public resource that should not be bought or sold.

Courtesy of Corporate Accountability International, whose "Think Outside the Bottle" campaign aims to educate the public about the dangers of corporate water privatization. For more information, visit StopCorporate Abuse.org.

Bottled water brands in North America owned by Nestlé:		
Acqua Panna	Ozarka	San Pellegrino
Arrowhead	Perrier	Trade Winds
Deer Park	Poland Spring	Zephyrhills
Ice Mountain	Resource	
Nestlé Pure Life	Sweet Leaf	

Bottled water brands in North America owned by Coca-Cola:		
Alhambra	Spring! by Dannon	Malvern Water
Aquarius Spring	Dasani	(Schweppes)
Ciel (Mexico)	Dasani Nutriwater	**Pepsi's big brand:**
Crystal	Evian	Aquafina

Centrifugal pumps are good for moving large volumes of water at relatively low pressure. As pressure rises, however, the water inside the centrifugal pump "slips" increasingly, until finally a pressure is reached at which no water is actually leaving the pump.

Positive-displacement pumps have some disadvantages. They tend to be noisier, as the water is expelled in lots of little spurts. They usually pump smaller volumes of water, they must start under full load, most require periodic maintenance, and most won't tolerate running dry. These are reasons that this class of pumps is not used more extensively in the AC-powered pumping industry.

Most AC-powered pumps are centrifugal types. This type of pump is preferred because of easy starting, low noise, smooth output, and minimal maintenance requirements. Centrifugal pumps are good for moving large volumes of water at relatively low pressure. As pressure rises, however, the water inside the centrifugal pump "slips" increasingly, until finally a pressure is reached at which no water is actually leaving the pump. This is 0% efficiency. Single-stage centrifugal pumps suffer at lifts over 60 feet. To manage higher lifts, as in a submersible well pump, multiple stages of centrifugal pump impellers are stacked up.

In the solar industry, centrifugal pumps are used for pool pumping and for some circulation duties in hot water systems. But in all applications where pressure exceeds 20 psi, you'll find us recommending the slightly noisier positive-displacement pumps, which require occasional maintenance but are vastly more efficient. For instance, an AC submersible pump running at 7%–10% efficiency is considered "good." The helical-rotor submersible pumps we promote run at close to 50% efficiency.

For Highest Efficiency, Run PV Direct

We often design solar pumping systems to run PV direct. That is, the pump is connected directly to the photovoltaic (PV) modules with no batteries involved in the system so that it pumps when the sun is shining, which is usually when it's most needed. The electrical-to-chemical conversion in a battery isn't 100% efficient. When we avoid batteries and deliver the energy directly to the pump, 20%–25% more water gets pumped. This kind of system is ideal when the water is being pumped into a large storage tank or is being used immediately for irrigation. It also saves the initial cost of the batteries, the maintenance and periodic replacement they require, and the charge controllers and the fusing/safety equipment that they demand. PV-direct pumping systems, which are designed to run all day long, make the most of your PV investment and help us get around the lower gallon-per-minute output of most positive-displacement pumps. Why make things more complicated if you don't have to?

However (every silver lining has its cloud), we like to use one piece of modern technology on PV-direct systems that isn't often found on battery-powered systems. A linear current booster, or LCB, is a solid-state marvel that will help get a PV-direct pump running earlier in the morning, keep it running later in the evening, and sometimes make running a possibility on hazy or cloudy days. An LCB will convert excess PV voltage into extra amperage when the modules

A Brief Glossary of Pump Jargon

Flow: The measure of a pump's capacity to move liquid volume. Given in gallons per hour (gph), gallons per minute (gpm), or liters per minute (Lpm).

Foot Valve: A check valve (one-way valve) with a strainer. Installed at the end of the pump intake line, it prevents loss of prime and keeps large debris from entering the pump.

Friction Loss: The loss in pressure due to friction of the water moving through a pipe. As flow rate increases and pipe diameter decreases, friction loss can result in significant flow and head loss.

Head: Two common uses: 1) the pressure or effective height a pump is capable of raising water; 2) the height a pump is actually raising the water in a particular installation.

Lift: Same as head. Contrary to the way this term sounds, pumps do not suck water, they push it.

Prime: A charge of water that fills the pump and the intake line, allowing pumping action to start. Centrifugal pumps will not self-prime. Positive-displacement pumps will usually self-prime if they have a free discharge—no pressure on the output.

Submersible Pump: A pump with a sealed motor assembly designed to be installed below the water surface. Most commonly used when the water level is more than 15 feet below the surface or when the pump must be protected from freezing.

Suction Lift: The difference between the source water level and the pump. Theoretical limit is 33 feet; practical limit is 10–15 feet. Suction lift capability of a pump decreases 1 foot for every 1,000 feet above sea level.

Surface Pump: Designed for pumping from surface water supplies such as springs, ponds, tanks, or shallow wells. The pump is mounted in a dry, weatherproof location less than 10–15 feet above the water surface. Surface pumps cannot be submerged and be expected to survive.

aren't producing quite enough current for the pump. The pump will run more slowly than if it had full power, but if a positive-displacement pump runs at all, it delivers water. LCBs will boost water delivery in most PV-direct systems by 20% or more, and we usually recommend a properly sized one with every system. It's well worth the small additional investment.

Direct Current (DC) Motors for Variable Power

Pumps that are designed for solar energy use DC electric motors. PV modules produce DC electricity, and all battery types store DC power. Direct current motors have the great advantage of accepting variable voltage input without distress. Common AC motors will overheat if supplied with low voltage. DC motors simply run slower when the voltage drops, slowing the flow of water but keeping it coming nonetheless. This makes them ideal partners for PV modules. Day and night, clouds and shadows; These all affect the PV output, and a DC motor simply "goes with the flow"!

Which Solar-powered Pump Do You Want?

That depends on what you're doing with it and what your climate is. We'll start with the most common and easiest choices and work our way through to the less common.

Pumping from a Well

Do you have a well that's cased with a 4-inch

Q: Will Solar Pumps Run Only During Direct Sunlight?

A. No. While we often run pumping systems directly off solar electric modules without batteries, battery-based systems can be used for round-the-clock pumping. Direct pumping is usually the better option (approximately 20% more efficient), but this works only during sunny times of day. PV direct is usually the preferred power delivery if you're pumping to a storage tank, doing direct irrigation, or running the backyard fountain. A PV-direct pump controller often is needed to help start the pump earlier in the day, keep it running later in the afternoon, or make any pumping possible on cloudy days.

or larger pipe, and a static water level that is no more than 750 feet below the surface? Perfect. We carry several brands of proven DC-powered submersible pumps with a range of prices, lift, and volume capabilities. The SHURflo Solar Sub is the lowest-cost system with lift up to 230 feet and sufficient volume for most residential homesteads. The bigger helical-rotor-type sub pumps like the Grundfos and SunPumps are available in over a dozen models, with lifts up to 750 feet or volume over 25 gallons per minute, depending on the model. Performance, prices, and PV requirements can be found on our website, realgoods.com. The SHURflo Solar Sub is a diaphragm-type pump, and unlike almost any other submersible pump, it can tolerate running dry. The manufacturer says just don't let it run dry for more than a month or two! This feature makes this pump ideal for many low-output wells. One that we highly recommend is ITT Goulds' three-phase 240-volt AC submersible pump with an ABB reactor and variable frequency drive (VFD). It's a simple matter to invert the DC to AC, and this way you get an industrial-quality pump for less money than a Grundfos, which cannot be serviced if it fails prematurely, which does happen.

Complete submersible pumping systems—PV modules, mounting structure, LCB, and pump— range from $1,900 to $12,000 depending on lift and volume required. Options such as float switches that will automatically turn the pump on and off to keep a distant storage tank full are inexpensive and easy to add when using an LCB with remote control, as we recommend.

Because many solar pumps are designed to work all day at a slow but steady output, they won't keep up directly with average household fixtures, like your typical AC sub pump. This often requires some adjustment in how your water supply system is put together. For household use, we usually recommend the following, in order of cost and desirability:

- **Option 1.** Pumping into a storage tank at least 50 feet higher than the house (21 psi by gravity feed), if terrain and climate allow.
- **Option 2.** Pumping into a storage tank at house level, if climate allows, and using a booster pump to supply household pressure.
- **Option 3.** Pumping into a storage tank built into the basement for hard-freeze climates, and using a booster pump to supply household pressure.
- **Option 4.** Using battery power from your household renewable energy system to run the submersible pump, and using a big pressure tank (80 gallons minimum).

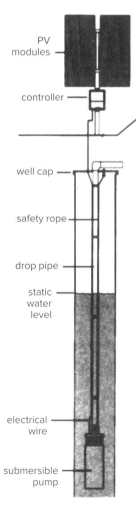

A simple PV-direct solar pumping system.

PV modules

controller

well cap

safety rope

drop pipe

static water level

electrical wire

submersible pump

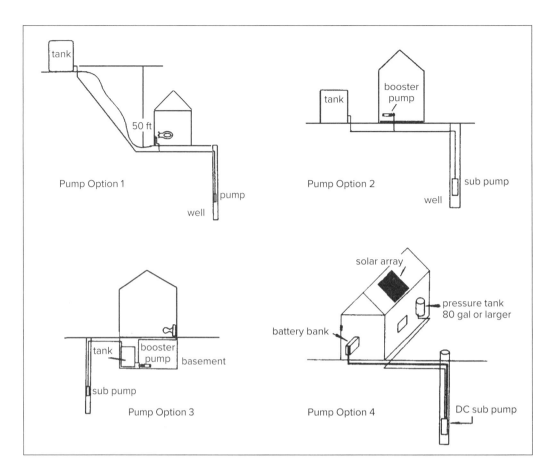

tank

50 ft

Pump Option 1

pump

well

tank

booster
pump

Pump Option 2

sub pump

well

tank

booster
pump

basement

sub pump

Pump Option 3

solar array

battery bank

pressure tank
80 gal or larger

Pump Option 4

DC sub pump

- **Option 5.** Using a conventional AC-powered submersible pump, large pressure tank(s), and your household renewable energy system with large inverter. There is some loss of efficiency in this setup, but it's the standard way to get the job done in freezing climates, and your plumber won't have any problems understanding the system. We strongly recommend the variable frequency drive agricultural pumps, which have no start-up surge, similar power use to a DC pump per amount of water pumped and pressure, and are long-lasting. Start-up surge will be kind to your inverter, output is 5–9 gpm, and power use is a quarter that of conventional AC pumps.

Many folks, for a variety of reasons, already have an AC-powered submersible pump in their well when they come to us but are real tired of having to run the generator to get water. For wells with 6-inch and larger casings, it's usually possible to install both the existing AC pump and a submersible DC pump. If your AC pump is 4 inches in diameter, the DC pump can be installed underneath it. The cabling, safety rope, and ½-inch poly delivery pipe from the DC submersible will slip around the side of the AC pump sitting above it.

Just slide both pumps down the hole together. The redundancy offers peace of mind and a little security for emergencies. It's often comforting to have backup for those times when you need it, like when it's been cloudy for three weeks straight, or the fire is coming up the hill and you want a lot of water fast!

Pumping from a Spring, Pond, or Other Surface Source

Your choices for pumping from ground level are a bit more varied, depending on how high you need to lift the water and how many gallons per minute you want. Surface-mounted pumps will not tolerate freezing temperatures. If you live in a freezing climate, make sure that your installation can be (and is!) completely drained before freezing weather sets in. If you need to pump through the hard-freeze season, we recommend a submersible pump as described above.

Pumps Don't Suck! (They Push)

Pumps don't like to pull water up from a source. Or, put simply, Real Goods' #1 Principle of Pumping: Pumps don't suck; pumps push. To operate reliably, your surface-mounted pump must be installed as close to the source as practical. In no case should the pump be more than 10 feet above the water level. With some positive-displacement pumps, higher suction lifts are possible but not recommended. You're simply begging for trouble. If you can get the pump closer to the source, such as the post for a pier or dock, for example, and still keep it dry and safe, do it! You'll be rewarded with more dependable service, longer pump life, more water delivery, and most importantly, less power consumption.

Low-cost Solutions

For modest lifts up to 50 or 60 feet and volumes of 1.5–3 gpm, we have found the SHURflo diaphragm-type pumps to be moderately priced and tolerant of abuses that would kill other pumps. They can tolerate silty water and sand without distress. They'll run dry for hours and hours without damage. But you get what you pay for. Life expectancy is usually two to five years, depending on how hard and how much the pump is working. Repairs in the field are easy, and disassembly is obvious. We carry a full stock of repair parts, but replacement motors don't cost much less than a new pump. Diaphragm pumps will tolerate sand, algae, and debris without damage, but these may stick in the internal check valves and reduce or stop output, necessitating disassembly to clean out the debris. Who needs the hassle? Filter your intake! Our customers love this pump, and people who don't mind doing periodic maintenance have used this low-cost solution for years.

Longer-life Solutions

For higher lifts, or more volume, we often go to a pump type called a rotary vane. Examples include the Slowpump and Flowlight Booster pumps. Rotary-vane pumps are capable of lifts up to 560 feet and volumes up to 5.59 gpm depending on the model. Of all the positive-displacement pumps, they are the quietest and smoothest. But they will not tolerate running dry, or abrasives of any kind in the water. It's very important to filter the input of these pumps with a 10-micron or finer filter in all applications. Rotary-vane pumps are very long-lived but will eventually require a pump head replacement or a rebuild.

Household Water Pressurization

We promote two pumps that are commonly used to pressurize household water: The Chevy and Mercedes models, if you will.

The Chevy model. SHURflo's 5000 series is our best-selling pressure pump. It comes with a built-in 54 to 60 psi pressure switch. With a 5 gpm flow rate, it will keep up with most household fixtures,

garden hoses excluded. The diaphragm pump is reliable and easy to repair, but it's somewhat noisy and has a limited life expectancy. We recommend 24-inch flexible plumbing connectors in a loop on both sides of this pump, and a pressure tank plumbed in as close as possible to absorb most of the buzz.

The Mercedes model. Flowlight's rotary-vane pump is our smoothest, quietest, largest-volume pressure pump. It delivers 3–4.5 gpm at full pressure and is very long-lived but quite expensive initially. This pump will keep up easily with garden hoses, sprinklers, and any other normal household use. Brushes are externally replaceable and will last 5 to 10 years. Pump life expectancy is 15 to 20 years.

Diaphragm-type pump.

Any household pressure system requires a pressure tank. A 20-gallon tank is the minimum size we recommend for a small cabin; full-size houses usually have 40-gallon or larger tanks. Pressure tanks are big, bulky, and expensive to ship. Get one at your local hardware or building supply store.

Solar Hot Water Circulation

Most of the older solar hot water systems installed during the tax-credit heydays of the early 1980s used AC pumps with complex controllers and multiple temperature sensors at the collector, tank, plumbing, ambient air, etc. This kind of complexity allows too many opportunities for Murphy's Law.

The smarter solar hot water systems simply use a small PV panel wired directly to a DC pump. When the sun shines a little bit, producing a small amount of heat, the pump runs slowly. When the sun shines bright and hot, producing

Hot water circulation pump.

lots of heat, the pump runs fast. Very simple, since the pump only runs when it needs to without temperature sensors. System control is achieved with an absolute minimum of gadgetry or points of failure. We carry several hot water circulation pumps for systems of various sizes.

For solar hot water systems with long, convoluted collection loops creating a lot of pipe friction, we can use multiple pumps. If some amount of lift is involved, we have to look at more robust pumps like the Dankoff SunCentric series with high-temperature pump heads, which will require substantially more PV power.

Swimming Pool Circulation

Yes, it's possible to live off the grid and still enjoy luxuries like a swimming pool. In fact, pool systems dovetail nicely with household systems in many climates. Houses generally require a minimum of PV energy during the summer because of the long daylight hours, yet the maximum of energy is available. By switching a number of PV modules to pool pumping in the summer, then back to battery charging in the winter, you get better utilization of resources. Currently we are recommending a high-quality AC variable frequency pump and controller (like the Pentair Intelliflo system) as the best choice for swimming pool circulation, whether on-grid or off. We've found customers to be unhappy with the quality of the available DC pool pumps.

Direct current pumps run somewhat more efficiently than AC pumps, so a slightly smaller DC pump can do the same amount of work as a larger AC pump. We also strongly recommend using a low-back-pressure cartridge-type pool filter. Diatomaceous earth filters are trouble. They have high back pressure and will greatly slow circulation or increase power use.

Water-powered Pumps

A few lucky folks have access to an excess supply of falling water. This falling-water energy can be used to pump water. Both the High Lifter and ram-type water pumps use the energy of falling water to force a portion of that water up the hill to a storage tank. That's right, no PV necessary, just let gravity do the work for you.

Ram Pumps

Ram pumps have been around for many decades, providing reliable water pumping at almost no cost. They are more commonly used in the east-

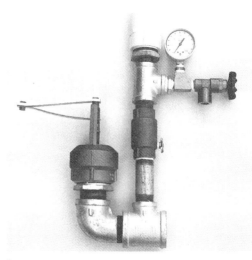

Ram pump.

ern US, where modest falls and large flow rates are the norm, but they will work happily almost anyplace their minimum flow rate can be satisfied. Rams will work with a minimum of 1.5 feet to a maximum of about 20 feet of fall feeding the pump. Minimum flow rates depend on the pump size.

Here's how ram pumps work: A flow is started down the drive pipe and then shut off suddenly. The momentum of moving water slams to a stop, creating a pressure surge that sends a little squirt of water up the hill. How much of a squirt depends on the pump size, the amount of fall, and the amount of lift. (Our technical staff can provide you with pump output data.) Each ram needs to be tuned carefully for its particular site. Ram pumps are not self-starting. If they run short of water, they will stop pumping and simply dump incoming water, so don't buy too big. Rams make some noise—a lot less than a gasoline-powered pump, but the constant 24-hours-a-day chunk-chunk-chunk is a consideration for some sites. Ram pumps deliver less than 5% of the water that passes through them, and the discharge must go into an unpressurized storage tank or pond. But they work for free, and you can expect them to last for decades. If you're lucky enough to have a site with falling water, go for it!

The High Lifter Pump

This pump is unique. It works by simple mechanical advantage. A large piston at low water pressure pushes a smaller piston at higher water pressure. High Lifters recover a much greater percentage of the available water than ram pumps do, but they generally require greater fall into the pump. This makes them better suited for more mountainous territory. They are available in two ratios, 4.5:1 and 9:1. Fall-to-lift ratios and waste-

Ram-type water pumps use the energy of falling water to force a portion of that water up the hill to a storage tank.

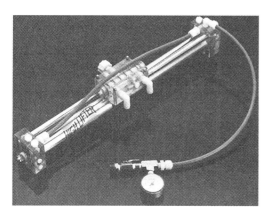

High Lifter pump.

water-to-pumped-water ratios are either 4.5:1, 9:1, or 22:1 (with 2,000 feet of net lift). Note, however, that as the lift ratio gets closer to theoretical maximum, the pump is going to slow down and deliver fewer gallons per day. High Lifters are self-starting. If they run out of water, they will simply stop and wait, or slow down to match what water is available. This is a very handy trait for unattended or difficult-to-attend sites. The only serious disadvantage of the High Lifter is wear caused by abrasives. O-rings are used to seal the pistons against the cylinder walls. Any abrasives in the water wear out the O-rings quickly, so filtering of the intake is strongly encouraged. High Lifter pumps can be overhauled fairly easily in the field, but the O-ring kit costs a lot more than filter cartridges. Output charts for this pump are included in the product section.

Selecting a Ram

Estimate amount of water available to operate the ram. This can be determined by the rate the source will fill a container. Make sure you've got more than enough water to satisfy the pump. If a ram runs short of water, it will stop pumping and simply dump all incoming water.

Estimate amount of fall available. The fall is the vertical distance between the surface of the water source and the selected ram site. Be sure the ram site has suitable drainage for the tailing water. Rams splash big-time when operating! Often a small stream can be dammed to provide the 1½ feet or more of head required to operate the ram.

Estimate amount of lift required. This is the vertical distance between the ram and the water storage tank or use point. The storage tank can be located on a hill or stand above the use point to provide pressurized water. Forty or 50 feet of water head will provide sufficient pressure for household or garden use.

Estimate amount of water required at the storage tank. This is the water needed for your use in gallons per day. As examples, a normal two- or three-person household uses 100–300 gallons per day, or much less with conservation. A 20 × 100-foot garden uses about 50 gallons per day. When supplying potable water, purity of the source must be considered.

These estimates will help you figure out which ram to select. The ram installation will also require pouring a small concrete pad, a drive pipe 5 to 10 times as long as the vertical fall, an inlet strainer, and a delivery pipe to the storage tank or use point. These can be obtained from your local hardware or plumbing supply store. Further questions regarding suitability and selection of a ram for your application will be promptly answered by our technical staff.

Wind-powered Water Pumps

If you've been reading this far because you want to buy a nostalgic old-time jack-pump windmill, we're going to disappoint you. They are still made but are very expensive, typically $3,000 and up. We don't sell them and don't recommend them. They are quite a big deal to set up and install into the well and require routine yearly service at the top of the tower. This is technology that largely has seen its day and is not worth the upkeep. Submersible pumps powered by PV panels are a much better choice for remote locations now.

Wind-powered pump.

There are a couple of effective ways to run a pump with wind power.

Compressed-air Water Pumps

Airlift pumps use compressed air. Three models are available, depending on lift and volume needs. They all use a simple pole-mounted turbine that

High Lifters recover a much greater percentage of the available water than ram pumps do, but they generally require greater fall into the pump.

In most areas of the country, freezing is a major consideration when installing plumbing and water storage systems. For outside pipe runs, the general rule is to bury the plumbing below frost level.

The usual rule for sizing PV-direct arrays is to add about 20% to the pump wattage in a mild climate, or 30% in a hot climate.

direct-drives an air compressor with a wind turbine. The air is piped down the well and runs through a carefully engineered air injector. As it rises back up the supply tube, it carries slugs of water in between the bubbles. The lift/submergence ratio of this pump is fairly critical. Lift is the vertical rise between the standing water level in the well and your tank. Approximately 30% of the lift is the recommended distance for the air injector to be submerged below the standing level. As lifts edge over 200 feet, the submergence ratio rises to a maximum of 50%. Too little submergence and the air will separate from the water; too much and the air will not lift the water, though considerable latitude exists between these performance extremes. This pump isn't bothered by running dry. Output depends on wind speed, naturally, but the largest model is capable of over 20 gpm at lower lifts, or can lift a maximum of 315 feet. The air compressor requires an oil change once a year.

Wind Generators Running Submersible Pumps

Some manufacturers offer wind generators that allow the three-phase turbine output to power a three-phase submersible pump directly. These aren't residential-scale systems and are mostly used for large agricultural projects, or village pumping systems in less industrialized countries. They are moderately expensive to buy. Contact our technical staff for more information on these options.

Freeze Protection

In most areas of the country, freezing is a major consideration when installing plumbing and water storage systems. For outside pipe runs, the general rule is to bury the plumbing below frost level. For large storage tanks, burial may not be feasible, unless you go with concrete. In moderate climatic zones, simply burying the bottom of the tank a foot or two along with the input/output piping is sufficient. In some locations, due to climate or lack of soil depth, outdoor storage tanks simply aren't feasible. In these situations, you can use a smaller storage tank built into a corner of the basement with a separate pressure-boosting pump, or you can pump directly into a large pressure-tank system.

Other Considerations and Common Questions

We hope that, by this point, you've zeroed in on a pump or pumps that seem to be applicable to your situation. If not, our technical staff will be happy to discuss your needs and recommend an appropriate pumping system—which may or may not be renewable-energy powered. Filling out the Solar Water Supply Questionnaire that follows will supply all the answers we're likely to need when we talk with you. At this point, another crop of questions usually appears, such as…

How Far Can I Put the Modules from the Pump?

Often the best water source will be deep in a heavily wooded ravine. It's important that your PV modules have clear, shadow-free access to the sun for as many hours as practical. Even a fist-size shadow will effectively turn off most PV modules. The hours of 10 AM to 2 PM are usually the minimum your modules want clear solar access, and if you can capture full sun from 9 AM to 3 PM, that's more power for you. If the pump is small, running off one or two modules, then distances up to 200 feet can be handled economically. Longer distances are always possible, especially when Linear Current Boosting MPPT pump controllers are installed, but consult with our tech staff or check out the wire-sizing formula on page 427 first, because longer distances require large (expensive!) wire. Many pumps routinely come as 24- or 48-volt units now, as the higher voltage makes long-distance transmission much easier. The Grundfos sub pump accepts DC input up to 300 volts. If you need to run more than 300 feet for sunshine, this may be just the ticket. Give us a call, and we'll make sure you're not selecting the wrong wire size and thereby depriving your pump of useful energy, wasted due to voltage drop.

What Size Wire Do I Need?

This depends on the distance and the amount of power you are trying to move. See the Wire Sizing Chart on page 427 and formula that can properly size wire for any distance, at any voltage, AC or DC. Or just give our tech staff a call—we do this kind of consultation all the time. Going down a well, 10-gauge copper submersible pump wire is the usual, although some of the larger sub pumps we're using now require 8-gauge, or even 6-gauge, for very deep wells. If you need anything other than common 10-gauge wire, we'll let you know.

What Size Pipe Do I Need?

Many of the pumps we offer have modest flow rates of 4 gpm or less. At this rate, it's okay to use smaller pipe sizes such as ¾ inch or 1 inch for pumping delivery without increasing friction loss. However, that's for pumping delivery only. There's no reason you can't use the same pipe to take the water up the hill to the storage tank and also bring it back down to the house or garden. Pipes don't care which way the water is flowing through them. But if you do this, you'll want a larger pipe to avoid friction loss and pressure drop when the higher flow rate of the household or garden fixtures comes into play. We usually recommend at least 1¼-inch pipe for household and garden use, and 2-inch for fire hose lines. See the pipe friction loss charts on page 245 to be sure.

What Size PV Modules Do I Need?

A number of the pumps Real Goods sells have accompanying performance and wattage tables. For instance, the Dankoff Solar Slowpump model 1403-24 lifting 440 feet will deliver 3.1 gpm and require 413 watts of PV. Because module ratings are based on ideal laboratory conditions, and you want your pump to work on hot or humid days, you should add 30% to the pump wattage. Heat reduces PV output; water vapor cuts available sunlight. So, in our example, we actually need 537 watts, which means buying two 280-watt modules. The usual rule for sizing PV-direct arrays is to add about 20% to the pump wattage in a mild climate, or 30% in a hot climate. And, of course, an LCB (linear current booster) of sufficient amperage capacity is practically standard equipment with any PV-direct system.

Can I Automate the System?

Absolutely! Life's little drudgeries should be automated at every opportunity. The LCBs that we so strongly recommend with all PV-direct systems help us in this task. These units are all supplied with wiring for a remote control option. This allows you to install a float switch at the holding tank, and a pair of tiny 18-gauge wires can be run as far as 5,000 feet back to the pump/controller/PV modules area. Float switches can be used either to pump to or pump from a holding tank and will automatically turn the system off when satisfied. The LCBs, float switches, and other pumping accessories can be found on our website, realgoods.com. LCBs can also be used on battery-powered systems when remote sensing and control would be handy.

Where Do We Go from Here?

If you haven't found all the answers to your remote water pumping needs yet, please give our experienced technical staff a call. They'll be happy to work with you to select the most appropriate pump, power source, and accessories for your needs. We run into situations occasionally where

Water Pumping Truths and Tips

1. Pumps prefer to *push* water, not *pull* it. In fact, most pumps are limited to 10 or 15 feet of lift on the suction side. Mother Nature has a theoretical suction lift of 33.9 feet at sea level, but only if the pump could produce a perfect vacuum. Suction lift drops 1 foot with every 1,000 feet rise in elevation. To put it simply, **Pumps Don't Suck, They Push. Choose your pump location wisely!**
2. Water is heavy, 8.33 pounds per gallon. It can require tremendous amounts of energy to lift and move.
3. Direct current electric motors are generally more efficient than AC motors. If you have a choice, use a DC motor to pump your water. Not only can it be powered directly by solar modules, but your precious wattage will go further.
4. Positive-displacement pumps are far more efficient than centrifugal pumps. Most of our pumps are positive-displacement types. AC-powered submersibles, jet pumps, and booster pumps are centrifugal types.
5. As much as possible, we try to avoid batteries in pumping systems. When energy is run into and out of a battery, 25% is lost. It's more efficient to take energy directly from your PV modules and feed it right into the pump. At the end of the day, you'll end up with 25% more water in the tank.
6. One pound per square inch (psi) of water pressure equals 2.31 feet of lift, a handy equation.

renewable energy sources and pumps simply may not be the best choice, and we'll let you know if that's the case. For 95% of the remote pumping scenarios, there is a simple, proven, cost-effective, long-lived renewable energy-powered solution, and we can help you develop it.

Water-saving Showerheads

Showers typically account for 32% of home water use

A standard old-style showerhead uses about 3 to 5 gallons of water per minute, so even a five-minute shower can consume 25 gallons. Since the 1990s, all showerheads sold in the US must use 2.5 gpm or less. The average for the low-flows that we sell (realgoods.com), however, is even less, a highly efficient 1.4 gpm! According to the US Department of Energy, heating water is the second-largest residential energy user. With a low-flow showerhead, water use, energy use, and costs for heating water for showers may drop as much as 50%. A recent study showed that changing to a low-flow showerhead saved 27¢ worth of water and 51¢ of electricity per day for a family of four. So, besides being good for Earth, a low-flow showerhead will pay for itself in about two months! Add one of our instantaneous water heaters for even greater savings.

Solar Water Supply Questionnaire

To thoroughly and accurately recommend a water supply system, we need to know the following information about your system. Please fill out the form as completely as possible. Please give us a daytime phone number, too.

Name: _____

Address: _____

City: _____ State: _____ Zip: _____

Email: _____ Phone: _____

DESCRIBE YOUR WATER SOURCE

Depth of well: _____

Depth to standing water surface: _____

If level varies, how much? _____

Estimated yield of well (gallons/minute): _____

Well casing size (inside diameter): _____

Any problems? (silt, sand, corrosives, etc.) _____

_____

WATER REQUIREMENTS

Is this a year-round home? _____ How many people full-time? _____

Is house already plumbed? _____ Conventional flush toilets? _____

Residential gallons/day estimated: _____

Is gravity pressurization acceptable? _____

Hard-freezing climate? _____

Irrigation gallons/day estimated: _____ Which months? _____

If you have a general budget in mind, how much? _____

Do you have a deadline for completion? _____

DESCRIBE YOUR SITE

Elevation: _____

Distance from well to point of use: _____

Vertical rise or drop from top of well to point of use: _____

Can you install a storage tank higher than point of use? _____

How much higher? _____ How far away? _____

Complex terrain or multiple usage? _____ (Please enclose map)

Do you have utility power available? _____ How far away? _____

Can well pump be connected to nearby home power system? _____

How far? _____ Home power system power battery voltage: _____

Mail your completed questionnaire to:

Tech Staff/Water Supply, Real Goods, 13771 South Highway 101, Hopland, CA 95449.

Questions? Call us at 888-567-6527.

Real Goods' Homestead Plumbing Recommendations

Types of Pipe
Metals

Black iron and galvanized iron—used for gas (propane or natural gas) plumbing but little else. It is slow and difficult to cut and thread, and requires special, expensive tools. (Consider renting tools if needed.) A special plastic-coated iron pipe is used for buried gas lines. The insides of galvanized pipe corrode when used for water supply and eventually restrict water flow. Older houses often suffer from this affliction, which causes ultrasensitivity of shower temperature and other exciting problems.

Cast iron—used to be the material of choice for waste plumbing, but ABS plastic thankfully has replaced it. Simple conversion adapters to go from cast iron to ABS are available if you find yourself having to repair an existing older system.

Copper—the most common material for water supply plumbing in new home construction. Most plumbing codes require copper for indoor work. Type M, the most common grade for interior work, comes in 20-foot rigid lengths. Houses typically use ¾-inch lines for feeders and ½-inch lines for fixtures. If gravity-fed water pressure will be lower than the 20 to 40 psi city standard, consider increasing your supply pipes one size. This cures many low-pressure woes by eliminating pressure loss within the house (and provides a shower that's nearly immune to temperature fluctuations). The thicker-walled flexible tubing L and K grades can be used legally in some localities, but they generally cost considerably more. Copper does not tolerate freezing at all! Copper is relatively easy to work with, and you need only simple, inexpensive tools. Joints must be "sweat soldered," which takes about 10 minutes to learn and is kind of fun afterward. Be sure to use lead-free solders for water supply piping! Some states allow the use of flexible copper tubing for gas supply lines, in which case special compression-type fittings are used.

Bronze/Brass—beautiful stuff, but too expensive for common use. Pre-cut, pre-threaded nipples are used occasionally for dielectric water tank protection (5 inches of brass qualifies as protection under code). Use brass only where plumbing will be exposed and you want it to look nice.

Plastics

Poly (polyethylene)—this black, flexible pipe is used widely for drip and irrigation systems. Available in utility or domestic grades. Never use utility grade for drinking water! It's made with recycled plastics, and you don't know what will leach out. Poly is almost totally freeze-proof. It's easy to work with (when warm) and only requires common hand tools. It's sold in 100-foot or occasionally longer rolls. Barbed fittings and hose clamp fittings *will* leak (why do you suppose it's used in *drip* systems?). Do not use this cheap plumbing for any permanently pressurized installations, unless you can afford to throw water away. Poly also degrades fairly rapidly in sunlight. Two to three years is the usual unprotected lifespan. One good use for poly is with submersible well pumps where the flexibility makes for easy installations. Poly pipe is the usual choice for sub pumps unless they're pumping hundreds of feet of lift at very high pressure. Special 100 and 160 psi high-pressure versions of poly pipe are used for sub pumps.

PVC (polyvinyl chloride)—one of the wonders of chemistry that makes homesteading easy. This is the white plastic pipe that should be used for almost everything outside the house itself. It's easy to work with and requires only common hand tools (although if you're doing a lot of it, a PVC cutter is a real timesaver over the hacksaw). PVC usually can survive mild freezing. Hard freezes will break joints, however, so use protection. It comes in 20-foot lengths. Sizes 1 inch and up generally are available with bell ends, saving you the expense and extra gluing of couplings. Use primer (the purple stuff), then glue (the clear stuff) when assembling. PVC must be buried or protected. If exposed to sunlight, it will degrade slowly from the UV, but it will also grow algae inside the pipe!

PVC grades: Schedule 40 is the standard PVC that is recommended for most uses. Some lighter-duty agricultural types are available. Class 125, class 160, and class 200 are graded by the psi rating of the pipe. Don't use these cheaper thin-wall classes if you're burying the pipe! Leaks from stones pushing through thinner walls are too hard to find and fix.

CPVC (chlorinated polyvinyl chloride)—a PVC formulation specifically made for hot water. The fittings and the pipe are a light tan color so you can tell the difference from ordinary PVC. Some counties and states allow the use of CPVC; most don't. It works just like PVC.

ABS (acrylonitrile butadiene styrene)—a black rigid pipe universally used for unpressurized waste and vent plumbing. It's easy to work with, requires common hand tools, and glues up quickly just like PVC. Make sure that the glue you use is rated for ABS. Some glues are universal and will work on both PVC and ABS; others are specific. Buy a big can of ABS glue—these larger pipes use a lot more glue. Also, the bigger cans come with bigger swabs, which you'll need on larger pipe.

PB (polybutylene)—a flexible plastic pipe that is used occasionally for home plumbing, but mostly for radiant-floor heating systems. It requires special compression rings and tools for fittings. PB will survive considerable abuse during construction without springing leaks. It makes no reaction with concrete floors. PB is the best choice for radiant floors and has nearly infinite life expectancy when buried in concrete.

> Try to arrange your house plan so that bathrooms, laundry room, and kitchen all fit back to back and can share a common plumbing wall. This saves thousands of dollars in materials costs, lots of time during construction, hours of time waiting for the hot water, and many gallons of expensive hot water.

Other Tips

Don't go small on supply piping. Long pipe runs from a supply tank up on the hill can produce substantial pressure drop. The typical garden needs 1¼-inch supply piping for watering and sprinklers. We recommend *at least* 1½-inch supply for most houses, and a 2-inch or larger line for a fire hose is often a very good idea.

Keep plumbing runs as short as possible. Try to arrange your house plan so that bathrooms, laundry room, and kitchen all fit back to back and can share a common plumbing wall. This saves thousands of dollars in materials costs, lots of time during construction, hours of time waiting for the hot water, and many gallons of expensive hot water.

Insulate *all* hot water pipes, even those inside interior walls before the walls are covered up. The savings and comfort over the life of the house quickly make up for the initial costs.

Store your glue cans upside down. This sounds wacky, but it works. Tighten the cap first, obviously. The glue seals any tiny air leaks and keeps all the glue inside and on the cap threads from drying solid. Don't store your primer upside down, though; it can't seal the tiny leaks and will disappear.

Friction Loss Charts for Water Pumping

How to Use Plumbing Friction Charts

If you try to push too much water through too small a pipe, you're going to get pipe friction. Don't worry, your pipes won't catch fire. But friction will make your pump work harder than it needs to, and it will reduce your available pressure at the outlets, so sprinklers and showers won't work very well. These charts can tell you if friction is going to be a problem. Here's how to use them:

PVC or black poly pipe? The rates vary, so first be sure you're looking at the chart for your type of supply pipe. Next, figure out how many gallons per minute you might need to move. For a normal house, 10–15 gpm is probably plenty. But gardens and hoses really add up. Give yourself about 5 gpm for each sprinkler or hose that might be running. Find your total (or something close to it) in the "Flow GPM" column. Read across to the column for your pipe diameter. This is how much pressure loss you'll suffer for every 100 feet of pipe. Smaller numbers are better.

Example: You need to pump or move 20 gpm through 500 feet of PVC between your storage tank and your house. Reading across, 1-inch pipe is obviously a problem. How about 1¼ inches? 9.7 psi times 5 (for your 500 feet) = 48.5 psi loss. Well, that won't work! With 1½-inch pipe, you'd lose 20 psi…still pretty bad. But with 2-inch pipe, you'd lose only 4 psi…ah! Happy garden sprinklers! Generally, you want to keep pressure losses under about 10 psi.

Friction Loss in PSI per 100 Feet of Scheduled 40 PVC Pipe

Flow GPM	Nominal Pipe Diameter in Inches							
	½"	¾"	1"	1¼"	1½"	2"	3"	4"
1	3.3	0.5	0.1					
2	11.9	1.7	0.4	0.1				
3	25.3	3.5	0.9	0.3	0.1			
4	43.0	6.0	1.5	0.5	0.2	0.1		
5	65.0	9.0	2.2	0.7	0.3	0.1		
10		32.5	8.0	2.7	1.1	0.3		
15		68.9	17.0	5.7	2.4	0.6	0.1	
20			28.9	9.7	4.0	1.0	0.1	
30			61.2	20.6	8.5	2.1	0.3	0.1
40				35.1	14.5	3.6	0.5	0.1
50				53.1	21.8	5.4	0.7	0.2
60				74.4	30.6	7.5	1.0	.03
70					40.7	10.0	1.4	0.3
80					52.1	12.8	1.8	0.4
90					64.8	16.0	2.2	0.5
100					78.7	19.4	2.7	0.7
150						41.1	5.7	1.4
200						69.9	9.7	2.4
250							14.7	3.6
300							20.6	5.1
400							35.0	8.6

Friction Loss in PSI per 100 Feet of Polyethylene (PE) SDR-Pressure Rated Pipe

Flow GPM	Nominal Pipe Diameter in Inches							
	½"	¾"	1"	1¼"	1½"	2"	2½"	3"
1	0.49	0.12	0.04	0.01				
2	1.76	0.45	0.14	0.04	0.02			
3	3.73	0.95	0.29	0.08	0.04	0.01		
4	**6.35**	1.62	0.50	0.13	0.06	0.02		
5	9.60	2.44	0.76	0.20	0.09	0.03		
6	13.46	3.43	1.06	0.28	0.13	0.04	0.02	
7	17.91	4.56	1.41	0.37	0.18	0.05	0.02	
8	22.93	**5.84**	1.80	0.47	0.22	0.07	0.03	
9		7.26	2.24	0.59	0.28	0.08	0.03	
10		8.82	2.73	0.72	0.34	0.10	0.04	0.01
12		12.37	**3.82**	1.01	0.48	0.14	0.06	0.02
14		16.46	5.08	1.34	0.63	0.19	0.08	0.03
16			6.51	1.71	0.81	0.24	0.10	0.04
18			8.10	2.13	1.01	0.30	0.13	0.04
20			9.84	2.59	1.22	0.36	0.15	0.05
22			11.74	**3.09**	1.46	0.43	0.18	0.06
24			13.79	3.63	1.72	0.51	0.21	0.07
26			16.00	4.21	1.99	0.59	0.25	0.09
28				4.83	2.28	0.68	0.29	0.10
30				5.49	**2.59**	0.77	0.32	0.11
35				7.31	3.45	1.02	0.43	0.15
40				9.36	4.42	1.31	0.55	0.19
45				11.64	5.50	1.63	0.69	0.24
50				14.14	6.68	**1.98**	0.83	0.29
55					7.97	2.36	0.85	0.35
60					9.36	2.78	1.17	0.41
65					10.36	3.22	1.36	0.47
70					12.46	3.69	**1.56**	0.54
75					14.16	4.20	1.77	0.61
80						4.73	1.99	0.69
85						5.29	2.23	0.77
90						5.88	2.48	0.86
95						6.50	2.74	0.95
100						7.15	3.01	**1.05**
150						15.15	6.38	2.22
200							10.87	3.78
300								8.01

CHAPTER 8

Water Heating

The Most Cost-effective Solar Alternative

Most of us take hot water at the turn of a tap for granted. It makes civilized life possible, and we get seriously annoyed by cold showers. Yet most people probably do not realize how much this convenience costs them. The average household spends an astonishing 20%–40% of its energy budget on water heating. And for most folks, all those energy dollars are given to an appliance that has a life expectancy of only 10–15 years and that throws away a steady 20% or more of the energy you feed it. Any efficiency improvements to your water heater reduce your energy consumption, lower your carbon footprint, and lessen the overall environmental impact of your home.

There are better, cheaper, and more durable ways to get hot water than using fossil energy water heaters. This chapter describes all the common water heater types, discusses the good and bad points of each, and offers suggestions for efficiency improvements. It also provides a detailed examination of solar hot water heating, which for most people is the simplest and most cost-effective renewable energy application around—indeed, it's possibly the best investment available in America today! If you're looking for a concrete action you can take to lower your carbon footprint and directly combat climate change, we suggest you seriously consider solar hot water.

We're indebted to our old friend Bob Ramlow for writing and revising this chapter. Bob is a former Real Goods storeowner in Amherst, Wisconsin, a longtime solar water system designer and installer, and author of *Solar Water Heating: A Comprehensive Guide to Solar Water and Space Heating Systems* (available at realgoods.com). We've developed some life-cycle cost comparison charts so you can examine the real operation costs over the lifetime of the appliance. Points to consider for each heater type are initial cost, cost of operation, recovery time (how fast does it heat water?), ease of installation or ability to retrofit, life expectancy, and life-cycle cost.

> The average household spends an astonishing 20%–40% of its energy budget on water heating.

> If you're looking for a concrete action you can take to lower your carbon footprint and directly combat climate change, we suggest you seriously consider solar hot water.

Common Water Heater Types

Storage or Tank-type Water Heaters

Storage/tank-type water heaters can be divided into two classifications: direct-fired and indirect-fired. Direct-fired tank-type water heaters consist of a storage tank with a built-in burner. Indirect-fired water heaters have a storage tank to hold the hot water, but there is no burner in the tank and therefore they require some other source of heat, typically a boiler of some sort (more on this below).

Direct-fired tank-type heaters are by far the most common kind of residential water heater used in North America. They typically range in size from 20 to 80 gallons and can be fueled by electricity, natural gas, propane, or oil.

Direct-fired storage heaters work by heating up water inside an insulated tank. Most tanks are steel with a baked-on epoxy liner. They have a burner at the bottom of the tank that resembles the gas burner on a gas kitchen stove. (An electric model will have an electric heating element.) A thermostat measures the temperature of the water in the tank, and when the temperature falls, the heating element or burner turns on and heats the water until it reaches the desired temperature, and then the heater turns off.

Traditional direct-fired storage-type water heaters have a flue that runs through the center of the tank. The simplest units are vented to a chimney. This design has not changed in over 60 years and is not particularly efficient—but it is cheap to manufacture. Other units have a blower on the exhaust allowing for a vent through the wall. Their good points are the modest initial cost, the ability to provide large amounts of hot water for a limited time, and the fact that they're well understood by plumbers and do-it-yourself homeowners everywhere. Their bad points include constant standby losses, slow recovery times, and somewhat low life expectancy. Because heat always escapes through the walls of the tank (standby heat loss), energy is consumed even when no hot water is being used. This wastes energy and raises operation costs.

Most manufacturers now offer "high-efficiency" storage heaters for a premium price. These use more insulation to reduce standby losses, and some models also incorporate burners with higher efficiency than standard models. Customer-installed water heater blankets will help reduce standby losses in any tank-type water heater (it is critical to follow the installation instructions that come with the insulation kits). Standby losses for traditional gas and oil heaters are higher because the air in the internal flue passages is constantly being warmed, rising, and pulling in fresh cold air from the bottom. If you must have a gas water heater like this, installing an automatic flue damper can save a lot of fuel and is always cost-effective.

Life expectancy for direct-fired tank-type water heaters averages 10–15 years. Usually the tank rusts through at this age, and the entire appliance has to be replaced. Life expectancy can be prolonged significantly if the sacrificial anode rod—installed by the tank manufacturer to keep the tank from rusting—is replaced at about five-year intervals. While most homeowners neglect anode rod replacement, it is a task every homeowner should have on their annual maintenance schedule. Anode rods should be checked every two years initially to monitor their deterioration. An anode rod is made of metals that corrode faster than the metal the tank is made of. This protects the tank from corrosion. When checking

the rod, you look to see if there is anything left of the rod. Most of these round rods are three to four feet long and are ½ inch thick. As long as there is some rod left, they will continue to protect your tank. When the rod is nearly gone it should be replaced. The cost of anode rods is minimal, and having a fresh rod will protect your tank from corrosion almost indefinitely. The rate of deterioration is dependent on a number of factors, the most significant one being water quality. While we are talking about regular water heater maintenance, the tank should be flushed annually. Attach a hose to the drain at the bottom of the water heater and drain around 20 gallons of water from the tank. Run this water down the drain. This removes sediment from the tank and helps prolong the life of the unit. This goes for all tank-type water heaters, regardless of the fuel type.

Direct-fired storage water heaters are fairly cost-effective if you use natural gas, which is still something of an energy bargain. If you're heating with electricity, oil, or propane, then a storage water heater, though cheaper initially, is your highest-cost choice in the long run. See the Life-Cycle Comparison Chart on page 265.

Condensing Gas Water Heaters

Among the problems with any type of water heater that has a burner is that a lot of heat is not used and goes up the chimney or out the side vent. A main reason for this is that you cannot cool the exhaust too much because that will damage the heater by condensing the water in the exhaust, so the heater rusts out from the inside. In addition, because for a long time the market was not very concerned with efficiency, improvements in burner efficiency were few and far between. It is cheaper to make an inefficient burner than an efficient one.

Over the last 15 or 20 years, the market for higher-efficiency heating units has grown and so has the technology. Driven first by the hydronic heating industry, vastly improved heating system efficiency has become more commonplace. These modern heating units are made of stainless steel and have improved burner efficiency, allowing for condensing conditions inside the heating units and resulting in vastly improved efficiencies—up to over 95%. The old standard vented water heaters and traditional cast-iron boilers were lucky to achieve a 50% efficiency.

Because this kind of heating appliance is so efficient at extracting the heat from the burner and delivering that heat to the water, very little heat is vented with the exhaust. Because the exhaust is cool, a high-temperature chimney is not required and the units can be vented using high-quality PVC pipe. The exhaust is forced out of the unit with a small blower, so the vent can terminate out a sidewall and does not have to vent out the roof. The combustion chambers are also sealed and do not use interior building air. That necessitates a second PVC pipe to bring in outside air to the unit for combustion. It also adds an extra safety factor that minimizes the chances of CO_2 entering the building.

There are two types of condensing water heaters on the market today: direct-fired and indirect-fired, which use a condensing boiler as the heat source. At this time all direct-fired condensing water heaters are fueled by gas (natural gas or propane). Indirect-fired water heaters can use condensing boilers fueled by either gas (by far the majority) or fuel oil.

Direct-fired Condensing Gas Water Heaters

Direct-fired condensing tank-type water heaters are one of the newest water heating technologies available for residential and commercial use. A direct-fired condensing water heater has a modern stainless steel high-efficiency gas burner installed in a very well-insulated storage tank. These units can achieve an efficiency of over 95%. *For most applications today, this technology is the best option for fossil-fueled water heating and solar water heating backup.* Because these appliances are designed from the ground up to be very efficient, the storage tanks are very well insulated, helping to minimize standby losses. Also, the whole unit is designed differently than the old standard tank-type water heater, which also decreases standby losses and increases overall efficiency.

A small microprocessor controls the operation of the heater. Because the appliance extracts the maximum amount of heat from the high-efficiency combustion process, PVC pipe can be used to vent the unit through a sidewall. Because of the ease of installation, this type of water heater is excellent for both new construction as well as retrofit installations.

As mentioned above, this technology is the new kid on the block. I like these water heaters. They address many of the problems typical of both tank-type and tankless water heaters. However, a direct-fired condensing water heater costs a bit more than a tankless water heater. When my older tankless heater needs to be replaced, I will buy a direct-fired condensing water heater to back up my solar water heater.

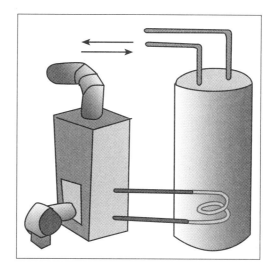

Indirect water heater.

Indirect-fired Water Heaters

An indirect water heater uses the home heating system's boiler to heat the domestic hot water. Hot water is stored in a separate insulated tank. Heat is transferred from the boiler to the storage tank using a small circulation pump and a heat exchanger. When used with the new high-efficiency gas- or oil-fired condensing boilers, indirect water heaters are a great option. If the heat source for your building is a condensing gas or oil boiler, then an indirect-fired water heating system is your best option for water heating or solar hot water backup. If you already have the high-efficiency heat source, you might as well use it for as many tasks as possible. The life expectancy of the heat exchanger and circulation pump is an exceptional 30 years, although the storage tank may need replacement a little sooner.

Heat Pump Water Heaters

If you use electricity to heat water, heat pumps are two to three times more efficient than conventional tank-type resistance electric water heaters. Heat pump water heaters use a compressor and refrigerant fluid to transfer heat from one place to another, like a refrigerator in reverse. Electricity is the only fuel option. The heat source is air in the heat pump vicinity, although some better models can duct in warm air from the attic or outdoors. The warmer the air, the better: Heat pumps work best in warm climates where they don't have to work as hard to extract heat from the air. In these applications, the upper element of the conventional electric water heater usually remains active for backup duty.

The advantages of heat pump water heaters are that they use only 33%–50% as much electricity as a conventional electric tank-type water heater, they will provide a small cooling benefit to

Over the last 15 or 20 years the market for higher-efficiency heating units has grown and so has the technology. These modern heating units are made of stainless steel and have improved burner efficiency, allowing for condensing conditions inside the heating units and resulting in vastly improved efficiencies—up to over 95%.

the immediate area, and life expectancy is a good 20 years. The disadvantages are that heat pumps are quite expensive initially (average installed cost is approximately $2,000), and recovery rates are modest, variable, and can be fairly low if the pump doesn't have a warm environment from which to pull heat. Like all storage tank systems, heat pumps have standby losses, and if installed in a heated room will rob heat from Peter (the furnace) to pay Paul (the water heater). Therefore this technology works best in warm climates where you do not have to pay for space heating.

Heat pumps make better use of electricity because it's much more efficient to use electricity to *move* heat than to *create* it. Heat pump water heaters are available with built-in water tanks, called integral units, or as add-on units to existing water heaters. Add-on units may be a smarter investment, as the heat pump will probably outlive the storage tank. If you live in a warmer climate and use electricity to heat water, a heat pump is your best choice. In fact, heat pump units stack up quite favorably against everything but natural gas and solar if you can afford the initial purchase cost (this is the case in warm or hot climates only).

Heat pump water heaters are a bit complex to install, particularly the better units that take their heat from a remote site or use a separate tank. You'll probably need a contractor to install one. Call your local heating and air conditioning contractor for more details on availability in your area, cost, and installation estimates.

Some people have promoted using the latest generation of heat pumps powered by a photovoltaic system as a solar water heater. The argument is that if you pair the high efficiency of a heat pump with the current low cost of a PV system sized to run the water heater, the economics may be favorable compared to a conventional solar water heating system. At this writing, however, the economics favor a conventional solar water heating system. In addition, at the present time, the physical size of the required PV array would be much greater than that of a conventional solar water heating array. Mechanically, a conventional solar water heating system is elegantly simple compared to a heat pump/PV system.

An "add-on" heat pump water heater.

An "intergral" heat pump water heater.

Tankless Coil Water Heaters

This type of heater is found only in older oil- or gas-fired boiler systems. A tankless coil heater operates directly off the house boiler; it does not have a storage tank, so every time there is a demand for hot water, the boiler must run. This may be fine in the winter, when the boiler is usu-

ally hot from household heating chores anyway, but during the rest of the year it results in a lot of start-and-stop boiler operation. Tankless coil boilers may consume 3 Btus of fuel for every 1 Btu of hot water they deliver. This type of water-heating system is not recommended.

On-demand, or Tankless, Water Heaters

Tankless water heaters have been around for a very long time. In Europe and Japan, they have been the primary type of water heater for decades, for two main reasons: because of their high efficiency (as mentioned below) and because they take up less space than tank-type water heaters. The tankless strategy is to produce the hot water only when needed, which eliminates the need for a storage tank and its associated standby losses. The advantages of demand water heaters are very low standby losses, low operation costs, unlimited amounts of hot water delivery, and a very long life expectancy. The main disadvantage is that the unlimited amounts of hot water can't be taken too rapidly, because flow rates are limited to the heater's abilities, so hot water use may need to be coordinated at times. (People tend to be *very* sensitive and excitable about shower temperatures.) Another disadvantage to on-demand water heaters is that in some models hard water can cause clogging problems inside the unit's heat exchanger.

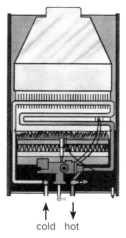

cold   hot

Basic demand water heater internal construction.

How Demand Water Heaters Work

Most on-demand water heaters are gas fired, either natural gas or propane. Demand water heaters turn on only when someone opens up a hot water faucet. When water flow reaches a minimum flow rate, the gas flame or heater elements come on, heating the water as it passes through a radiator-like heat exchanger. Tankless heaters do not store any hot water for later use but heat water only as demanded at the faucet. The minimum flow rates required for turn-on prevent any possibility of overheating at very low flow and ensure that the unit turns off when the faucet is turned off. Minimum flow rates vary from model to model but are generally about 0.5–0.75 gallon per minute for household units. Other safety devices include the standard pressure/temperature relief valve that all water heaters in North America are required to carry. Tankless heaters also use an additional overheat sensor or two on the heat exchanger.

The great advantage of tankless heaters is that you run out of hot water only when either the gas or the water runs out. On the other hand, tankless heaters will meter out the hot water at just so many gallons per minute. Excess water flow will result in lower temperature output. Some tankless heaters are limited to running just one fixture at a time. Larger household-sized heaters can run multiple fixtures simultaneously. Showers are a touchy issue, so what complaints we hear usually revolve around showers and multiple-fixture uses. Tankless heaters are probably the best choice for smaller homesteads of two people or fewer, where hot water use can be coordinated easily. Larger homes with an intermittent use that's a long distance from the rest of the household hot water plumbing, such as a master bedroom at the end of a long wing, are also good candidates for a tankless heater just to supply that isolated area.

Tankless With Solar

As we mentioned above, the standard tankless units sense the outgoing water temperature and adjust the heat input accordingly to maintain a steady output temperature. As flow rate or incoming water temperature varies, the heat input will be modulated up or down to compensate. The standard units can modulate heat input down to only about 20,000 Btus and are not particularly recommended for solar backup. Some gas-fired units and all electric units can modulate all the way down to zero Btu and are highly recommended if there's any possibility you may use preheated solar water in the future. Some units will also shut off at their preset temperatures if the solar preheated water comes in at that temperature. These units do have high temperature limits, so it is best to limit (with a tempering valve) the incoming temperature to just under the tankless set-points. If the incoming water is already preheated to your selected output temperature, then the tankless heater will come on only briefly. If your preheated water is at 90°F and you've got a 120°F output set, then the water heater will come on just enough to give 120°F output.

Can I Get an Electric Demand Water Heater?

Yes, you can, but first a warning or two. Electricity is easily the most expensive power source for heating applications. At average North American 2014 energy prices, natural gas is the least expensive choice for water heating, propane is 30% to 50% more expensive, and electricity is about another 15% above that. It takes a shocking amount of electricity to heat water on the fly. Most households will need a 200-amp electric service at a minimum to support an instant electric water heater. That said, there are high-quality electric water heaters with precise digital control and multiple temperature sensors that will do a great job with no standby losses.

Ask Real Goods

Is it possible to use one of the tankless heaters on my hot tub?
With some ingenuity and creativity, this has been done successfully. You need to bear in mind that these heaters are designed for installation into pressurized water systems. For safety reasons, they won't turn on until a certain minimum flow rate is achieved. The flow rate is sensed by a pressure differential between the cold inlet and the hot outlet. Tankless heater flow rates are 3 gallons per minute and less, which is a much lower flow rate than hot tubs. A hot tub system is very low pressure but high volume. What usually works is to tee the heater inlet into the hot tub pump outlet, and the heater outlet into the pump inlet. This diverts a portion of the pump output through the heater, and the pressure differential across the pump usually is sufficient to keep the heater happy. In addition, you'll need an aquastat to regulate the temperature by turning the pump on and off (your tub may already have one if a heating system already was installed). You'll also need a willingness to experiment and a good dose of ingenuity. CHECK WITH YOUR LOCAL BUILDING INSPECTOR BEFORE ATTEMPTING THIS INSTALLATION.

The advantages of demand water heaters are very low standby losses, low operation costs, unlimited amounts of hot water delivery, and a very long life expectancy. The main disadvantage is that the unlimited amounts of hot water can't be taken too rapidly, so hot water use may need to be coordinated at times.

Solar Water Heaters

In the realm of water heating, solar offers a great alternative to replace most of our traditional fossil fuel use. There was a time in this country when the *only* option for hot running water was solar water heating.

Indeed, solar water heating has a long history that our society has generally ignored. The first solar water heating system was patented in 1891: a simple black water tank mounted in a box that had glass on one side, what we call today an ICS (integral collector storage) or batch water heater. In 1909, the first flat-plate collector was patented, and a few years later, the first closed-loop anti-freeze system was introduced. This all happened before the development of electric and gas water heaters, meaning that the very first type of automatic water heating system was solar. Between 1935 and 1941, more than half the population of Miami, Florida, used solar water heaters.

Solar water heaters enjoyed a huge surge of interest following the Arab oil embargo of the mid-1970s and early 1980s. Solar water heating is a technology that nearly everyone can use toward the goal of reducing fossil fuel consumption. Heating domestic hot water uses 20%–40% of a home's annual energy budget. A solar water heating system can supply 60%–90% of that energy load. The return on your investment is higher than the return for any other renewable energy investment you can make. Depending on your particular situation, the savings in conventional fuel can pay for the cost of the solar water heating system in as little as three years. Most often, the payback is around 5–10 years, still a great investment, equating to a 10%–20% return on investment (ROI), which is way better than any safe investment available in America today. A better way to look at the financial implications of investing in a solar water heating system is to ask, "How does this investment affect my cash flow?" If you buy a quality solar water heating system, your cost for fossil fuel will go down. Even if you borrow the money for the solar investment, your monthly payment will be less than your energy savings. The result is positive cash flow—more money in your pocket every month, starting the minute the solar water heater turns on. From a climate change perspective, a typical two-panel system will offset between 1,500 and 2,000 pounds of CO_2 per year. That is the equivalent of driving between 1,300 and 1,750 miles in a car getting 22 miles per gallon. So how do you get started?

Getting Started with Solar Hot Water

The first step in the direction of solar hot water is the same first step in going solar for space heating or electricity: Conserve energy! As always, the three general principles are: reduce losses, increase efficiency, and reduce consumption.

To start, examine your heating system from top to bottom, and look for places where heat might leak out. When it comes to water, heat loss is nothing but waste. A little cheap insulation can go a long way and make a noticeable difference. Thorough insulation of hot water pipes is easiest in new construction, but it's a good idea, and usually not too hard, to insulate as much of the piping as you have access to. You should also insulate your existing water heater with a jacket made for that purpose. (Newer models are often well insulated to begin with.) Heat loss can also come from leaks. A faucet that leaks 30 drops of water a minute will waste almost 100 gallons a month. Fix leaky faucets promptly.

Next, try to increase the efficiency of everything in your home that uses hot water. Generally, you'll be limited to the washing machine and the dishwasher. If you upgrade these appliances to more energy-efficient models, you will see sig-

> A solar water heating system can supply 60%–90% of your domestic hot water energy load. The return on your investment is higher than the return for any other renewable energy investment you can make. Depending on your particular situation, the savings in conventional fuel can pay for the cost of the solar water heating system in as little as three years.

Bob Ramlow

nificant reductions in energy consumption. For instance, a front-loading washing machine uses half as much hot water as a standard top-loading model. This increase in efficiency saves 10–20 gallons of hot water per load, which can amount to thousands of gallons of hot water a year.

Finally, you can conserve energy by simply using less. A lot can be accomplished without significantly changing your daily habits. For example, when washing dishes in the sink, don't let the water run while rinsing; instead fill one sink with wash water and the other with rinse water and/or turn off the hot water between rinses. Soak pots and pans instead of letting the water run while you scrape them clean. If you use a dishwasher, run only full loads. Use cold water with the garbage disposal—it solidifies grease, allowing the disposal to get rid of it more effectively. Showers use less water than baths, but make sure you've installed low-flow showerheads. Standard showerheads use 3–4 gallons per minute; even a brief five-minute shower can consume 20 gallons of hot water. Low-flow showerheads will use less than half that amount, so that a family of four can save well over 1,000 gallons a month.

How Solar Water Heating Systems Work

The obvious answer is that the sun heats the water—but exactly how this happens depends on many factors, including your climate, how much hot water you use, and the solar access and conditions at your site. It is fundamental to realize that a solar water heating system will not provide 100% of your hot water needs, so you also have to have a backup heater if you want hot water 24

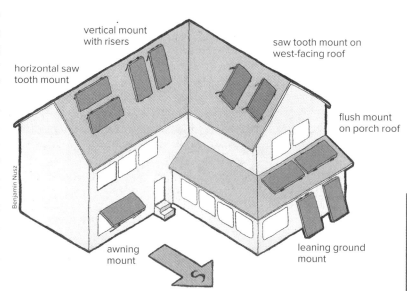

Collector mounting locations.

hours a day, 7 days a week, 12 months of the year. The reason is that the sun does not shine every day, and when the sun does not shine, you get no solar heated water. For both economic and practical reasons, solar water heating systems are sized to provide 100% of the daily hot water load for ONE day. The bottom line is that when it is a sunny day the solar system will provide all your hot water, and during that day your backup water heater will not come on and you will use no fossil fuel. Exactly what percentage of your annual hot water the solar heater will provide depends on a number of factors, including climate, weather patterns, and size of the collector array and balance of system type. Most systems provide 50% or more of the annual hot water load (the federal incentive requires a 50% minimum

solar contribution to the annual load). The ultimate limiting factor is the number of sunny days per year, which varies across different areas of the country and the world. But even in climates that experience considerable cloudy or short day conditions, a solar water heating system will supply some hot water and displace some fossil fuel use. Most solar installers use computer programs containing NASA weather data when estimating solar production from a particular system at a specific location.

A solar water heating system is comprised of two parts, the collector type and the system type. There are several types of collectors and several types of systems. They are described more fully below. Many, but not all collector types can be used in most, but not all system types. This affords you choices, and one of the best ways to get advice about what system and collector types are appropriate for your particular climate is to talk to local solar installers.

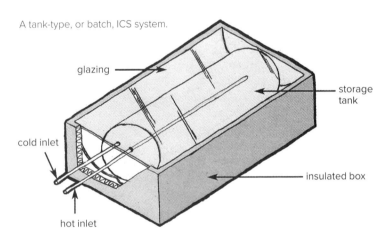

A tank-type, or batch, ICS system.

glazing

storage tank

cold inlet

insulated box

hot inlet

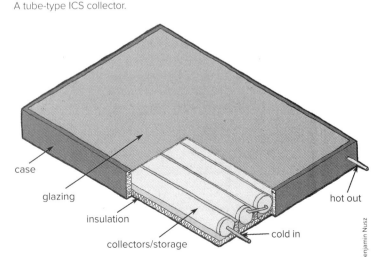

A tube-type ICS collector.

case

glazing

insulation

collectors/storage

hot out

cold in

Benjamin Nusz

Types of Solar Water Collectors

ICS Collectors

ICS stands for integral collector storage, also known as a batch heater. In a batch heater, the hot water storage tank *is* the solar absorber. The tank or tanks are painted black or coated with a selective surface and mounted in an insulated box that has glazing on one side. The sun shines through the glazing and hits the black tank, warming the water inside it. Some models feature a single large tank (30–50 gallons) while others feature a number of metal tubes plumbed in series (30- to 50-gallon total capacity). The single tanks are typically made of steel, while the tubes are typically made of copper. These collectors weigh 275–450 pounds when full, so wherever they are mounted, the structure has to be strong enough to carry this significant weight; you may have to reinforce an existing roof. You should always mount the collectors tilted so that the system will drain properly.

Batch heaters are widely used around the world where the climate *never* experiences freezing conditions. They work great, given the climatic restrictions. They are a type of direct system, as the water heated in the collector is the water you actually use in your home. (Indirect systems are described below.) Batch heaters are also perfect in seasonal applications such as campgrounds and summer homes where they are used only during the warm months of the year and are drained before freezing occurs.

The tube type of batch heater will outperform the tank type because more surface area is exposed to the sun. Another advantage of the tube type is that their profile is much smaller, thereby minimizing their aesthetic impact on a building. On the other hand, tube collectors cool off more quickly at night because of the larger surface area, so they lose efficiency, especially when the nights are cool. You can maximize the efficiency of this kind of system by using as much hot water as possible during the day and early evening hours.

Flat-plate Collectors

Flat-plate collectors are the most widely used type of collector in the world for domestic solar water heating and solar space heating applications. These collectors have an operating range from well below 0°F to around 200°F, which is precisely the operating range required for these applications. They are durable and effective. Flat-plate collectors also have a distinct advantage over other types because they shed snow very well, which makes them a good choice in colder

REAL GOODS

climates. They are the standard to which all other kinds of collectors are compared.

Flat-plate collectors are shallow rectangular boxes that typically are 4 feet wide by 8 feet long and 4–6 inches deep, though they also come in 4 × 10-foot and 3 × 8-foot sizes. They consist of a strong frame, glazing fastened to the front of the collector, and a solid back. Just beneath the glazing lies an absorber plate. This absorber plate has manifolds that run across the top and bottom of the collector, just inside the frame. These manifolds are usually 1-inch-diameter copper pipe and extend out both sides of the collector through large rubber grommets. These internally manifolded collectors can be easily ganged together to make large arrays. Smaller riser tubes, typically ½-inch copper pipes, run vertically and are welded to the manifolds above and below, spaced 3–6 inches apart (the closer the better). Attaching a flat copper fin to each riser completes the absorber plate. The fin must make intimate contact with the riser tube to facilitate effective heat transfer from the fin to the tube; soldering or welding makes the best connection. The fins are usually made of copper and either plated with a selective surface or painted to maximize solar absorptivity. They also are usually dimpled or corrugated to increase absorptivity. The absorber plates are not attached to the frame; they rest inside of it and can expand or contract as they are heated or cooled and not be restricted by the frame. The fins and tubes must be made of the same metal to reduce the chance of corrosion. (One absorber on the market has copper waterways with aluminum fins. They get away with this by laminating the metals together in such a way that galvanic reactions do not take place.)

The strength of a flat-plate collector is all in the frame, and strength is very important because the collectors must be able to withstand high wind conditions without breaking apart. These frames are almost exclusively made of extruded aluminum, although some are made of rolled aluminum; we recommend the extruded aluminum. Mounting hardware fastens to channels or flanges built into the frame. Because these flanges go completely around the collector, great flexibility in mounting options is available.

Another important component to consider are the fasteners used to assemble the collector. All fasteners should be made of stainless steel. It is critical to always use compatible metals when attaching different kinds together. Aluminum and stainless steel are compatible, while aluminum and plain or galvanized steel are not. This prin-

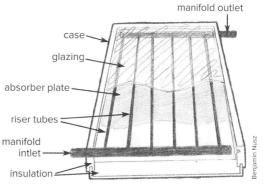

A flat-plate collector.

ciple must follow through to the mounting hardware as well as the collector construction. Because each manufacturer makes its own mounting hardware, and because that collector is tested with its specific hardware, you should always purchase the mounting hardware to match the collector.

While many materials have been used for the glazing of flat-plate collectors, only low-iron tempered glass has stood the test of time. This glass is usually patterned on the outside to reduce glare and reflection and to increase absorptivity. An EPDM rubber gasket is fitted to the edges of the glass plate both to protect the edge and to create a good seal where it sets against the collector frame. Note that if you ever have to take the glazing off of a collector, the edge of tempered glass is very fragile. If you even tap the edge/side of a tempered glass pane, it can literally explode apart, so be very careful and always wear safety glasses and gloves when handling glass. All kinds of plastics have been used as glazing material, but they have all failed under direct and constant exposure to the sun.

Evacuated-tube Collectors

While flat-plate collectors are all made essentially the same way and perform similarly from one brand to another, evacuated-tube collectors vary widely in their construction and operation. Evacuated-tube collectors consist of a number of annealed glass tubes that each contain an absorber plate. During the manufacturing process, a vacuum is created inside the glass tube. The absence of air creates excellent insulation, allowing higher temperatures to be achieved at the absorber plate.

The similarity between different types ends there. Some evacuated-tube collectors have a riser tube fastened to the absorber plate that sticks out of each end of the tube and is attached to a manifold, much like that of a flat-plate collector absorber. The solar fluid circulates through each

Flat-plate collectors are the most widely used type of collector in the world for domestic solar water heating and solar space heating applications. They are the standard to which all other kinds of collectors are compared.

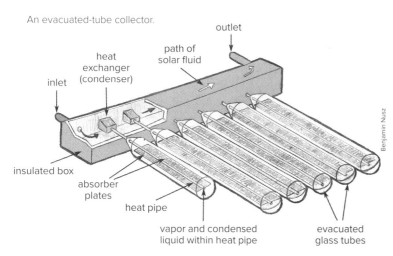

An evacuated-tube collector.

outlet

path of
solar fluid

heat
exchanger
(condenser)

inlet

insulated box

absorber
plates

heat pipe

vapor and condensed
liquid within heat pipe

evacuated
glass tubes

While flat-plate collectors are all made essentially the same way and perform similarly from one brand to another, evacuated-tube collectors vary widely in their construction and operation.

tube and is heated. Others use a hollow pipe attached to the absorber plate that is closed at one end and exits the glass tube on the top. A special liquid inside the pipe evaporates when heated, and the hot vapor rises to a heat exchanger manifold located along the top of the tubes. Solar fluid is heated in the exchanger and circulated throughout the system. As the solar fluid cools the vapor, it condenses and drops back down into the pipe. These liquid/vapor tubes often require a valve on each pipe. Still other models use a solid metal rod attached to the absorber that sticks out the top end of the glass tube and is inserted into a manifold. Sun shining on the absorber heats the rod, and when solar fluid is circulated through the manifold, it picks up heat from the rod ends.

Evacuated-tube collectors have characteristics different from flat-plate collectors. First, these collectors can get hotter than flat-plate collectors and can generate temperatures above the boiling point of water. This can cause significant problems in a solar water heating or space heating

system so it is therefore critically important to always make sure there is an adequate load on the system to keep the temperatures below 210°F within the system. One way you can keep the temperatures under control is by oversizing the solar storage tank or undersizing the collector, which is what is typically done in Asia and Europe where these collectors are most popular. It is also prudent to avoid using this type of collector if the system will sit idle for a period of time—for example, if you usually take long vacations. To avoid potential overheating with evacuated-tube collectors, it is best to use them in drainback systems, which are described below.

Second, the tubes are fragile. They are made of annealed glass, which is much more delicate than tempered glass. Care must be taken when transporting and handling the glass tubes. Another issue is that collector performance depends upon the vacuum inside the tubes. Because a rod or pipe exits the tube on one end or both, a seal must be maintained at this junction. If the seal is broken, the collector performance is no better than that of a flat-plate collector, and probably worse. Some tubes, called folded tubes, are more like a thermal pane window with two panes of glass close together and the vacuum between them. These folded tubes do not suffer from vacuum loss. Another problem with evacuated-tube collectors is that they do not shed snow or melt frost. Because the evacuated tube is such a good insulator, little heat escapes from it, and the snow or frost that accumulates on the tubes can stick for a long time. Their surface is also irregular, so snow packs between the tubes as well. It is not uncommon for roof-mounted evacuated-tube collector arrays to become packed with snow in the early winter and stay that way until spring, which renders them completely useless for a good portion of the year. The problem is compounded because the fragility of the glass makes it difficult to scrape accumulated snow off the tubes.

In climates that do not experience snow that lingers for more than a week, systems using evacuated tube collectors can work very well, especially if used in drainback systems. Be sure to choose an installer who has experience with this technology to ensure you are getting a system that addresses the issues noted above. Properly designed, an evacuated-tube collector can work well.

Concentrating Collectors

A concentrating collector utilizes a reflective parabolic surface to reflect and concentrate the sun's energy to a focal point where the absorber is located. To work effectively, the reflectors must

track the sun. These collectors can achieve very high temperatures because the diffuse solar resource is concentrated on a small area. In fact, the hottest temperatures ever measured on Earth's surface have been at the focal point of a massive concentrating solar collector. Concentrating collectors have been used to make steam that spins an electric generator in a solar power station. This is sort of like starting a fire with a magnifying glass on a sunny day.

Today there are many solar-powered electric generating plants in operation in the US and worldwide that use concentrating collectors. They are among the least expensive electric generating plants to build and operate, and they create little (if any) pollution. They are always installed in desert climates where there is little cloud cover annually. More systems like this are being installed every year.

Attempts have been made to use concentrating collectors in domestic water heating systems, but to date no successful or durable products have been developed. Problems have been encountered with the tracking mechanisms, the precision needed in the mechanisms, the whole mechanism freezing up, and the durability of the reflectors and linkages. They also tend to make the water too hot on a regular basis.

Pool Collectors

The single largest application of active solar water systems is to heat outdoor swimming pools. Special collectors have been developed for this purpose, made of a special copolymer plastic. The collectors don't have to be glazed because they are used only when it is warm outside. Like any other solar water collector, they cannot withstand freezing conditions.

Solar pool heating collectors are direct systems because the pool water itself circulates through the collectors. This is the most efficient configuration, as there are no heat exchanger losses. Most pool water contains additives like chlorine, which is highly corrosive to copper, and this is one of the reasons the collectors and piping are made of plastic .

Plastic pool heating collectors are typically mounted flat on a roof. The collectors are held in place with a set of straps that go over the collectors but are not actually attached to them. The straps are often plastic-coated stainless steel and are threaded through special clips that are bolted to the roof. This method of holding down the collectors allows them to expand and contract on the roof without binding.

Pool heating collectors are not appropriate for heating domestic water.

The single largest application of active solar water systems is to heat outdoor swimming pools.

Types of Solar Water Heating Systems: Operation and Installation

As we've been discussing, solar water systems can be either direct or indirect. In a *direct* solar water system, domestic water—the actual water you use in your home—enters the solar collector, where it is heated before being circulated into the house. In an *indirect* system, a heat transfer fluid (also called solar fluid) is heated in the collectors and then circulated to a liquid-to-liquid heat exchanger where the solar heat is transferred from the solar fluid to your domestic water supply.

All systems except batch systems require some kind of tank to store the water that was heated by the sun. Most systems use a storage tank for the solar heated water, and also a backup heater that can be either a tank type or an on-demand (tankless) type. These are called *two-tank systems*. Some system designers attempt to use the solar storage tank as the backup heating system as well by including an electric heating element or a gas burner to heat the water when insufficient solar energy is available to bring the water up to the desired temperature. These are called one-

tank systems. We recommend two-tank systems because you usually need a second tank to have suitable storage capacity for the collectors. All the illustrations in this section are of two-tank systems. The only situation where a one-tank system may be appropriate is in a warm climate where it is sunny most days.

Over the years, a number of different domestic solar water heating systems have been devised, but there are only three we would ever recommend: ICS or batch, drainback, and closed-loop antifreeze. These three are the only system types that have proven their reliability over the past 20 years.

ICS or Batch

Batch systems are considered passive because they require no pumps of any kind to operate. They are direct systems because the domestic water actually enters the collector. The batch unit is typically plumbed in series between the cold-water supply and the conventional water

Because of their simplicity, batch collectors are among the least expensive solar water heating systems available.

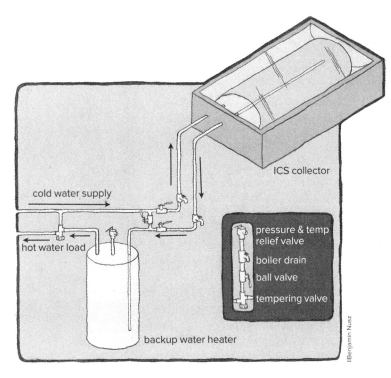

cold water supply

hot water load

ICS collector

pressure & temp relief valve

boiler drain

ball valve

tempering valve

backup water heater

IIBenjamin Nusz

An ICS system.

heater. Whenever a hot water tap in the dwelling is opened, cold water from the supply enters the batch collector and forces the solar heated water stored in the batch collector into the conventional or backup water heater. If the water from the batch collector is hotter than the setting on the backup heater, that heater will not activate. If the water from the collector is warm but below the temperature setting of the backup heater, the backup will have to add only enough heat to raise the temperature to the preset level. If no solar heating has taken place and all the water in the batch collector is cold, the backup heater will have to deliver the whole load.

Because water is plumbed through the batch collector, this type of system is suitable only for climates or seasons where freezing conditions will *not* occur. Batch collectors are very popular along the extreme southern parts of the US and in tropical climates. They have also been successfully installed in vacation homes and recreational facilities like parks, where they are used only during the summer months and are drained the rest of the year.

Because of their simplicity, batch collectors are among the least expensive solar water heating systems available. A complete system consists of the batch collector and piping: No pumps, extra storage tanks, controllers, or other components are needed. Unfortunately, because it is simple and inexpensive, people who live in climates where freezing conditions do occur are sometimes tempted to use a batch system. They often

think that if freezing conditions are forecast, they will simply drain the water out of the collector and piping and everything will be all right. In theory, this assumption is correct. In reality, Murphy's Law often comes into play (Murphy's Law states that if anything can go wrong, it will). Actually, all batch collectors can withstand mild freezing conditions. The mass of the water in the collector will keep it from freezing for quite some time (depending on the temperature). It is the piping going to and from the collector that is vulnerable. A well-insulated ¾-inch copper pipe will freeze in less than five hours at 29°F. Unless you plan to use it only seasonally, you're better off not installing a batch system if you live in a cold climate.

Drainback Systems

Drainback systems use flat-plate or evacuated-tube collectors to heat a solar fluid, usually distilled water or a weak antifreeze/water solution, that circulates through them. Two insulated pipes connect the collectors to a specialized tank called a drainback tank. A high-head pump is installed on the pipe that feeds the collector array. A differential temperature controller turns the circulating pump on and off. When on, the solar fluid is pumped from the drainback tank to the collectors, where it is heated. The solar fluid then drops back down to the drainback tank, completing the circuit. When the system turns off, the pump stops and all the solar fluid in the collectors and the piping drains back into the tank. Drainback tanks are relatively small, so they do not store much heat. When the system is operating, various methods (detailed below) are used to transfer the heat from the drainback tank to a solar storage tank.

Drainback systems are one of the three most popular types of solar energy systems installed worldwide. They are an excellent choice for all climates except those that experience severe or extended cold, or where a significant amount of snow is expected annually. In hot climates, they are an excellent choice.

Drainback systems work very well in warm climates, because when the water in the storage tank gets heated to its maximum desired temperature (its high limit), the system turns off and all the fluid drains out of the collector, which prevents the solar fluid from degrading due to overheating. This is especially important when systems experience idle periods like vacations during the summer months in warm or hot climates.

Drainback systems are classified as indirect

solar water heaters because the domestic water is heated by a solar fluid and heat exchanger and not heated in the collector. They are classified as closed-loop because the solar fluid remains within a single circuit at all times. They are classified as active systems because they always use a pump or pumps to circulate fluids through the system.

The major limitation of drainback systems is their inability to prevent freeze-ups in climates that experience extended cold or snowy conditions. Some installers like to add antifreeze to the solar fluid in drainback systems to extend their range northward.

A limitation of drainback systems is that the collectors have to be located above the drainback tank. This eliminates the option of ground-mounted arrays. Pump selection is very critical in drainback systems, and there are height restrictions on how high a high-head pump can lift water. Unless the circulating pump has a motor over ½ hp, the distance between the water level in the drainback tank and the top of the collectors must be less than 28 feet.

Most manufacturers offer drainback kits that work well for the majority of installations. It is certainly true that most solar water heating systems are the same size and are installed the same way, so kits are fine for most situations. These kits are pre-engineered to take the guesswork out of designing a system. We highly recommend such kits if purchased from a reputable manufacturer. Sizing a drainback system collector array is exactly the same as sizing a closed-loop antifreeze system. To size a drainback tank, calculate all the liquid that would fill the collectors and all the piping above the tank and add 4 gallons.

The operation of the solar loop is straightforward and was described above. To make sure that a system functions properly, you need to follow a few simple rules. The system must be installed to facilitate complete and fast drainage when the system turns off. The collectors must be mounted so that they drain toward the inlet of the array. This would be the bottom manifold where the solar fluid enters the collectors. They should be mounted at a 15-degree angle sloping toward the feed inlet (¼" per foot). Collectors should never be mounted so the riser tubes of the absorber plate are horizontal, as they will sag over time and prevent proper drainage. If sagging occurs, water can be trapped in the pipe and burst during freezing conditions. All piping that is not in conditioned space must be sloped at a 15-degree angle (¼" per foot) toward the drainback tank, and all horizontal pipe runs must be supported at least

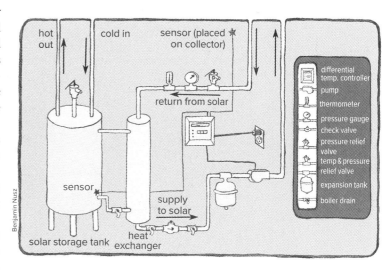

every 3 feet to prevent any sagging, which would inhibit proper drainage. Use a pair of 45-degree elbows instead of a 90-degree elbow whenever practical to facilitate faster drainage. The minimum pipe size for drainback systems is ¾-inch hard copper. Never install any valves of any kind on the solar loop.

Drainback tanks should always be located inside the conditioned space and can be located at the highest point possible to reduce head pressure. This is an important consideration when the system is installed on a two-story house and the storage tank is in the basement. Drainback tanks should be unvented (except on large space heating systems or large commercial systems). They should be fitted with a sight glass to monitor fluid levels within the tank. These tanks should be well insulated to prevent heat loss.

The solar loop pump must be a high-head pump of sufficient size to raise the water from the drainback tank to the top of the collector array. When calculating the head, measure from the bottom of the drainback tank to the highest point of the collector array and add 4 feet. Your pump must be able to exceed that head. Once all the piping is full of solar fluid, the pump does not have to work very hard, because gravity pulling the fluid back down the return line helps pull fluid up the feed line. It will take a 120-volt AC pump to do this job, so the system must be powered by 120 volts AC and use a differential temperature controller to turn the pump on and off. These systems cannot be powered by photovoltaics, because when the system turns on, the pump must start with full force to overcome the head pressure, and PV-powered pumps don't work that way. The pump should be located at least 2 feet below the bottom of the drainback tank and set in a vertical pipe, pumping up to the collectors.

A drainback system with antifreeze, AC powered.

The major limitation of drainback systems is their inability to prevent freeze-ups in climates that experience extended cold or snowy conditions.

A drainback system.

Most solar water heating systems are the same size and are installed the same way, so kits are fine for most situations.

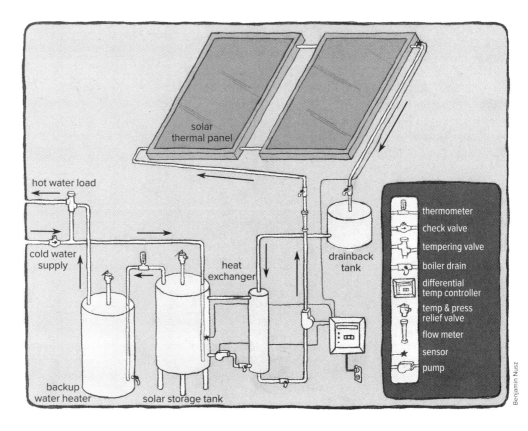

Benjamin Nusz

Drainback systems include a tank to store the solar heated water for later use. There are various methods of getting the heat from the drainback tank to the storage tank, but all use a liquid-to-liquid heat exchanger. One method circulates the hot solar fluid through an in-tank heat exchanger (including the wrap-around type) where the heat exchanger is located in the solar hot water storage tank. In this design, one pump circulates the hot solar fluid throughout the system; hence, it's called a single-pumped system. Another method uses an external heat exchanger mounted to the storage tank. In this case, a second pump may be needed to circulate the water from the storage tank, through the heat exchanger, and back to the storage tank—this is called a double-pumped system. Both pumps are controlled by the differential temperature controller and turn on and off at the same time. The heat exchanger pump should be very small, because the flow through that circuit should be slow, and because there is very little total head to overcome. The third method of getting heat from the drainback tank to the storage tank is to have a heat exchanger below the minimum fluid level inside the drainback tank. A second pump is required with this system to circulate water from the storage tank, through the heat exchanger, and back to the tank. All these systems work very well when properly installed.

Closed-Loop Antifreeze Systems

Closed-loop antifreeze systems are the most versatile of all solar water heating systems. They can be installed in virtually all climates and will not freeze or overheat when properly sized and installed; collector arrays can be mounted in almost any imaginable way and can be located at considerable horizontal distances from the exchanger.

Closed-loop antifreeze systems are similar to drainback systems in many ways, but they have a few significant differences. Antifreeze systems use flat-plate or evacuated-tube collectors to heat a solar fluid, usually a high-temperature propylene glycol/water mixture that is circulated through them. Two insulated copper pipes connect the collectors to a heat exchanger. A relatively small circulating pump is installed on the pipe that feeds the collector array, along with an expansion tank, some drain valves, a safety valve, and a gauge. A loop of piping starts at the heat exchanger, travels through the collector array, and then completes the loop traveling back to the heat exchanger. This piping loop is completely sealed and kept completely full of the solar fluid at all times.

These systems can be controlled and powered by a PV panel and a DC pump, or they can use a differential temperature controller operating a 120-volt AC pump. When the pump starts, the solar fluid circulates through the collector array, through the hot supply pipe to the heat exchanger, and then back to the collectors through the return line. When the pump stops, the solar fluid simply stops moving within the closed loop. Because the solar fluid stays in the entire closed loop, from the collector array to the piping that travels through unconditioned space, it must be able to act as a heat transfer fluid and also protect the system from freezing. Heat transfer from the solar fluid to the domestic water can be accomplished a couple of ways, which are detailed below.

Closed-loop antifreeze systems are the most widely distributed type of solar thermal system worldwide. They are an excellent choice for all climates except where it is hot and the application does not present a reliable and consistent load every day. They are not necessary in climates that do not experience freezing, but they are the only failsafe system for climates where freezing occurs and should be the only choice for climates that experience prolonged or severe cold weather and/ or heavy snowfalls. As with drainback systems, most manufacturers make complete system kits that work just fine when properly installed.

The major limitation of closed-loop antifreeze systems is the limitation of the solar fluid. The best antifreeze solution available today is high-temperature formulated propylene glycol. This fluid can eventually break down and form a corrosive solution that can harm system components. High temperatures degrade the fluid, and the rate of degradation is directly proportional to the intensity of the overheating, so the hotter it gets, the quicker it will degrade. All overheating scenarios can be avoided, so this should not be considered a fatal flaw. The best way to reduce overheating is to make sure the system circulates fluid whenever it is sunny. If you have this type of system, it is critical that you check the condition of the solar fluid regularly. Propylene glycol solar fluid solutions need to be checked after any prolonged overheating episode. This would occur if the circulating pump failed and went undetected for a month or more, for example.

Taking note of a few important points will ensure that your closed-loop antifreeze system will perform effectively. The hot supply pipe between the collectors and the heat exchanger must always be made of copper and covered with high-temperature pipe insulation, because under rare circumstances it can get very hot for short periods of time. Copper is the only practical material that can withstand this heat. On residential solar water systems, the return line from the exchanger to the collectors should also be copper, but standard high-quality pipe insulation can be

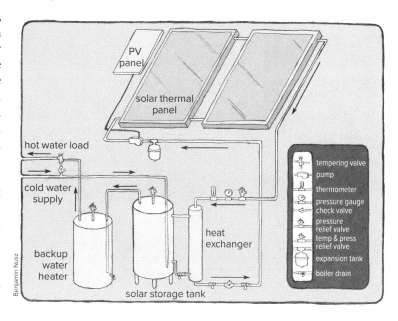

A closed-loop antifreeze system.

used on that line. (In large solar space heating systems, the return line can be made of PEX tubing, except for the final 10 feet where it attaches to the collectors.) Collector arrays should be mounted with a slight slope toward the return inlet where the solar fluid enters the array, and all piping should slope slightly toward the heat exchanger. For arrays that are remote, a drain should be placed at the inlet where the solar fluid enters the array to facilitate drainage. All horizontal copper pipe runs should be supported every 5–6 feet, and all vertical runs should be supported every 10 feet. Large space heating systems or any systems that are seasonal or may experience periodic idle times where no load is present require a heat diversion load, sometimes called the shunt load. This shunt diversion load is necessary to prevent the solar fluid from overheating. A shunt loop is typically a buried length of uninsulated pipe or a radiator located outside, preferably in a cool or windy spot. Outdoor hot tubs are also a common shunt load—not a bad idea!

Because fluid remains in the solar loop at all times, a method to prevent thermosiphoning must be included in the solar loop. Thermosiphoning can happen when collectors that are located above the storage tank are colder than the tank, as is the case with roof-mounted arrays. Heat from the storage tank and heat exchanger could rise up the supply pipe to the top of the array, and cold-heavy solar fluid could drop down the return pipe to the heat exchanger from the bottom of the array. Under this undesirable scenario, the fluid is circulating in the opposite direction from when it is heating. A check valve will stop this from happening.

The check valve should be located between the two boiler drains that are used to charge the system. The check valve will also ensure that the fluid flows correctly when charging the system. If your collectors are not located above the storage tank, and no check valve is required, you may need to install a full-port ball valve between the drains to facilitate charging.

All that said, there are occasions when reverse thermosiphoning is a good thing. For instance, if you are going on an extended vacation (over two weeks), you may want to keep your tank cool to prevent overheating and speeding the degradation of the solar fluid. You can install a vacation bypass around the check valve to allow reverse thermosiphoning to occur. (This method would not be effective if your collectors are located below the storage tank.)

The solar fluid within the solar loop is put under pressure when the system is charged. The pressure is typically set at 32 pounds at 60°F fluid temperature. When the fluid is heated, it expands, raising the pressure within the closed loop. Eventually, the pressure would increase enough to burst the piping. To prevent a catastrophic event, an expansion tank is installed on the loop. The expansion tank will also provide some compensation when pressure in the loop drops as the fluid cools. It is not unusual to see the pressure in the solar loop fluctuate between 10 and 45 pounds seasonally, and sometimes daily. It is critically important to properly size the expansion tank used in a closed-loop system. Formulas are available for sizing these expansion tanks. It is also important to use a "solar rated" expansion tank, because regular expansion tanks are not robust enough for use in a solar water heating system. Set the pressure of the expansion tank 3–5 pounds below the pressure in the system at 60°F.

The best location for the expansion tank within the closed loop is directly before the circulating pump in relation to fluid flow. This maximizes flow by compensating for the negative pressure that is created on the suction side of the circulating pump. The expansion tank is the point of no pressure change within the closed loop, so it eliminates the negative pressure or vacuum caused by the pump. It is not absolutely necessary to follow this suggestion, but it will ensure optimum circulation in your system. The expansion tank should always hang below the pipe it is attached to. Heat will accelerate the deterioration of the bladder inside the tank, and by hanging it below the pipe, the tank will stay cooler.

A pressure gauge located in the loop should hang down from the pipe and be visible from the charging drain valves. By hanging below the pipe, no air will get trapped in the fitting or gauge. Unfortunately, most pressure gauges are set up to be above the pipe, so when you install the gauge hanging below it, the scale on the gauge will be upside down. Don't panic, you'll get used to it.

A pressure relief valve is also fitted to the closed loop. Notice that this is a pressure-only relief valve. The relief valve should be set to open at 80–90 pounds. Be sure to install a drainpipe on the relief valve and terminate the drain near the floor.

There is no static head pressure that the solar loop circulating pump must overcome because all the pipes are filled with fluid at all times so the only head pressure the pump has to overcome is friction head. Therefore, low-head circulating pumps can be used to circulate the solar fluid through the system. PV-powered pumps can be used here, because the pump can start slow and circulation will start immediately. In fact, a PV-powered system is optimal for several reasons. Most important, PV pumps run whenever the sun is shining, so there is always circulation in the solar loop when the collectors are hot. This extends the life of the solar fluid by eliminating stagnation when the sun is shining. PV-powered pumps are also naturally variable speed, so when the sun's resource is low, the pump runs slower, but the thermal collector's output is also lower under this condition, so the pump speed is perfectly matched to the collector output. On the other side of the coin, under full sun conditions the pump runs faster and matches the temperature of the thermal collector by increasing the flow.

A differential temperature controller can also control closed-loop systems. With such a controller, a low-head 120-volt AC pump is used in the solar loop. If a differential temperature controller is used, it's a good idea to disable the high-limit function in the controller to assure that the pump will circulate at all times that the collector is hot. If the power goes out during a sunny period and a controller is used to regulate the pump, the

Inside an expansion tank.

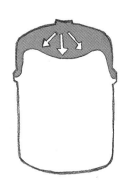
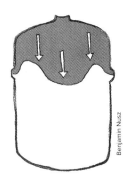

Benjamin Nusz

Chamomile Ramlow

pump will stop and stagnation can occur, which will eventually lead to deterioration of the solar fluid if it happens regularly. This is a reason to periodically check the condition of the solar fluid if this scenario happens regularly.

Extra care needs to be taken when filling the system with solar fluid. It is essential to get all the air out of the system at that time. As already noted, the solar fluid, once installed into the closed loop, will be in there for at least 10 years under normal circumstances, and air can cause circulation problems, so a good job done at the start will result in many years of trouble-free service. Many installers over the years have specified placing an automatic air vent at the highest point in the closed loop. This practice is not recommended. Automatic air vents fail on a regular basis and should not be used on closed-loop systems. If you still feel that you require a port to bleed air from the system, install a short riser at the high point of your system on the collectors. Install a ball valve on that riser and a Schrader valve on the end of the riser. Air will accumulate in the riser, because it is the high point in the loop. You can open the ball valve, open the Schrader valve to expel any accumulated air, and then close the ball valve.

Some contractors that are used to installing traditional hydronic heating systems will want to put an automatic fill valve on a solar closed-loop system. They do this out of habit because most hydronic heating systems require this valve to keep the system pressurized. *Never install an automatic fill valve on a closed-loop system!* It will destroy a system by diluting the solar fluid, which can lead to decreased freeze protection.

Closed-loop systems can use a tank-integrated heat exchanger or an external heat exchanger. Tank-integrated heat exchanger systems are the simplest to install and are suited for situations where space is limited. These are always single-pumped systems, as the only pump required is the solar-loop pump. External heat exchanger systems are more efficient and can also be single pumped if the heat exchanger is of the thermo-siphon type. A plate-type heat exchanger or a coiled tube-in-shell heat exchanger can also be used. In these cases, the system must be double pumped. Double-pumped systems can either be PV powered or use a controller and two 120-volt AC pumps. Remember to calculate the correct wire size when running long low-voltage wires between the PV panel and the DC pumps.

Large space-heating systems and large commercial systems often use a closed-loop design. Because of their size, this may be the only option where the collectors cannot be located above the storage tank and heat exchanger. A shunt load must be provided on large systems to reduce overheating during nonload conditions. The world's largest flat-plate solar water heating system, in Green Bay, Wisconsin, utilized more than 5,250 4×8-foot collectors (157,689 sq. ft. of collector area). This closed-loop system performed extremely well, producing over 37,500 million Btus annually. Unfortunately, when natural gas prices fell to record lows during the 1990s, the system was shut down and dismantled. Solar water heating makes tremendous ecological and financial sense, now more than ever.

Collectors and balance of system components can be purchased from a local dealer. Whether you are installing your system yourself or having the system installed by a solar professional, be sure to always check references for the companies you are dealing with. We also suggest that you read Bob Ramlow's book *Solar Water Heating: A Comprehensive Guide to Solar Water and Space Heating Systems* (New Society, updated edition), which you can buy at realgoods.com

Solar water heating makes tremendous ecological and financial sense, now more than ever.

Solar Water Heating Economics

As an introduction to this section, I want to provide a reality check.

Reality Check #1. Nobody has a crystal ball to forecast what the price for any fuel will be in the future (even tomorrow). Projecting future fossil fuel prices is entirely speculative because we never know how the future will play out. For instance, 7 years ago the US Department of Energy predicted that natural gas prices would increase rapidly in the near future. That didn't happen because of fracking. The lower than expected natural gas prices also affected the price of electricity, because most new power generating stations operate on natural gas. Nonetheless, it still seems reasonable to assume that fossil fuel prices will increase yearly, and that the rate of increase will likely increase too, as we continue to use them up at the current rate of consumption and as populations increase and climate change continues to shape our world. As I write this, natural gas prices are up 32% from a year ago and propane prices have more than doubled in the last 2 months. When consulting the chart below, please also keep in mind that fossil fuel prices depend on one's location and vary significantly from one region of the country to another. For the following examples, I have chosen conservative fuel inflation prices and rates. Your situation may differ.

Reality Check #2. To make an accurate and true comparison between the cost of a fossil fuel and the cost of a form of renewable energy, you have to include the cost of "externalities" for both. Externalities, as you probably know, are the costs associated with obtaining, processing, and using the fuel. Externalities are not included in the price we pay "at the pump," but they are real costs borne by society. Some examples of externalities include, but are not limited to: acid rain, air pollution in general (asthma, heart conditions, physical damage to structures), water pollution, and climate change. These costs are not trivial and are often ignored, because it is hard to put a price tag on them, and because it is inconvenient to acknowledge them as true costs: If we acknowledge these costs, then we must take responsibility for them. Some economists have argued that including externalities related to the fossil fuel cycle (mine/well through combustion) would at least double the price of fossil fuels. A fair comparison must add externalities to the price of renewable energy as well. While there certainly are some, because renewable energy equipment is made of materials, some of which are extracted and processed with some environmental costs, the externalities related to renewable energy systems pale in comparison to those of fossil fuels. I have not added any costs for externalities in the following calculations. We leave it up to you to consider these values along with the economic analyses.

Many people choose solar water heating because they want to reduce their carbon footprint and be responsible earthlings. Others choose solar hot water because it is a good economic decision. You get two for the price of one with this kind of investment. A solar water heating system is considered a long-term investment and is amortized over its lifetime. While the business sector regularly uses what is called "life-cycle costing," homeowners typically do not. This may be a change of perspective for those not accustomed to this type of analysis, but it is the only accurate financial perspective to use when analyzing the true economics of a solar water heating system investment.

The first point to underscore is that when you install a solar water heater you make an investment that will increase the value of your home. You gain in equity what you spent on the cost of installation. Solar water heaters typically have a lifespan of at least 30–40 years. In most cases, the collectors will outlast your roof. So if you decide to sell your house, you should get back most of what you paid to buy and install the system from the increased value of your home. Under this scenario, you recoup your investment at the time of sale and you have enjoyed many years of pollution-free hot water at no cost.

The second critical point is life-cycle costing. People often ask, "Why would I consider purchasing a solar water heater that costs several thousand dollars when I can purchase a gas or electric water heater for around a thousand dollars?" The answer lies in life-cycle costing, which adds the original cost of a piece of equipment to its operating cost over its lifetime. Life-cycle cost analysis gives an accurate picture of the total cost of a purchase over time, and takes into account long-term trends in energy prices. The following chart uses a realistic but conservative estimate that the rate of inflation for fossil fuels will average 3.6% for electricity and 5.9% for natural gas (the historical averages for 1970–2013) over the next 30 years. Solar hot water compares favorably with its competitors.

Note that in this comparison of solar with natural gas and electric water heating, the solar heater has no fuel cost Most solar water heating systems use a small circulating pump that, when operating, draws about as much energy as a 25-watt light bulb that is on 5 hours per day. Like any other piece of mechanical equipment, it does require some maintenance, but this amounts to only 1% of the system cost per year. As you can see, viewing the systems over the long term makes for a fair comparison.

Like other renewable energy investments, solar water heating is different from a conventional investment because the free energy it harvests reduces a monthly bill that you pay to someone else. The savings gained from the solar water heater pays for the solar investment.

One more thought about the economics of solar water heating. People often ask the question, "What's my payback?" This is interesting because they likely have never asked that question about any other investment they ever made, but we have been influenced to ask it about renewable energy. When the calculation equates to 10 or 12 years, we get the wrong impression that this is not such a good investment. We fail to recall that few, if any, of our other investments perform this well. A 10- or 12-year payback equals an impressive 8% to 10% return on investment (this is like getting an 8%–10% interest rate on your savings account

or CD), not including the fact that after avoided fossil fuel costs pay for the system, everything else is free.

Cash flow makes the world go round, and having a solar water heating system affects your cash flow in a positive way. If you borrow the money to pay for the system, two things impact your cash flow: 1) Your fossil fuel bill for water heating will go down, and 2) You will have a monthly payment on the system. In many cases (it all depends on the actual cost of your water heating fuel and the efficiency of your fossil-fueled water heater), the monthly payment for the solar water heater will be less than your fuel reduction value, resulting in more money at your disposal at the end of the month—a positive cash flow impact. The greater the length of the loan, the better the cash flow impact. Many people include such a loan in their mortgage, which is an excellent strategy.

When you install a solar water heating system, you join millions of people worldwide. In many countries, nearly every house has a solar heater as their primary source of hot water. While many of these countries are in warm climates, Germany, which is not, has achieved 50% solar water heating, and they are still cranking.

Life Cycle Comparison Chart			
	Electric Water Heater @ 90% efficiency	Natural Gas Water Heater @ 65% efficiency	Solar Water Heater
Energy Required to Produce 11 million BTU	5,200 kWh	246 therms	11 Million BTU
Cost per Energy Unit	$0.15/kWh	$1.25/therm	$0
Energy Inflation Rate (40-year average)	3.6%	5.9%	—
Installed Cost Minus Federal Tax Credit	$1,500	$1,500	$6,720 ($9,600 – 30% federal tax credit)
Maintenance	$50/year	$50/year	$100/year
Cost to Operate per Year + Initial Cost			
1	$830	$358.00	$100
2	$858.51	$376.17	$100
3	$888.07	$395.42	$100
4	$918.70	$415.80	$100
5	$950.45	$437.38	$100
6	$983.37	$460.23	$100
7	$1,017.49	$484.44	$100
8	$1,052.86	$510.07	$100
9	$1,089.52	$537.21	$100
10	$1,127.52	$565.96	$100
10-Year Life-Cycle Cost	$10,216.52	$6,040.67	$7,720

Water Heating System Costs

The following chart is a life-cycle economic analysis over 10 years. It compares the yearly cost to operate an electric water heater and a natural gas water heater to a solar water heater under the following assumptions:

- 64 sq. ft. flat-plate closed-loop solar water heating system, family of 4 located in Columbia, MO (the middle of the country).
- Electric water heater efficiency = 90%,[1] gas water heater efficiency = 65%,[2] solar fraction = 69%

The following is a cash flow analysis comparing the cost of owning a solar water heating system (same system as above) to the cost of operating an electric water heater. This analysis assumes you add the final installed cost of the solar system to your existing mortgage. The chart could be extended for another 20 years to reflect the length of the loan. As you can see, there is positive cash flow starting the first year, and it remains positive throughout the life of the loan. Note that a quality solar water heater that is properly installed will last over 30 years.

Monthly Bill	Electric Water Heater Monthly Savings (reduction in electric bill after adding solar water heater)	Solar Water Heater Monthly Payment (installed cost of $6,720 added to a 30-year mortgage)	Cash Flow Impact per Month
1st year	44.85	$27.79	+$17.06
2nd year	46.49	$27.79	$18.70
3rd year	51.06	$27.79	$23.27
4th year	52.83	$27.79	$25.04
5th year	54.65	$27.79	$26.86
6th year	56.54	$27.79	$28.75
7th year	58.51	$27.79	$30.72
8th year	60.54	$27.79	$32.75
9th year	62.65	$27.79	$34.86
10th year	64.83	$27.79	$37.04

Energy Conversions

Btu: British Thermal Unit, an archaic measurement, the energy required to raise 1 lb. of water 1°F

1 gal. liquid propane	= 91,500 Btus
1 gal. liquid propane	= 4 lb. (if you buy propane by the pound)
1 gal. liquid propane	= 36.3 cu. ft. propane gas @ sea level
1 therm natural gas	= 100,000 Btus
1 cu. ft. natural gas	= 1,000 Btus
1 kW electricity	= 3,414.4 Btus/hr.
1 hp	= 2,547 Btus/hr.

REAL GOODS

CHAPTER 9

Water and Air Purification

Clean Water and Clean Air Are Human Rights!

IF YOU ARE CONCERNED about the health of our planet and the degradation of our environment, you are no doubt also conscientious about the foods you eat. If you are conscientious about the foods you eat, then you likely pay attention to the quality of the air you breathe and the water you drink. Like food, air and water can carry into your body the nasty contaminants that saturate the natural world. But you might not realize the extent to which your source of domestic water and the air inside your house may be polluted. This chapter examines the bad stuff that may be lurking in your air and water, the conditions that contribute to potentially unhealthful levels of these contaminants in your living space, and what you can do about these problems. Some good resources are listed at the end of the chapter, and you can find recommended products at our online store, realgoods.com or by calling our knowledgeable technicians at 1-800-919-2400.

The Need for Water Purification

The Environmental Protection Agency (EPA) states that no matter where you live in the United States, some toxic substances are likely to be found in the groundwater. Indeed, the agency estimates that one in five Americans, supplied by one-quarter of the nation's drinking water systems, consume tap water that violates safety standards under the *Clean Water Act*. Even some of the substances that are added to our drinking water to protect us, like chlorine, can form toxic compounds. In the case of chlorine, trihalomethanes (THMs) have been linked to certain cancers. Most chlorines have been long banned in Europe. The EPA has established enforceable standards for more than 100 contaminants. However, credible studies have identified more than 2,000 contaminants in the nation's water supplies.

The simplest way of dealing with water pollution is a point-of-use purification device. The tap is the end of the road for water consumed by our families, so this is the logical and most efficient place to focus water treatment. Different water purification technologies each have strengths and weaknesses and are particularly effective against specific kinds of impurities or toxins. So the system you need depends first and foremost on the nature of the problem you have, which in turn requires testing and diagnosis.

We will describe these systems in detail after discussing the contaminants you might encounter. Most treatment systems are point-of-use and deal with only drinking and cooking water, which accounts for less than 5% of typical home use. Full treatment of all household water is a very expensive undertaking and is usually reserved for water sources with serious, health-threatening problems.

Contaminants in Water

Presumably your drinking water comes from a municipal system, a shallow dug well, a deep drilled well, or a spring. If you drink bottled water, you've already taken steps to control what you consume—but you should also read about

> The Environmental Protection Agency estimates that one in five Americans, supplied by one-quarter of the nation's drinking water systems, consume tap water that violates safety standards under the Clean Water Act.

The most pressing and widespread water contamination problem results from the organic chemicals (those containing carbon) created by industry.

the problems and perversions of bottled water in Chapter 7. The fact is, most good purification systems provide tasty potable water for far less money than the cost of regularly drinking bottled water, and you won't be supporting the insidious movement perpetrated by Coca-Cola, Pepsi, and Nestlé to privatize water—a human right! Each type of water supply is more or less vulnerable to different kinds of pollutants, because surface water (rivers, lakes, reservoirs), groundwater (underground aquifers), and treated water each are exposed to environmental contaminants in different ways. The hydrogeological characteristics of the area you live in, and localized activities and sites that create pollution—factories, agricultural spraying, a landfill—potentially will impact your water quality. And some impurities, such as lead and other metals, can be introduced by the piping that delivers water to your tap.

Reliable and inexpensive tests are available to identify the biological and chemical contaminants that may be in your water. Here's what you may find.

Biological Impurities: Bacteria, Viruses, and Parasites

Microorganisms originating from human and animal feces, or other sources, can cause waterborne diseases. Approximately 19,500 cases of waterborne illness are reported each year in the US (ncbi.nlm.nih.gov/pubmed/18020305). Additionally, many of the minor illnesses and gastrointestinal disorders that go unreported can be traced to organisms found in water supplies.

Biological impurities largely have been eliminated in municipal water systems with chlorine treatment. However, such treated water can still become biologically contaminated. Residual chlorine throughout the system may not be ad-

equate, and therefore microorganisms can grow in stagnant water sitting in storage facilities or at the ends of pipes.

Water from private wells and small public systems is more vulnerable to biological contamination. These systems generally use untreated groundwater supplies, which could be polluted due to septic tank leakage or poor construction.

Organic Impurities: Tastes and Odors

If water has a disagreeable taste or odor, the likely cause is one or more organic substances, ranging from decaying vegetation and algae to organic chemicals (organic chemicals are compounds containing carbon).

Inorganic Impurities: Dirt and Sediment, or Turbidity

Most water contains suspended particles of fine sand, clay, silt, and precipitated salts. This cloudiness or muddiness is called turbidity. Turbidity is unsightly, and it can serve as food and lodging for bacteria. Turbidity can also interfere with effective disinfection and purification of water.

Total Dissolved Solids (TDS)

Total dissolved solids consist of rock and numerous other compounds from the earth. The significance of TDS in water is a point of controversy among water purveyors, but here are some facts about the consequences of higher levels of TDS:

1. High TDS results in undesirable taste, which can be salty, bitter, or metallic.
2. Certain mineral salts may pose health hazards. The most problematic are nitrates, sodium, barium, copper sulfates, and fluoride.
3. High TDS interferes with the taste of foods and beverages.
4. High TDS makes ice cubes cloudy and soft, and they melt faster.
5. High TDS causes scaling on showers, tubs, and sinks, and inside pipes and water heaters.

Toxic Metals or Heavy Metals

The presence of toxic metals in drinking water is one of the greatest threats to human health. The major culprits include lead, arsenic, cadmium, mercury, and silver. Maximum limits for each of these metals are established by the EPA's Primary Drinking Water Regulations (epa.gov/safewater /consumer/pdf/mcl.pdf.

Toxic Organic Chemicals

The most pressing and widespread water contamination problem results from the organic chemicals (those containing carbon) created by industry. The American Chemical Society lists more than 4 million distinct chemical compounds, most of which are synthetic (manmade) organic chemicals, and industry creates new ones every week. Production of these chemicals exceeds a billion pounds per year. Synthetic organic chemicals have been detected in many water supplies throughout the country. They get into the groundwater from improper disposal of industrial waste (including discharge into waterways), poorly designed and sited industrial lagoons, wastewater discharge from sewage treatment plants, unlined landfills, and chemical spills.

Studies since the mid-1970s have linked organic chemicals in drinking water to specific adverse health effects. However, only a fraction of these compounds have been tested for such effects. More than three-quarters of the substances identified by the EPA as priority pollutants are synthetic organic chemicals.

Volatile Organic Compounds (VOCs)

Volatile organic compounds are very lightweight organic chemicals that easily evaporate into the air. VOCs are the most prevalent chemicals found in drinking water, and they comprise a large proportion of the substances regulated as priority pollutants by the EPA.

Chlorine

Chlorine is part of the solution and part of the problem. In 1974, scientists discovered that VOCs known as trihalomethanes (THMs) are formed when the chlorine added to water to kill bacteria and viruses reacts with other organic substances in the water. Chlorinated water has been linked to cancer, high blood pressure, and anemia. Chlorine has been banned for years in Europe and most industrialized countries due to its status as a known carcinogen, but for some insidious reason, it is still legal in the US!

The scientific research linking various synthetic organic chemicals to specific adverse health effects is not conclusive and remains the subject of considerable debate. However, given the ubiquity of these pollutants and their known presence in water supplies, and given some demonstrated toxicity associations, it would be reasonable to assume that chronic exposure to high levels of synthetic organic chemicals in water could be harmful. If they're in your water, you want to get them out.

Pesticides and Herbicides

The increased use of pesticides and herbicides in American agriculture since World War II

has had a profound effect on water quality. Rain and irrigation carry these deadly chemicals into groundwater as well as into surface waters. These are poisonous, plain and simple. Roundup weed killer, made by Monsanto, is the best known and most commonly used herbicide in the world. According to the most recent EPA estimates (from 2006 but published in 2011), in the US 185 millions pounds are used annually in agriculture and another 11 million pounds are used on lawns and gardens; those figures are probably higher today. Another way to look at Roundup's ubiquity is that in 2013, 83% of all corn and 93% of all soybeans grown in the US were genetically modified to be Roundup resistant. Numerous studies have shown Roundup to cause birth defects in laboratory animals, and there is some evidence that exposure to glyphosate (the active ingredient) might lead to conditions like Parkinson's disease, infertility, and cancer (huffingtonpost.com /2011/06/24/roundup-scientists-birth-defects_n _883578.html; naturalnews.com/041150_Monsan to_Roundup_glyphosate.html).

Asbestos

Asbestos exists in water as microscopic suspended fibers. Its primary source is asbestos-cement pipe, which was commonly used after World War II for city water systems. It has been estimated that some 200,000 miles of asbestos pipe are currently in use delivering drinking water. Because pipes wear as water flows through them, asbestos shows up with increasing frequency in municipal water supplies. It has been linked to gastrointestinal cancer.

Radionuclides

The Earth contains naturally occurring radioactive substances. Certain areas of the US exhibit relatively high background levels of radioactivity due to their geological characteristics. The three substances of concern to human health that show up in drinking water are uranium, radium, and radon (a gas). Various purification techniques can be effective at reducing the levels of radionuclides in water.

Testing Your Water

The first step in choosing a treatment method is to find out what contaminants are in your water. Different levels of testing are commercially available, including a comprehensive screening for nearly 100 substances.

Even before you test your water, a simple comparison can help you figure out what level of treatment your domestic water supply may need.

Chlorinated water has been linked to cancer, high blood pressure, and anemia. Chlorine has been banned for years in Europe and most industrialized countries due to its status as a known carcinogen.

It has been estimated that some 200,000 miles of asbestos pipe are currently in use delivering drinking water. Because pipes wear as water flows through them, asbestos shows up with increasing frequency in municipal water supplies. It has been linked to gastrointestinal cancer.

Why pay for bottled water forever when treatment will be cheaper? Bottled water typically costs $2 per gallon or more. Many purifiers produce clean water for just pennies a gallon.

Find some bottled water that you like. Note the normal levels of total dissolved solids (TDS), hardness, and pH, and also how high these levels range. If you don't find this information on the label, call the bottler and ask. Then get the same information about your tap water, which can be obtained by calling your municipal supplier. If your tap water has lower levels of these things than the bottled water, a good filter should satisfy your needs if you decide you want to treat it. If your tap water has higher levels than the bottled water you prefer, you may need more extensive purification.

If you're drinking water from a private well or spring, you'll have to get it tested to know what's in it. Some common problems with well water, such as staining, sediment, hydrogen sulfide (rotten egg smell), and excess iron or manganese, should be corrected before you purchase a point-of-use treatment system, because they will interfere with the effective operation of the system. These problems can be identified easily with a low-priced test; many state water-quality agencies will perform such tests for a nominal fee. If you have any reason to worry that your water may be polluted, we strongly advise you to conduct a more comprehensive test. You can get a water test kit from Real Goods at realgoods.com.

Water Choices: Bottled, Filtered, Purified

If you're not satisfied with your drinking water, you suspect it may be contaminated, or you've had it tested and you know it's polluted, you have two choices: You can buy bottled water, or you can install a point-of-use water treatment system in your home.

There are three kinds of bottled water: distilled, purified, and spring. Distillation evaporates the water, then recondenses it, thereby theoretically leaving all impurities behind (although some VOCs can pass right through this process with the water). Purified water is usually prepared by reverse osmosis, deionization, or a combination of both processes (see below for explanations). Spring water is usually acquired from a mountain spring or an artesian well, but it may be no more than processed tap water. Spring water generally will have higher total dissolved solids than purified water has. Distilled water and purified water are better for batteries and steam irons because of their lower content of TDS.

But why pay for bottled water forever when treatment will be cheaper? Bottled water typically costs $2 per gallon or more. Many purifiers produce clean water for just pennies a gallon. For example, the cost per gallon of water produced by the Multipure Water Guardian filter that Real Goods sells starts at about 35 cents and decreases the longer you own it. Certainly these systems will pay for themselves in cost savings within a few years.

The two basic processes used to clean water are filtration and purification. The word "filter" usually refers to a mechanical filter, which strains the water, and/or a carbon filter system, which reduces certain impurities by chemically bonding them—especially chlorine, lead, and many organic molecules. Purification refers to a slower process, such as reverse osmosis, that greatly reduces dissolved solids, hardness, and certain other impurities, as well as many organics. Many systems combine both processes. Make sure that the claims of any filter or purifier you consider have been verified by independent testing.

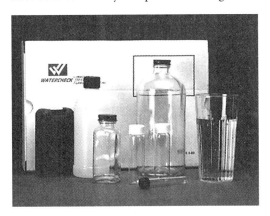

Methods of Water Filtration and Purification

The most efficient and cost-effective solution for water purity is to treat only the water you plan to consume. A point-of-use water treatment system eliminates the middleman costs associated with bottled water and can provide purified water for pennies per gallon. Devices for point-of-use water treatment are available in a variety of sizes, designs, and capabilities. Some systems improve only the water's taste and odor. Other systems reduce the various contaminants of health concern. Different systems work most effectively against certain contaminants. A system that utilizes more than one technology will protect you against a broader spectrum of biological pathogens and

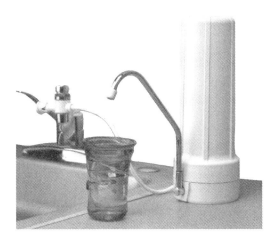

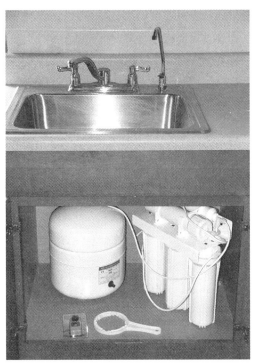

chemical impurities. Combining activated carbon filtration and reverse osmosis generally is considered the most complete and effective treatment.

When considering a treatment device, always pay careful attention to the independent documentation of the performance of the system for a broad range of contaminants. You should read the data sheets provided by the manufacturer carefully to verify its claims. Many companies are certified with the National Sanitation Foundation (NSF), whose circular logo appears on its data sheets.

The following is a brief analysis of the strengths and weaknesses of each option. Various options are available from Real Goods: real goods.com.

Mechanical Filtration

Mechanical filtration can be divided into two categories. Strainers are fine mesh screens that generally remove only the largest particles present in water. Sediment filters (or prefilters) remove smaller particles such as suspended dirt, sand, rust, and scale—in other words, turbidity. When enough of this particulate matter has accumulated, the filter is discarded. Sediment filters greatly improve the clarity and appeal of the water. They also reduce the load on any more expensive filters downstream, extending their useful life. Sediment filters will not remove the smallest particles or biological pathogens.

Activated Carbon Filtration

Carbon adsorption is the most widely sold method for home water treatment because of its ability to improve water by removing disagreeable tastes and odors, including objectionable chlorine. Activated carbon filters are an important piece of the purification process, but they are only one piece. Activated carbon effectively removes many chemicals and gases, and in some cases, it can be effective against microorganisms. However, generally it will not affect total dissolved solids, hardness, or heavy metals. Only a few carbon filter systems have been certified for the removal of lead, asbestos, VOCs, cysts, and coliform.

Each type of carbon filter system—granular activated carbon and solid block carbon—has advantages and disadvantages (see below).

Activated carbon is created from a variety of carbon-based materials in a high-temperature process that creates a matrix of millions of microscopic pores and crevices. One pound of activated carbon provides anywhere from 60 to 150 acres of surface area. The pores trap microscopic particles and large organic molecules, while the activated surface areas cling to, or adsorb, small organic molecules.

The ability of an activated carbon filter to remove certain microorganisms and certain organic chemicals—especially pesticides, THMs (the chlorine by-product), trichloroethylene (TCE), and PCBs—depends upon several factors:

1. The type of carbon and the amount used.
2. The design of the filter and the rate of water flow (contact time).
3. How long the filter has been in use.
4. The types of impurities the filter has removed previously.
5. Water conditions (turbidity, temperature, etc.).

Carbon adsorption is the most widely sold method for home water treatment because of its ability to improve water by removing disagreeable tastes and odors, including objectionable chlorine.

Granular Activated Carbon

Any granular activated carbon filter has three inherent problems. First, it can provide a base for the growth of bacteria. When the carbon is fresh, practically all organic impurities (not organic chemicals) and even some bacteria are removed. Accumulated impurities, though, can become food for bacteria, enabling them to multiply within the filter. A high concentration of bacteria is considered by some people to be a health hazard.

Second, chemical recontamination of granular activated carbon filters can occur in a similar way. If the filter is used beyond the point at which it becomes saturated with the organic impurities it has adsorbed, the trapped organics can release from the surface and recontaminate the water, with even higher concentrations of impurities than in the untreated water. This saturation point is impossible to predict.

Third, granular carbon filters are susceptible to channeling. Because the carbon grains are held (relatively) loosely in a bed, open paths can result from the buildup of impurities in the filter and rapid water movement under pressure through the unit. In this situation, contact time between the carbon and the water is reduced, and filtration is less effective.

To maximize the effectiveness of a granular activated carbon filter and avoid the possibility of biological or chemical recontamination, it must be kept scrupulously clean. That generally means routine replacement of the filter element at 6- to 12-month intervals, depending on usage.

Solid Block Carbon

These filters are created by compressing very fine pulverized activated carbon with a binding medium and fusing the composite into a solid block. The intricate maze developed within the block ensures complete contact with organic impurities and, therefore, effective removal. Solid block carbon filters avoid the drawbacks of granular carbon filters.

Block filters can be fabricated with a porous structure fine enough to filter out coliform and other disease bacteria, pathogenic cysts such as giardia, and lighter-weight volatile organic compounds such as THMs. Block filters eliminate the problem of channeling. They are also dense enough to prevent the growth of bacteria within the filter.

Compressed carbon filters have two primary disadvantages compared with granular carbon filters. They have smaller capacity for a given size, because some of the adsorption surface is taken up by the inert binding agent, and they tend to plug up with particulate matter.

Thus, block filters may need to be replaced more frequently. In addition, block filters are substantially more expensive than granular carbon filters.

Limitations of Carbon Filters

To summarize, a properly designed carbon filter is capable of removing many toxic organic contaminants, but it will fall short of providing protection against a wide spectrum of impurities.

1. Carbon filters are not capable of removing excess total dissolved solids (TDS). To gloss over this deficiency, many manufacturers and sellers of these systems assert that minerals in drinking water are essential for good health. Such claims are debatable. However, some scientific evidence suggests that minerals associated with water hardness may have some preventive effect against cardiovascular disease.
2. Only a few systems have been certified for the removal of cysts, coliform, and other bacteria.
3. Carbon filters have no effect on harmful nitrates or on high sodium or fluoride levels.
4. With both granular and block filters, the water must pass through the carbon slowly enough to ensure complete contact and filtration. An effective system therefore has to have an appropriate balance between a useful flow rate and adequate contact time.

Reverse Osmosis (Ultrafiltration)

Reverse osmosis (RO) is a water purification technology that utilizes normal household water pressure to force water through a selective semipermeable membrane that separates contaminants from the water. Treated water emerges from the other side of the membrane, and the accumulated impurities left behind are washed away. Sediment eventually builds up along the membrane, and then it needs to be replaced.

Reverse osmosis (RO) is highly effective in removing several impurities from water: total dissolved solids (TDS), turbidity, asbestos, lead and other heavy metals, radium, and many dissolved organics. RO is less effective against other substances. The process will remove some pesticides (chlorinated ones and organophosphates, but not others) and most of the heavier VOCs. However, RO is not effective at removing lightweight VOCs such as THMs (the chlorine by-product) and TCE (trichloroethylene), and certain pesticides.

These compounds are either too small, too light, or of the wrong chemical structure to be screened out by an RO membrane. RO is also the best way to remove fluoride from drinking water, although this is not 100% effective.

Reverse osmosis and activated carbon filtration are complementary processes. Combining them results in the most effective treatment against the broadest range of water impurities and contaminants. Many RO systems incorporate both a pre-filter of some sort and an activated carbon post-filter.

RO systems have two major drawbacks. First, they waste a large amount of water. They'll use

anywhere from 3 to 9 gallons of water per gallon of purified water produced. This could be a problem in areas where conservation is a concern, and it may be slightly expensive if you're paying for municipal water. On the other hand, this wastewater can be recovered or redirected for purposes other than drinking, such as watering the garden or washing the car. Second, reverse osmosis treats water slowly: It takes about three to four hours for a residential RO unit to produce one gallon of purified water. Treated water can be removed and stored for later use.

Other Treatment Processes

Three other water treatment processes are worth knowing about. Distillation is a process that creates clean water by evaporation and condensation. Distillation is effective against microorganisms, sediment, particulate matter, and heavy metals; it will not treat organic chemicals. Good distillers will have a carbon filter to remove organic chemicals.

Ultraviolet (UV) systems use UV light to kill microorganisms. These systems can be highly effective against bacteria and other organisms; however, they may not be effective against giardia and other cysts, so any UV system you buy should also include a 0.5-micron filter. Other than some moderately expensive solar-powered distillers on the market, any distiller or UV unit will require a power source.

Ozone is a strong oxidizer that is also used to remove microorganisms and other organic matter from water. Ozonators are usually combined with carbon or other filters in a water cleaning system. These systems are much more common in commercial rather than residential applications.

A Final Word on Water

Our best advice for those of you considering a drinking water treatment system consists of three simple points.

1. Test your water so you know precisely what impurities and contaminants you're dealing with.
2. Buy a system that is designed to treat the problems you have.
3. Carefully check the data sheets provided by the manufacturer to make sure that claims about what the system treats effectively have been verified by the National Sanitation Foundation (NSF) or another third party.

These actions will enable you to purchase a system that will most effectively meet your particular needs for safe drinking water. Products for testing and filtering your water are available at realgoods.com.

Reverse osmosis (RO) is highly effective in removing several impurities from water: total dissolved solids (TDS), turbidity, asbestos, lead and other heavy metals, radium, and many dissolved organics.

Air Purification

Most of us are familiar with the idea that the water we drink at home might be polluted or contaminated. But did you realize that the air inside your home (or place of work) potentially could be even more hazardous to your health than the water?

The Problem with Indoor Air

The problem of indoor air quality has attracted attention only in the last few decades. One reason is that the sources of indoor air pollution are more mundane and subtle than the sources of outdoor air and water pollution. In addition,

The combination of high-performance buildings that allow minimal infiltration of outside air and the continual increase of synthetic products that we bring into our homes has added up to a possible public health problem of significant proportions.

before highly insulated, tightly constructed buildings became common in the 1970s, most American homes were drafty enough that the buildup of harmful gases, particles, and biological irritants indoors was unlikely. However, the combination of high-performance buildings that allow minimal infiltration of outside air and the continual increase of synthetic products that we bring into our homes has added up to a possible public health problem of significant proportions.

Little is known with relative certainty about the unhealthful consequences of constant exposure to polluted indoor air. Nonetheless, the EPA has concluded that many of us receive greater exposure to pollutants indoors than outdoors. EPA studies have found concentrations of a dozen common organic pollutants to be two to five times higher inside homes than outside, in both rural and industrialized areas. Furthermore, people who spend more of their time indoors and are thus exposed for longer periods to airborne contaminants are often the very people considered most susceptible to poor indoor air quality: very young children, the elderly, and the chronically ill.

Sources of Indoor Air Pollution

You can probably identify some of the sources of indoor air pollution: new carpeting, adhesives, cigarette smoke, dust mites. But many things that contribute potentially irritating or harmful substances to indoor air may not be obvious. Indoor air contaminants typically fall into one or more of the following categories:

- Combustion products
- Volatile chemicals and mixtures
- Respirable particulates
- Respiratory products
- Biological agents
- Radionuclides (radon and its by-products)
- Odors

What kinds of substances are we talking about? Let's start with chemicals and synthetics. Building materials and interior furnishings emit a broad array of gaseous volatile organic compounds (VOCs), especially when they are new. These materials include adhesives, carpeting, fabrics, vinyl floor tiles, some ceiling tiles, upholstery, vinyl wallpaper, particleboard, drapery, caulking compounds, paints and stains, and solvents. In addition to producing emissions, many of these materials also act as "sponges," absorbing VOCs and other gases that can then be reintroduced into the air when the "sponge" is saturated. These kinds of building and furnishing materials also

Breathe Easier Without Indoor Pollutants

A five-year EPA study found the concentrations of 20 toxic compounds to be as much as 200 times higher in the air inside homes and offices than outdoors. Poor air circulation is partly to blame. Real Goods' air filters can remove hazardous pollutants, as well as dust, pollen, and pet dander, so the air you breathe is cleaner (realgoods.com).

can be the source of respirable particulates, such as asbestos, fiberglass, and dusts.

Appliances, office equipment, and office supplies are another major source of VOCs and particulates. Among the culprits are leaky or unvented heating and cooking appliances; computers and televisions; laser printers, copiers, and other devices that use chemical supplies; preprinted paper forms and rubber cement; and routine cleaning and maintenance supplies, including carpet shampoos, detergents, floor waxes, furniture polishes, and room deodorizers.

Your own routine "cleaning and maintenance supplies" are not above suspicion, either. Many of the personal care and home cleaning products we use emit various chemicals (especially from the components that create fragrance), some of which may contribute to poor indoor air quality. Clothes returning from the dry cleaner may retain solvent residues, and studies have shown that people do breathe low levels of these fumes when they wear dry-cleaned clothing. Even worse, the pesticides we use to control roaches, termites, ants, fleas, and other insects are by definition toxic. Sometimes these poisons are tracked indoors on shoes and clothing. Do you have any of this stuff stored underneath the sink? One EPA study has suggested that up to 80% of most people's exposure to airborne pesticides occurs indoors.

Other human activities can also have a dramatic impact on indoor air quality. Tobacco smoke from cigarettes, cigars, and pipes is obviously an unhealthy pollutant. Household heating and cooking activities may introduce carbon monoxide, carbon dioxide, nitrogen oxide and dioxide, and sulfur dioxide into the air. Candles and wood fires contribute CO and CO_2 as well as various particulates.

Many biological agents contribute to indoor pollution: We are surrounded by a sea of microorganisms. Bacteria, insects, molds, fungi, protozoans, viruses, plants, and pets generate a number of substances that contaminate the air. These range from whole organisms themselves to feces, dust, skin particles, pollens, spores, and even some toxins. Humans contribute too: We exhale microbes, and our sloughed skin is the primary source of food for dust mites, which can cause asthma and allergies for many people. Furthermore, certain indoor environments, such as ducts and vents, or humid areas such as bathrooms and basements, provide ideal conditions for microbial growth.

Radon (and its decay products, known as "radon daughters") is one type of indoor pollutant that has received widespread notice. As noted above, indoor radon buildup can be a problem in certain parts of the US where it is present in rocks and soil. A radioactive gas, radon has been implicated as a cause of lung cancer. Simple and inexpensive radon test kits are available at hardware stores. Contact your regional EPA office—you can look it up in the phone book or online—for more information about effective radon remediation strategies.

Odors are always present. Some are pleasant, others are unpleasant; some signal the presence of irritating or harmful substances, others do not. The perception of odors is highly subjective and varies from person to person. Most odors are harmless, more of a threat to state of mind than to health, but eliminating them from an indoor environment may be a high priority to keep people content and productive.

Health Effects of Indoor Air Pollution

Before you become too alarmed about the health dangers that may be floating around in the air you breathe at home, remember the following proviso. As with most environmental pollutants, adverse health effects depend upon both the dose received and the duration of exposure. The combination of these factors, plus an individual's particular sensitivity to specific substances, will determine the potential health effect of the contaminant on that individual.

With that in mind, the health effects of indoor air pollution can be short term or long term and can range from mildly irritating to severe, including respiratory illness, heart disease, and cancer. Most often the effects of such pollutants are acute, meaning they occur only in the presence of the substance; but in some cases, and perhaps in many cases, the effects may be cumulative over time and may result in a chronic condition or illness.

The clinical effects of indoor air pollution can take many forms. The most common clinical signs include eye irritation, sneezing or coughing, asthma attacks, ear-nose-throat infections, allergies, and migraines. More seriously, prolonged exposure to radon, tobacco smoke, and other carcinogens in the air can cause cancer. Most volatile organic compounds can be respiratory irritants, and many are toxic, although little applicable toxicity data exists about low levels of indoor VOCs. One VOC, formaldehyde, is now considered a likely human carcinogen. Needless to say, the best way to protect your health is to avoid or reduce your exposure to indoor air contaminants.

Nearly all observers agree that more research is needed to better understand which health effects occur after exposure to which indoor air pollutants, at what levels, and for how long.

Improving Indoor Air Quality

There are three basic strategies for improving indoor air quality: source control, ventilation, and air cleaning.

Source control is the simplest, cheapest, and most obvious strategy for improving indoor air quality. Remove sources of air pollution from your home. Store paint, pesticides, and the like in a garage or other outbuilding. Read the labels on the products you bring into your home, and buy personal care, cleaning, and office products that contain fewer synthetic volatile compounds or other suspicious and potentially irritating substances.

Other sources can be modified to reduce their emissions. Pipes, ducts, and surfaces can be enclosed or sealed off. Leaky or inefficient appliances, such as a gas stove, can be fixed.

Regular house cleaning is effective: Sweeping, vacuuming, and dusting can keep a broad spectrum of dust, insects, and allergens under control. You should be conscientious about cleaning out ducts and replacing furnace filters on a regular basis. Also, reduce excess moisture and humidity, which encourage the growth of microorganisms, by emptying the evaporation trays of your refrigerator and dehumidifier.

Ventilation is another obvious and effective way to reduce indoor concentrations of air pollutants. It can be as simple as opening windows and doors, operating a fan or air conditioner, and making sure your attic and crawl spaces are properly ventilated. Fans in the kitchen or bathroom that exhaust to the outdoors remove moisture and contaminants directly from those rooms. It is especially important to provide adequate ventilation during short-term activities that generate

A five-year EPA study found the concentrations of 20 toxic compounds to be as much as 200 times higher in the air inside homes and offices than outdoors.

Nearly all observers agree that more research is needed to better understand which health effects occur after exposure to which indoor air pollutants, at what levels, and for how long.

It is especially important to provide adequate ventilation during short-term activities that generate high levels of pollutants, such as painting, sanding, or cooking.

high levels of pollutants, such as painting, sanding, or cooking. Energy-efficient air-to-air heat exchangers that increase the flow of outdoor air into a home without undesirable infiltration have been available for several years. Such a system is especially valuable in a tight, weatherized house. See Chapter 6, page 188, for more information.

Unfortunately, indoor air cleaning or purification is more complicated than purifying water. Not many technologies are available for household use, most of them have limited effectiveness, and some of the more promising methods are controversial. Air filtration is problematic because it's only really effective if you draw all the indoor air through the filter. Even then, such filters reduce only particulate matter. Any furnace or whole-house air conditioning system has a filter. Some of them, such as electrostatic filters or HEPA (High Efficiency Particulate Air) filters, can reduce levels of even relatively small particulates significantly. Portable devices incorporating these technologies have also been developed, which can be used to clean the air in one room. Some of them have EnergyStar ratings, which means they'll consume less power than other models. Follow the manufacturer's instructions about how frequently to change the filter on your furnace or stand-alone air cleaner. However, be aware that air filters are not designed to deal with volatile organic compounds or other gaseous pollutants.

Ionization is a second method of cleaning particulates out of the air. A point-source device generates a stream of charged ions (via radio frequencies or some other method), which disperse through the air and meet up with oppositely charged particles. As tiny particles clump together, they eventually become heavy enough to settle out of the air, and you won't breathe them in. But like filtration, ionization is not effective against VOCs. Check out the air purification systems available on the Real Goods website: realgoods.com.

What About Ozone?

Another method of air purification uses ozone, in combination with ionization. Ozone (O_3) is a highly reactive and corrosive form of oxygen. You know about the ozone layer in the upper atmosphere, which prevents a lot of ultraviolet light from reaching Earth's surface where it can damage living organisms. Ozone also exists naturally in small concentrations at ground level, where it has various effects. It is a component of smog, produced by the reaction of sunlight upon auto exhaust and industrial pollution. Ozone is also produced by lightning and is partially responsible for that fresh, clean smell you notice outside after a thunderstorm. Some studies suggest that slightly greater-than-natural outdoor levels of ozone can reduce respiratory, allergy, and headache problems, and possibly even enhance one's ability to concentrate. However, scientists agree that at slightly elevated levels in the air, ozone can be dangerous and unhealthful to people. In concentrations that exceed safe levels as determined by the EPA, ozone is considered an air pollutant.

Manufacturers of air cleaners that operate as

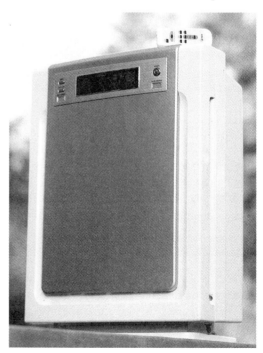

ozone generators make strong claims about their products. They assert that these devices kill biological agents, such as molds and bacteria, that contribute to indoor air pollution; that the ozone penetrates into drapes and furniture and filters down to the floor into carpets and crevices to oxidize and destroy absorbed contaminants and those that fall out of the air; and that the ozone breaks down potentially harmful or irritating gases, including many VOCs, into carbon dioxide, oxygen, water vapor, and other harmless substances. They claim that their machines accomplish all this with levels of ozone that are safe, and that ozone production can be monitored, either automatically or manually, to ensure that ozone concentration does not become excessive.

The EPA strongly disputes many of these claims and considers them misleading. The agency's position is that "'available' scientific evidence shows that at concentrations that do not exceed public health standards, ozone has little potential to remove indoor air contaminants [and] does not effectively remove viruses, bacteria, mold, or other biological pollutants." While the EPA acknowledges that some evidence shows that ozone can combat VOCs, it points out that these reactions produce potentially irritating or harmful by-products. In addition, the EPA is skeptical of manufacturer's claims that the amount of ozone produced by such devices can be controlled effectively so as not to exceed safe levels. The bottom line for the EPA is that "no

> Manufacturers of air cleaners that operate as ozone generators claim ozone kills biological agents, such as molds and bacteria, that contribute to indoor air pollution; oxidizes and destroys contaminants absorbed by drapes and furniture; and breaks down potentially harmful or irritating gases, including many VOCs—all this with levels of ozone that are safe and can be monitored. The EPA strongly disputes many of these claims and considers them misleading.

agency of the federal government has approved ozone generation devices for use in occupied spaces. The same chemical properties that allow high concentrations of ozone to react with organic material outside the body give it the ability to react with similar organic material that makes up the body and potentially cause harmful health consequences. When inhaled, ozone will damage the lungs. Relatively low amounts can cause chest pain, coughing, shortness of breath, and throat irritation." (This information is posted on the EPA website, at epa.gov/iaq/pubs/ozonegen.html.)

The ozone issue is still being debated, and at present, Real Goods does not carry ozone generators.

Resources

The best source of up-to-date information about indoor air pollution and residential air cleaning systems is the federal Environmental Protection Agency. The EPA maintains several useful telephone services and distributes, free of charge, a variety of helpful publications. You can view many of these publications online at epa.gov/iaq/pubs/. You can also request these publications from the regional EPA office in your area, which you can find in your local telephone book under US Government, Environmental Protection Agency, or online. The following resources may be particularly valuable.

The Inside Story: A Guide to Indoor Air Quality, publication #402K93007 (almost 15 years old, but still relevant according to the EPA).

Guide to Air Cleaners in the Home, publication #402-F-08-004 (2008).

Residential Air Cleaning Devices: A Summary of Available Information, publication #402-F-09-002 (2009).

National Service Center for Environmental Publications and Information
epa.gov/ncepihom/

Indoor Air Quality Information Clearinghouse
epa.gov/iaq/iaqxline.html

National Radon Hotline
800-SOS-RADON (767-7236); information recording operates 24 hours a day
epa.gov/radon/rnxlines.html

CHAPTER 10

Composting Toilets and Greywater Systems

Novel, Safe, and Clean Solutions to an Age-old Human Issue

Actually, all pollution is simply an unused resource.
Garbage is the only raw material that we're too stupid to use.
ARTHUR C. CLARKE

Every house should be surrounded by an oasis of biological productivity
nourished by the flow of nutrients and water from the home.
ARTHUR C. LUDWIG

Get your shit together!
UNKNOWN HIPPIE FROM THE 1970S

The greywater produced by an average family (from the sink, shower, tub, and washer) is about enough to irrigate all the fruit trees that can fit around a house on a suburban lot. In a cold climate, heat from greywater can dramatically increase production in an attached passive solar greenhouse. The excreta (urine and feces) from a family contains nearly the same amount of nutrients as all the fertilizer used to grow the food to feed that family.

Combined and concentrated, greywater and excreta can become an expensive waste disposal problem when treated in the conventional manner. In most communities, sewage disposal is more expensive and energy intensive than providing a potable water supply. This high economic and ecological expense is revenue for the sewage-industrial complex, which lobbies heavily for codes to mandate the most inherently expensive solutions for all plumbing problems, without regard to their external costs.

Regulators, for the most part, are unaware that they are being used as victims to shake down the citizenry. They sincerely believe they are protecting health. (If your system requires a permit, please treat regulators with respect and compassion, no matter how exasperating the rules are.)

Well managed, greywater and excreta are actually valuable assets. Implementing decent management of greywater and excreta is quite easy. Really good management can be more complex.

Well managed, greywater and excreta are actually valuable assets.

This chapter is newly revised by Art Ludwig, author of *Create An Oasis with Greywater* and the *Builder's Greywater Guide*. Art is an ecological designer who consults internationally on the design of water and wastewater systems, emphasizing integrated "systems of systems." He has designed four new types of greywater systems, helped to write greywater codes in several states, and formulated the first plant and soil biocompatible laundry detergents.

A real two-story outhouse from the early 1900s.

Ecological systems design is about context, and integration between systems. The entirety of integrated, ecological design can be reduced to one sentence:

Do what's appropriate for the context.

Ecological systems—rainwater harvesting, runoff management, passive solar, composting toilets, edible landscaping—are more context-sensitive than their counterparts in conventional practice; that's most of what makes them more ecological. They are also more interconnected.

Greywater systems in particular are more context-sensitive than any other manmade ecological system, and more connected to more other systems. They are like the keystone species of household systems.

What this means is that, get the greywater just right and you've got the whole package right. Which also means that if you take the trouble to get the greywater just right, along the way you'll have gotten right your water supply, water efficient fixtures, hot water, rainwater and runoff management, use of yard waste and compost, sanitation, composting toilets, irrigation and salt management in the soil, edible landscaping, microclimate modification, permaculture zonation, and solar orientation—all this will be just right as well.

Clearly an in-depth examination of all of these interconnected systems is beyond the scope of this short article, but we'll connect you with information and hardware that help you get your system of systems working together synergistically.

Composting toilets can close the nutrient cycle, turning a dangerous waste product into safe compost, without smell, hassle, or fly problems. They are usually less expensive than installing a conventional septic system, and they will reduce household water consumption by at least 25%.

Reclaiming Valuable Resources

In our rush to sanitize everything in sight, we have ended up throwing away a potentially valuable and money-saving resource, and we over-design expensive and energy-intensive disposal systems that pollute surface and groundwater. Our conventional plumbing systems mix a few pounds of valuable nutrients and a few micrograms of potentially dangerous pathogens with hundreds of gallons of very lightly polluted greywater from our sink, shower, tub, and washer. By going for the highest revenue solution, we've taken a small problem and made it much larger, more difficult, and more expensive, especially in long term, external costs.

Outhouses can smell bad in the summer and feel too close to the house, and allow flies to spread filth and disease. In the winter, they don't smell, but feel too far away.

We're going to talk about what happens if we separate the nutrients from the water, and compost the excreta until the pathogens are gone. Greywater can be used directly for landscape and garden watering. Human excreta can be composted safely and odorlessly to kill pathogens, then used as a highly beneficial compost.

Composting toilets can close the nutrient cycle, turning a dangerous waste product into safe compost, without smell, hassle, or fly problems. They are usually less expensive than installing a conventional septic system, and they will reduce household water consumption by at least 25%.

What Is a Composting Toilet?

A composting toilet in the broadest sense of the term is any system used for managing human feces and urine with zero or very minimal water for flushing, and no connection to a sewer or septic system. There are dozens of different types, each with its particular advantages and disadvantages.

We're going to focus on modular, manufactured composting toilets.

Composting toilets (for brevity we're going to refer to modular manufactured composting toilets simply as "composting toilets") have several advantages: small size, a pre-engineered design, reliable operation, and the creation of high-quality compost. The disadvantages include, in higher-capacity units, the use of electricity for ventilation and warming, as well as the use of plastic for the toilet body. Low use is not a problem with this kind of toilet (except if the composting chamber is cold). Adequate sizing is important: Sustained overuse will be a problem.

Overall, modular manufactured composting toilets are ideally suited for light or occasional use, and less suited to heavy continuous use.

This type of composting toilet is basically a warm well-ventilated container with a diverse community of aerobic microbes living inside that break down the urine and feces. (We are trying to roll back the mischaracterization of excreta as human "waste" or greywater as "waste" water in our language and culture.)

People are invariably amazed at the quality of the finished compost. The process creates a rich, fluffy compost similar to that from a well-maintained garden compost pile or the composted leaf litter from an old-growth forest. If you are someone who can appreciate rich soil, you may find your senses calling you to lift a double handful to your face and inhale while letting it crumble deliciously through your fingers…even as your mind is saying, "hang on a second: a year ago this was.…"

Flowers and fruit trees love it, though we don't recommend using this compost on kitchen gardens, as some human pathogens possibly could survive the composting process.

The composting process in such a toilet does not smell. Rapid aerobic decomposition—active composting—which takes place in the presence of oxygen, at relatively high temperatures, is the opposite of the slow, sometimes smelly process that takes place in an outhouse, which works by anaerobic decomposition and/or slow aerobic moldering. Anaerobic microbes cannot survive in the presence of oxygen and the more energetic microbes that flourish in an oxygen-rich environment. If a composting toilet smells bad, it means something is wrong. Usually smells indicate pockets of anaerobic activity caused by inadequate mixing.

How Do Composting Toilets Work?

A composting toilet has three basic elements: a place to sit, a composting chamber, and an evaporation tray. Many models combine all three elements in a single enclosure, although some models have separate seating, with the composting chamber installed in the basement or under the house. In either case, the evaporation tray is

A modern composting toilet.

Composting Toilet Basics

The manufactured modular composting toilets we sell use rapid aerobic decomposition, like a well-turned garden compost pile, to break down wastes. Almost all the material that goes into the composting toilet disappears up the vent as water vapor or gases.

Why would I want one?
Composting toilets can provide safe waste processing in locations where conventional septic systems are impossible or ill advised, such as lakeside cabins, sites with very slow or very fast perk, high water tables, or other issues. They are often the most environmentally friendly method of managing excreta.

Do they smell?
Forget your outhouse experiences. If a composting toilet smells, it's telling you something's wrong.

What's left, and what do I do with it at the end of composting?
There's a dry, fluffy, odorless material in the finishing drawer. Once every few months you pull out the entire drawer, and carry it out to your fruit trees or ornamental plants. It's compost! It's good for them!

When is a composter not a good idea?
Apartment dwellers, this isn't for you. Intermittent use in cold environments is a problem, too. The biological community in a composter likes 60°F or warmer. Below 50°F the little critters go into hibernation and all composting activity stops. Composting toilets can freeze without harm, but obviously you can't use them then, except for very occasional use as a storage tank.

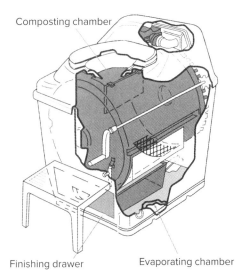

A typical Sun-Mar composting toilet.

Composting chamber

Finishing drawer

Evaporating chamber

positioned under the composting chamber to catch any liquids, and some sort of removable finishing drawer is supplied to carry off the finished compost material.

In order for the most effective composting to occur, the right mix of oxygen, water, temperature range, and nutrients is required, just like a compost pile. These factors all affect each other. We'll consider each in turn.

Oxygen
Smaller composters usually employ some kind of mixing or stirring mechanism to ensure adequate oxygen to all parts of the pile and faster composting action, which raises the temperature.

Water
Too much or too little water will stop the composting process. In these kinds of toilets, too

much water is the one to worry about. Ninety percent of what goes into a composting toilet is water. Compost piles need to be damp to work well, but many composting toilets suffer from too much water. Evaporation is the primary way to rid a composting toilet of excess water. If evaporation can't keep up, then many units have an overflow that is plumbed to the household greywater or septic system. Heat and air flowing through the unit assist the evaporation process. Every composting toilet has a vertical vent pipe to carry off moisture. Air flows across the drying trays, around and through the pile, then up the vent to the outside of the building.

The low-grade heat produced by composting helps to provide sufficient updraft to carry vapor up the vent. However, like any passive vent with minimal heat, these are subject to downdrafts. Electric composters include vent fans as standard equipment. All manufacturers include, or offer, an optional vent fan with their nonelectric models that can be battery- or solar-driven.

Temperature
Composting toilets work best at temperatures of 70°F or higher; at temperatures below 60°F the biological process slows to a crawl, and at temperatures below 50°F, it comes to a stop. The composting action itself will provide some low-level heat, but not enough to keep the process going in a cold environment. Electric models include a small heating element as standard equipment. It is okay to let a composting toilet freeze, for example, in a summer cottage over the winter, although it shouldn't be used when cold or frozen.

A Historic Composting Toilet: The Clivus Multrum

The earliest composter designs, such as the Clivus Multrum system, use a lower-temperature decomposition process known as moldering, which takes place slowly over several years. These composters have air channels and fan-driven vents, but they lack supplemental heat or the capability of mixing and stirring. Therefore, these very large composters don't promote highly active composting. They employ a different strategy for liquids. These are periodically removed by hand or pumped out. In these units a large proportion of the nutrients are in this liquid, which is fairly innocuous, relatively raw urine. This type of large, slow composter is most effective in public access sites, and many are currently in successful use at state or national parks. Their large bulk lets them both absorb sudden surges of use, and weather long periods of disuse without upset.

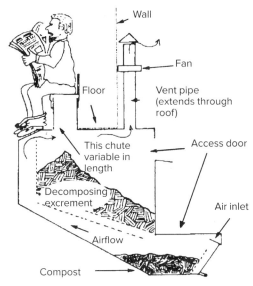

Wall

Fan

Floor

Vent pipe (extends through roof)

This chute variable in length

Access door

Decomposing excrement

Air inlet

Airflow

Compost

An early Multrum toilet—the first of the modern generation of composting toilets.

REAL GOODS

Normal biological activity will resume when the temperature rises again.

Nutrient Mix

The main factor is the ratio of carbon to nitrogen. Thirty to one will produce the fastest composting. Other ratios will work, but more slowly. Every one to three days, the owner should add a cup of high-carbon-content bulking agent, such as peat moss, wood chips, or dry-popped popcorn. This helps soak up excess moisture, makes lots of little wicks to aid in evaporation, and creates air passages that prevent anaerobic pockets from forming.

Manufacturer Variations

Every compost toilet manufacturer will be delighted to tell you why their unit is the best. We'll attempt to take an impartial attitude and give the pluses and minuses of each design.

First, some general comments. Smaller composters certainly cost less, but because the pile is smaller they are more susceptible than larger models to all the problems that can plague any compost pile, such as liquid accumulation, insect infestations, low temperatures, and an unbalanced carbon/nitrogen ratio. Smaller composters require the user to take a more active role in the day-to-day maintenance of the unit. We have found that the smaller units with electric fans and thermostatically controlled heaters have far fewer problems than the nonelectric units. Choosing the best system for your context will reward you with better performance.

Sun-Mar Composting Toilets

Sun-Mar is the largest and most experienced of the small residential-size composting toilet manufacturers. Their composting toilets use a rotating drum design, like a clothes dryer, that allows the entire compost chamber to be turned and mixed easily—and remotely. This ensures that all parts of the pile get enough oxygen, and that no anaerobic pockets form. Routine once- or twice-a-week mixing also tumbles the newer material in with older material, so the microorganisms can get to work on recent additions more quickly. Periodically, as the drum approaches one-half to two-thirds full, a lock is manually released and the drum is rotated backwards. This action lets a portion of the composting material drop from the drum into a finishing drawer at the bottom of the unit. The next time the drum needs a portion moved along, the finishing drawer, which now contains a load of finished compost, is emptied on your orchard, ornamental plants, or selected parts of your garden, reinstalled, and a fresh load dumped in.

Sun-Mar has a large Centrex 3000 model that has a composting drum twice as long as anything we've seen, with stacked drying trays. This design doubles the solids capacity, triples the drying tray area, and doubles the time that material can spend fully composting, alleviating many of the more common complaints we've heard about Sun-Mar units. (See realgoods.com.)

The main disadvantage of the Sun-Mar drum system is that fresh material can be included with the material dumped into the finishing drawer, and as a result, some pathogens may survive the composting process. Also, the limited size of most Sun-Mar composters means they work best for intermittent-use cabins or small households. With the exception of the large Centrex 3000 model, these composters can on occasion be overwhelmed by full-time use in larger households.

Sun-Mar produces both fully self-contained composting toilets and centralized units, in which the compost chamber is located outside the bathroom and connected to either an air-flush toilet or an ultra-low-flush toilet with standard 3-inch waste pipe. Electric models all feature thermostatically controlled heating elements to keep the microbe community happily active and assist with evaporation, and a small fan to ensure fresh airflow and negative air pressure inside the compost chamber. We strongly recommend the electric models for customers who have utility power available. For those with intermittent AC power, AC/DC models are available that take advantage of AC power when it's available but can operate adequately without it as well. Sun-Mar also makes nonelectric models that don't need any power, although use of these models should be limited to one or two people or a low-use cabin.

EcoTech Carousel Composting Toilets

The EcoTech Carousel is a very large composter for full-time residential use. It features a round carousel design with four separate composting chambers. Like most composters, it has drying trays under the composting chamber and a fan-assisted venting system to ensure odorless operations. Each of the four chambers is filled in turn, which can take from two to six months; then a new chamber is rotated under the dry toilet.

This gives each of the four batches up to two years to finish composting without being disturbed. Batch composting has the advantage of not mixing new material with older, more advanced compost, allowing the natural ecological

It is okay to let a composting toilet freeze, for example, in a summer cottage over the winter, although it shouldn't be used when cold or frozen. Normal biological activity will resume when the temperature rises again.

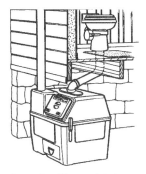

Some Sun-Mar models have the toilet separated from the composting chamber.

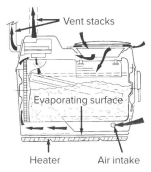

Cross-section of a typical Sun-Mar composter.

Sun-Mar Composting Toilets: Saving the Planet One Toilet at a Time

With toilets on all seven continents —including sites in Antarctica and Mount Everest base camp—the Sun-Mar company has single-handedly conserved more than 2 billion gallons of water.

That's why, since 1985, Real Goods has been proud to feature products by Sun-Mar, the world leader in consumer composting. This remarkable company was founded in the mid-1960s in Sweden with a mission to create waterless toilets off the traditional septic grid, using Mother Nature instead of hazardous chemicals to compost human waste into a healthy and usable product.

Based in Canada, Sun-Mar has been in the composting toilet business longer than almost anyone. Sun-Mar designs are all based on a rotating drum, like a clothes dryer. Turning a crank on the outside of the composter rotates the drum, which ensures complete mixing and good aeration of the "pile." A well-turned compost pile works faster and more effectively. Mixing should happen once or twice a week whenever the composter is in active use. When the drum is rotated in the normal mixing direction, the inlet flap swings shut, and all material stays in the drum. Periodically, when the drum approaches one-half to two-thirds full, a lock is released that allows the drum to rotate backwards. This lets some compost drop into the finishing tray at the bottom. It finishes composting here until the next time the drum needs to dump some material.

The fully self-contained Sun-Mar composters, the Excel and Compact models, are good choices for summer camps, sites with intermittent use, sites without electricity (excel NE only), or full-time residences with winter heating (composting drums must be in a 65°F or warmer space to function properly).

The Centrex series puts the composting drum outside or in the basement for those who want a more traditional-looking toilet or toilets in the bathroom. Centrex models are available for a full range of uses from intermittent summer camps to full-time residential use. The entire line is ANSI/NSF listed now, which will make it easier for local health officials to accept their use.

If you have utility electricity available at your site, we strongly recommend using one of the standard AC models with fan and heater. They compost faster, are more tolerant of variable uses, and are much less likely to suffer any performance problems. This electrical advice applies to all composting toilet models.

Smelling Like a Rose

As you might imagine, Sun-Mar's customer service team gets some interesting queries and comments—most notably,"what about the smell?" Answer: It's a non-issue with Sun-Mar composting toilets. Aerobic decomposition produces no sewer gases, and a partial vacuum is maintained by the venting system to ensure odor-free operation.

Sun-Mar composting toilets are available at real goods.com.

Santerra Composting Toilets

Santerra Green Systems are some of the most advanced composting toilet systems available today. A patented Automatic Six-Way Aeration™ process differentiates Santerra Green from other composting toilets. The automatic aeration process speeds up the composting and evaporation process significantly through forced-air, heat, and natural microbe action. And, unlike others, there is no drum to manually turn.

Santerra Green Systems come "plug & play," ready to install out of the box. And, to make it easier, the toilet (waterless, low water, or vacuum flush) is included in the system price. Installation is easy and is usually completed in a few hours.

Santerra Green offers two slab-level installation options: 1) X Series models are waterless and all-in-one. They install right on the bathroom floor; 2) FlushSmart V Series models can flush up, down, or sideways, avoiding any issues with at- or below-grade installations.

Santerra Green can provide an economical and environment-friendly sanitation solution to countless applications. These composting toilets are perfect for use in cottages, cabins, chalets, farms and barns, homes, pool cabanas, modern homes, prefabricated structures, public facilities, warehouses, and yurts.

The Carousel Composter for full-time and larger family use.

Carbon/Nitrogen Ratios

Proper composting requires a balance of carbon and nitrogen in the organic material being composted. The ideal balance is 30:1 carbon to nitrogen. Human excreta are not properly balanced as they are too high in nitrogen. They require a carbon material to be added for the encouragement of rapid and thorough microbial decomposition. In the mid-1800s, the concept of balancing carbon and nitrogen was not known, and the high nitrogen content of humanure in dry toilets prevented the organic material from efficiently decomposing. The result was a foul, fly-attracting stench. It was thought that this problem could be alleviated by segregating urine from feces (which thereby reduced the nitrogen content of the fecal material), and dry toilets were devised to do just that. Today, one popular class of composting toilets segregates urine from feces. In the other classes of composting toilets, the simple addition of a carbonaceous material to the feces/urine mix will balance the nitrogen of the material and render the segregation of urine unnecessary

—from *The Humanure Handbook*, 2nd edition, Joe Jenkins. Used with permission.

cascading of composting processes. This results in more complete composting, and greatly reduces the risk that disease organisms will survive. The EcoTech Carousel can handle full-time use and large families or groups. Because it does not have a heating element, it needs to be installed in a conditioned indoor space. But because there is only the small vent fan to run, the power use will be very low. Its disadvantages are obvious: It's big, and it's expensive initially.

Will Your Health Department Love It as Much as You Do?

Probably not, but this depends greatly on the enlightenment quotient of your local health official. Some health departments will welcome composting toilets with open arms, and even encourage their use, while others will deny their very existence. In Northern California where Real Goods is located, many counties accept composting toilets, including our own Mendocino County. We have often found that in lakeside summer cabin situations, health officials prefer to see composting toilets rather than pit privies or poorly working septic systems that leach into the lake. Several composters now carry the NSF (National Sanitation Foundation) seal of approval, which makes acceptance easier for local officials but does not mandate it. It is really up to your local sanitarian. He or she can play God within their own district, and there is nothing that you, Real Goods, or the manufacturer can do about it. Our advice if you're trying to persuade your authorities? Be courteous. Health officials have to consider that a composting toilet might suit you to a tee, and

you'll take good care of it, but what if you sell the house? Will the next homeowner be willing, or able, to take care of it, too?

Also, be aware that composting toilets only deal with toilet wastes, the blackwater. You still need a way to treat your greywater wastes—all the shower, tub, sink, and washer water. Unconventional greywater systems will face similar approval hurdles to composting toilets. Some localities happily will allow alternatives that use the greywater for landscape watering, but others may require a standard full-size septic system. Greywater alternatives are covered in more detail below.

Composting Toilet Installation Tips
Keep It Warm
Compost piles like warmth, because it makes all the little microbes work faster. Thus, composting toilets work best in warm environments. If your installation is in a summer-use cabin, then a composting toilet is ideal. A Montana outdoor installation of a modular manufactured toilet will not work at all in wintertime. If you plan to use the composter year-round, then install it in a heated space on an insulated floor. The small electric heater inside the compost chamber will not keep it warm in an outdoor winter environment, but it will run your electric bill up considerably trying. A temperature boost can be gained by following the same passive solar principles as for space

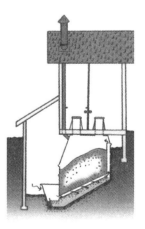

The CTS System for high-use public sites.

Compost piles like warmth, because it makes all the little microbes work faster. Thus, composting toilets work best in warm environments.

heating: protected south exposure, insulation, etc. Talk to one of Real Goods' technicians if you have any doubts about proper model selection or installation.

Plug It In

If you have utility power, by all means use one of the electrically assisted units; they have far fewer problems overall. If you have intermittent AC power from a generator or other source, use one of the AC/DC or hybrid units. Anything that adds warmth or increases ventilation will help these units do their work. We strongly recommend adding the optional vent fans to nonelectric units. Consider that running a generator to make electricity to evaporate excess water is going to undo a fair bit of the ecological benefit of using a composting toilet.

Let It Breathe

The compost chamber draws in fresh air and exhausts it to the outside. If the composter is inside a house with a wood stove and/or gas water heater that are also competing for inside air, you may have lower air pressure inside the house than outside. This can pull cold air down the composter vent and into the house. Composters shouldn't smell bad, and something is wrong if they do, but you don't want to flavor your house with their gaseous by-products either. Do the smart thing: Give your wood stove outside air for combustion. This will give an energy efficiency and comfort boost as well, by turning down the airspeed on all the cold drafts through every crevice that must

otherwise provide the air supply for your wood stove.

Don't Condense

Insulate the composter's vent stack where it passes through any unheated spaces. The warm, humid air passing up the stack will condense on cold vent walls, and moisture will run back down into the compost chamber. Most manufacturers include vent pipe and some insulation with their kits. Add more insulation if needed for your particular installation.

Keep Everything Flowing

With central units such as the Sun-Mar Centrex models, where the composting chamber is separate from the ultra-low-flush toilet, the slope of any horizontal waste pipe run is critical. The standard 3-inch ABS pipe needs to have ⅛ to ¼ inch of drop per foot of horizontal run. More drop per foot than that allows the liquids to run off too quickly, leaving the solids high and dry.

Vertical runs are no problem, so if you want a toilet on the second floor, go ahead. Sun-Mar recommends horizontal runs of 18 feet maximum, but customers have successfully used runs of well over 20 feet. Play it safe and give yourself cleanout plugs at any elbow when assembling the ABS pipe. If you are using an air-flush unit, it must be installed directly over the composter inlet.

Like the venerable outhouse, composting toilets only deal with human excreta. Unlike a modern septic system, they won't provide greywater treatment.

Incinerating Toilets

We love the composting toilet's ability to manage waste (and even recycle nutrients!) very efficiently. But, let's face it, they just don't work everywhere you might need sanitary facilities, for instance in an unheated cabin that sees only a few weekends use every winter. ECOJOHN™ provides efficient incinerating toilets as an alternative to composting toilets. They are designed to handle most types of climates and can run off the grid. By use of propane, these units incinerate waste into a sterile ash that only needs to be emptied a few times per year. The ECOJOHN™ products provide environmental, logistical and economical benefits.

WE DO IT EVERY DAY, BUT DO WE EVER THINK ABOUT IT!

I DEFECATE, THEREFORE I AM . . .

With thanks to Joseph Jenkins and *The Humanure Handbook.*

Greywater Systems

What Is Greywater?

Any wash water that has been used in the home, except water from toilets, is called greywater. Dish, shower, sink, and laundry water comprise 50%–80% of residential "waste" water. This generally may be reused for other purposes, especially landscape irrigation. (This is the definition common in Europe and Australia. Some jurisdictions in the US exclude kitchen sink water and diaper wash water from their definition of greywater. These are most accurately defined as "dark grey" water.) Most homes produce 20 to 40 gallons of greywater per person, per day. Greywater contains oil, grease, hair, food bits, bacteria, and possibly traces of pathogens. None of these are a problem for plants, though they must be considered in the plumbing design. Toxins from cleaners can't be processed by plants and soil. The simplest solution is to ban them from the house and instead use alternatives that are biocompatible with plants and soil. Note that the elements sodium, chlorine, and boron, though common in non-toxic cleaners, are not biocompatible with plants and soil, and should be banned from the house as well.

Why Would You Want to Use Greywater?

Two main reasons: Ecological treatment, or reuse of water for irrigation (or both). It's a waste to irrigate with great quantities of drinking water when plants thrive on used water containing small bits of compost. Unlike a lot of ecological stopgap measures, greywater reuse is a part of the fundamental solution to many ecological problems and will probably remain essentially unchanged in the distant future. The benefits of greywater recycling include:

- Lower freshwater use
- Less strain on failing septic tank or treatment plant
- Better treatment (topsoil is many times more effective than subsoil or treatment plant)
- Less energy and chemical use
- Groundwater recharge
- Plant growth
- Reclamation of otherwise wasted nutrients
- Increased awareness of and sensitivity to natural cycles

Greywater is used primarily for landscape irrigation, because soil is an excellent water purifier. By

> Unlike a lot of ecological stopgap measures, greywater reuse is a part of the fundamental solution to many ecological problems and will probably remain essentially unchanged in the distant future.

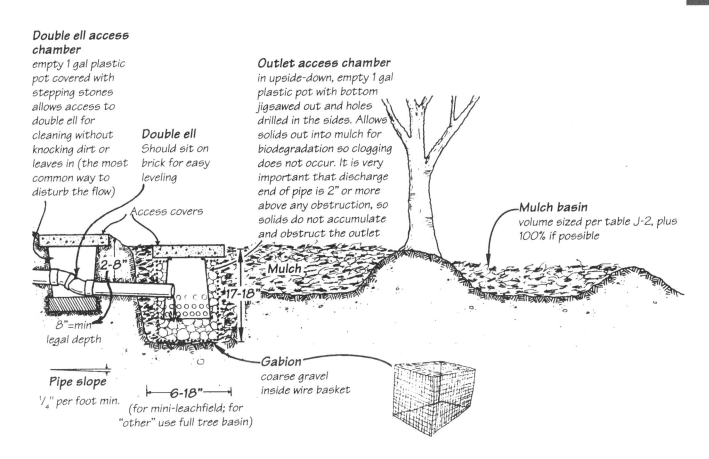

Double ell access chamber
empty 1 gal plastic pot covered with stepping stones allows access to double ell for cleaning without knocking dirt or leaves in (the most common way to disturb the flow)

Double ell
Should sit on brick for easy leveling

Access covers

Outlet access chamber
in upside-down, empty 1 gal plastic pot with bottom jigsawed out and holes drilled in the sides. Allows solids out into mulch for biodegradation so clogging does not occur. It is very important that discharge end of pipe is 2" or more above any obstruction, so solids do not accumulate and obstruct the outlet

Mulch basin
volume sized per table J-2, plus 100% if possible

Mulch

2-8"

17-18"

8"=min legal depth

Pipe slope
¼" per foot min.

6-18"
(for mini-leachfield; for "other" use full tree basin)

Gabion
coarse gravel inside wire basket

All responsible greywater designs stem from two basic principles:
• Natural purification occurs while greywater passes slowly through healthy topsoil.
• No human contact should occur before purification.

As with composting toilets, acceptance of greywater depends entirely on the folks at your local health department. Some will give you a choice, some won't.

substituting for freshwater, greywater can make landscaping possible in dry areas. It may also keep your landscaping alive during drought periods. Sometimes a greywater system can make a house permit possible in sites that are unsuitable for a septic system.

When Would Greywater Be a Poor Choice?

Greywater recycling isn't for everybody. High-rise apartment dwellers would be seriously challenged. But let's assume we're speaking to folks with at least a suburban plot around them. Even then, you may have insufficient yard and landscaping space, major bits of your drain plumbing may be encased in concrete, your climate may be too wet or too frozen, costs may outweigh benefits, or it might just be illegal where you live.

How Does a Greywater System Work?

Although the trace contaminants in greywater are potentially useful, even highly beneficial, for plants and landscaping, their presence demands some modest caution in handling and disposal. How do you recycle it safely? All responsible greywater designs stem from two basic principles:
- Natural purification occurs while greywater passes slowly through healthy topsoil.
- No human contact should occur before purification.

Greywater system designs display great diversity. At the simplest end of the spectrum, greywater can be carried over to plants by hand, or dispersed through the humble "drain out back" (a pipe pointed down the nearest hillside, the way every rural structure in the US was plumbed until recently). At the most complex, it can be finely filtered and automatically distributed as needed through subsurface drip irrigation tubing. For most contexts the recommended systems are:

Laundry: laundry to landscape system (described in the Oasis Design instructional video: see oasisdesign.net or realgoods.com). This consists of connecting the washing machine to an irrigation line with several outlets at various mulch basins.

Mulch basins (or swales) are tiny watersheds formed around trees with the root crown high and dry on an island in the middle 6–20" high, a flat-bottomed moat to contain greywater, rough mulch to cover it, and a low wall around the perimeter to keep mulch and greywater in and surface runoff out.

Shower, bath, sinks: branched drain to mulch basins. Branched drain systems take one flow and split it into several permanent outlets.

Kitchen sink: branched drain to subsoil infiltrators. Subsoil infiltrators are completely subsurface like a septic leach field and are sanitary for dark greywater (or even blackwater).

Feces: composting toilet or green septic. Green septic is like the offspring of a branched drain greywater system and a septic system. It uses very low-flow toilets and distributes the water predictably underground to a number of fruit trees

Urine: composting toilet, direct to plants, or to a disposal field. Urine is tricky, as it has as much salt (poison) as nitrogen and phosphorous (gold, basically). The best destination depends on the climate, soil, and plants. If urine is applied at a rate faster than the salt is leached out of the root zone, the salt will, over time, ruin the soil and kill most food plants.

All that being said, the one general rule is that there are no general rules for greywater systems. Even more than other aspects of ecological design, greywater systems are highly context-specific. Each system needs to suit the site, climate, the owner, and users (and perhaps regulators as well). Other major factors are how much water you've got, how much irrigation you need, the soil permeability, your budget, and legal constraints.

Will Your Health Department Love Greywater as Much as You Do?

As with composting toilets, acceptance of greywater depends entirely on the folks at your local health department. Some will give you a choice, some won't. Some may allow greywater systems so long as you also install a full-size approved septic system. Greywater regulation is being rationalized worldwide, often in giant leaps, but some areas still have archaic ideas about this resource. Since most people don't bother to get a permit, it is common to find whole populations who are unaware that the greywater reuse they've practiced for generations has been made illegal. Also common are regulators in other areas who are unaware that greywater is now permissible in their jurisdiction. Generally, official attitudes are moving toward a more accepting and realistic stance, and authorities will often turn a blind eye toward greywater use.

In the late 1970s, during a drought period, the state of California published a pamphlet that explained the illegality of greywater use, while showing detailed instructions of how to do it! Since then, California has become one of many places where you can install a greywater system legally. The rules are better with each revision, but still only one greywater system in 10,000 has gone through the permitting process. Happily, there is a new trend toward reasonableness and practicality in greywater regulation. Since 2001, the state of Arizona has allowed the installation of greywater systems of less than 300 gallons a day that meet a list of reasonable requirements without having to get a permit. In California, the same is now true for laundry-only systems: no permit required. For up-to-date information, and much more info on presenting a greywater system to your friendly local health department, consult the *Builder's Greywater Guide* or the Greywater Policy Center at greywater.net.

Recycling and greywater use may suit you perfectly, but someday you're likely to sell your perfect house. Will a greywater system suit the next owners as well? It's the people who will buy the house from you that the health department is thinking about. The best advice is simply to call your health department and ask. But ask as a hypothetical example: "A friend of ours is considering installing a greywater system...." Health officials recognize the "hypothetical" question readily. This allows you to ask specific pointed questions, and them to answer fully and honestly, without anyone admitting that any crime or bending of the rules has occurred or is likely to occur. In all cases, be unfailingly nice with your local administrative authority. This strategy pays dividends both locally, system by system, and regionally. Us polite greywater outlaws have been welcomed into the crafting of policies and codes—in some cases, our suggested wording has been used verbatim.

How to Make a Greywater System

Legal greywater systems tend toward engineering overkill, while many of the simple and economical methods that folks actually use are still technically illegal. Start with the greywater guides mentioned in the Resources section. These provide the best overview of the subject and detail every type of greywater system, from the simple dishpan dump to fully automated designs. Aim for the best possible execution with the simplest possible system. During new home construction, it's fairly easy to separate the washer, tub, sink, and shower drains from the toilet drains. Retrofits are a bit more trouble, and in some cases impossible, as when the plumbing is encased in a concrete slab floor. Kitchen sinks are often excluded from greywater systems because of antiquated legal definitions, and/or the high amount of oil, grease, and food particles that require accommodation with the hardware and maintenance.

In a world where it feels like most infrastructure we use is doing far, far more damage than necessary, greywater reuse can be tremendously satisfying.

"Spring water flows across the canyon and to our solar collectors by gravity. The now piping-hot spring water comes out the low-flow showerheads, and pours deliciously over us in our his-and-hers

> In a world where it feels like most infrastructure we use is doing far, far more damage than necessary, greywater reuse can be tremendously satisfying.

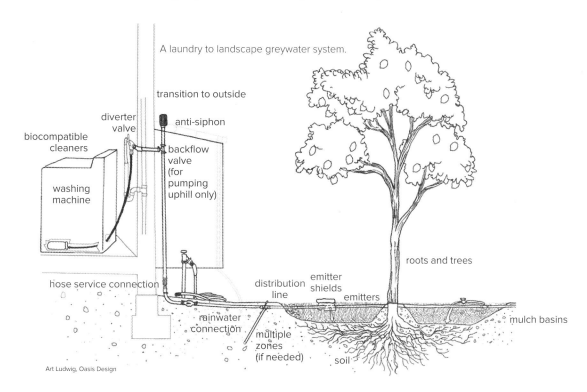

A laundry to landscape greywater system.

biocompatible cleaners

washing machine

diverter valve

transition to outside

anti-siphon

backflow valve (for pumping uphill only)

hose service connection

rainwater connection

multiple zones (if needed)

distribution line

emitter shields

emitters

roots and trees

mulch basins

soil

Art Ludwig, Oasis Design

Art Ludwig, Oasis Design

double outdoor shower. Spread and divided, the greywater flow trickles through mulch into the rich soil around a thirsty tangerine tree. You can tell the tree is happier for this rivulet of spring flow that we have diverted its way. From the shower, we reach up and grab a couple ripe fruit, and can't help but feel that this is the way we're meant to live."

—*Art Ludwig*

Greywater Resources

All the resources listed here are available at oasis design.net or realgoods.com.

Complete greywater book and video set

All books listed below plus a free copy of *Principles of Ecological Design* by Art Ludwig (*Create an Oasis with Greywater, Builder's Greywater Guide, Laundry to Landscape* instructional DVD).

Create an Oasis with Greywater (book)

Create an Oasis with Greywater describes how to choose, build, and use twenty different types of greywater systems. It thoroughly covers all greywater basics, and will benefit everyone who is using or contemplating the use of greywater. This 5th edition of the world's best-selling greywater book includes 50 pages of new text, photos, and figures, as well as complete, detailed design and installation information on branched drain greywater systems. By Art Ludwig (Oasis Design, 2006).

"Greywater for dummies and greywater encyclopedia in one information goldmine." —Dan Chiras, author, *The New Ecological Home, The Solar House*

Laundry to Landscape instructional video (DVD)

An instructional video explaining how to design and build Laundry to Landscape greywater systems. These are the simplest, most economical greywater systems to install yourself as an owner or renter, and a good green collar job opportunity (as a self-employed installer or as part of your landscape, plumbing, or construction business). Produced by Art Ludwig (Oasis Design, 2010), 90 minutes.

Builder's Greywater Guide (book)

Installation of greywater systems in new construction and remodeling. This supplement to *The New Create an Oasis with Greywater* will help building professionals or homeowners work within or around building codes to successfully

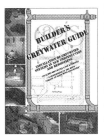

include greywater systems in new construction or remodeling. Includes information on treatment effectiveness, sample US codes, and permit submissions. You need *Create an Oasis with Greywater* to make sense of the *Builder's Greywater Guide*.

NOTE: *Builder's Greywater Guide* is currently undergoing extensive expansion. The currently available edition is accurate and up to date except it does not have the current version of the California Greywater Standards, which the authors helped write; these can be found at the website below. Physically, the current *Builder's Guide* is our simplest formatted book; it is black and white, saddle-stitched. The next edition will have a color cover, be perfect bound, and cost more. There isn't a release date for the new edition, but it will be late 2014 at the earliest. Art Ludwig (Oasis Design, 2014).

Oasisdesign.net

At oasisdesign.net/greywater, you can find:

- Excerpts from *Create an Oasis with Greywater*, including the system selection chart, which lists the attributes of twenty different greywater system options
- The common mistakes and preferred practices chapter from *Create an Oasis*
- Calculations and references about greywater safety
- Greywater Policy Center
- Trailer for the *Laundry to Landscape* video.

CHAPTER 11

Regenerative Homesteading and Farming

Growing Your Own Organically and Biodynamically Using

the Permaculture Principles While Living the Good Life

AS WE HAVE EMPHASIZED IN PREVIOUS CHAPTERS on relocalization, shelter, and water, the way we treat the land on which we live is an integral part of the stewardship ethic that characterizes the practice of sustainable and regenerative living. Land use and energy use are two sides of the same coin, and a responsible approach to each is based on the same values. The more we can do to build the soil and improve our land for future generations, the better and healthier our legacy will be. With substantial help from two practitioners—Benjamin Fahrer (Permaculture) and Jim Fullmer (Biodynamics)—we are pleased to offer this introduction to the concepts and techniques of Permaculture and Biodynamic Agriculture. Both are holistic systems for designing growing spaces and integrating food production into larger ecological, social, and spiritual contexts. Both are fundamentally grounded in principles and practices drawn from close observation and deep understanding of the natural world. Permaculture and Biodynamics offer us profound guidelines for the right way to practice agriculture on any scale in a world scorched by destructive agriculture in the 20th century and increasingly defined by peak oil and global climate change.

Permaculture and Biodynamics are conceptual tools that, while of interest to anyone seeking a low-environmental-impact lifestyle, are particularly appropriate for off-grid living and any kind of homesteading. Real Goods can provide you with many of the physical tools that will enable you to do the "hard work of simple living" efficiently and effectively. Humans are a tool-using, tool-loving species, and good tools improve the quality of our lives and contribute to making them more fulfilling. The better the quality of your tools, the better your craftsmanship is likely to be. Jobs will also be easier and more enjoyable.

Many of the tools and resources for sustainable living offered by Real Goods are categorized under labels like solar power, water pumping, or composting toilets. But we've always carried a selection of useful gizmos that defy nice, neat categorization. You can check out our wild, diverse, and fun collection of things that defy categorization on our website: real goods.com.

> Permaculture and Biodynamics are conceptual tools that, while of interest to anyone seeking a low-environmental-impact lifestyle, are particularly appropriate for off-grid living and any kind of homesteading.

Permaculture: A Holistic Design System for Your Site

Permaculture is about designing like nature would, where everything is a part of the whole. Permaculture embraces the totality of a place.

You may have heard of Permaculture, and chances are what comes to mind is a style of gardening. That's true—but Permaculture is not a style or technique used to grow one's food, it is a design system, a philosophy, in which all kinds of techniques are used to implement a food production system as well as all the elements that sustain us and all of life. Permaculture is about designing like nature would, where everything is a part of the whole. Permaculture embraces the totality of a place.

Grounded in three main ethical intentions—Earth Care, People Care, and Fair Share—Permaculture design is a system of assembling conceptual, material, and strategic components in patterns that provide mutually beneficial, regenerative, and secure places for all forms of life on this Earth.[1] The Permaculture ethos comes from the depths of indigenous cultures and from patterns found in nature. So permaculture is how everything is connected through a design. It is an approach to gardening and a way to manage land as well as a way to build shelter. But even more, Permaculture as design is a system that can be also applied to economic and social aspects of our society. For a discussion of how permaculture principles can be applied to living in urban environments, see Chapter 12, Urban Homesteading, pages 305–32.

History and Background

The word *Permaculture* comes from marrying the two words *permanent* and *culture*. Our species for the last 10,000 years has developed a culture that is the most destructive our planet has ever seen. Based on an extractive process, an industrial capitalistic complex has gotten us into a terrible mess. Our soils are depressed and depleted, water and air are polluted, and natural resources are peaking in their supply vs. demand. Permaculture provides us some solutions, and limiting

Permaculture to simply a way of gardening would be like limiting the concept of energy production to the collection of solar power. The concept is much broader, invoking all aspects of a healthy food system and a holistic way of thinking and living. Permaculture is truly about design, connectivity, and relationships.

Two Australians, Bill Mollison and a younger David Holmgren, who were studying the unstable and unsustainable characteristics of Western industrialized culture, coined the term Permaculture in the late 1970s. Holmgren was asking questions about how the human species had existed for so long in harmony and balance with nature yet was now contributing so much to the destruction of Earth. They were drawn to indigenous worldviews and the self-sustaining and regenerative qualities of natural processes and systems. Holmgren began to think about how human processes and systems could be patterned after the designs found in nature. Other people were doing similar work at the same time, and Permaculture drew from and contributed to a larger conversation among pioneers of ecological thinking: P. A. Yeomans' Keyline Plan, J. Russell Smith's work on tree crops, Masonobu Fukuoka's Natural Farming techniques, James Lovelock's Gaia Hypothesis, Buckminster Fuller's radical design thinking, and many others who were breaking into the field of ecological design, such as Lawrence Halprin. Holmgren brought all these currents together in a thesis he wrote under Mollison's tutelage, which was published in 1978 as *Permaculture One*. Mollison published *Permaculture Two* (1979), and then the book that became the definitive text of Permaculture design, *Permaculture: A Designer's Manual* (1988), which covers the ecology of all living systems and how to design our human place in them.

While Mollison traveled the world learning from and integrating practices into the Permaculture theory, Holmgren focused on applying and testing the theories and concepts that he and Mollison had developed. At his farm and homestead Melledora, Holmgren established a successful design firm that has now created projects all over the world over the past 30 years, the majority of them in Australia and Europe. In the seminal *Permaculture: Principles and Pathways Beyond Sustainability* (2003), Holmgren elaborates a comprehensive, integrated set of Permaculture principles. His most recent book is *Future Scenarios: How Communities Can Adapt to Peak Oil and Climate Change* (2009).

An herb spiral embodies the spirit of Permaculture.

Benjamin Fahrer

Permaculture Ethics, Principles, and Practices

Permaculture begins with an ethic of personal responsibility for taking care of the Earth by our actions and the choices we make. As Mollison says, "The only ethical decision is to take responsibility for our own existence and that of our children. **Make it now.**" This foundational ethic is elaborated in three simple concepts:

- **Earth Care.** Allowing provisions and resources for all life systems to continue and multiply.
- **People Care.** Allowing provisions for people to access those resources necessary to their existence.
- **Fair Share.** Return of surplus and setting limits to our consumption. By governing our own individual needs, we can set resources aside for Earth Care and People Care.

In *Permaculture: A Designer's Manual*, Mollison articulated five primary organizing principles for the practice of Permaculture. Other practitioners have since articulated related sets of ideas. The core design ideas of Permaculture can be distilled into the following 12 principles:

1. **Work with nature,** rather than against natural elements, forces, processes, dynamics, and evolutions, so that we can assist rather than impede natural developments. On a practical level, this translates into using gravity, native species, and the sun and wind, for example, in designing homescapes, landscapes, gardens, and other environments.
2. **The problem is the solution;** everything works both ways. How we see things determines whether situations or elements are advantageous or not. Everything is a valuable resource.
3. **Make the least change for the greatest possible effect.** Allow work to be a source of your energy, not a sink.
4. **The yield of the system is theoretically unlimited.** The limits on the number of uses for a given resource within a system are the information available and the imagination of the designer.
5. **Everything gardens,** or has an effect on its environment; everything is connected.
6. **Relinquishing power.** The role of beneficial authority is to return function and responsibility to life and people.
7. **Unknown good benefit.** If we start with good intentions, other good things follow naturally.
8. **Succession of evolution.** Natural design inherently evolves toward stability and resiliency in a system.
9. **Cyclical opportunity.** Every cyclical event increases the opportunity for increased yield; therefore, increased cycling of resources increases yield.
10. **Functional design.** All functions are supported by many elements, while each element performs many functions.
11. **Stability** is created by a number of beneficial connections between diverse beings.
12. **Information as a resource.** Information is the critical potential resource. Bad information can result in a poor design, whereas good information increases the opportunity for a good design.

Observation, Cycles, Relationships, and Patterning

While these ethics and principles supply the framework for the Permaculture design process and govern the choices designers make, careful observation is the crucial tool that informs those choices.

Observation is by definition the attentive watching of somebody or something. Permaculture goes deeper in calling for Protracted And Thoughtful Observation (PATO). It is thoughtful because, as a designer, you are observing the relationships and cycles that are happening before you, and it is protracted because you are observing them over a period of time. Upon first impressions of a situation or site, one can only assume and project patterns and flows. Observations over time are necessary to glean information that can reveal probable patterns of ecological succession and evolution. Many Permaculturalists will tell you that that the trees are the best translators of the environment and by developing observational

> Permaculture begins with an ethic of personal responsibility for taking care of the Earth by our actions and the choices we make.

Permaculture is whole-system design, in which the social or invisible aspects are just as integral as the physical and visible ones.

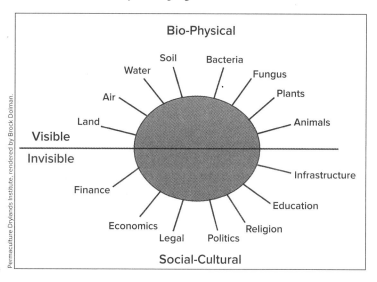

Permaculture Drylands Institute, rendered by Brock Dolman.

Holmgren's Permaculture Principles

David Holmgren articulated a set of principles in his 2002 book, *Permaculture: Principles and Pathways beyond Sustainability*. Although different than Mollison's, these principles are also guided by the three main permaculture ethics of Earth Care, People Care, and Fair Share. They are based on ecological concepts and use familiar terms. And they make a lot of sense when put into practice. As the thread that weaves together the design, these principles are what make permaculture often feel like second nature to those who practice.

1. **Observe and interact**
   "Beauty is in the eye of the beholder"
   By taking the time to engage with nature we can design solutions that suit our particular situation.

2. **Catch and store energy**
   "Make hay while the sun shines"
   By developing systems that collect resources when they are abundant, we can use them in times of need.

3. **Obtain a yield**
   "You can't work on an empty stomach"
   Ensure that you are getting truly useful rewards as part of the work that you are doing.

4. **Apply self-regulation and accept feedback**
   "The sins of the fathers are visited on the children unto the seventh generation"
   We need to discourage inappropriate activity to ensure that systems can continue to function well.

5. **Use and value renewable resources and services**
   "Let nature take its course"
   Make the best use of nature's abundance to reduce our consumptive behavior and dependence on nonrenewable resources.

6. **Produce no waste**
   "A stitch in time saves nine." "Waste not, want not."

By valuing and making use of all the resources that are available to us, nothing goes to waste.

7. **Design from patterns to details**
   "Can't see the forest for the trees"
   By stepping back, we can observe patterns in nature and society. These can form the backbone of our designs, with the details filled in as we go.

8. **Integrate rather than segregate**
   "Many hands make light work"
   By putting the right things in the right place, relationships develop between them and they support each other.

9. **Use small and slow solutions**
   "The bigger they are, the harder they fall." "Slow and steady wins the race."
   Small and slow systems are easier to maintain than big ones, making better use of local resources and producing more sustainable outcomes.

10. **Use and value diversity**
    "Don't put all your eggs in one basket"
    Diversity reduces vulnerability to a variety of threats and takes advantage of the unique nature of the environment in which it resides.

11. **Use edges and value the marginal**
    "Don't think you are on the right track just because it's a well-beaten path"
    The interface between things is where the most interesting events take place. These are often the most valuable, diverse, and productive elements in the system.

12. **Creatively use and respond to change**
    "Vision is not seeing things as they are but as they will be"
    We can have a positive impact on inevitable change by carefully observing, and then intervening at the right time.

Integrate rather than segregate. Gardening with relative location in mind. A bed of lettuce is planted in the shade of corn, next to a summer cover crop of buckwheat, Oceansong.

Use and value renewable resources and services: "Let nature take its course." Water integration in the cityscape, a curb-cut in Portland, Oregon.

Catch and store energy: "Make hay while the sun shines." By developing systems that collect resources when they are abundant, we can use them in times of need. Water spilling into pond catchments, Big Sur.

skills, a seasoned designer can read the landscape and draw conclusions about flows and patterns based on the way the trees lean and grow as well as on the species that are present. Another approach is to use a hammock or a raft floating on a pond as design tools. By melting into the surroundings and spending time in focused observation, nature begins to reveal itself to you, and this unveiling is an invaluable resource.

Considering cycles, relationships, and patterns brings us to the idea that Permaculture as a whole design science has both visible and invisible structures. A simple homestead, for example, encompasses many cycles and patterns that are visible: the nutrient cycle, the life cycle, the hydrological cycle, plant cycles, seasonal cycles, and so on. A huge variety of patterns are present within the elements of these cycles. Whether it is the web of a garden spider, the spiral of a sunflower, or the branching of a leaf or tree, recognizing those patterns is essential to creating integral concepts that work with nature. Form is the envelope within which life pulsates. And in nature's design, form follows function. Permaculture relies on biomimicry, or emulating nature's patterns and processes, to create efficient and elegant designs for human use that also enable other living things to flourish.

In the realm of human relations and social institutions, the principles of Permaculture can help to orient us to the power of invisible structures. Take self-governance within a community, for example. The cycling of roles and responsibilities, the flow of money within larger financial cycles, yearly cycles of meetings and fundraisers: The character of these kinds of interactions is strongly influenced by people's habitual patterns of behavior, emotional and physical, that surround them in a working environment—and they are largely invisible to the people involved. The foundational practice of Protracted And Thoughtful Observation enables conscious participants to sustain ethical intention and help keep everyone positively connected in pursuit of the tasks and goals at hand. Permaculture is about mutually beneficial relationships, in human affairs as well as in the ways humans relate to nature. The following chapter on Urban Homesteading considers these issues in the context of permaculture in city environments.

Zone and Sector Analysis

Permaculture design fundamentally works with nature rather than against it, and therefore a solid understanding of the energy potential of the site is crucial for its success. The approach to assessing these energies is the concept of zone and sector analysis, which is an energy conservation placement pattern for a site. Whatever the scale of the site, this baseline assessment tool is invaluable in setting yourself up for success. Analyzing zones and sectors within an environment literally energizes the principle noted above: making the least change for the greatest possible effect, working with nature and relative location.

Zones refer to the energy resources and components that are located onsite and are somewhat controllable through human interaction. Sectors refer to the external energy resources that flow through the site and are more or less uncontrollable because they are produced at the command of nature.

Zones are areas and pathways that are in reference to a point of origin from which we work, a focal point of the system where the design starts. These zones are determined and defined by frequency of use. That point of origin is called Zone 0 and is typically the inside of your house. In a more contemporary sense, Zone 0 can also be thought of as the interior, subjective place where your being and spirit reside—the Self.

Zone 1 begins when you step outside of your house or personal space. Items or areas within Zone 1 are visited multiple times throughout the day and usually require continual observation and work, as they are areas of intensely cultivated space, such as a kitchen garden. Here nature is arranged in such a way as to fit specific human needs.

Zone 2 is not as intensely cultivated and is visited one or two times a day. The components of Zone 2 do not need as much attention as those within Zone 1, yet they still require daily interaction. Animals such as chickens or goats are

Zones reflect frequency of use, or how often an element within the system is visited and needs attention. Zone 0 is the point of origin, such as a house. Things placed in Zone 1 are visited intensively, and frequency decreases through Zone 5, a wild place that provides inspiration. The concept is cyclical, so that energy from Zone 5 is brought back into the other zones.

Zones – Onsite energies; Frquency of Use

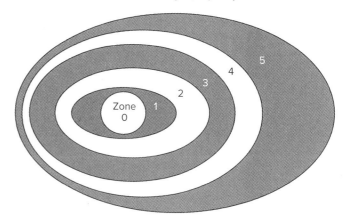

Benjamin Fahrer, adapted from Bill Mollison, *Permaculture: A Design Manual*

Pulsation of Zonation. What is our site?
What level are we working at? Sense of scale.

Integration of sheep into a walnut orchard, Full Belly Farm.

Sector analysis considers the wild, or external, energies that flow through a site. These energies may be harnessed, channeled, deflected, or blocked.

housed here along with the more robust garden and small fruit trees. Zone 2 is designed to express a greater balance between human needs and nature.

Zone 3 is considered the "farm zone," and objects or elements in this area are meant to supply the site with an abundance of surplus to either sell or reinvest. Larger orchards, row crops, and grazing pasture located in Zone 3 require less than daily attention, perhaps visitation only one or two times a month.

Zone 4 is a managed area of land that borders on wilderness. Including tree crops grown for fuel and food, large-scale water collection in ponds and lakes, and wildcrafting and foraging of native plants, Zone 4 is managed about once a year.

Zone 5 is the area considered wilderness and has no management strategies except those imposed by nature. In this zone, we are visitors, not managers. Zone 5 is where we look for examples to emulate and the true teaching of natural design. Although you might think this zone is the least used area of land on the site, and thus fre-

quency of interaction would be very low, that is not necessarily so. In fact, Zone 5, as with all the zones, can actually cross over to or border one's own home zone. Observations within Zone 5 can happen every morning over tea as one looks out over pristine forest and beholds nature's masterpiece. Zone 5 actually feeds back into Zone 0 and provides constant inspiration and rejuvenation. It is this feedback loop that adds a pattern of connectivity and wholeness between the core of your system and the rest of the natural world.

Sectors are external forces like the sun, wind, water, fire, and sound, to name a few, that flow through the site. Depending on the characteristics of the site and the goals of the design, there are three main ways to work with the incoming energy from various sectors: to open the sector up, block it, or channel the energy for better use. Many approaches to landscape or site design utilize the blocking or channeling of sector energies. What's unique about Permaculture is its holistic and integrated view that seeks to optimize the use of energy flows throughout the environment of a site. In many cases, the energy of a sector is used to do work that would normally be performed with an onsite energy resource. Examples would be siting a house to utilize passive solar heating, thereby reducing the demand for firewood from Zone 4, and planting deciduous trees on the south side of the house to provide summer shade and passive cooling. Windscreens are a common tool, for example trees or fences built to help shelter a field or home from a prevailing wind, or to hide a road from view. Trees can also be planted to channel wind certain ways and use it as a resource, for example, to provide evaporative cooling to a house or to create a Venturi effect for a wind turbine.

Once zones and sectors are identified and outlined, they are mapped out on the site. Such mapping pays particular attention to how various forces and elements overlap and interact. Determining an appropriate Permaculture design usually requires a lot of jostling around of buildings, plantings, and elements. Ultimately, an effective energy conservation placement pattern takes form that efficiently utilizes the resources at hand.

Permaculture Design Process

The general core model is a graphic aid that demonstrates how all the components of Permaculture fit and work together in the design process. As just noted, designing a site requires the participants to be open to change, trial and error, and revision, as their understanding of energy flows and resource interactions evolves and ma-

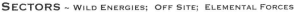

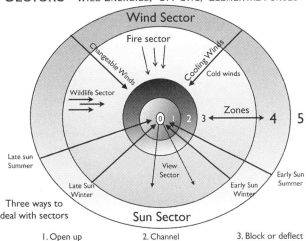

SECTORS ~ Wild Energies; Off Site; Elemental Forces

Wind Sector

Fire sector

Changeable Winds

Cooling Winds

Cold winds

Wildlife Sector

Zones

0 1 2 3 4 5

Late sun Summer

View Sector

Late Sun Winter

Early Sun Winter

Early Sun Summer

Three ways to deal with sectors

Sun Sector

1. Open up

2. Channel

3. Block or deflect

Benjamin Fahrer, adapted from Bill Mollison, *Permaculture: A Design Manual*

REAL GOODS

tures through sustained observation. Representation as a mandala suggests that the process is a never-ending loop of assessment, visioning, conceptual planning, and master planning. This mandala forms a lens of natural-systems thinking that is used to view all dimensions of the site in question. What works on paper does not necessarily work in the field, and small changes in one aspect can ripple into dramatic changes in overall structure and design.

This process can drive some conventional designers and clients crazy. But more and more are embracing a way of whole-systems thinking. Permaculture practitioners accept the ever-evolving design as a living, creative entity. Designer Benjamin Fahrer tells the story of observations made over the winter rains that revealed previously unrecognized aspects of a client's site and prompted a complete rethinking of the original homestead and garden design. What seemed like a problem at first soon became a wonderful opportunity to fulfill the client's deepest desires—but only because the people involved were willing to observe natural dynamics, reassess initial assumptions, and reimagine their plans. Flooding and drainage problems became a pond, gardens were relocated to a more advantageous spot, and those changes provided more sun for the house.

The essential purpose of design is to create the possibility for events to happen—or an "imperfect design." A perfect design is fixed and stagnant, and requires no further input, human

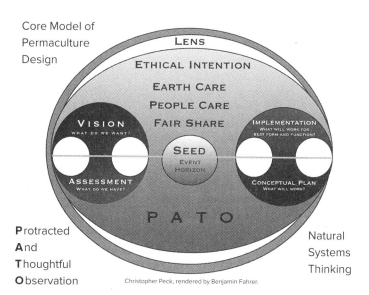

Core Model of Permaculture Design

LENS

ETHICAL INTENTION

EARTH CARE
PEOPLE CARE
FAIR SHARE

VISION
WHAT DO WE WANT?

ASSESSMENT
WHAT DO WE HAVE?

SEED
EVENT
HORIZON

IMPLEMENTATION
WHAT WILL WORK FOR
BEST FORM AND FUNCTION?

CONCEPTUAL PLAN
WHAT WILL WORK?

P A T O

Protracted
And
Thoughtful
Observation

Natural
Systems
Thinking

Christopher Peck, rendered by Benjamin Fahrer.

or natural. An imperfect or incomplete design accepts change and is enhanced by chance occurrences—enriched by weeds and edge, by the changing patterns of sunlight and shade, even by a branch falling on the paths or terrace. A habitation is a natural system, and it should be designed and treated as such.

In its most profound sense, the Permaculture design process is not limited to land-based projects: It can be applied to any aspect of life. It is inclusive, it is participatory, it is a navigational compass for designing one's entire life—and it is Permaculture when it lives up to the ethical intentions of Earth Care, People Care, and Fair Share.

The seed is the idea or site that is being designed. Permaculture uses the method of Protracted And Thoughtful Observation, grounded in ethical intentions, to generate a process of assessment, visioning, conceptual planning, implementation, and further assessment.

Backyard swales for rainwater harvesting, the initial infrastructure for a future orchard.

A keyline design that allows water to spread responsibly through a landscape is done with a yeoman's plow at Orella Ranch.

Erik Ohlsen

Benjamin Fahrer

Biodynamic® Agriculture: Food and the Cosmos

Current relevance

Conceived in 1924 by Dr. Rudolf Steiner (also the founder of the Waldorf education movement), Biodynamic Agriculture is one of the original foundations of what we today call "organic" and "sustainable" agriculture. Steiner laid out the principles of Biodynamic Agriculture in a series of lectures, in response to requests from European farmers who were alarmed both at the growing use of synthetic chemical fertilizers and pesticides, and at what they perceived as a noticeable and relatively rapid decline in crop and animal vitality. Demeter certification was initiated in 1927 and has been in existence since, now active in over 29 countries worldwide with a Standard that has its base developed via the democratic input of all these countries and cultures.

The organic food industry has grown significantly over the past three decades. The various national organic standards in effect around the world have been focused more on the materials used in organic agriculture and less on the interconnected biological systems that deliver it. The Demeter Biodynamic Farm Standard focuses strongly on the biological systems aspect of agriculture and falls back on the national organic regulations to define appropriate materials to be imported, but also with limitations on their importation. An intrinsic result of such a systems approach is close attention to what has to be imported onto the farm with the aim of generating the needed inputs out of the living dynamics of the farm itself. Importing materials to an organic agricultural system reintroduces some of the same problems that synthetics-based industrial agriculture presents, namely dependence on Earth's natural resources to mine, refine, and transport a myriad of products that are shipped all over the world. By its nature, this practice puts pressure on natural resources and the natural systems where these materials are mined or harvested. The goal of a Biodynamic agricultural system is to be as regenerative as possible, thus generating the inputs onsite out of the living dynamics of the agricultural system itself.

The Philosophy of Biodynamic Agriculture: A Farm as a Living Organism

Biodynamic Agriculture is a holistic approach that aims to generate an agricultural ecosystem/individuality that operates as a self-contained system in its most holistic sense. As explained by Dr. Ehrenfried Pfeiffer, a student of Steiner's who introduced the Biodynamic method to the United States in 1938, the name *Biodynamic* refers to "working with the energies which create and maintain life."

Biodynamic farm management requires close attention to the living and dynamic interrelation of the constituent parts of the agricultural system, rather than only concentrating on the individual parts in isolation from each other. A critical aim of this practice is to provide necessary agricultural inputs out of the living dynamics of the farm itself, rather than importing them from the outside. Such an approach to agriculture requires cooperation with living systems archetypically inherent to the identity of this planet. Biodynamic farming involves managing a farm utilizing the principles of a living organism. A concise model of a living organism ideal would be a wilderness forest, where there is a high degree of self-sufficiency in all realms of biological survival. Fertility and feed arise out of the recycling of the organic material the system generates. Avoidance of pest species is based on biological vigor and its intrinsic biological and genetic diversity. Water is efficiently cycled through the system.

Such systems operate within an archetypal biological concept of "time" that is followed by the natural world on Earth, such as the progression of the seasons, related weather patterns, sunrise and sunset, and the rhythmic processes of growth and decay that result. A farm by nature is once removed from a wild setting, but the aim of a Biodynamic farm is to observe and utilize the principles wisely inherent to such a setting. This biological timeframe dictates how quickly a farm can reach a level of fine-tuned efficiency and maximum productivity. The local biological networks that support a farm or garden can be intensified and accelerated to a degree, but the system itself cannot be pushed beyond its means without bringing in help in the form of imported materials. Such imported materials are temporary measures, and the ultimate goal is to generate the agricultural system's ways and means internally.

Approaching a farm or garden as a living organism involves four general and coexisting realms. These are the *mineral*, such as the constituents of physical matter; the phenomenon of *life* that flows through the mineral—essentially giving it form and expression, such as we see in the plant world; a sense of *consciousness/self-awareness* that also permeates the living forms, such as we see in the animal world; and finally a sense of *will*

Plant growth forms, spiraling and enveloping.

and intent that consciously works with the inter-relation of all these realms. Human beings are one of the species on Earth that can consciously mold the other realms with will and intention. The molding of a landscape in this manner is the ancient art and science of agriculture. Thus, generating a Biodynamic agricultural system is not unlike the relationship of the conductor to the symphony or the sculptor to the clay. The human will and intent molds a living, nutritious landscape utilizing holistic tools and practices that are ancient and highly evolved.

Merriam-Webster defines an organism as a system with many parts that depend on each other and work together. Observing the natural world makes it apparent that this is indeed true and that the expansiveness of the connections involved are not limited to artificially created borders. Important in this regard is the use, and future evolution, of the Biodynamic preparations introduced by Rudolf Steiner in 1924 (see below). In a sense, their role is to keep these wider connections grounded with the Earth, treating the Earth medicinally/homeopathically, not unlike a naturopath would approach you or me as living organisms with wider rhythmic connections. In principle, a Biodynamic farm is formed in the image of a living organism.

Consider the relationship between the plant world and the sun, for example. The light and warmth the sun generates drives all life on Earth. The plant world gathers the sun and turns it into chlorophyll, and from that process literally comes a huge diversity of plant forms manifest into physical form. From this archetypal fact, all else in the world of matter evolves. That is miraculous! Sunlight and warmth literally pull plant forms into existence. If you watch closely the growing tips and resulting morphology of most green plants, you will see an expression of levity, defying gravity, grasping at sunlight, spiraling, enveloping—a pulling of a plant species into physical form, not a pushing out into space.

In your imagination of the scope of a Biodynamic farm, consider the images below, and on the following page.

Demeter Association, Inc.

These aerial photos (left, and at the top and bottom of the following page) show how a single farm is embedded in and a part of a series of ecosystems, watersheds, and bioregions.

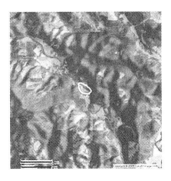

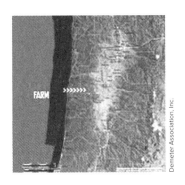

Demeter Association, Inc.

The first three photos are aerial views (increasing in altitude) of a Biodynamic farm in western Oregon. Within the boundaries of this farm (circled) is a dynamic system complete with biodiverse habitat (as forest, wetland, river riparian, hedgerow, diverse cropping, and developed insectaries) and internal humus-based fertility dynamics (livestock/composting manure, careful crop rotation, green manures). The development of this system also provides the farm with its own crop nutrition, pest control, and water conservation inputs without having to import them on a truck or in a bag. Instead of seeking this or that remedy for specific management problems, single inputs that address numerous management concerns are implemented to more or less kick-start the biological wisdom that is already inherent to the farm ecology. For instance, by implementing a fertility program to develop soil humus, both water management (by increasing the soil's ability to hold water) and pest control (through balanced crop growth that receives nutrition from the dynamic relationship between roots and soil life) are also addressed.

As one ascends skyward, it is clear that this organism within the fence line is itself part of a wider ecosystem. The farm is part of the Douglas fir forest ecosystem that surrounds it. A multiplicity of birds, amphibians, mammals, and insects move out of the surrounding forest and flow through the farm; some move in and contribute.

The next three photos carry the theme further. In the first, it can be seen that the farm is situated on the flood plain of a river valley. On a clear night, the cold air from the highlands pours down onto the valley floor where the farm resides and creates a chilly microclimate that is quite cooler than areas a few miles downstream. Next, that valley can be seen in relation to the other drainages that shed water down the east slope of the Willamette Valley in western Oregon. The farm is an element of an even broader system that drains copious amounts of water eventually back to the sea.

As the next images show, the farm and its valley are situated in the Pacific Northwest of North America, an environment that is conditioned by the ocean to the west and the winds and weather it brews. A great river of air ebbs and flows across the face of the North American continent—the Jet Stream. The farm, always a microcosm within a greater macrocosm, is an element of the identity of Earth, herself a living organism with a cold top and bottom and a warm middle. It spins, and thus the farm (and the farmer) contends with the forces of gravity.

Phenomena such as the 2004 tsunami shown below remind us of the dynamic nature of the Earth as a living organism. At her core, the Earth is extreme fire: the temperature there is thought to be as hot as the surface of the sun. This sun-like intensity yields to varying states of molten matter upon which tectonic plates float and drift. This tsunami resulted from the Earth's Indian Plate being subducted by the Burma Plate, which caused the entire planet to vibrate as much as 1 centi-

Demeter Association, Inc.

meter (0.4 inches) and triggered other earthquakes as far away as Alaska. At the subduction point, water was displaced by earth. The resulting swirling flow of water that came upon land in its path was both majestic and tragic.

Earth is an element of a wider solar system. Astronomy tells us that Earth and its moon are in a rhythmic dance with each other and also with other celestial bodies, circling the sun and each other. This dance goes on with the utmost mathematic precision and predictability. The star that Earth and its comrades dance around is one of countless others in a larger system, all a microcosm of yet a greater macrocosm, the Milky Way galaxy. And galaxies abound in the universe.

And so it goes—everything is interconnected. A critical point in the explanation of Biodynamic Agriculture is that even the tiniest aspect of an agricultural system, a seed or a bud for instance, contains the archetypal imprint of the widest celestial sphere, essentially the infinite.

Left to right: Sri Lankan flood; Earth's relation to the moon; our solar system; the Milky Way galaxy, of which our solar system is an element.

Biodynamic Preparations

(courtesy of Hugh Courtney)

The good agronomy practices that are at the base of Biodynamic Farming are not sufficient to achieve the process of healing. Steiner developed a set of unique "Biodynamic preparations," which are a critical element of Biodynamic Farming. In the US, the Biodynamic preparations have been most successfully created and distributed by Hugh Courtney at the Josephine Porter Institute, for many years now. Some regional groups and individual farmers have also been making the preparations.

In total, there are nine BD "preps." Horn manure (BD #500) and horn silica (BD #501) are field sprays, diluted and stirred vigorously for one hour in water, which relate to earth and light forces, respectively. The yarrow (BD #502), chamomile (BD #503), stinging nettle (BD #504), oak bark (BD #505), dandelion (BD #506), and valerian (BD #507) preparations are known as the compost preps and are applied to compost and manure piles. The horsetail herb, *Equisetum arvense* (BD #508), is sprayed in liquid form and has a regulatory function with respect to fungus and other manifestations of the watery element.

A different way of thinking is required to approach agriculture from the Biodynamic perspective. In our "conventional chemical" or even in the "organic" approach, we are conditioned to think in terms of substances—or, rather, in terms of chemical requirements that can be met by particular substances. In chemical agriculture, we bring nitrogen (N) to the soil via ammonia or urea, and in organic agriculture, we do so via manure. For phosphorus (P), the choice is superphosphate or rock phosphate. Regardless, we are still thinking in terms of chemical substances or, as many understand it, in NPK terms. Biodynamic Agriculture and its preparations ask us to think in terms of *forces* rather than substances. One need not throw out all knowledge of soil chemistry to think in Biodynamic terms—but we must go beyond the chemical point of view. Just as the effects of the forces of gravity or magnetism can be observed without actually being able to *see* these forces, so too we can recognize through their effects the forces that are released through use of the soil-enlivening Biodynamic preparations. For instance, Steiner describes the plant as a being that exists between sun and Earth, or between the polarities of silica and limestone with clay as the mediating factor. In present-day agriculture, silica is totally ignored, and the variables associated with limestone are barely recognized. When Steiner refers to silica, limestone, or clay, he doesn't mean the substances but rather the cosmic forces that permeate the substances. The beneficial effects of cosmic forces are noticeable

A critical point in the explanation of Biodynamic Agriculture is that even the tiniest aspect of an agricultural system, a seed or a bud for instance, contains the archetypal imprint of the widest celestial sphere, essentially the infinite.

in greater seed viability, improved food quality, and healthier livestock and crops.

Practicing Biodynamic Agriculture

To understand the forces behind the minute amounts of substances used in Biodynamic preparations, here is Hugh Courtney's analysis of the preps as they relate to silica/limestone/clay:

Limestone	Clay (or clay-humus)	Silica
Polarity	Mediating factor	Polarity
BD #500	BD #502-507	BD #501 & #508

To enable the enlivening cosmic forces standing behind the silica and limestone to work with full effect, the first step is to introduce the mediating forces of the BD Compost Preparations to the soil. This is usually done via properly made BD compost using BD #502-507. Proper composting using the preps is a key feature of Biodynamic Agriculture. Another means for applying the compost preps (BD# 502-507), particularly when beginning a conversion, is by use of the BC (Biodynamic Compound Preparation), a product containing BD #502-507 that can be applied as a spray. BD-made compost itself, however, has the virtue of imparting a longer-lasting effect to the soil than does the BC. Once the forces of BD

The Biodynamic Principles of Soil Fertility

1. To restore to the soil the organic matter it needs to hold its fertility in the form of the very best humus.
2. To restore to the soil a balanced system of functions. This requires looking at the soil not only as a mixture or aggregation of chemicals, mineral or organic, but as a living system. Biodynamics concerns itself with the conditions under which soil microlife can be fully established, maintained, and increased.
3. While the Biodynamic method does not deny the role and importance of the mineral constituents of the soil, especially the so-called fertilizer elements and compounds—including nitrogen, phosphate, potash, lime, magnesium, and the trace minerals—it sponsors the most skillful use of organic matter as the basic factor for soil life. (Interestingly, the importance of trace minerals for normal and healthy plant growth was actually pointed out by Rudolf Steiner as early as 1924.) Advocates of organic farming were among the first champions of the value of manure and compost.
4. In the Biodynamic method, life and health depend on the interaction of matter and energies as well as chemical substances. A plant grows under the influence of light and warmth—that is, energies—and transforms them into chemically active energies by way of photosynthesis. A plant consists not only of mineral elements (inorganic matter that makes up only 2%–5% of its substance) but also of organic matter such as protein, carbohydrates, cellulose, and starch, all of which derive from the air (carbon dioxide, nitrogen, oxygen) and make up the major part of the plant mass aside from water (15%–20%). The greater part of plant mass, some 70% or more, consists of water.
5. The interaction of the substances and the energy factors forms a balanced system. Only when a soil is balanced can a healthy plant grow and transmit both substance and energy as food. The Biodynamic method aims to establish a system that brings into balance all factors that maintain life.
6. Were we to concentrate only on nitrogen, phosphate, and potash, we would neglect the important role of biocatalysts (trace minerals), enzymes, growth hormones, and other transmitters of energy. The Biodynamic way of treating manure and composts includes the knowledge of enzymatic, hormonal, and other energy factors.
7. Restoring and maintaining balance in a soil requires proper crop rotation. Soil-exhausting crops with heavy demands on fertilizing elements should alternate with neutral or even fertility-restoring crops—on the farm as well as in the garden and even in the forest. A soil that has been put to maximum effort—producing corn, potatoes, tomatoes, peppers, and cabbage (all greedy crops), for example—should have a rest period with restorative soil-building crops such as legumes. Temporary cover with grass and clover helps to improve humus and nitrogen levels. Greedy crops and arable cultivation consume humus.
8. The entire environment of a farm or garden is important. One must pay attention to the air, the functioning biological system, the quality of the hillsides, and the water balance.
9. The soil also has a physical structure. The maintenance of a crumbly, friable, deep, well-aerated structure is an absolute must if one wants to have a fertile soil. All factors that lead to structural disintegration of the soil (like plowing soil that is too wet) must be considered.

#502-507 are working in the soil, the BD #500 (horn manure) is applied, followed by BD #501 (horn silica) and BD #508 (horsetail herb/*Equisetum arvense*) according to the need.

The preparations must be made with exactitude, following Steiner's directions, so they are able to convey forces to enliven the soil and heal the Earth. Equally careful attention must be paid to the processes of stirring, spraying, and storing the preps by the person intending to put them to use. Some of the preparations require stirring in a particular way for specific lengths of time. *The stirring is the most important step* and cannot be circumvented if the forces of the preparations are to be transferred to the water and brought to the soil as a spray. Proper equipment to do these tasks should be on hand before obtaining large quantities of preps.

Ultimately, the Biodynamic farmer should plan to make his or her own preparations so as to regain control over the economic destiny of the farm. Further help in making preps is available from the Biodynamic Farming and Gardening Association (see below) and through workshops offered at the Josephine Porter Institute for Applied Bio-Dynamics in Woolwine, Virginia (see below). JPI also sells preparations to those new to Biodynamic Agriculture, or otherwise unable to make their own.

If you're serious about practicing the art of Biodynamic Agriculture, it takes a serious commitment. The Earth, a living being, can be healed, and our foods can be endowed with spiritual and cosmic forces, which are otherwise lacking in foods grown using NPK thinking. The nine BD preparations need to be applied as a totality and should be orchestrated by the farmer or gardener out of an attempt to understand the dynamic forces and influences working into nature out of the cosmos.

Resources

Demeter Association
PO Box 1390
Philomath, OR 97370
541-929-7148
demeter-usa.org
(Education and certification of Biodynamic Agriculture)

The Josephine Porter Institute for Applied Bio-Dynamics
PO Box 133
Woolwine, VA 24185
276-930-2463
info@jpibiodynamics.org
jpibiodynamics.org
(Product catalog available for biodynamic preps)

Bio-Dynamic Farming and Gardening Association
1661 N. Water St., Suite 307
Milwaukee, WI 53202
262-649-9212
info@biodynamics.com
biodynamics.com

Bio-Dynamic Farming and Gardening Association of Northern California
PO Box 453
Fair Oaks, CA 95628
916-965-0389
biodynamic@aol.com
biodynamics.com

CHAPTER 12

Urban Homesteading

Heirloom Skills for Sustainable Living

WHEN REAL GOODS BEGAN MORE THAN 35 YEARS AGO, our primary audience was people living off the grid, often in remote rural areas. But times have really changed! Today we recognize that people from all walks of life, living in all kinds of environments, are concerned about sustainability and climate change and interested in the kind of lifestyle that Real Goods aims to facilitate. It's not just rural anymore. The Urban Homesteading movement is thriving, and more and more people want to incorporate renewable energy, sustainable water management, food growing and permaculture, and other "heirloom skills" into city living. That's why we've asked Rachel Kaplan and K. Ruby Blume to contribute this new chapter on Urban Homesteading for our *Solar Living Sourcebook*. Rachel is a homesteader living and working in Sonoma County, California. She teaches and writes widely about homescale resilience, and she and her family homestead a small rental property they call Tiny Town Farm. Ruby runs the Institute for Urban Homesteading in Oakland, California, which offers over 60 classes a year. This piece is adapted from their book, *Urban Homesteading: Heirloom Skills for Sustainable Living*.

The weed growing up through the cracks in a city sidewalk—that sharp green shard of life persisting against all odds—reflects nature's resilience. It's also a metaphor for the uprising earth consciousness growing in our cities—small, surprising, commonplace. Spreading. Across this country, citizens are looking for solutions to the seemingly intractable problems of our time, and evolving new ways to live. And it's not about moving *back* to the land anymore—it's about tending to the land where you live. Picking up the shovel and the hoe, turning their closets and roofs and backyard decks into places to grow food and their yards into chicken coops, urban farmers are reclaiming heirloom agrarian practices as strategies for artful living. All of this is in direct response to dire resource depletion, climate chaos, and the political stalemate that characterize our time.

Urban homesteading is happening in small and large cities across the country; practitioners are relearning skills that have been abandoned in the relentless march toward convenience; valuing thrift and community self-reliance; and tending to our home places in an intentional repudiation of the cultural forces of speed, need, and greed. This is all part of a global movement for change

rooted in respect for indigenous peoples and values, a cadre of environmental first responders and a network of progressive social change organizations seeking peace and reconciliation at every level. All of these together forge an opportunity to rewrite the story of our relationship to the earth and the possibility of remaking culture around an ethic of care and stewardship for this place that is our shared home. This series of earth-based actions make an immediate difference in the places we call home, and are creative, inventive, and delicious ways to redesign our cities on the template of nature's resilience.

All the systems that sustain us—food, water, shelter, medicine, family, and community—are at risk from the ongoing disintegration of life brought about by global capitalism's profound disrespect for natural limits. It's past time for us to redesign our cities and our lives with an ethic of care at its core, remaking local systems on the model of the earth itself—adaptive, lush with diversity, and fertile with possibility. Rather than continuing to direct our life energies toward a system that is degenerate on every level—personal, social, and environmental—urban homesteading offers urban folks a chance to reskill, a

urban-homesteading.org

The urban homesteading way seeks a local life-serving economy that creates, as David Korten artfully said, "a living for all, rather than a killing for a few."

strategy for maximizing interdependence, resilience, and a sense of sufficiency in living locally.

Restructuring local economies to protect the earth and evolve our culture is central to the homesteading path. We are currently enmeshed in an extractive economy, where corporate wealth is regarded as the foundation for economic health; where mining our Earth's resources and exploiting our citizens and international neighbors is accepted as the cost of doing business. The urban homesteading way seeks a local life-serving economy that creates, as David Korten artfully said, "a living for all, rather than a killing for a few." These practices protect our common inheritance of clean water, breathable air, and a life of joy and meaning for our families.

One of the central ethics of homesteading is a sense of bioregionalism, an awareness of, and commitment to, place. Bioregionalism teaches us about the specific ecological and cultural relationships happening around us, engaging a process of asking simple questions about moonrise and moonset, about soil, about air and wind, about where our water comes from and where our waste goes. This way of becoming native to place, of living within nature's limits and gifts, is a way of creating a life that can be shared by all and passed on to future generations.

Some of the central urban homesteading practices are the same as homesteading practices everywhere—growing and preserving food, caring for and harvesting animals, foraging, making medicine, tending to the resources of water and waste and energy. While they are all scaled to meet the opportunities and limits of city living, the skills are largely the same. But a city's unique and abundant resource is human energy—the intelligence, creativity, needs, hurts, history, and futures of a city's people converging in exciting and sometimes destructive ways. Learning to harvest this energy and direct it toward projects meeting the needs of our local and regional communities

Bioregional Quiz: Where Are You?

Urban homesteading is grounded in place. How familiar are you with the place you call home? Get curious. If you don't have all the answers, take some time to find them. Knowing these details about your home will help you become a more responsive steward of your place.

1. Can you trace the water you drink from precipitation to tap?
2. How many days from today until the moon is full and/or new?
3. Describe the soil around your house.
4. What were the primary subsistence techniques of the cultures(s) that lived in your area before you?
5. Name five edible plants in your bioregion and their seasons of availability.
6. From what direction do winter storms generally come?
7. Where does your garbage go?
8. Where does your sewage go?
9. How long is your growing season?
10. Name five resident and migratory birds in your area.
11. Name five resident and migratory human beings in your area.
12. What is the land use history by humans in your bioregion in the past century?
13. What primary geological events and processes influenced the landforms of your bioregion?
14. What animal or plant species have become extinct in your region?
15. From where you are, point to the north.
16. Name one of the first spring wildflowers to bloom in your area.
17. What kinds of rocks and minerals are found in your area?
18. Were the stars out last night?
19. Name some non-human being with whom you share your space.
20. Do you celebrate the turning of the winter and summer solstice?
21. How many people live next door to you? What are their names?
22. How much gasoline do you use, on average, in a week?
23. What form of energy costs you the most money?
24. What is the largest wilderness area in your bioregion?
25. What are the greatest threats to the integrity of the ecosystem in your bioregion?
26. What is the name of the creek or river that defines your watershed?
27. What geographic and/or biotic features define your bioregion?
28. What particular place or places have special meaning for you? *

* Bioregional quiz shared with me by my first permaculture teacher, Penny Livingston-Stark. Not certain of original provenance.

will be a central survival strategy of the 21st century, and is a big part of the urban homesteading ethic.

In many places throughout this chapter, you will notice that the homesteading strategy we describe is rooted in community building and cooperation, skills that are as important as, for example, any specific knowledge of land management or food growing. (And sometimes more difficult to master.) The land frontiers have all been conquered; our final frontier is learning how to live in harmony with one another and the world around us. Rebuilding a network of relationships between the Earth and all its inhabitants will be key to human evolution and survival.

Urban Permaculture Practice

Urban homesteading is often rooted in the design science of permaculture, which has an elegant way of looking at a whole system or problem: observing how the parts relate; mending sick structures by applying ideas learned from long-term working ones; and maximizing connections between key parts in every design. Permaculturists observe and imitate the working systems of nature to mend the damaged landscapes of human agriculture and cities. This same thinking can be applied to the design of your backyard, the organization of your kitchen, getting around town, relating to people at home or work, or managing the water that falls on your house. Permaculture design is discussed in detail in other parts of this book (see Chapter 11, especially pages 292–97), so will not be dealt with in depth here. What is important to this chapter is the application of permaculture to the urban landscape, how it tends to the small scale of each homestead, and the need to extend beyond the bounds of your property and connect with other areas of your community to access and share resources.

Permaculture design attempts to replicate nature by developing edible ecosystems closely resembling their wild counterparts. The prime example of this is the food forest, a seven-level perennial planting design closely mimicking the intelligence of the forest ecology. Starting at the root level and rising to the tree canopy level, this design works well for the rural home site in temperate and tropical regions and has been widely documented. We can plant a modified food forest in our smaller urban places, and this is often a good use of small spaces, but the application more unique to urban centers is a redesign of city space using these same ecological principles.

Efficient design systems based on the resilient model of the forest ecology can be developed for city infrastructure, food distribution, and human relations. This means looking at the different elements that make up the city's landscape and beginning to forge new relationships among them. A city is made up of streets, and buildings, and a sewage system, and various laws and regulations and governmental entities. It is also made up of communities of people who have different alliances, histories, and cultural traditions. Cities house people's dreams, their economy, their children's education, their entertainment, and their health.

Urban permaculture design asks how can we bring each of these elements into better working relationships with one another. How can we bring each element of what makes up a city—the visible and invisible structures that make up our lives—into a synergistic working alignment? Each

Urban homesteading is grounded in place. How familiar are you with the place you call home? Get curious. If you don't have all the answers, take some time to find them.

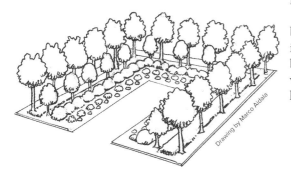

In the middle of the city, an urban food forest can maximize production in small spaces, and create oases of fertility, biodiversity, food production, pollination opportunities, and green space. The seven levels of the permaculture food forest are: A) Canopy layer, B) Smaller trees to large shrubs, C) Small shrubs, D) Herbaceous layer, E) Root layer, F) Low ground cover, G) Vining or vertically climbing plants.

Permaculturists observe and imitate the working systems of nature to mend the damaged landscapes of human agriculture and cities. This same thinking can be applied to the design of your backyard, the organization of your kitchen, getting around town, relating to people at home or work, or managing the water that falls on your house.

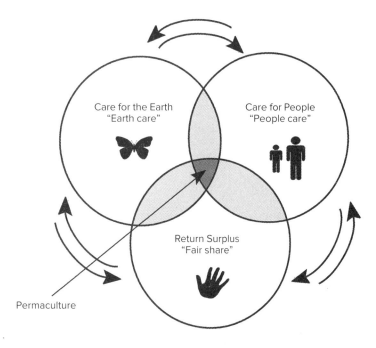

Permaculture

Cities have a vital resource that is less abundant in rural environments: the engines of human creativity, ingenuity, and diversity, resulting in an increased possibility of evolving different solutions to problems that affect us all.

element in the redesign of our cities can be looked at in terms of the input of energy and resource, and possible return, just as we do when we look at the forest ecology and notice the closed-loop system of nature's design.

Although urban density is sometimes hard to endure, it does have some ecological advantages. Human beings living in clustered settlements do less damage to our remaining wild lands; cities, with their already developed infrastructure, are prime targets for intelligent redesign; and because of people's proximity in cities, energy output, especially in the realms of transportation and home heating, can be significantly reduced. Cities have a vital resource that is less abundant in rural environments: the engines of human creativity, ingenuity, and diversity, resulting in an increased possibility of evolving different solutions to problems that affect us all. The city is where most of us are and will continue to be in the 21st century. We obviously have a lot of work to do if we are to recreate our cities on nature's model. Permaculture design is one tool for regenerating urban systems toward greater integration and productivity, reduced waste, and finding a place for everyone at the table.

Zones and Sectors in an Urban Setting

Zonation is an important concept in permaculture, a way of understanding and organizing the objects, plants, and animals you have most interaction with closest to the center of your life.[1] Bill Mollison's original model organized different

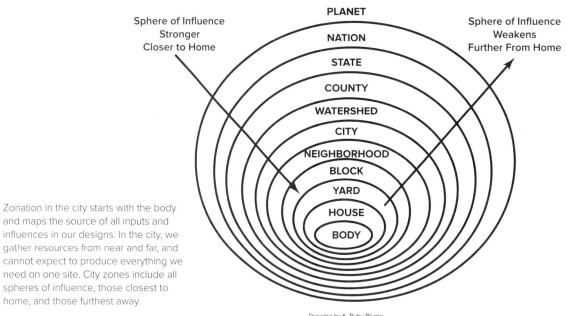

Zonation in the city starts with the body and maps the source of all inputs and influences in our designs. In the city, we gather resources from near and far, and cannot expect to produce everything we need on one site. City zones include all spheres of influence, those closest to home, and those furthest away.

Drawing by K. Ruby Blume

REAL GOODS

areas in the homestead as Zones 1–5, with Zone 1 being the area right outside your door and Zone 5 being the outlying areas of your farm. For Mollison, the golden rule is to develop the nearest area first; in our small city homesteads, it is likely that everything will be pretty close in, and our zones will extend far beyond the property line because we cannot produce everything we need onsite as we sometimes can in a rural homestead.

We can focus on zones when we plan our gardens and figure out the best place for our chickens and bees, but another way to think about zones in the city is to start with the territory closest in— the body. The body is Zone 00, the place where you are always at home. Beyond the body, zones spread in concentric circles to include your house or apartment, your garden, the street where you live, your neighborhood, your communities of necessity and affinity, your city, local government, watershed, bioregion, and so on. Delineating these zones can give us a sense of where resources come from, and how we might make new choices to localize our living. (For a discussion of zones and sectors in a more rural environment, see Chapter 11, pages 295–96.)

All of these areas offer us different resources— the comfort of home, the diversity of the marketplace, the proximity of parks and recreational areas, the availability of wild land. Notice that as the forces *increase* in scale and distance, our power to influence them *decreases*. With its emphasis on living a local life, permaculture focuses on the zones we can most easily touch and influence, but it is also important to understand how the larger circles of influence affect the choices we make in our intimate spaces.

Because a conversation about zones is always about conserving energy, another way to think about zones in a city has to do with the amount of fossil fuel energy it takes to get to a certain place. Walking then becomes the center of the concentric circles, followed by cycling, public transit, automobile travel, and air travel as the outer zone.

When we see how much space we occupy in each of these zones, we learn something about our own participation in the fossil fuel debacle. Make your own map by placing the different things you do over the course of a day, week, or month in relationship to the energy it takes to access these resources. For example, a home garden will be in a walking zone; a visit to a national park is likely to be a car ride, unless you live adjacent to one. Can you walk to your grocery store? Work? School? In this way, you will begin to understand how much fossil fuel energy you are using, and how you might begin to curtail your use.

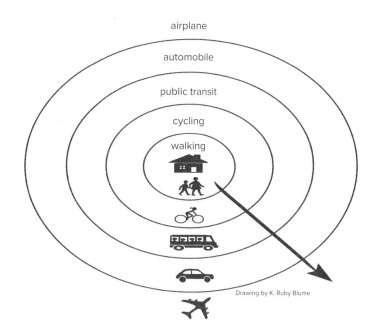

Drawing by K. Ruby Blume

Transit Zones map elements in the city by proximity and the energy it takes to get to each zone. Fill out the map with the different things you do each day or week or month. It will provide information about how much renewable and non-renewable energy you use to power your life.

Zones help us understand how to manage onsite energy; sectors are a way to look at natural or wild energy as it flows across the land. In a classic permaculture analysis, we observe wind, water, fire, sun, and weather as they affect our site. In an urban sector analysis, these natural elements are observed, but we also assess the cultural and economic forces and flows affecting our lives. Urban zone and sector analysis reflects the interplay between small personal spaces and communities and the larger social networks that impact us all.

Private property is one of the biggest socioeconomic forces defining sectors in a city. We can delineate seven sectors defined by different types of ownership:

- **Personal:** the household—rent or own.
- **Family and friends:** informal but strong relationships.
- **Associations:** clubs, churches, volunteer groups, etc.
- **Community:** neighborhood, city, county, state, federal.
- **Local businesses:** retailers, professionals, farmers, and crafts people.
- **Mega-corporations:** conglomerates, chain stores, the Fortune 500.
- **Undefined:** resource of lands without clear ownership, such as vacant lots, underpasses, abandoned houses.

These sectors can be mapped like this:

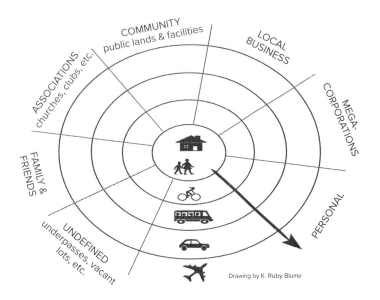

Drawing by K. Ruby Blume

This diagram shows the overlap of city zones and sectors, divided by distance and lines of property. Fill it in for yourself to map your own urban experience.

Using this new model, map the fossil fuel-based zones as they are influenced by these cultural sectors. Fill in this map for yourself. Where do your resources come from? What kind of energy do you need to access them? What do you notice about your energy and resource use that could change to meet the permaculture values of Earth Care, People Care, and Fair Share? Hopefully, your map demonstrates that there are many ways to meet your needs other than through personal ownership. For example, if you don't have the space to grow your own food at home, you can get nutritious food from a community garden plot or the farmer's market. Use these maps to assess how you might begin to streamline and localize your lifestyle.

Gardening in the Urban Habitat

Growing food is one of the central actions of any homesteader, rural or urban. Volumes have been written on the topic of organic growing. For the purposes of this chapter, we will focus on issues and projects relevant for the small and urban-scale garden. The basic gardening techniques, from seed to stem, don't change when we do them in the city, but where we do them and how we manage the scarce resource of land certainly does.

Got Land?

Access to land is not a simple issue, and the widening gap between rich and poor continues to

Left: Once a driveway, the space has been depaved, sheet mulched, and liberated to grow herbs, flowers, trees, and vegetables.
Right: Urban orchard growing in containers along the edge of a backyard.

Photo by K. Ruby Blume

REAL GOODS

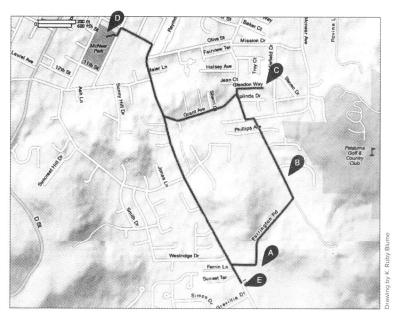

Sharing spaces: Rachel's Walking Gardens

A. Home Garden B. Chickens & Bees C. Garden Shared with Friend D. Community Garden E. Bees

Sharecropping in the city expands usable space, and turns the whole neighborhood into a farm.

Drawing by K. Ruby Blume

The median in front of your house or the narrow strip between your place and your neighbors, empty lots, alleyways, rooftops, backyards, front yards, against a wall, and decks are all places where you can grow good food.

reinforce the problem of access to private property in all of our cities. Because everyone does not have access to land, and many urban people are renters, much urban homesteading happens off-site, in alliance with others. Especially in marginalized and impoverished communities, many people do not have access to any kind of land. Available space is often used for shared projects, including community gardens, guerrilla gardening sites, food security projects, gleaning opportunities, youth leadership and job training programs, as well as school gardens.

The front lawns of municipal buildings, office parks, and businesses are other places to look for available land—they are often ready for an upgrade, especially if you do it on the cheap and commit to taking care of it. And using a municipal and public building has multiple functions—not only do you get to garden, but you also create a project that inspires others to do the same. The flat roof of your neighborhood movie theater or grocery store might be ready and waiting for a hive of bees and some fruit trees waiting to be pollinated.

Gardens can be constructed in raised beds, or in moveable barrels and pots. The median in front of your house or the narrow strip between your place and your neighbors, empty lots, alleyways, rooftops, backyards, front yards, against a wall, and decks are all places where you can grow good food. You can use a bale of straw on the median between sidewalk and street to grow your vegetables. Another way around the space dilemma is to borrow some from your neighbors and garden in their backyard. If you find yourself looking over the fence and lusting over your neighbor's unkempt yard, get down off that ladder, walk next door and offer to turn it into a productive garden. Urban homesteading has a quality of scrappy resourcefulness to it—a can do, DIY attitude. Once you've passed over the idea that a city isn't any place for a garden, you'll be amazed at how your senses get tuned in to all the places that can be useful for growing food.

Growing a wide diversity of plants in a small space on a back deck. At the height of spring, this 27-square-foot container garden contained 26 different kinds of herbs, vegetables, and flowers.

Rachel Kaplan

Self-watering Containers

Self-watering containers save on watering time and are useful throughout even the driest summers. Here's how to make one yourself.

Gather your materials.

- Two 5-gallon buckets (or any other container that can stack)
- 1 lid
- 1 plastic tub OR drain grate (The height of the tub/drain grate should be approximately the same height as the gap between the two buckets when stacked)
- 1 2-foot-long, 1-inch-diameter plastic pipe (make sure it is longer than the height of the buckets when stacked)
- 1 mesh baggie (find them as packaging for fruit or veggies)
- Drill with 1-inch bit and 1-inch masonry bit
- Utility knife with extra blades
- Rounded file
- Saw
- Permanent marker

1. **Mark the buckets.**

    a) Hole for wicking basket: on the bottom of the first bucket, trace your drain grate or plastic tub and mark a circle on the bottom of the first bucket. Be sure your circle is smaller than the lip of the container.

    b) Hole for pipe: on the same bucket, mark a hole for the pipe, also ½ inch from the wall of the bucket.

    c) Side drainage holes: measure and mark drainage holes on the side of the second bucket. Just place the buckets one next to the other and figure out how much of a gap there is between them when they stack together. Mark two drainage holes, one on each side, just below that line.

    d) Second hole for pipe: on the lid, mark a hole for the pipe (½ inch from the edge).

    e) Holes for plants: next mark holes for the seedlings on the lid, or one big hole for an established plant.

2. **Cut the holes in the buckets.**

    a) Cutting plastic kicks up a lot of little plastic bits. Protect your eyes and nose and mouth accordingly.

    b) For the big holes on the first bucket and the lid, start them with a drill, using a 1-inch masonry bit. Use the utility knife to widen the holes.

    c) Cut drainage holes in the bottom of your first bucket, using a ¼-inch-diameter drill bit. Next, cut the side drainage holes on the second bucket.

    d) Do not cut the side drainage holes in the bucket with the holes in the bottom.

3. **Prepare the pipe.** Cut an angled segment from the bottom of the pipe, using your hacksaw. The reason you're doing this is so that water can flow out of the pipe when it's at the bottom of the buckets.

4. **Assemble the wicking basket.** Either line the drain grate with mesh, or cut holes in your solid plastic container. You can also use food containers, as long as there is enough of a lip and they are the right height. The drain cover, though more expensive, seems sturdier and better for this project.

5. **Assemble the bucket.** Place the assembled wicking basket in the bottom of the bucket. Push the pipe through the holes in the lid and the bottom of the inner bucket. Stack two buckets, with the basket hanging between the two. Place the top of the bucket underneath the whole setup to catch extra water. Fill your buckets up with soil, and you're ready to grow.*

* instructables.com/id/The-Dearthbox-A-low-cost-self-water ing-planter/

The self-watering container, from top to bottom, includes:
2 buckets that fit into one another.
Top tub is filled with soil.
Connecting wick between two tubs.
Drain grate between two tubs.
Holes in bottom of top tub for connecting wick and drain grate.
Bottom tub has an air layer with an overflow hole and a water layer.

Self-watering Container

top tub with soil
connecting wick
bottom tub
air layer with
overflow hole →
water layer →

Marco Aidala

Containers and Raised Beds

Container gardening is a good way to maximize space when you don't have too much of it, and containers can be placed in parts of the yard or deck or patio that have the most sun. You can also put a raised bed that's at least 12 inches deep right on top of your driveway and garden without any problem. Many vegetables and herbs will grow in small containers and don't need to be placed directly into the ground. You can also grow a lot of vegetables in a 5-gallon bucket. As most vegetables need at least 12 to 18 inches of root space, a 5-gallon pot is the minimum size for growing food successfully, though a compact lettuce needs less depth than a wandering 5-foot-high tomato.

Community Tool Shed

Urban homesteaders take the lead on community-based projects that make our gardens grow better. Here's one: a community tool shed. If you're in good communication with your neighbors (or even if you're not), you can each buy certain tools you'll need and commit to sharing them around. That cuts down on costs for all, and brings the level of neighborly collaboration up a notch. Does everyone really need to own a ladder or a rototiller? A group of gardening friends can also invest in tools that are too expensive to buy alone, and for which you only have sporadic need: chain saw, wood chipper, back hoe, and so on. Find a common shed in which to store tools; your community tool co-op is a great urban homesteading asset.

Another great option is a tool lending library. Some cities sponsor such libraries that function just like regular libraries. People borrow tools for a short time for free and return them when done. Volunteers or nonprofit organizations often run these lending libraries. If your city or town doesn't have one yet, start one yourself.

Healing the Urban Earth: Compost, Vermicompost, and Mycelium

As in all gardens, urban gardens need soil that has been cared for and amended. City soil is especially degraded by decades of building and development, lead, and asbestos. You definitely do not want to plant food directly in a bed of untended urban soil. Fortunately, all soil can be amended by the addition of compost and organic matter, so some of our first projects in creating a city garden are to build a compost bin or worm bin and get to work composting our kitchen scraps. (Vermin are an issue in the city, so you'll want to create containers that allow for the decomposition of or-

Self-watering Container

1. Planks   2. Pallet   3. Hardware Cloth

Marco Aidala

This compost bin has two sections, both made by easy-to-find transport pallets, and is lined with hardware cloth to keep out vermin.

ganic matter and the limitation of rodent access. Cover everything you build to keep the rats out.)

You can build a compost bin in an afternoon. You need a three-sided box ideally 3 × 3 feet in diameter. One of the simplest designs involves three pallets nailed together into the shape of a C. If you have the room, make a compost bin with two sections, in the shape of an E. Make a gate with some chicken wire in the front of it, but latch it in such a way that you can move it for the times when you turn over the compost in the pile. Once you've got the bin built, it's easy to get started composting your kitchen and yard wastes.

If you don't have enough space to host a compost bin, a worm bin is the best way to compost your kitchen scraps. Building and maintaining a worm bin is a great urban gardening project—it doesn't take up a lot of space, uses available kitchen scraps and turns them into soil, and can be easily placed on a deck, back porch, or small yard setting. Once the worms eat your garbage, you'll end up with vermicompost, one of the best soil amendments available.

The Magic Mushroom

Cultivating mushrooms and mycelium grows the fertility of your homestead, and provides another food source. Mushrooms are important keys to human and planetary health, living in a symbiotic relationship with trees and other plants of the forest by breaking down matter and distributing nutrients. We can grow mushrooms for environmental remediation projects as well as growing edible mushrooms, adding another taste treat to the long list of foods easily grown in small urban spaces.

Building and maintaining a worm bin is a great urban gardening project—it doesn't take up a lot of space, uses available kitchen scraps and turns them into soil, and can be easily placed on a deck, back porch, or small yard setting.

Build a Worm Bin

Materials Needed

- Two 8- to 10-gallon dark-colored plastic storage boxes
- Drill (with ¼-inch and ¹⁄₁₆-inch bits)
- Newspaper
- One pound of red worms

1. Drill about 20 evenly spaced ¼-inch holes in the bottom of each bin. These holes will provide drainage and allow the worms to crawl into the second bin when you are ready to harvest the castings.

2. Drill ventilation holes about 1½ inches apart on each side of the bin near the top edge using the ¹⁄₁₆-inch bit. Also drill about 30 small holes in the top of one of the lids.

3. Prepare bedding for the worms by shredding newspaper into 1-inch strips. Moisten the newspaper by soaking it in water and then squeezing out the excess. Worms need moist bedding. Cover the bottom of the bin with 3 to 4 inches of fluffed up moist newspaper. Old leaves or leaf litter can also be added. Throw in a handful of dirt for "grit" to help the worms digest their food.

4. Add your worms to the bedding. One way to gather red worms is to put out a large piece of wet cardboard on your lawn or garden at night. The red worms live in the top three inches of organic material, and like to come up and feast on the wet cardboard. Lift up cardboard to gather the worms. An earthworm can consume about half of its weight each day. If your food waste averages ½ pound per day, you will need 1 pound of worms or a 2:1 ratio. There are roughly 500 worms in one pound. Don't worry if you start out with less than one pound—they multiply very quickly. Just adjust the amount that you feed them to fit your worm population.

5. Cut a piece of cardboard to fit over the bedding, and get it wet. Then cover the bedding and the worms with the cardboard. (Worms love cardboard, and it breaks down within months.)

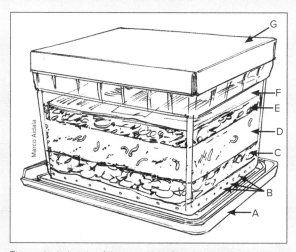

Marco Aidala

The multiple levels of the worm bin, from bottom to top, include: A: Original container lid; B: Drilled holes in the bottom of the container; C: Drainage materials; D: Worm bedding; E: Organic wastes; F: Newspaper; G: Lid on top of bin. Worms live in the worm bedding and transform this organic matter into vermicompost. Any extra liquid ends up in the container lid and can be used to fertilize plants in the garden.

Construct a cylinder of chicken wire and fill it with substrate to feed mushroom spores. Add water. Oyster mushrooms are fast and simple to grow and offer a plentiful yield.

K. Ruby Blume

Mycelium, the root structure of mushrooms, "runs" or spreads throughout the soil in a fibrous film-like network. To generate connectivity and soil health, we want to spread mycelium through the garden. A simple way to do this is to build a propagation bed that will contain and grow mycelium, as well as sprouting edible mushrooms. Some simple-to-grow edible mushrooms are the oyster mushroom, the garden giant (wine cap stropharia), shiitakes, or shaggy mane. Do a taste test before starting this project to decide which mushrooms you want to cultivate in your yard and start with the one you like best.

Within a month, the mycelium will have spread throughout the box. When the first mushroom sprouts, leave it to refruit into the propagation bed. Wait for the next mushroom to appear before you start to harvest your food. If you want to generate more mushroom patches through-

Building a Propagation Bed

Materials Needed

- Wood 4–5 inches wide, between 1–2 feet 1¼-inch hardware cloth, at least 2 foot square
- Hardwood chips, mulch, or wood shavings (alder or oak). (If hardwood is not available, use a softwood like pine or eucalyptus.)
- Mycelium or spores for the mushrooms you wish to cultivate. A common choice to cultivate in this manner is the wine cap stropharia.
- Screws
- Drill

1. Build a square box between 10 and 24 inches wide, 4 to 5 inches deep. You can also use pre-made plastic boxes with ventilation holes on the bottom.
2. Staple ¼-inch hardware cloth on the bottom.
3. Soak the substrate of hardwood chips, mulch, or wood shavings for about 30 minutes before you put them in the propagation bed. If you're using wood shavings like pine or eucalyptus with volatile oils in them, they will need to be soaked overnight.
4. Drain the substrate, and fill the box with the materials.
5. "Seed" the substrate with mycelium spores. You can get them from a friendly mycologist or online at one of these sites: fungiperfecti.com or field forest.net.
6. Divide the mycelium spawn into five sections. If you are using a package of inoculated wooden dowels, use five per propagation bed.
7. Place them in the box with some distance between them to encourage the spreading of mycelium through the bed.
8. Put the box in a shady area of the garden. Keep this box well watered, but not drenched.

out the garden or spread the mycelium around to support plant life in the garden, take a portion of the mycelium-infused substrate from the propagation bed and place it under an apple tree, or even into a garden bed underneath a shady plant, where it will get enough moisture and humidity to continue to fruit and spread. This mycelium and its fruiting body will enrich your growing soil and can be left to grow and decompose in the garden, or be eaten, whichever you prefer.

Growing Mushrooms in Straw Bales

You can also grow mushrooms in straw bales placed in cool, shady places in the garden. Oyster mushrooms are easily cultivated in this manner. Wet the straw bales well, and stuff the spawn deeply throughout the straw bale. Keep the bale watered and wet. It is best to inoculate just before the major rainy season in your area so you don't have to do much of the watering yourself. It should take a few months for the bale to start sprouting mushrooms. Once you have harvested the mushrooms and the straw bale begins to decompose, you can mulch it into the garden, which will also spread the beneficial mycelium.

Other Space-saving Garden Projects

The following two sidebars give instructions for two other simple projects that will help you optimize the use of space in your urban garden in ways that also nurture your plants: a cold frame and a mini hoop house.

Urban Orchard: Growing Fruit in the City

Fruit trees are perennial and provide a lot of food for many people over many years. They also provide forage for beneficial insects, shade, and beauty in the garden. An urban orchard is surprisingly easy to tend. Many fruit trees grow happily in pots for many years, or thrive in small, out-of-the-way places. Columnar trees (pruned to grow straight, rather than branch out) are easy to set into small places at the top of a driveway or the corner of a deck. An espaliered tree won't take up much space, provides shade and privacy when flattened up against a fence, and yields a bounty of fruit. Planting fruit trees among other plants

The central stem of an espaliered tree is trained vertically against a wall, and lateral branches grow out at right angles to the central trunk.

K. Ruby Blume

Build-it-Yourself Cold Frame

Materials Needed
- Salvaged window, any dimension
- Scrap wood, at least 1" × 6"
- Screws
- Drill
- Hinge (optional)

1. Measure the window frame.
2. Use scrap wood at least 1" × 6" to construct a box that fits the measurement of the window. Construct it so the window can rest comfortably on this wooden frame.
3. Screw together your frame into the shape of the window. Set the window over the frame.
4. Optional: Attach the window to the frame with a hinge. (This will make it possible to prop the cold frame up on hot days to provide ventilation for plants, but the cold frame will function with or without the hinge.)
5. Put it in a sunny location and you're ready to go.

You can skip steps 1–3 if you have an already constructed box—an old cabinet or a set of shelves you find at a garage sale or thrift store. Simply turn this into a cold frame by filling it with dirt and covering it with a properly sized glass window.

Start plants in separate flats underneath the glass, or fill the cold frame with dirt and grow them right in there. The dirt will need to be periodically replenished as you lift the seedlings out and transplant them into the garden. Attach the window to the frame with a hinge, and use a stick to prop open the cold frame on warm spring days. Remember to prop it open when the days warm up, or you'll fry your starts. Site your cold frame so it gets maximum sunlight.

or throughout the garden is part of a perennial strategy that evolves into lush urban gardens. Trees need water, but not as much as an annual vegetable or herb, and are usually satisfied with a good drenching once or twice a month. Fruit trees planted in the ground do well with less water than trees planted in pots. Once established, a tree in the ground will live and produce for many years. Check with your neighbors and find out which fruit trees fluorish in your neighborhood and microclimate.

Caring for Trees in Containers
Some fruit trees are grafted onto dwarf stock, which means they won't grow above a certain height—these are often ideal for smaller spaces, as well as for long lives in containers, barrels, or large pots. When keeping trees in pots, fertilize

them and prune them each year to keep them happy in their small spaces. Trees in pots need more fertilizer, more water, and should be kept smaller than trees planted in the ground. A tree in a container should not be allowed to grow out of proportion to the container itself. A 12-foot tree in a 3-foot container will not thrive. Be vigorous and regimented about fertilizing and cutting back your trees when they live in containers. You may need to work with their roots every few years or so, tipping them out of the pots, loosening the roots as much as possible, and replanting them in a bigger pot.

Another great urban project is a community orchard, managed by a few devoted arborists, but planted throughout a neighborhood or city. Sometimes it just takes a few motivated people to turn an empty yard, street median, or abandoned alley into a food-growing zone. "Borrowing" people's yards and tending them in exchange for some of the shared bounty is a strategy for planting an urban orchard that could eventually feed many people locally and provide an urban arborist with a living.

Check out commonvision.org for orchard planting resources. These folks will come to your town in their painted bus loaded down with fruit trees, and plant 20 to 50 trees in an afternoon in places where they will flourish and grow. These events are affordable and great community builders, drawing in children, adults, and elders.

> "Borrowing" people's yards and tending them in exchange for some of the shared bounty is a strategy for planting an urban orchard that could eventually feed many people locally and provide an urban arborist with a living.

Easy Mini Hoop House

One of the best ways to stretch your growing season, especially in four-season climates, is by using a greenhouse. In most urban settings, you just won't have room for a true greenhouse structure, but you can simulate greenhouse conditions easily and provide another zone for starting sprouts in a small space. If you do have space for a greenhouse, we envy you, and encourage you to go online to find a design that suits you. Some of the best ones are narrow structures placed against the sides of buildings and constructed out of PVC pipe and greenhouse plastic. For the rest of us, simulating greenhouse conditions will have to suffice.

To make your mini hoop house, you're going to be building a tent made out of bamboo, rattan, and PVC pipe or some other flexible material, and covering it with plastic sheeting. You can make a freestanding setup that you place on an outside deck, patio, or yard space, or you can attach the greenhouse "tent" to a table and place it in a sunny place in your yard.

1. Find a long table that won't mind living outside (something metal or plastic will work).
2. Create a frame around the table by attaching bamboo or wire in an arched shape to the edges of the table. If you are using PVC, get 3-way fittings to make the corners. This will create a square base around the table and allow you to fit the hoops into the corner pieces. If you opt for material like bamboo or rattan, attach the corners of the tent with strong wire to hold them in place.
3. Make the form high enough to provide space for starter flats and roomy enough above the flats for the starts to grow.
4. Cover the frame with greenhouse-grade plastic or Reemay. You'll have to purchase this new, and it

may cost a bit, but it lasts for years and provides the protection and sunlight plants need. Attach it with clamps, or clothespins if the framing material is narrow enough.

5. Attach the cover to the "greenhouse" on three sides, and leave a fourth side moveable. This way your plants have protection, and you have access to your plants.
6. Place your flats with seeds in them on the table and cover with the plastic. Make sure you situate your "greenhouse" in a sunny part of the yard. Check frequently to make sure moisture, heat, and light needs are being met.

You can take this contraption apart in the winter when you won't use it. If you set it up on a folding table, it should be easy to stash somewhere until the next growing season.

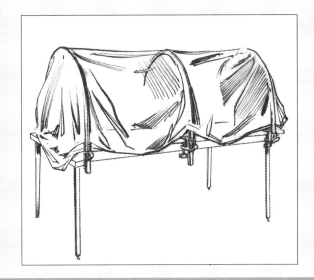

Animals on the Urban Homestead

Animals turn your garden into a farm. We are delighted to see chickens, rabbits, ducks, quail, bees, and the occasional goat on urban farms from coast to coast. When thinking about taking responsibility for any animal on an urban homestead, please consider their basic needs and whether you can meet them in the context of a busy urban life.

Shelter/Housing/Protection
All backyard animals need predator-proof homes that are dry, clean, and spacious enough to meet their needs. Chickens need more space than quail or rabbits, and goats will need more than chickens. If you are breeding animals for food, you may have to provide separate housing for differently sexed animals until you bring them together to breed. Most animals prefer to have their houses out of direct sunlight, in close proximity to places where they can roam freely for part of the day, and lifted off the ground to provide protection from predators and weather.

Nutritional Needs and Food Sources
On most urban homesteads, it will be unrealistic to imagine you could produce all the food your animals will need, though some of them, like ducks and chickens, supplement their food with

Think about the people who live around you when turning your yard into a farm. It's going to impact them, and that will have ripple effects on your neighborhood relations. But for the most part, neighbors are excited and interested in animals, and they are a great way to make friends.

An animal cage on wheels can be placed in different parts of the yard that need mowing. Rabbits keep the lawn trimmed and deposit high-quality fertilizer as they go.

snails and slugs and bugs from the garden. Other animals, like goats, chickens, and rabbits, will eat kitchen and garden scraps and convert them into fertilizer for the garden, but this won't be enough nutrition for them overall. Some urban farmers feed their animals by dumpster diving. This may be a fine short-term solution for feeding an animal you are going to butcher at the end of a season, but maybe not the best choice for a long-term relationship. Make sure you have access to the proper kind of feed at a price you can afford before you take on an animal that's going to eat every day. Different animals have different nutritional needs—providing the animals the food they need to keep them healthy and productive is key.

Composting

With any animal, you're going to have to figure out what to do with their waste. Fortunately, the waste of many backyard barnyard animals provides excellent fertilizer for the garden. Chicken poop needs to be composted for some time before being added to the garden; rabbit poop can be integrated into the garden without a stint on the compost pile, and goat manure is also excellent for the garden as is. Ducks poop in their watery habitats; when the pond needs to be cleaned, this water can be added to the compost pile to increase its fertility.

The ease of mucking out your animal's home should be considered. A rabbit hutch with a wire bottom is simple: Just let the poop drop on the ground underneath the hutch, rake it up, and put it in the garden. A chicken coop can also be designed with wire flooring to serve the same function. You will be glad to limit the number of times per year that you have to walk into your chicken coop to clean it out. Pre-built structures can also

be adapted to make cleaning and composting an easier task.

Legality

Before you start keeping farm animals in the city, get familiar with city ordinances regarding small livestock in your area. Some places allow chickens, but not roosters; some places forbid bees. Many urban centers balk at goats or pigs but are happy enough to house the rabbit, inside or outside the house. Check with your municipality to find out what's allowed in your area. Use your best judgment coupled with local ordinances to make decisions about how many animals to keep. More important than the law, though, are the needs of the animals themselves. Too many animals shoved into a small space makes for unhealthy conditions and unhappy animals, the exact opposite of kind of quality living we want to provide.

Neighbors

Keep your neighbors in mind when getting backyard animals. Chances are that a rooster under a neighbor's window won't go over too well. Animals with a strong smell aren't usually much appreciated either. Don't be cavalier about this—think about the people who live around you when turning your yard into a farm. It's going to impact them, and that will have ripple effects on your neighborhood relations. But for the most part, neighbors are excited and interested in animals, and they are a great way to make friends. Very few people say no to a fresh egg, or a jar of honey, or the chance to watch a goat wandering through the neighborhood.

Alliances with Other Community Members.

Keeping animals is a pleasure, but it is also a daily chore. Some animals, like goats, require more care, and sometimes the best way to provide it is to share the work with others. To meet this need, goat-sharing cooperatives are springing up and providing the people who own goats some help, while giving members the opportunity to learn how to take care of goats and enjoy their milk products.

Time and Attention the Animals Will Need

Once an animal's home is established, its care doesn't take much more than minutes a day. Make sure you assess your own time limitations before you commit to a new animal on your homestead. If you travel a lot, a good way to manage your animals is to share their care, feeding, and produce with neighbors and friends. Three rabbits really don't take much more time than one rabbit,

K. Ruby Blume

but breeding rabbits and taking care of the babies takes extra time. The same is true with chickens—the size of flock doesn't add time to your day, but if you are taking care of small chicks that need more attention to survive, factor that in when you start your animal-keeping project.

Predators and Pets

Some city neighborhoods have an abundance of predators that can be dangerous for your animals. A large gathering of feral cats will threaten small chickens, ducks, or other fowl. Dogs can also be a problem for smaller barnyard animals, and some are wired to grab a chicken by the neck and shake it, even when they know you love the chicken too. As mentioned before, the design of the animal's shelter should reflect an awareness of opportunistic predators like raccoons and possums, as should your habits for securing your animals at night.

Chickens in the Garden

We don't really need to sing the praises of the chicken here—chickens are very popular on the urban homestead, for good reasons: They provide eggs, high-quality nitrogen-rich poop, and endless hours of entertainment. However, we would like to encourage people to really think about whether they need their *own* flock, or are able to share the love with their neighbors. While people enjoy the eggs, and some intrepid homesteaders love the homegrown meat, keep in mind that a chicken will lay for a certain number of months each year and only for a certain number of years, and then is just eating imported feed in your backyard and pooping up a storm. They take up space, make noise, and have a strong odor. Are they worthwhile for your homestead? Do you want a *pet* chicken? Think it through before you invest money in a coop and a flock, and assess whether or not doing this as a multi-family project might not be the most efficient use of space and energy. And be mindful of space. You don't want to have "free range" chickens in an area so small they can barely get around.

Urban Rabbits for Fun, Food, and Fertilizer

Rabbits are assets to the backyard homestead for some of the same reasons as chickens—they produce the best manure of all the backyard animals, they'll help you keep your lawn trimmed, they breed like, well…rabbits, are a sustainable source of meat, and are much beloved by children on the homestead. Taking care of rabbits is easy, but your work load will be different if you are raising

rabbits for fun, food, or fertilizer. Make sure you think out all the dimensions of your rabbits and the uses they serve as you design the place they will occupy on your homestead.

The Duck: The Perfect Permaculture Pet

Ducks in the backyard homestead also have a number of uses and benefits. Like chickens, they scratch and peck, and they are extremely good at picking snails, bugs, and insects off your vegetables. (The permaculture cliché goes like this: "You don't have a snail surplus. You have a duck deficiency.") Ducks also like to eat little sprouts, so you want to set them out in the garden after your plants have reached a pretty good size, or make a contained run for them in places where you want them to go. They happily waddle through the garden in a little flock, seeking out food and nesting under any available straw or leaf pile. They have great personalities and are fun to watch, much like their avian cousins. They lay eggs, and can also be turned into fancy food products if you are so inclined.

You can house a small flock of two to four ducks in a large doghouse (easy to find at a recycling center or a lucky street giveaway) because they don't need much inside space to be happy. You can use an old box crate or desk, predator-proof, and out of the hot sun for much of the day. It should have good ventilation, and be a bit off the ground. Ducks will wander through the garden and seek shelter when needed, but are mostly out and about during the day. Unlike the chicken and the rabbit, the duck needs a water habitat to be happy. You can set up an old bathtub filled with water-loving plants for use as a duck habitat; you can grow plants in it, and store some rain as well, creating a multi-purpose feature in your yard.

Raising Quail for Eggs and Meat

Coturnix, or Japanese quail, are a delight to raise. Their space requirements are small, they don't eat a lot, they convert feed into protein efficiently, and they are more congenial by far than the chicken. The modern coturnix has been bred to begin producing eggs when less than two months old. Once she starts laying, the hen will produce an egg daily for at least a year. The males are equally rapid growers, being ready for the table at six to eight weeks of age.

Coturnix eggs are nearly identical in taste and nutritional quality to chicken eggs. Coturnix hens, however, need less than two pounds of feed to produce a pound of eggs. Chickens need almost three pounds of feed to make that same

Bees are a true addition to the homestead because they offer pollination to plants and food to humans; in exchange, beekeepers and gardeners can offer healthy opportunities for home and forage to an endangered and essential pollinator.

pound of eggs. Five coturnix eggs equal one chicken egg. Because of their small size, coturnix can be kept in small pens, such as a wire cage, rabbit hutch, or even a small dog kennel. Cages can be raised or can rest on the ground. If the cage is raised, it will be easier to clean, the birds will never be standing in manure, and their eggs will remain clean.

Bee Kind

Bees are a true addition to the homestead because they offer pollination to plants and food to humans; in exchange, beekeepers and gardeners can offer healthy opportunities for home and forage to an endangered and essential pollinator. Bees forage within a three-mile radius of their hive, so they pollinate your neighbor's gardens as well. Bees are feral and untamed; they bring wildness into the garden. Beehives can be sited in a backyard, on a roof, or in an empty lot. Because bees use the sun to direct their flights into and out of the hive, they thrive in a place with morning sun and afternoon shade.

In most places, it is perfectly legal to keep bees, though in others an absence of laws about beekeeping can be taken for consent. When people are concerned about bees, it is usually because they think of them as stinging hazards, and do not understand the benefits they provide. You do the bees a service by educating people about the gentle, non-aggressive nature of the bee, who only stings in defense of her brood and her food. If you have a neighbor who is truly allergic to bees, some caution may be warranted when placing your hive, but otherwise education and the sharing of local honey is often enough to help

the neighbors relax and enjoy the hive. Bees have few predators, unless you live in a city with bears, and dogs and cats tend to stay away, especially if they've been stung once.

While there are many types of beehives, the two most common ones are the Langstroth—those white boxes we are all familiar with—and the Top Bar Hive, which can be made in a variety of shapes, but is generally long and narrow. The Langstroth boxes are the dominant style in the US and can be easily purchased from any beekeeping supply company. They are well understood and there is a lot of support available for beginning beekeepers.

The Top Bar Hive, developed by the Peace Corps as a cottage industry with low start-up cost for African villagers, is gaining popularity among natural beekeepers in the US. Not yet readily available commercially, they are inexpensive and easy to construct with minimal carpentry skills. In this system, the bees build free-form comb off bars that sit across the top of the opening of the box.

Goats...In the City?

We see goat herding on the outer edges of what's possible in most backyard homesteads, and are amazed to see goats gaining ground in cities around the country. Goats are curious, friendly, smart, and deeply relational creatures; they need attention, diverse forage, and one another. They are also territorial, especially around their children, and can be noisy and belligerent when crossed, threatened, or in unfamiliar territory. Goats are generally kept for the milk and compost they provide. They are ideally suited for sloping areas that may be challenging for gardens, and can speedily clear a yard of bamboo or blackberry brambles. Goats are happy in marginal spaces and will quickly take over an abandoned lot. But they demand a commitment of both time and money that needs to be assessed before taking the plunge. While forage goats are hardy creatures that may struggle along on whatever is available, a goat that is expected to produce milk every day needs good-quality feed, housing, and medical care.

For these reasons, a goat-keeping collective is an excellent way to care for an urban herd. Members share the tasks of tending, walking, milking, hoof trimming, and cleaning up after the goats in exchange for milk, meat, compost amendments, and good company. If you can join a goat collective or find someone who has goats who will teach you, becoming an apprentice is a good way to begin keeping goats.

Walking a goat in the neighborhood. Good for the goat, good for the girl.

K. Ruby Blume

How to Build a Top Bar Hive

One of the appeals of the Top Bar Hive is its affordability and the fact that with some basic carpentry skills, you can build one yourself. Here's a description of how to the build the TBH at home. The total cost of building a Top Bar Hive should be less than $100. You can generally buy a full complement of bees for your hive, including the queen, for around $100 as well.

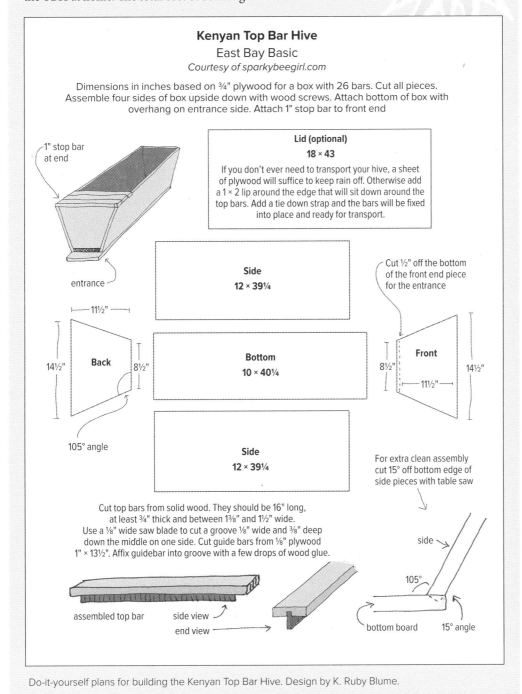

Kenyan Top Bar Hive

East Bay Basic

Courtesy of sparkybeegirl.com

Dimensions in inches based on ¾" plywood for a box with 26 bars. Cut all pieces. Assemble four sides of box upside down with wood screws. Attach bottom of box with overhang on entrance side. Attach 1" stop bar to front end

1" stop bar at end

entrance

Lid (optional)

18 × 43

If you don't ever need to transport your hive, a sheet of plywood will suffice to keep rain off. Otherwise add a 1 × 2 lip around the edge that will sit down around the top bars. Add a tie down strap and the bars will be fixed into place and ready for transport.

Side
12 × 39¼

Cut ½" off the bottom of the front end piece for the entrance

11½"

14½"   **Back**   8½"

Bottom
10 × 40¼

8½"   **Front**   14½"

11½"

105° angle

Side
12 × 39¼

For extra clean assembly cut 15° off bottom edge of side pieces with table saw

Cut top bars from solid wood. They should be 16" long, at least ¾" thick and between 1⅜" and 1½" wide. Use a ⅛" wide saw blade to cut a groove ⅛" wide and ⅜" deep down the middle on one side. Cut guide bars from ⅛" plywood 1" × 13½". Affix guidebar into groove with a few drops of wood glue.

side

105°

assembled top bar   side view   end view   bottom board   15° angle

Do-it-yourself plans for building the Kenyan Top Bar Hive. Design by K. Ruby Blume.

Even without growing your own food, it is possible to realign your food buying and eating habits within a sustainable urban framework, by frequenting local food co-ops, buying in bulk, shopping at farmer's markets, or getting your veggies from a community supported agriculture farm.

Full-up and fantastic forage from a local orchard.

Harvesting Your Animal Friends

Learning to slaughter your backyard animals in a compassionate and timely manner is good homesteading practice. We recognize that slaughtering animals is not for everyone—some people won't even consider it, and would rather support their barnyard flocks into their dotage. And some people raise animals because of what they offer on intangible levels, and would never consider eating these feathered or furry friends. We respect that, but feel that as we prioritize maximum yield on our homesteads, selective and humane butchering of our barnyard animals is a good idea.

If you're going to enter into the harvesting of your animals and do it in a responsible way, we highly recommend getting a few lessons from an expert. You can learn fairly simply how to harvest a chicken, rabbit, or quail. When it comes to the larger mammals—goats and pigs—we suggest doing an apprenticeship with a master before taking it on yourself. Keep in mind that animal slaughter in cities is definitely regulated by health ordinances; make sure you know what they are before you start killing animals in the backyard. For this homesteading task, you may need to make friends with someone who lives outside your neighborhood and has more land where this kind of action is more appropriate.

Remaking Our Relationship to Food

Remaking our relationship to food—where it comes from, how it's grown, how it's processed, how we eat it, and how to make it available for everyone's benefit—is an important part of the homesteading lifestyle. Simple and banal as it may seem, changing your relationship to food is one of the most powerful ways to cut loose from an unsustainable and dangerous industrial system and participate in making a better one. Even without growing your own food, it is possible to realign your food buying and eating habits within a sustainable urban framework, by frequenting local food co-ops, buying in bulk, shopping at farmer's markets, or getting your veggies from a community supported agriculture farm.

Beyond gardening, here are a couple of other options for food gathering in the city.

Gleaning

By far, the most reliable source of food in the city's gleaning web is those fruit trees standing in front of people's houses waiting for someone to pick them. Don't delay. Walk up to that house and knock on the door. Ask your neighbor if you can harvest their fruit. Sometimes people who don't tend their trees are happy to share some of the excess, especially if you're willing to pick as much as you can and leave some in baskets on their front porch. A fruit foraging practice yields hundreds of pounds of fruit for your family each year, an excellent return on the time you'll spend wandering the streets and picking it up.

Petaluma Bounty, a local northern California nonprofit, has a program called the Bounty Hunters. After identifying trees whose fruit is going to waste, the organization makes relationships with the owners of the trees, and then heads out to pick up the extra produce. It's then given to local residents, or sold at a low fee at the community farm that was established to serve food-insecure folks in our area. The Bounty Hunters picked three tons of food in their first year. In their second

Slice fruits and vegetables about ¼-inch thick and let the sun do the work of drying them into tasty treats that will last into the next season.

Building a Solar Drying Rack

Materials Needed
- 1" × 1" wood
- Metal brackets
- Food-grade plastic mesh
- Staple gun and staples
- Drill and screws

This solar drying rack was built in a few afternoons out of leftover 1" × 1" wood.

1. Construct three squares out of wood. Brace them with metal brackets.
2. Cover squares with plastic mesh by stapling mesh to the frames. Overlap mesh around wood frames to secure it. These are your drying trays.
3. Build two more squares of wood big enough to go around the drying trays. Brace them with brackets as well.
4. Turn these two squares into a cube by connecting them together with four pieces of wood in the corners, approximately 6 inches high.
5. Once you have made the rectangular cube, affix pieces of wood the length of the frame to the short sides of the cube. If you want to have three drying racks, affix three pieces of wood to the frame. These will support the trays as they sit in the rack.

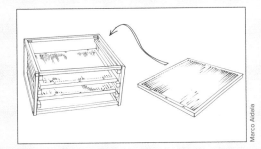

Marco Aldala

A simple-to-build solar food dryer is a rectangular box with shelves holding a set of drying racks. The bottom is open, which allows for airflow; the top can be covered with a screen to keep off insects. This dryer took a few afternoons for an imperfect carpenter to design and build.

You can also build and use a solar oven in the city, much as you would on any rural homestead. These nifty tools allow us to access the power of the sun, no matter where we live. Or you can purchase a solar oven from Real Goods, see realgoods.com.

Urban homesteaders ferment a wide variety of foods in their kitchens—cheese, kombucha, sourdough bread, sauerkraut, kimchee, kefir. We particularly enjoy lacto-fermented sodas and find it simple to cultivate in the urban homestead kitchen.

year, four tons of food, which otherwise would have gone to waste and rotted on people's lawns, was foraged and distributed to people who need it. This program resourcefully uses local bounty to strengthen community. You can start a project like this your neighborhood, too. (Petaluma Bounty.org)

The Dumpster: The Last American Commons

In 1988, the Supreme Court declared that once something is thrown away in a dumpster, it becomes common property, thus establishing the dumpster as possibly the last and largest commons in the US. Most people think eating or foraging out of a dumpster is nasty, but when you really get in there, you'd be amazed at how much perfectly fine food, good books, clean clothing, furniture, and kitchenware the dumpster holds. Some find that dumpster-diving is a perfectly abundant way to find dinner. Most of us won't go that hardcore route, but a moderate approach to the riches abandoned in a dumpster never hurt anyone. Dumpster diving is a truly conservative approach to resource use. Dumpsters can feed

you and your farm animals quite well, while saving massive amounts of perfectly good stuff from going right into the landfill.

Mixing up cabbage for sauerkraut.

Diane Dew Photography

The Heirloom Kitchen Skills

Urban homesteaders, like homesteaders everywhere, spend a lot of time in the kitchen during harvest season, turning fresh food into preserved products that can last through the winter. The processes for food preservation and storage—canning, drying, fermenting, root cellaring, and freezing—are the same in the urban kitchen, and are discussed in depth in this and many other books and websites. All you need to know from an urban perspective is that these skills give us an opportunity to reclaim some self-sufficiency around our food sources. Urban homesteaders are rediscovering that it is fun to play with your food.

Drying the Harvest

Drying the surplus harvest is a simple way to preserve it for the months when things aren't growing in the garden. A solar food dehydrator will maintain a nutritious and tasty supply of high-quality, locally grown foods all year long. If you live in a part of the country where summers are hot and dry, a solar dryer is a wonderful tool. When assessing your conservation goals, the choice for a solar dryer over an electric one is clear.

Fermentation

People have co-evolved and partnered with countless microbial life forms in mutually beneficial relationships for a long time. The long list of fermented foods that many of us know and love includes olives, cheese, sourdough bread, dry meats, wine, fish sauce, ketchup, vinegar, coffee, tea, vanilla, ginger beer, tofu, soy sauce, miso, natto, tempeh, sauerkraut, and chocolate. Other simple-to-cultivate treats include yogurt, kefir, kombucha, kimchee, and lacto-fermented vegetables and sodas. When we are culturing and fermenting foods, we recruit the microbes and beneficial bacteria from our own kitchens in the process. These traditional foods also enhance our health and the proper functioning of our bodies. Unseen, beneficial microbes enhance the flavors and nutritional quality of our foods, transforming fresh food into something tasty, healthful, and new.

Urban homesteaders ferment a wide variety of foods in their kitchens—cheese, kombucha, sourdough bread, sauerkraut, kimchee, kefir, etc.… We particularly enjoy the recipe on the following page for lacto-fermented sodas and find it simple to cultivate in the urban homestead kitchen.

Natural Building

Reimagining our homes through better design and the use of more renewable resources is one of the central themes of this *Sourcebook* (see Chapter 2), and of the products sold by Real Goods (realgoods.com). City folks are redesigning their homes with alternative building strategies, including reused and upcycled materials, and pushing the edges of permitting processes to bring them to the municipalities that make the rules controlling our building practices. Many books about natural building methods are available from Real Goods (see realgoods.com). Meet the activists at Villa Sobrante, a collective home in a suburb of Oakland, California, who are working this home front.

On the Ground: Urban Retrofit

Villa Sobrante is a collective household in El Sobrante, California. The resident owners are four young women—Sasha, Massey, Trilby, and Lindsey—joined in a shared love of natural building and homesteading. They all met at the Real Goods Solar Living Institute, where as interns they were introduced to all of these homesteading arts. Once done with that training, they decided to stake their claim "outside the borders of Berkeley" (i.e., a community in the "know"), and bring the good news of natural building to the outlying suburbs. "When we decided to do this,

Mixing the cob by foot.

Miguel Micah Elliott

Lacto-Fermented Sodas

Infusions for homemade sodas can be made from the wide range of what's available, including sassafras, sarsaparilla, ginger, elderflower, elderberry, valerian, cinnamon, spruce, juniper, spearmint, wintergreen, hops, star anise, yarrow, rosehips, hibiscus, lemon, orange, grapefruit, raspberry, strawberry, peach, and pear. Use whatever's good and growing in your neighborhood.

For all the lacto-fermented sodas described below, you'll need the following equipment to get started.

Equipment
- Pot
- Funnel or sieve
- Fermentation vessels—jars, crocks, carboys, buckets, etc.
- Measuring cups and spoons
- Bottles and caps

The ingredients for all lacto-fermented sodas are the same, and vary only in the kinds of herbs or fruits with which you infuse your beverage, and your choice of sweetener.

Ingredients
- Non-chlorinated water
- Fruit or herbs
- Sweeteners—agave, cane, maple, honey
- Starter cultures—this can be whey, a commercial culture, your last batch of soda, or make your own.
- Ginger (optional)

Making Your Own Starter Culture
Lacto-fermented sodas need a starter culture to begin the process of fermentation. Every beverage will need either a starter culture or whey to begin the fermentation process.

How to
1. Grate or finely dice fresh ginger root and put a tablespoon of it into a mason jar ¾ full of water, along with 2 tsp white sugar.
2. Add another 2 tsp each sugar and ginger every day for a week, at which time it should become bubbly with a pleasant odor.
3. If it gets moldy, dump it and start over.

Once you have your starter, use it to make a small quart-sized batch. Then use two cups of this for a gallon batch. (You can drink the rest.) Save a pint from each batch you make to inoculate the next batch. Use two cups of this starter culture or one cup of whey per gallon of soda. Both whey and starter can be stored in the refrigerator for a few weeks before the cultures use up all available food and start to die off. You can stave off having to "catch" a new one by adding a tablespoon each of sugar and ginger to your starter.

Basic Lacto-Fermented Soda
- 1 cup sweetener per gallon
- 1 cup whey per gallon OR 2 cups starter per gallon
- ½ tsp salt per gallon
- Infuse to taste

Putting It Together
1. Wash your fermenting vessels in hot soapy water.
2a. Make an herbal infusion by placing herbs into hot water, as you would for tea. Let the herbs steep to taste, then strain them out of the water; or
2b. Place fresh fruit or fresh fruit juice, up to 2 cups per half gallon, into your fermenting vessel.
3. Add approximately 1 cup sweetener per gallon.
4. Add 2 cups starter culture or 1 cup whey per gallon to fermentation vessel.
5. Add infusion to jar until close to the rim of the jar or, if making a fruit soda, top the jar off with pure non-chlorinated water.
6. Add two or three slices of fresh ginger and/or a pinch of salt. (This is optional, but it helps with the fermenting process.)
7. Cover loosely (you can use a canning lid, just don't screw it on all the way).
8. Let sit for two to six days, until the mixture in the jar starts to get fizzy. Timelines will vary with temperature and cultures. The only way to tell how the culturing process is going is to check your ferments every day by tasting them until they've reached the taste you like.
9. Bottle, refrigerate, and drink. If you seal in airtight containers, you can let it sit out for a few hours so the carbonation builds up, but be careful—a potent lacto-fermented soda can be quite explosive! If your container is airtight, it is best to release extra carbonation, even if refrigerated, if you don't drink the soda within the first few days.

Playing with plaster during an earthen plastering workshop. Natural building made from cob, earthen plaster, and recycled glass.

Sasha Rabin/Vertical Clay

"We were seen as regenerating the neighborhood just by being here."

Students mixing earthen plaster for adobe building.

Sasha Rabin/Vertical Clay

we wanted to find out how much we can do for ourselves as urban homesteaders," Sasha explains. "We wanted to learn how to put things into practice, how to live with them, observe them, and learn to do it with other people.

"We wanted to do this in a place where it isn't being done much, and to see what kind of influence that has. Partly we chose this site because it was a good place to regenerate. It needed so much, and was such a problematic property, so when we arrived we were welcomed here. If this place had had a perfect lawn in the front of the house when we bought it, the neighbors would not have welcomed us as much. As it was, even a sheet-mulched lawn was better than a foreclosed property that was a target for homeless people. We were seen as regenerating the neighborhood just by being here."

Villa Sobrante, which houses the first permitted natural building retrofit in Contra Costa County, is the most substantial natural building

urban retrofit we found in the Bay Area. "We're trying to experiment with alternative housing in an urban setting in a way that's legitimate," Massey says. "We're taking this 800-square-foot house and making it livable for four or more people, within code. When people who have power over the permitting process get to see that these kinds of structures are sound and safe and a viable alternative to the waste and expense of building, and the problems of creating affordable housing especially in very dense urban areas like this one, the laws will change. We care about the legality issue because we're trying to do this in a way that can be a model. Working with the county and the existing laws is the only way to get them to change."

Other small buildings on the property are made from natural materials like cob, straw, earth bags, and locally sourced bamboo. Materials are reused and recycled from the dump, Craigslist, other building projects, and the overflow of their retrofit. Alongside the building projects, residents are creating gardens, digging swales for water catchment, and learning to live together. "The swales and earthworks were a natural thing as we started working with the dirt—when you dig a hole, there's a lot of extra dirt, so what we didn't use in the buildings, we used in shaping the land," Massey explains. "And since growing food was one of our goals and there are so many issues with water in our area, managing the dirt so that it helps sink the water into the ground was a smart design. It helped us really see our property in a different way—how does the water flow? Where does it go? What's the best way to use the earth we have to the most effect? How do you go about designing what you have in a way that's efficient and doesn't waste a lot of resources or energy?"

Villa Sobrante residents share living space, garden space, financial commitment, and work load for the project. "Working with the county is a learning experience, and learning to make a great garden bed is a challenge too, but probably the biggest learning edge is in living together. In the beginning, we were living pretty much on top of one another," Lindsey recalls. "That helped us decide pretty quickly to build structures where we could have some privacy while living together so intensively. We hope the natural building we do here serves as an inspiration for people, but sometimes we think the way we've learned to live together is just as important as a model for another way of living in our culture."

Homesteading has become meaningful for Sasha, Massey, Trilby, and Lindsey in ways that are personal and political. Sasha summed it up when she said, "When people are more involved in the systems that sustain their own lives, something more freeing happens than when we live in a more constricted way. When we are involved in how our food grows, where it comes from, how our structures are built, where we live, some deeper part of our humanity comes through.

There is so much to learn from being connected to your environment in this way, spending a lot of time outside with different systems and being more connected to the root of everything that provides for our basic human needs. If you are more connected to them, you feel more alive."

Powering Down

Like our new building practices, urban homesteaders are also revisioning the uses of energy in the city. Cities have an energetic advantage over rural regions—you can live without a car in many cities, using bicycle and public transport, and the occasional rented car to get around. Now that electric vehicles are proliferating and coming down in cost, they're the perfect vehicle for city dwellers. (For more information, see Chapter 13, Sustainable Transportation.) All of the usual fixes for truing up our houses to use less energy are happening in the city. And energy use is often lower in multiple-tenancy buildings, and can be more affordable.

Here's one good idea for aggregating energy use and distribution in a city or suburban area: district heating, a neighborhood-based energy-sharing collective to lower the use and cost of natural gas. In a district heating scheme, each household of close neighbors buys shares in an efficient system designed to conduct hot water to different buildings. Neighbors agree about the placement of the energy-efficient boiler, the infrastructure for piping the energy from house to house, and a cost-sharing arrangement. Each home gets energy from the boiler, and all expenses are shared.

District heating is a good way to install energy-efficient systems at lower cost, generate interdependent and resilient structures between neighbors, and begin to significantly lower the carbon footprint at home. Costs always go down when people share them, fewer resources are used, and the workload and cost for maintaining the system is spread among many. Similar neighborhood arrangements take advantage of shared solar arrays for generating power.

Not surprisingly, large utility companies have already passed laws restricting such neighborhood co-ops for energy sharing. As it stands in some places, a group would have to become an independent utility company to make this work, and that's pretty rough going. In California in 2010, the local industrial energy corporation placed on the ballot a measure that would have strengthened the utility's monopoly on energy provision and made it nearly impossible for counties to figure out ways to provide their own alternative energy. Fortunately, this ballot measure was overturned by a majority of Californians, but political initiatives like this must be carefully scrutinized, and vigorously opposed.

For full detailed discussions of residential renewable energy systems, especially photovoltaics, see Chapters 3 and 4; for a full discussion of solar hot water, see Chapter 8. Every urban dweller should know how easy it is to go solar for no money down and a smaller electric bill than they currently pay, through "third-party ownership" like leases and power purchase agreements. A full range of renewable energy products is available from Real Goods at realgoods.com.

Before embarking on this kind of collective enterprise, make sure you know the laws in your municipality and within your utility district. This is an obvious place for legislative agitation: Either utility companies need to provide people with renewable energy alternatives, or they need to allow people to provide it for themselves. We advocate such community-based solutions over personal and private property solutions at all times. Working with others in these kinds of small-scale collectives is a direct way to take action on the issue of significantly reducing nonrenewable energy use at home.

Every urban dweller should know how easy it is to go solar for no money down and a smaller electric bill than they currently pay, through "third-party ownership" like leases and power purchase agreements.

Renting or living in an apartment doesn't mean giving up water sustainability. Greywater harvesting from sinks, showers, tubs, and laundry can be done at low cost, and the systems are low- to no-tech, or fully reversible when you move.

Greywater and Rainwater Use

As drought and water issues become increasingly more common, it is essential that we all reduce our use, and participate in reusing and recycling as much water as we can. Greywater and rainwater catchment are important not only in rural areas, but in our cities as well. In many municipalities, it is not legal to divert greywater to the landscape, but this is beginning to change as our challenges with water are exacerbated by population growth, agricultural excess, and dwindling water resources, especially in the arid West. Arizona, New Mexico, Texas, and Wyoming have very simple guidelines to follow for legal greywater reuse that do not require a permit. California recently passed a new code allowing washing machine greywater systems to be simply and legally installed, following state guidelines.

There are no laws (yet!) against using buckets and hoses to divert water to the landscape, but the diversion of laundry water and bathwater to the landscape remains controversial in many cities. Kitchen sink water is considered "blackwater"

Rainwater Storage: The Daisy Chain

In a city homestead, it is unlikely that you will have enough space to use a cistern to store rainwater and a single rain barrel often does not hold enough water to make much difference. But using the narrow spaces between buildings or side alleys allows us to set up a chain of water barrels that increases the storage capacity of your system. The engineering of a classic downspout and storage system is the same, but the volume you catch will be much greater. Separate from the cost of the barrels, the total cost of the system should not exceed $50.

Materials

- Electrical male adapters (1 per barrel)
- Rubber washers (not an O-ring)
- Female threaded barbed tee fittings (1 per barrel)
- 1-inch vinyl tubing to connect the barrels (1 per pair of barrels)
- 1 brass shutoff valve

Rainwater diverted from roof into rainwater barrels. To increase storage capacity, barrels are connected to form a daisy chain. Barrel on the far right has a spigot for water distribution when needed. Pipe coming from the top of the right-hand barrel allows the water to overflow when barrels have reached maximum capacity. Direct this pipe toward the landscape and away from the foundation of the building.

1. The first barrel has a hose connection to the second barrel. Install a female threaded tee to this first barrel.
2. On the side that points out and away from the barrel, attach a brass male hose thread to male pipe thread (MHT to MPT) connector, with a hose shut-off valve connection. The side that points toward the next barrel uses a male pipe thread with barbed fitting and the vinyl tube connecting the two barrels is hose clamped over the barb.
3. When the rainwater is needed for irrigation, a garden hose is attached to the first part of the female tee and the shut-off valve is turned "on." The middle barrels have a female barbed tee with a 1" vinyl tube hose clamped on each end.
4. The last barrel in a string of barrels ends with the tube tied off with wire. This allows for easy connection of more barrels in the future.
5. If no further connections are desired, the last barrel could end with a female-threaded 90-degree bend, with a MPT by barbed coupling. The 1" vinyl tubing fits over the barbed fittings and is hose clamped. It is important to use electrical male adapters and not plumbing adapters because the plumbing fittings have rounded edges and it's harder to create a watertight seal with them. When troubleshooting and maintaining the system, check the screens and make sure they are intact.

Barrels are connected with tee fittings, 1-inch distribution line, electrical male adapters, and O-rings inserted into the barrel itself. Spigot at the end of the line attaches to garden hose and water can flow from there into the landscape.

in some, but not all states, which designates it as unsafe to be reused and toxic for land and humans. It is therefore illegal to pipe this water into the landscape, even into a place where the water would be filtered and cleaned (through an easily constructed biofilter). In some states, it is illegal to harvest the rainwater falling on your roof.

You will have to decide for yourself if your conservation ethic trumps your law and order ethic. Know the law in your state, and if you feel uneasy about the legal aspects and it keeps you from being able to conserve water in a way you'd like, work to help local agencies understand the values, benefits, and low health risks associated with greywater reuse and rainwater harvesting. Support them in changing laws to allow the intelligent reuse of these available and plentiful waters.

Strategies for greywater and rainwater reuse are covered in Chapters 7 and 10 of this book. Resources are available at harvestingrainwater.com, oasisdesign.net, and realgoods.com.

Some Specific City Solutions: Greywater for the Renter

Renting or living in an apartment doesn't mean giving up water sustainability. Greywater harvesting from sinks, showers, tubs, and laundry can be done at low cost, and the systems are low- to no-tech, or fully reversible when you move. In an apartment where you have access to a backyard and garden space, you can set up a simple composting toilet and use the urine as fertilizer in the garden, divert the water from the washing machine, or siphon water from the bathtub into the garden without spending much more than $20. You can also harvest the water from the bathroom sink for reuse.

Without significantly changing your (landlord's) house you can also install a shower shut-off valve, a $5 piece of hardware that will reduce shower water use by 50%. You can wash dishes using a dishpan, and use the water on your garden. Another simple way to save water is to add bulk (a rock or a brick) to displace water in the toilet tank. This will lower the number of gallons that get flushed each time you do flush the toilet.

Laundry to Landscape Greywater Reuse

"Laundry to landscape" greywater reuse is becoming more common and is a good way to recycle a lot of water into the garden. In some states, this is legally permitted; in others, you are a hacker until you press your municipality to see the benefit of this simple reuse strategy.

This outdoor washing machine provides an easy way to divert water to the landscape. Note the red valve on left side of washer—it allows for choice about whether water will divert to landscape or drain. Run a single or multi-trunk line, with or without valve or branches, to the landscape. This system can also be constructed on an indoor washing machine. The mechanism is the same.

Distribution plumbing of 1-inch should be used to get water to different parts of the landscape. The outlets are directed toward mulch basins filled with organic matter that further sinks the water into the ground.

Covered water outlets help sink water into the landscape.

Direction of greywater can also be controlled from a branched drain in the garden. Each time the system is used, water can be diverted to a different part of the garden. Branches can be 1-inch, 2-inch, or ½-inch.

Waste

We need to move beyond Reduce, Reuse, and Recycle to the five Rs of Rethink, Redesign, Reduce, Repair, and Recover.

Zero waste is the Holy Grail of the sustainable city. You probably have "Reduce, Reuse, and Recycle" tattooed on the inside of your head by now. If you're a compost devotee, Rot has been tagged onto the three Rs as well. This is all good, but it doesn't really address the small-world reality: There's simply nowhere for the waste to go. We need to move beyond Reduce, Reuse, and Recycle to the five Rs of Rethink ("waste" is a *resource*), Redesign (from open to closed system), Reduce (consumption and use), Repair (systems and objects themselves), and Recover (compost what's organic, and return it to the earth as nutrients).[2] Once we've internalized the permaculture ethics of Earth Care, People Care, and Fair Share, zero waste practices become second nature, a central strategy for meeting our need to be of service to the Earth.

Reduce

Of the three Rs, Reduce is absolutely the most important. Our country's over-consumptive buying habits are literally trashing the globe. Committing to the Reduce ethic is solid homesteading, a way of taking your energy (and money) away from the consumer economy and making peace with what you've got (which we venture to bet is an awful lot anyway). About stuff we think we need to buy but really don't, we have a simple mantra: Just Say No.

Recycle and Reuse

Recycling is great, but reusing is better. What are alternative uses for your recyclables? We advocate for the transformation of recycling into upcycling: Make something you need out of materials you formerly thought of as "garbage." Pretend the engines of production have shut down in your city, and notice how much you have around you that can be repurposed and reused.

Humanure: The Ultimate Recycling Project

Human excrement is an unused resource, not a waste product, and recycling it is good for our gardens, the waters, and the Earth. Our current

Recover: A Kitchen Sink Recycling System

Here's a solid reuse project for your kitchen. Dirty sink water can be toxic for plants and humans, especially when it contains meat and dairy products whose fats and oils are not beneficial for plants or for the soil in your garden. But rather than sending the water down the drain to further contaminate the waterways, we advocate constructing a biofilter to remediate this water and direct it to further uses on your homestead. This system will use the worm bin you built when you were making a vermicompost setup for your back porch (see above, page 314).

1. Add a third level to the two existing levels of your worm bin by getting another container that fits into the already constructed worm bin.
2. Drill holes in the bottom of the box for ventilation.
3. Place straw in this box and inoculate it with oyster mushroom spawn.
4. Pour the blackwater from your kitchen sink into this chamber.
5. Continue to place kitchen scraps in the second level of the worm bin as before.
6. Mushrooms will grow in the upper box. You can eat them, or plant them out in your garden to spread the mycelium to the trees and plants you are growing.

7. The straw in this level of the system will absorb some blackwater. The rest will drop into the next level of the box where the worms live and are fed by kitchen scraps. The kitchen sink water keeps this part of the system well watered, and brings more nutrients for worms and other microorganisms to eat.
8. The water will continue to filter down to the final bottom box of the worm bin.
9. The water has now gone through two levels of remediation. What drains from this final level of the worm bin is a very fine vermicompost tea, a golden elixir for the garden.
10. Drain this water and return it to the garden.

This triple stacking function system converts the blackwater from the kitchen sink into a great organic fertilizer for your plants in about two square feet on your back patio. The mushroom spawn and worms have transformed a waste product into food for themselves and the garden.*

* Kevin Bayuk, personal conversation, 3/6/10.

How to Build a Simple Composting Toilet

What You'll Need
- Medical toilet seat
- Two 5-gallon buckets
- One plant pot that fits inside 5-gallon bucket
- Subsoil
- Compost
- Red worms

The simplest way to make a composting toilet for your home is to purchase a medical toilet seat (the kind where you can easily sit down to pee), and keep two buckets nearby. One bucket is for pee, the other bucket is for poop. Both buckets are layered with sawdust in between deposits. You can also purchase the Separette Privy Kit, which separates urine and feces for easy collection. This lowers the work load, eliminates odors, and makes waste composting simple.

Place the toilet seat over the buckets, and do your business as usual. Sprinkle a handful of sawdust into the bucket after each use. The urine your household produces is safe to use without treatment. When the pee bucket is full, dump it on the compost pile, or layer it in the garden as mulch. The nutrient-rich sawdust will enhance either garden or compost pile. Urine contains nitrogen, phosphorus, and potassium and makes an amazing plant fertilizer. It is typically sterile, and if separated from feces, can be easily and safely reused.

In the simple two-bucket system, the poop bucket is constructed slightly differently than the pee bucket. For the poop bucket, take another five-gallon bucket (which has a cover) and find a plant pot that fits within it. Usually the black plastic pots that come with small trees or large plants will fit perfectly. Line the bottom of that pot with about one inch of subsoil, packing it in so that it keeps the fecal matter from leaking out. This creates a block that prevents any disease vectors from escaping, but allows for a little bit of drainage. Add a layer of compost with live worms in it, about two inches deep. You can then start making deposits in the bucket, covering each deposit with a bulk agent, like wood chips or sawdust. Keep the bucket covered when not in use.

When the plant pot is nearly full (about three inches from the top), remove it from the five-gallon bucket, and add another layer of worm-rich compost. Plant a tree in the center of the pot, and let the pot sit in the shade under a tree for a few months. Keep it watered, but not drenched. This will keep any contaminants from leaking into the soil. Rodents don't get in because they won't dig through the subsoil. The trees love this soil, which has no pathogens because the worms have been processing it for the month or so it takes for one person to fill up the pot.

After you let it sit for a few months, you can transplant the tree into the ground, or give it to someone you love.*

Please note that Real Goods also sells composting toilets: realgoods.com.

* Kevin Bayuk in personal conversation, 3/6/10, as taught to him by Ben Jordan.

---

sewage system is broken, poisoning the waters with bodily wastes that could be used to rebuild our diminishing topsoil. You can build yourself a composting toilet in the backyard in an afternoon. Most ecological toilets are built so that poop and pee go in different places (one bucket for each). There are also urine-diverting inserts that can be integrated into the system that split the pee and poop into different places, making it possible to sit and do your business without having to separate liquid from solid waste on your own. Some homesteaders site their composting toilets outside in a little nook in the garden, protected from prying eyes by bamboo stands or other beautifully smelling plants. It's nice to make the privy private, even when it's outside, as cultural taboos about excrement are extremely powerful and inhibiting to the goal of zero waste. For more information on humanure and composting toilets, refer to Chapter 10. Or check out *The Humanure Handbook*, by Joe Jenkins (available at realgoods.com).

Personal Ecology

Our bodies, our minds, our emotions, and our spirits together are our first homesteads—unique, demanding, mountainous, oceanic, rocky, fertile, wind blown, and ultimately compostable (see Chapter 14, Natural Burial, for a different, sustainable take on this topic). It's smart to sustain the most precious resource of the self. Rather than thinking of self-care as an indulgence, we see

> Rather than thinking of self-care as an indulgence, we see it as the beginning of cultivating the ground from which we take positive actions in the world, and a reflection of how we care for the world around us.

it as the beginning of cultivating the ground from which we take positive actions in the world, and a reflection of how we care for the world around us. Continuing to separate our personal bodies from the body of the Earth just doesn't work. Instead of getting to ourselves last, or sacrificing ourselves to "save the world," cultivating self-love and care as an outgrowth of our concern for our world and our future is a priority.

We talk about energy a lot when we talk about sustainable living, but focus mostly on efficiency or solar energy or biofuels or bicycles. Managing human energy needs to be part of the conversation if homesteaders are to be successful over time. Our physical energy—the ability to do, to build, to dig, to eat—is supported by exercise, rest, and play. Our emotional energy—happiness, or at least contentment and satisfaction—needs support and appreciation to flourish. Our mental energy or clarity—the ability to plan, make decisions, and solve problems—is often balanced through non-linear processes like art making, ritual, and ceremony. Our spiritual energy—our awareness of the life force flowing within us—is fed by relationships with nature, meditation, and other spiritual practices, and our engagement in community. Tending to these aspects of our personal lives is like tending to the garden in different seasons, or the chickens in the morning and the bees on a sunny afternoon. The work we do on ourselves will have an impact on many levels, and affect all of our practices in the art of living.

Resources

There are many resources pertinent to the Urban Homesteading lifestyle, and homesteading in general, and this short resource guide is only meant to direct you toward some of our favorites at this time. If you find yourself interested in any aspect of the homesteading way, like water, or food production, or waste management, make sure you check out the abundance of resources that exist, in book, media and online form.

Rachel Kaplan, with K. Ruby Blume, *Urban Homesteading: Heirloom Skills for Sustainable Living* (Skyhorse, 2011). Available at urban-homesteading.org

Daily Acts
dailyacts.org

The Institute for Urban Homesteading
iuh.oakland.com

Kelly Coyne and Eric Knutsen, *The Urban Homestead* (Process, 2010). Also see their website, Rootsimple.com

Chris Shein, *Vegetable Gardener's Guide to Permaculture* (Timber Press, 2013).

John Jeavons, *How to Grow More Vegetables*, 8th ed. (Ten Speed, 2012).

H. C. Flores, *Food Not Lawns* (Chelsea Green, 2006).

Wendy Tremayne, *The Good Life Lab: Radical Experiments in Hands-on Living* (Storey, 2013). See also her blog, Holy Scrap: blog.holyscrap hotsprings.com

BioFuels Oasis
biofueloasis.com

Novella Carpenter and Willow Rosenthal, *The Essential Urban Farmer* (Penguin, 2011).

Novella Carpenter, *Farm City: The Education of an Urban Farmer* (Penguin, 2010).

Urban Farm Magazine
facebook.com/urbanfarm

Creating sacred space out of a garage wall. Cob art by Miguel Micah Elliott.

Miguel Micah Elliott

CHAPTER 13

Sustainable Transportation

Toward a Future of Carbon-free Mobility

YOU MIGHT BE SURPRISED to find a chapter on sustainable transportation in the *Solar Living Sourcebook*, because Real Goods is not in the business of selling vehicles. We offer some excellent resources but only a few relevant products. No matter. Let's confront the reality head on: For most Americans, our personal use of motor vehicles is responsible for as much or more of our carbon footprint as operating a household. Nationwide, the transportation sector accounts for 28% of greenhouse gas emissions. Cars and light trucks account for more than half of that total. Our cultural addiction to fossil fuels creates problems beyond climate change, most notably air pollution, sprawl, a large military establishment predicated on protecting access to oil, and vulnerability to volatile prices, supply interruptions, and terrorism. For all these reasons—because the energy and climate change implications of transportation for long-term sustainable living on Earth are so critical—we think it is important to discuss the subject in the *Sourcebook*.

The concept of sustainable transportation should be understood holistically, just like the concepts of sustainable shelter, energy, and agriculture. As defined by the Global Development Research Center, "Sustainable transportation concerns systems, policies, and technologies. It aims for the efficient transit of goods and services and for sustainable freight and delivery systems. The design of vehicle-free city planning, along with pedestrian- and bicycle-friendly design of neighborhoods, is a critical aspect for grassroots activities, as are telework and teleconferencing" (gdrc.org/uem/sustran/sustran.html). For most of us who live in automobile-centered societies, where personal mobility is considered a civil right, the key conceptual shift is to understand that transportation should be about access to people and goods, rather than freedom of mobility. Urban design and transportation policy should favor people and healthy communities over motorized vehicles. (Ideas about relocalization and resilience explored in Chapter 1 are relevant too.) However, with more than 300 million gas-guzzling cars and trucks filling roadways in the US, it is clear we need to transform the core concepts behind providing fuel and systems of fuel production while these vehicles still rule our streets.

> For most Americans, our personal use of motor vehicles is responsible for as much or more of our carbon footprint as operating a household. Nationwide, the transportation sector accounts for 28% of greenhouse gas emissions.

Lower-carbon and more energy-efficient alternatives to fossil fuels for automotive transportation come in three basic flavors. Powering motor vehicles with electricity or biofuels is realistic, affordable, and increasingly popular right now. Fuel cells that burn hydrogen rather than hydrocarbons are a technology still in development and not on the near horizon for consumer vehicles. We've asked long-time Real Goods associates and experts in their respective fields David Blume (biofuels) and Steve Heckeroth (electric vehicles) to update the state of affairs with these alternatives in 2014. As you'll see, they have differing perspectives about the relative merits of biofuels and electricity as the best choice for addressing climate change in the short term and building a world of sustainable transportation in the longer term—and they're both skeptical about the promise and utility of fuel cells. At Real Goods, our best choice is to let David and Steve each have their say and let you, the reader and decision maker in your own life, assess the options that might work

for you. It's clear to us that *any* shift away from fossil fuels that results in an immediate reduction of greenhouse gas emissions is a good thing. Ultimately, building a sustainable transportation system will require bold public policy decisions, investment in the necessary infrastructure, and incentives that encourage and assist consumers to kick the fossil fuel habit.

> Ultimately, building a sustainable transportation system will require bold public policy decisions, investment in the necessary infrastructure, and incentives that encourage and assist consumers to kick the fossil fuel habit.

The Problem with Fossil Fuels

The biggest problem with fossil fuels, of course, is that burning them produces carbon dioxide and other air pollutants. The carbon emissions from internal combustion engines are a major component of the greenhouse gases that are causing climate change. (For more on the specifics of climate change, see Chapter 1.)

A second big problem with fossil fuels is that producing them is not energy-efficient. Compared with the alternatives, for fossil fuels the energy return on investment (EROI)—the amount of energy produced per unit of energy consumed in the process of production—is unfavorable. And when the entire production process is taken into account, energy consumption and greenhouse gas emissions are higher. This is especially true in our current era of peak oil, when fossil fuels are becoming depleted and tapping the remaining sources is difficult, expensive, and energy intensive.

The average EROI for conventional oil before 1970 was roughly 25:1. In other words, 25 units of oil-based energy are obtained for every one unit of other energy that was used to extract it. The EROI for deep-water drilling can be as little as 5:1. For heavy oil that needs to be heated before it can be pumped out of the ground, the EROI can be as low as 4:1, and for oil shale and tar sands, it can be less than 3:1. More energy used to get less oil out of the ground means more carbon emissions and other pollution for less production. (Rachel Nuwer, *Inside Climate News*, February 2013.) Furthermore, the natural gas that is used to extract oil out of tar sands also has an EROI that is spiraling out of control as fracking becomes a common practice, contaminating ground water and causing earthquakes across America. When all the externalities are accounted for, including restoration of miles of lifeless scars in what used to be pristine wilderness, the EROI will probably be negative for tar sands.

The environmental risks of extracting unconventional sources of petroleum, via deep-water drilling or strip mining of tar sands, can be dev-

astating, as we witnessed with the Deepwater Horizon spill in the Gulf of Mexico in 2010. The cheapest way to move oil is by pipeline, but pipelines can take years to build, so as once marginal oilfields become profitable, thousands of rail tanker cars are pushed into service. In 2013, more oil spilled in the US from trains than the total spilled since the federal government began collecting data on spills in the early 1970s. There have been major derailments in Alabama and North Dakota, not to mention the crude oil spilled in Quebec, Canada, on July 6, 2013, when a runaway train derailed and exploded, destroying most of a town and killing 47 people.

A final problem created by our fossil fuel addiction is geopolitical. Protecting US access to petroleum resources around the world remains a guiding principle of American foreign policy and a rationale for a large military establishment. Even though those very problematic unconventional sources in North America are enabling the US to reduce oil imports from other parts of the world, like the Middle East, that trend is unlikely to alter the desire of American business and political leaders to maintain that high level of military readiness. Replacing fossil fuels in our transportation system with renewably sourced electricity or biofuels will help to alter that logic. For example, according to David Blume, if the US were to move to the 30% alcohol fuel blend mandate referenced in the EPA's Blend Report (*New York Times*, May 3, 2013), we would no longer need to import any petroleum from the Middle East.

In this era of climate change and our quest for fossil fuel alternatives, perhaps the best way of assessing the relative merits of motor vehicle fuel systems is to consider the concept of energy efficiency from sun to wheel. In other words, since the sun is the primary source for all types of energy, it is important to compare the overall conversion efficiency of how do different vehicle fuels—fossil, biomass, electricity, hydrogen—compare in their overall conversion efficiency from solar energy to transportation energy.

The Problem with Today's Motor Vehicles

There are over 200 million licensed drivers in the US who each burn an average of 800 gallons of fuel annually to go a total of more than 3 trillion miles. And while small steps have been made recently to mandate increased efficiency, in 2013 the US reached an all-time high in the average fuel economy of American cars. Are you ready?—24.6 miles per gallon. That's compared to a fleet average of 45 mpg in the European Union, and higher in Japan. Only one word is appropriate to describe this "achievement": Shameful. The American automobile and petroleum industries and their allies in government continue to stand in the way of real, concrete progress that is urgently needed to deal with the reality of climate change.

One of the main reasons for the low average fuel efficiency of the American fleet is the continuing popularity of SUVs and light trucks. According to federal regulations, light-duty trucks are primarily designed for the transport of property or for off-road operation and have a gross vehicle weight of less than 14,000 lbs. The "light truck" classification was created in the early 1970s to acknowledge that vehicles used for work or on the farm would have difficulty meeting the same standards as cars. In 1978, Congress enacted the Gas Guzzler Tax and exempted light trucks, so that vehicles in this category do not need to meet the same safety, fuel economy, or emission standards as cars. Over the years, the auto industry has exploited this fact by introducing luxury trucks, vans, and 4 Wheel Drive Sport Utility Vehicles (SUVs) that are mainly used for the on-road transport of people. A study by Friends of the Earth found that, since 1999, automakers have avoided paying over $150 billion in Gas Guzzler taxes by calling passenger vehicles "light trucks." Because trucks don't need to meet the same standards as cars, they are much cheaper to manufacture and their profit margin can be more than ten times greater than that of the more fuel-efficient cars that serve the same purpose. The billions of dollars spent on advertising these grossly inefficient vehicles has taken "light truck" sales from 10% of sales in 1990 to over 50% of the nearly 20 million vehicles sold in the US in 2013. Indeed, the top-selling vehicle in America for the past 32 years is the Ford F Series pickup truck, and in 2013, three out of the five best-selling autos in America were pickups that get less than 20 mpg. As a result, the fuel economy of the US vehicle fleet as a whole has remained relatively constant even though the fuel efficiency of "cars" has been increasing.

Increasing the average fuel economy of cars to 40 miles per gallon—which is easily accomplished with existing hybrid technology—would save 5 million barrels of oil a day. That would represent an enormous savings in CO_2 emissions, too: 5 million barrels × 55 gal./barrel × 365 days/yr. × 10 years × 23.8 lb. CO_2/gal. = nearly 12 billion tons of CO_2 over a decade. (See Paul Roberts, *The End of Oil: On the Edge of a Perilous New World.*)

Trends during the past several years have proven beyond a doubt that alternative-fuel vehicles are here to stay. Automakers are scrambling to keep up with growing consumer demand for hybrid electric cars. Constantly evolving battery technologies and emerging plug-in hybrids (PHEVs) promise to make super-low-emission driving for commuting and short trips—by far the majority of driving that Americans do—a reality in the foreseeable future. All-electric vehicles are finally coming onto the market, and they offer the promise of a vehicles fueled by virtually zero-carbon solar power. And biofuels are catching on in a big way, with the infrastructure for cost-effective and energy-efficient production of biodiesel and ethanol being put into place.

Switching to zero-emission vehicles that don't burn any fossil fuels will eventually make a big difference, and in some scenarios our economy might someday be powered primarily by hydrogen, which can itself be produced without greenhouse gas emissions using renewable energy sources. But those days are far in the future, and there is considerable urgency to act now to begin withdrawing from our destructive fossil fuel habit. What can a mobility-loving, energy-conscientious individual do?

- Drive a vehicle that gets good fuel mileage.
- Drive a flex-fuel vehicle that can burn bioethanol (alcohol fuel).
- Purchase a fuel conversion kit that transforms gas engines into flex-fuel engines with the intelligence to know what type fuel is being used and how to use it most efficiently.
- Stop buying and using diesel fuel—diesel is implicated in numerous studies as directly linked to causing lung cancer. Or switch to biodiesel.
- Buy a hybrid gas-electric vehicle, and consider using bioethanol instead of gasoline.
- Buy an electric vehicle, and plan to charge it with solar panels. (If it's charged with grid power, it will have a considerable carbon and pollution footprint.)

In this era of climate change and our quest for fossil fuel alternatives, perhaps the best way of assessing the relative merits of motor vehicle fuel systems is to consider the concept of energy efficiency from sun to wheel.

Increasing the average fuel economy of cars to 40 miles per gallon—which is easily accomplished with existing hybrid technology—would save 5 million barrels of oil a day. That would represent an enormous savings in CO_2 emissions, too.

- Use mass transit as much as you can, and lobby your service provider to switch away from diesel fuel.
- Take advantage of carpooling or car sharing in communities where that system exists.
- Walk or ride your bike for a commuting change.
- Plant trees to offset your personal CO_2 emissions.

Organizations such as American Forests (americanforests.org) and Trees for the Future (treesfor thefuture.org) have excellent reforestation programs in the United States and around the world; all you need to do is plunk down a little cash. The American Forests website even has an easy-to-use Personal Climate Change Calculator that tells you how many trees have to be planted to offset your annual contribution to global warming.

Let's take a look at the near and far horizons of vehicle development and see what choices are now on the market and likely to be available in the future.

Hybrids and Other Low-emission Vehicles

The fact that hybrid electric vehicles are becoming wildly popular with consumers and that the auto manufacturers are having trouble keeping up with demand is good news all around, because gas-electric hybrids represent the simplest, most effective, most affordable short-term way to reduce fossil fuel consumption in passenger cars and the environmental impact of driving motor vehicles. In response to more stringent state and federal emissions standards, the manufacturers are also marketing Ultra-Low Emission and Super Ultra-Low Emission gasoline vehicles (ULEVs and SULEVs) that feature redesigned, more efficient engines and more effective tailpipe emission controls.

The Toyota Prius hybrid.

A hybrid combines the best features of internal combustion and electric drive, while eliminating the worst parts of each. Hybrid vehicles have both an internal combustion engine and an electric motor, with a highly sophisticated automatic control system that chooses which one is running under what conditions. The electric motor is used at low speeds and for acceleration boost; the internal combustion motor delivers cruising speed (above 35 miles an hour) and long range but automatically shuts off at stoplights. Deceleration and braking actually capture energy by generating electricity that recharges the batteries.

These cars are fuel-efficient and affordable. The 2014 Toyota Prius gets 51 mpg in the city and 48 mpg on the highway; the 2014 Honda Civic hybrid gets 44 and 47 mpg, respectively. The Ford C-Max hybrid gets 45 and 40, respectively. The lowest-priced models each sell for between $24,000–$25,000.

Biodiesel, Ethanol, and Other Biofuels

By David Blume, founder of the International Institute for Ecological Agriculture and author of Alcohol Can Be a Gas!

Like gas-electric hybrids, biodiesel is a readily available transportation choice that is more environmentally friendly than driving a conventional gasoline-powered car. The increasing public awareness and use of biodiesel actually represents a "back to the future" development. Believe it or not, the original engines invented by Rudolph Diesel in 1895 were designed to run on vegetable oil. That is the promise of biodiesel and other biofuels, such as ethanol, in the 21st century and beyond. These are clean burning, relatively nonpolluting fuels made from renewable resources.

Biodiesel is a biodegradable, nontoxic fuel derived either from vegetable oils such as soybean or canola, or from recycled waste cooking oil. Biodiesel is made in a refinery process that removes glycerin from the oil. The process is not highly energy intensive, and the glycerin by-product can be sold for use in the manufacturing of soap and cosmetics. Biodiesel can be burned in compression-ignition (diesel) engines, as a pure fuel or blended with petroleum diesel (aka dino-diesel), with little or no modification to the vehicle.

Any use of biodiesel instead of petroleum diesel reduces emissions of pollutants and greenhouse gases. As demonstrated in Environmental Protection Agency tests, burning pure biodiesel reduces particulate emissions by 47%, polycyclic aromatic hydrocarbons by 75%–85%, unburned hydrocarbons by 68%, carbon monoxide by 48%, and sulfur oxides and sulfates by nearly 100% compared with petroleum diesel. The only exception is nitrogen oxide emissions, which may increase slightly with biodiesel. The positive impact on climate change is even better. Not only does burning biodiesel reduce carbon dioxide emissions by 78%, but biodiesel is carbon-neutral (as are other biofuels): While combustion of any biofuel releases carbon dioxide into the atmosphere, growing the agricultural crop used in that fuel captures a similar amount of CO_2 through the process of photosynthesis. More and more people today run B20, a blend of 20% biodiesel with 80% petroleum diesel. Burning B20 reduces tailpipe emissions by one-fifth of the percentages just noted. All the Volkswagen TDI (turbo diesel injection) models of the last several years run just fine on B100 or 100% biodiesel.

Biodiesel also has a positive energy balance. For every unit of energy consumed in the process of growing and manufacturing it, the resulting fuel contains anywhere from 2.5 to 3.2 units of energy. Fossil fuels have a negative energy balance, because extracting and refining them is highly energy intensive. While a given unit of biodiesel contains slightly less energy than the same unit of petroleum diesel, it has a slightly higher combustion efficiency, so using biodiesel in place of petroleum does not noticeably affect vehicle performance. You can even have a hybrid electric vehicle that uses a diesel motor instead of a conventional gasoline motor, and several major automakers including Volkswagen, Honda, Toyota, and Chevrolet currently offer diesel hybrids.

While it's true that you can use biodiesel without making any engine modifications, it does have a solvent effect that may release accumulated deposits on tank and pipe walls, which may, in some vehicle models, initially clog filters and necessitate their replacement. Some experts also recommend replacing rubber hoses with ones made of synthetic materials in certain makes of vehicles. However, B20 does not appear to present a problem to those rubber components. Biodiesel also clouds and gels at higher temperatures than petroleum diesel, which often presents a problem in colder climates. As with petroleum, however, winterizing agents added to biodiesel enable it to be used trouble-free to at least −10°F. Here at the Solar Living Center, we dispense B100—100% biodiesel from 100% recycled restaurant vegetable oil—to numerous cars traveling the northern California Highway 101 corridor. Several of our Hopland Real Goods employees currently drive biodiesel vehicles using B100.

Biomass Energy Content

The highest yield feedstock for biodiesel is algae, which can produce 250 times the amount of oil per acre as soybeans.

1 acre of algae yields 10,000 gallons of biodiesel
1 bushel of soybeans (60 lb.) yields 11 pounds of soybean oil, which makes 1.5 gallons of biodiesel
1 acre of soybeans yields about 65 gallons of biodiesel

1 bushel of corn (56 lb.) yields about 2.5 gallons of ethanol
1 ton of corn stover yields 80–90 gallons of ethanol
1 ton of switch grass yields 75–100 gallons of ethanol

Biogas from animal waste (methane) The number of animals it takes to produce methane with an energy content of 1 gallon of gas per day: 46 pigs or 5.5 cows or 810 chickens

A number of town vehicle and school bus fleets in the Northeast have successfully piloted using B20 year-round. The momentum of the fuel switch is growing, especially since the price of B20 continues to be competitive with petroleum diesel. As of this writing, the retail price of B20 is about 15 cents higher than conventional (dino) diesel. Many school districts and municipalities are able to achieve even greater savings because they purchase their diesel in bulk. Given that the price of oil is destined to rise drastically in our lifetime, every price increase makes biodiesel that much more competitive in the marketplace. The list of cities, towns, and schools that are turning to biodiesel, out of concern for the health of

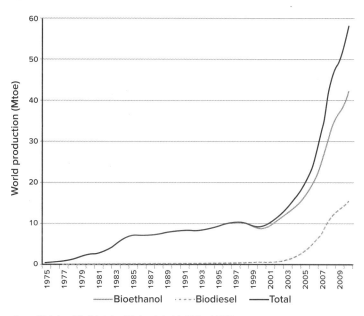

Source: "Biofuels and World Agricultural Markets: Outlook for 2020 and 2050" by Guyomard, Forslund, and Dronne (2011), cdn.intechopen.com/pdfs-wm/18291.pdf

schoolchildren and municipal employees as well as the global environment, continues to grow. Keene, New Hampshire, uses it in their truck fleet, as do several towns in Vermont; Warwick, Rhode Island, uses it in 60 school buses (and to heat school buildings), and the idea is spreading to other cities in the state; Deer Valley School District in Phoenix, Arizona, uses it in 140 buses and 5 maintenance vehicles; and the states of Michigan and Minnesota have created incentive programs to encourage the switch. Companies such as L. L. Bean and Alcoa are experimenting with biodiesel, and even the federal government is getting into the act, successfully testing a B100 truck in Yellowstone National Park.

Other biofuels may have even more promise than biodiesel produced from oil seeds or recycled cooking oil. Ethanol is familiar to many people, a fuel primarily made from corn today. It's controversial in some quarters because it is expensive to make from corn; this is the process most commonly used in the US, and it yields a small positive energy balance. But ethanol can be produced more efficiently and sustainably from other biomass feedstocks, including a variety of nontraditional crops like buffalo gourd, prickly pear cactus, agave, fodder beets, sorghum cattails, mesquite pods, and the one billion tons of food-processing waste generated each year. Just converting the lawn clippings currently going into landfills would yield 11 billion gallons of fuel per year. In Brazil and India, the energy input is not fossil fuel but sugar cane bagasse (waste plant fiber) or self-produced methane. As alcohol plants in the US retrofit to burn biomass or self-produced methane—this is beginning to happen—the energy return on energy input will rise dramatically, and parallel the 8 units of energy out for one unit in achieved by sugar cane. Scientists working to improve the production of ethanol from cellulose predict that foreseeable technological advances implemented on a large scale could lower the cost of ethanol to 50¢ a gallon. Cellulosic ethanol projects in North America are moving from the drawing board to commercial-scale pilot plants, but so far the promise of this technologically complex process has not materialized. Given the abundance of food waste and cultivated feedstocks available to produce biofuels, large-scale production of ethanol from cellulose may not become an important part of our fuel energy mix.

The potential environmental benefits are enormous. Burning ethanol instead of gasoline reduces carbon dioxide emissions by more than 80% and eliminates sulfur dioxide (which causes emphysema and acid rain). Ethanol also reduces emissions of carbon monoxide, nitrous oxides, and hydrocarbons by over 90%. And unlike biodiesel, ethanol can be used in both gasoline and diesel vehicles.

In 2011, the Environmental Protection Agency began to update its fuel blend rulings. It specified the increased blending of alcohol fuel in unleaded gasoline from 10% to 15%. This decision would reduce carbon emissions and ostensibly lower the cost of fuel. The decision was fought in the media and in court by the America Petroleum Institute (a nonprofit arm of the petroleum industry). Using specious arguments and half-truths about performance impacts, and concerns that older cars would suffer significant damage from using the increased amount of alcohol fuel, the API forced the EPA to delay its decision to implement until it could fully document the benefits and lack of hazards associated with the use of e15 in the US fuel mix. While the controversy stirred, David Blume, working with actress/activist Daryl Hannah, converted her 1979 Pontiac Firebird (famous from the movie *Kill Bill*) to run on alcohol fuel. They took the car to a California Certified Smog Station and ran a series of smog tests. The EPA was claiming at the time that it had insufficient data that alcohol fuel was safe for older cars, so Blume and Hannah showed them in a videotaped test of fuels that a 1979 car performed quite normally and far cleaner with each incremental increase in alcohol fuel added to the gas mix. The test covered unleaded E10, E15, E30, and E85 fuels; you can view it at youtube.com/watch?v=OmbqulFnSqk. In 2014, the API lawsuit against the EPA to limit states' rights to increase alcohol fuel percentages was lost in the Supreme Court. It is now legal for any state to specify E30 (similar to Brazil) or any blend it chooses.

Brazil offers an intriguing case study of the potential positive impact of biofuel use on a massive scale. In the 1970s, Brazil's military dictatorship spurred the creation of an ethanol industry, based on sugarcane, to fuel the country's transportation needs. Today all motor fuel is at least a blend of 30% ethanol with 70% gasoline, and since 2009, 92.3% of all cars and light vehicles run exclusively on alcohol.

Flex-fuel vehicles, which can switch back and forth between different types of fuel carried onboard, are one key. Though there are some reports of concern about serious mileage loss using clean and renewable alcohol fuel, a number of studies, including the EPA's own research, indicate that an optimized or flex-fuel vehicle running alcohol fuel will perform at better than manufacturer specifications. The US could supply all its transportation fuel from ethanol on less than 2% of its agricultural land if our municipal sewage was diverted to cattail-based energy farms (cattail is a temperate crop similar to sugar cane), producing ethanol without petroleum-based fertilizers. Ethanol production in the US currently stands at 16 billion gallons annually (plus another 18 million gallons from cellulosic sources). Nearly 2,300 filling stations (out of a total of 175,000) now sell E85. And the number of flex-fuel vehicles on the road has gone from zero in 1993 to more than 10 million today, though consumers don't always know they have them. Any gasoline car can be converted to alcohol for as little as $50–$300 in parts. More important, all fuel-injected cars can run on at least 50% alcohol without any modifications! Most hybrids can run on E85 without changes even though they are not advertised as being E85 capable.

Scientists, including some at the National Renewable Energy Laboratory, are investigating the possibility of biofuel made from algae. Algae are the most efficient organisms on the planet at consuming carbon dioxide and producing oil. According to NREL estimates, a 1,000-square-meter pond of algae can produce more than 2,000 gallons of oil per year. In comparison, the same crop area of high-yield canola can produce 50 gallons of oil. Even if only 10% of the algal oil can be extracted for use as fuel, it still outperforms the canola by a factor of 4. A half-million acres of algae ponds (that's equivalent to less than 1% of the fallow cropland in the US) could produce something on the order of 10 billion gallons of oil annually, which could then be processed into biodiesel. In the grandest scenario, marine algae farms off the California coast and in the Dead Zone in the Gulf of Mexico could supply all the transportation fuel for the entire country, without planting a single seed in the ground. Talk about renewable energy! Stay tuned.

Biofuels, Electric Cars, and a Clean Energy Future

How do electric cars and alcohol coexist for a clean future? Electric vehicles have their difficulties. If electric cars proliferate without clean fuel charging, reliance on both coal electricity and nuclear energy could be exacerbated. Advocates say this situation is only temporary and the answer will eventually be photovoltaic panels on everyone's roof to charge their cars cleanly. The amount of electricity it takes to charge an electric car is comparable to the amount of electricity used by an average American home. Therefore, depending on the efficiency of your home, electricity rates where you live, and how much you drive, the cost to install the PV panels needed to charge your car could be substantial. Another problem is the use of nonrenewable metals like copper in electric motors. The world currently has difficulty producing enough copper from diminishing supplies just to provide plumbing and wiring for a growing population. Repowering 1.5 billion vehicles with electric motors may not be physically possible, and the price of copper could skyrocket beyond affordability with even a modest attempt to convert the world fleet.

For those enamored of their electric cars, there is a positive solution. Again depending on one's circumstances, it might be cheaper to install a small generator at home running on alcohol to recharge batteries, compared to installing PV. The resulting hot water from the generator engine can provide all the hot water a home needs for a week, dramatically reducing the electricity an all-electric home needs for hot water and heat. For some, this technology might remain a cheaper way of recharging electric vehicle batteries.

Sources of Further Information about Biofuels

International Institute for Ecological Agriculture, permaculture.com.

When Cooking Fuel Kills by Jonathan Alter. The transition to clean indoor cooking and air is one of the critical issues Blume Distillation and IIEA are involved in, along with organizations including the UN and EPA, through the Global Alliance for Clean Cookstoves and the Partnership for Clean Indoor Air. See: nationalmemo.com /content/when-cooking-fuel-kills

Regarding Food vs Fuel Myth:
Listen to David's interview with Ira Flatow of NPR Science Friday: npr.org/templates/story /story.php?storyId=93636627&ft=2&f=510221

Endorsements of bioethanol from Ed Begley Jr. and Hunter Lovins (September 2009): alcohol canbeagas.com/node/1345

E15 Test: youtube.com/watch?v=OmbqulFnSqk

Model T—First Flex-Fuel Car: youtube.com/watch ?v=5qDYoEupI28

Any gasoline car can be converted to alcohol for as little as $50–$300 in parts. More important, all fuel-injected cars can run on at least 50% alcohol without any modifications! Most hybrids can run on E85 without changes even though they are not advertised as being E85 capable.

Electric Vehicles

By Steve Heckeroth.

On the first Earth Day in 1970, Stephen Heckeroth committed his life to replacing fossil energy with direct solar energy. He has designed and built over 20 passive solar homes and 30 electric vehicles, including Electric Porsche Spyders and Electric Agricultural Tractors. He was Director of Building Integrated Photovoltaics for Uni-Solar from 2000–2008 and has written dozens of articles and given over a hundred presentations on solar design and sustainable transportation. He is a contributing editor to *Mother Earth News* and was chair of the American Solar Energy Society Sustainable Transportation Division for 10 years.

The technologies exist to clean the air, stabilize the climate, and maintain our standard of living all at the same time. By relying on clean renewable technologies, we can eliminate much of the US trade deficit and the reason for war while achieving energy independence. Battery electric vehicles (EV) are zero emission, and plug-in hybrids (PHEV) have ultra-low carbon emissions. They can be charged from zero-emission renewable energy sources like the sun and wind. By adding more batteries to hybrid electric vehicles (HEV), plug-in hybrids can be built that offer the range of gas vehicles (400 miles) with the zero-emission and cost-saving benefits that battery electric vehicles provide for short trips.

EVs have a history that stretches back before gas cars were on the road, and even before there were paved roads. They have always been the clean and quite solution, but until recently the weight and size of batteries made long-distance travel impractical. The development of nickel metal hydride (NiMH) batteries in the 1990s made ranges of up to 100 miles per charge possible in a compact car. Just in the last few years, the development of lithium batteries now allows full-sized luxury cars to go up to 300 miles on a charge. These batteries are still expensive, but as the volume of EVs increases, the battery cost continues to drop.

From Sun to Wheel

Internal combustion vehicles are only about 15% efficient at moving a typical 2-ton vehicle and far less than 1% efficient at moving the 100–200 lb. people driving them. Vehicle efficiency is usually measured in miles per gallon (mpg), assuming that the fuel somehow magically appears in the tank. Recently the US Department of Energy (DOE) has started measuring fuel efficiencies more accurately from well to wheel. But as the conventional wells run dry, we need a better way to measure transportation efficiency. The standard to use from a sustainability perspective is solar energy, because the sun is the source of all renewable fuels.

Producing electricity from solar energy using photovoltaics (PV) is from 15%–20% efficient. Current battery charge-discharge efficiency varies from 80%–95%. Electric motors can be over 90% efficient including line and controller losses. Total efficiency from sun to wheel for electric cars is between 4%–6%.

PV panels placed in the sun produce enough clean renewable energy to pay back the energy that was used to make them in as little as a couple of months. They will keep producing free energy for many decades with very little maintenance.

Photosynthesis by plants is about 1% efficient at converting solar energy into carbohydrates. The efficiency of producing biofuels from carbohydrates and then getting fuel into a vehicle's tank varies widely from 10%–35%, depending on the process and the distance to the use. The internal combustion engine (ICE) and drive train have a low 10%–15% efficiency. This calculates to an overall sun-to-wheel efficiency for burning biofuels of about .01%–.05%.

When measured from sun to wheel, solar-charged electric vehicles are 80 to 600 times more efficient than vehicles burning biofuels. Solar-charged EVs also have a very minimal impact on soil, water, and air when compared to biofuels. Of course, burning biofuels is much better than burning fossil fuels—but why settle for less pollution when you can have zero emissions?

Biofuelishness

Fueling cars on recycled veggie oil had a great EROEI because the oil was free, but fast-food grease pits could only supply a limited number of early adopters. As biofuels became more mainstream, drivers wanted to have their green fuel at the gas pump. Soon there was not enough recycled oil to keep up with demand, and oil crops were planted. As demand increased, federal programs encouraged the diversion of food crops to produce fuel with billions of dollars in funding. When the majority of the US corn crop was used to make sufficient ethanol to comply with these federal programs, the price of tortillas in Mexico quadrupled. The corn used to make enough ethanol to fill up the tank of a SUV one time could

---

SUSTAINABLE TRANSPORTATION

> Battery electric vehicles (EV) are zero-emission and plug-in hybrids (PHEV) have ultra-low carbon emissions. They can be charged from zero-emission renewable energy sources like the sun and wind. By adding more batteries to hybrid electric vehicles (HEV), plug-in hybrids can be built that offer the range of gas vehicles (400 miles) with the zero-emission and cost-saving benefits that battery electric vehicles provide for short trips.

provide enough calories to feed a person for an entire year.

Because it takes about 10 units of fossil energy to produce 1 unit of carbohydrate energy when plants are forced to grow in monocrop conditions on the massive scale of corporate agribusiness, the EROEI went negative. (Some advocates say that the ratio may be closer to 6:1, but that's not much better. Sugar cane, the other world monocrop grown for fuel, does have a positive EROEI.) Controversy about ethanol production, against the background of the world financial crisis and rising food prices, triggered a reevaluation of federal subsidies. Many observers believe that it will be hard for the ethanol industry to maintain profitability without government subsidies. [Editor's note: David Blume does not advocate using corn to produce ethanol.]

Hydrogen Hoax

The pure hydrogen required to power a fuel cell does not exist in nature. Almost all the hydrogen used in the US is extracted from nonrenewable natural gas. Separating hydrogen atoms from natural gas uses additional fossil energy. Hydrogen can also be extracted from water, but the process uses more energy than the resulting hydrogen contains. Safely storing enough hydrogen to go more than 100 miles before needing to refuel is another issue that needs to be resolved. Furthermore, hydrogen refilling infrastructure is very expensive and currently nonexistent. Finally, a fuel cell vehicle still costs over $200,000 and has a very limited range, even after 15 years and billions

of dollars of federal funding. The planet cannot afford to wait for hydrogen fuel cells to become economically viable.

The Plug-in Solution

As already noted, electric vehicle technology exists today that can provide us with zero-emission transportation when battery charging is provided by renewable energy sources. A quick study of the following chart shows the overwhelming advantages of plug-in hybrid (PHEV) and battery electric vehicles (EV).

In 2002, I purchased one of about 200 RAV4 EVs that were sold as part of Toyota's compliance with the California Air Resources Board (CARB) Zero Emission Vehicle (ZEV) Program. Toyota was the only manufacturer to offer EVs for sale. All the other automakers only offered leases to comply with the program. Shortly after I bought my EV, General Motors and the Bush administration sued the State of California, claiming that only the federal government had the right to set fuel economy standards. Never mind that the federal government wasn't doing its job, or that California was regulating emissions, not fuel economy standards. In April 2003, CARB abandoned the ZEV Program and adopted a Hydrogen Program that was supposed to require auto manufacturers to build 250 fuel cell vehicles by 2008. So instead of the hundreds of thousands of competitively priced EVs required by 2003, in the original ZEV Program, only a few dozen multimillion dollar fuel cell vehicles were tested for a short time, and zero emission was declared

Vehicle Type	$ Gas 25 mi./ day [update]	kWh 25 mi./day	$/year 25 mi./day [update]	Gal./year 25 mi./day	Tons of CO₂/year Tailpipe	* + Tons of Upstream CO₂/year
10 mpg Gas	8.75	100	$3,200	915	10.5	13.7
20 mpg Gas	4.37	50	$1,600	460	5.3	6.8
30 mpg Gas	2.93	34	$1,050	305	3.5	4.5
40 mpg HEV	2.20	25	$800	230	2.6	3.4
50 mpg HEV	1.75	20	$640	180	2.1	2.8
Plug-in HEV 25-mi. Range	0	10	$200	0	0	0.7
Battery EV	0	6	$100	0	0	0.4
Solar/Pedal/Electric	0	1	0	0	0	ZERO

*This column includes upstream CO_2 emissions for exploration, extraction, transport, refining, and distribution of gasoline, as well as CO_2 emissions from the California mix of power plants that produce electricity to charge electric vehicles.

The main assumptions used to produce the values in the chart are:

1. The average cost of gasoline over the next year will be approximately $3.50/gallon.
2. The Time of Use (TOU) rate for nighttime charging is approximately $0.07/kWh.
3. Including the production energy there are over 40 kWh of energy in a gallon of gasoline.
4. Including the production exhaust burning 1 gallon of gasoline creates over 25 lbs. of CO_2.

An electric vehicle, outstanding in its field.

"I use about 20 kWh to commute about 60 miles a day and charge the E-RAV at night using the low off-peak rate of $.07/kWh. So a full charge costs about $1.50, or about 2½ cents a mile. Adding windshield washer fluid and changing the tires once are the only other expenses the car has required in 12 years."

to be an unrealistic goal. Meanwhile, GM and all the other manufacturers recalled and crushed all the EVs they had leased when the ZEV Program was scrapped.

Fortunately, those of us who purchased RAV4 EVs were able to hold on to our cars. We now have 160,000 trouble-free miles on ours and can testify that zero-emission vehicles are measurably superior in almost every way to internal combustion engine cars. We installed a 3 kW PV roof to offset the electricity used to charge our EVs and power our homestead in 1999. I use about 20 kWh to commute about 60 miles a day and charge the E-RAV at night using the low off-peak rate of $.07/kWh. So a full charge costs about $1.50, or about 2½ cents a mile. Adding windshield washer fluid and changing the tires once are the only other expenses the car has required in 12 years. Because the E-RAV is equipped with regenerative breaking that charges the batteries when I slow down or stop, the mechanical brakes are still like new. During the day, my solar array pumps energy back into the grid, and I get compensated at peak rates of up to $.35/kWh, so my solar roof paid for itself long ago and now gives us free power.

The RAV4 EVs originally sold for $29,000, after rebate, but because they were the only commercially produced EVs available for many years, some were resold for as much as $80,000. Until recently, the auto industry continued to claim that there was not enough demand for EVs for them to be commercially manufactured. Then Tesla changed everything, first with their roadster that goes from 0–60 mph in 2.8 seconds and has a 235-mile range between charges. Then in 2010, they introduced the Model S7 passenger luxury sedan that goes 0–60 in 4 seconds with up to a 300-mile range. That same year, Nissan came out with the all-electric Leaf. In 2011, the Chevy Volt plug-in hybrid went on sale, with a 40-mile all-electric range and a backup ICE to take the car up to 400 miles on a tank. In 2013, every major auto

manufacturer announced the introduction of at least one plug-in model. US plug-in sales have gone from a few thousand in 2010 to 100,000 in 2013. These numbers are encouraging until you consider that plug-ins only represent .07% of the 240 million licensed cars and trucks in America. In addition, many of them are probably charged using conventional utility power.

If you are one of the millions of Americans who has been compelled by corporate advertising to buy a truck that rarely has anything in the bed or a 4WD SUV that has never been off the road, please give future generations a fighting chance and get your gas-guzzler off the road permanently. Even if you are driving an ICE car that gets supposedly "good" mileage, go to pluginamerica .org/vehicle-tracker and make your next car one of the 25 different plug-in models now available. Then consider having solar panels installed on your roof and never going to a gas station again.

Designing Communities for People Not Cars

Even if we were able to totally switch to solar-charged electric vehicles tomorrow, we would still have a long list of problems to solve. Traffic accidents are the leading cause of death in the US for people between the ages of 4 and 45, and diseases linked to lack of exercise are the leading cause of death for everyone over 45. Most American towns and cities are covered with asphalt for roads and parking lots, and the places where people can walk are narrow strips of dirty concrete. It doesn't have to be this way. There are towns all over the world that were laid out as walkable communities before cars existed. The most successful of these towns have several things in common:

- **Adequate but Not Too Much Water:** Access to a sustainable freshwater supply was the primary determining factor in the location of most preindustrial towns (but building above the floodplain was another very important consideration).
- **Green Belts:** Without massive inputs of fossil energy, towns were surrounded by protected farms and wild land to assure a sustainable source of food and building materials. This scenario should reign again.
- **Narrow Roads:** Only a few wider roads for commerce come into the town center, and all other roads exist primarily for foot traffic and narrow carts.
- **Density:** Living and working spaces are tightly packed with shop owners often living above their shops. Ideally, daily activities are all within walking distance.

These basic guidelines for walkable towns have not changed over millennia, but recent advances in technology and renewable power generation add new opportunities to make communities even more vibrant and sustainable:

- **Solar Access:** The increased efficiency and reduced cost of photovoltaics and solar thermal systems now makes it possible for anyone with solar access to power most of their energy needs from the sun.
- **Internet Access:** Cell phones and connected computers allow communication and many work-related tasks without the need to move our bodies or vehicles.
- **Mass Transit:** Technologies like Pod Cars and Evacuated Tube Transport will be able to move people and goods from town to town using a small fraction of the land and energy needed for roads and conventional vehicles.
- **Ultralights:** Very lightweight pedal, pedal/electric, and solar/electric 1- and 2-person vehicles will make personal transportation orders of magnitude more efficient than the autos of today.
- **V2G:** Vehicle to Grid (V2G) technology allows EV batteries to interact with whatever they are plugged into. For example, when your EV is plugged into the grid and there is a power failure, the batteries in your car could power your home. Or when a power outage threatens a critical care facility, instead of running diesel generators, the EVs in combination with solar shade structures in the parking lot could keep the power on. Or even better, when thousands of EVs are plugged into the grid, power outages can be avoided by tapping into power from batteries when electric demand is high and charging batteries when demand is low.
- **GPS and GIS:** Global Positioning Systems (GPS) work with Geographic Information Systems (GIS) to locate where we are at any given time or where a point of interest is located. Navigation systems using GPS and GIS have all but eliminated the need for printed maps in just a few years. These systems can also be used to locate the best places to install solar panels or locate housing within walking or biking distance from work and schools.
- **Lighter Than Air and Water:** Blimps, dirigibles and zeppelins covered with solar panels could offer luxury long-distance travel through the air, and solar ships could make ecotourism really eco.

Burn or Breathe

People can survive about 4 weeks without food, 4 days without water, and 4 minutes without oxygen. Oxygen is arguably the most precious resource on Earth. Burning a gallon of gasoline consumes enough oxygen to keep a baby alive for about 2 weeks.

The longer we continue burning fossil fuel, the more dire the warnings from scientists about the irreversible consequences of CO_2 emissions. As white heat-reflecting polar ice gets replaced by dark heat-absorbing open seawater, warming accelerates. As the warming continues, the permafrost thaws, releasing more carbon dioxide and methane, which in turn causes more melting and so on. It is unclear whether we have already driven these processes to the point where the Earth will be unable to support life, but it is clear that we are accelerating in that direction.

The stakes couldn't be higher. If you really need to drive, get an electric car and install a solar array on your roof to charge it. Better yet, live where your home is within walking or cycling distance of employment, schools, and services. Work at home if you can. Support local businesses that attempt to move toward sustainability. Remember, what we buy, we empower. If we stop empowering life-threatening systems, they will change or fade away.

Futuristic Vehicles

When we first wrote about hydrogen fuel cells in the 12th edition of this *Sourcebook* in 2005, the hype about hydrogen fuel cell vehicles was mounting. Well, that hype has become muted now, and our skepticism about fuel cells has deepened. Hydrogen—the most common element in the universe—may be the wave of the future. It might become a primary energy storage medium sometime during the 21st century, and that would be a good thing for people and the planet. However, even more today than ten years ago, we are persuaded by the analysts who argue that fuel cell and hydrogen vehicles have a long way to go before they're commercially viable, and therefore represent a distant long-term solution when more realistic short-term fixes are urgently needed now. A strategically targeted, well-funded crash hydrogen development program is perhaps advisable, but other steps can and should be taken immediately to reduce the greenhouse gases and other pollutants generated by the transportation sector of the economy. For a description of how a hydrogen fuel cell works, and why they're not a viable alternative for the foreseeable fugure, see Chapter 3, pages 118–19.

In previous editions of the *Sourcebook*, we profiled the concept of the Hypercar being developed by the Rocky Mountain Institute. While RMI is no longer working on that particular project, they are focusing on driving the transition to a fossil-fuel-free US transportation system by 2050, specifically through the lightweighting of vehicles by the integration of carbon fiber composite parts. For more information, see rmi.org/autocomposites.

The bottom line: If you feel compelled to act now to reduce your greenhouse gas emissions from transportation, switch to biofuels or some form of electric vehicle.

Additional Resources about EVs

Plug-In America
pluginamerica.org

Inside EVs
insideevs.com

Electric Auto Association (EAA)
electricauto.org

Department of Energy Vehicle Technologies Office
www1.eere.energy.gov/vehiclesandfuels/

Estimated Alternative Fueled Vehicles in Use in the US, 1992–2010[1]

Year	Liquefied Petroleum Gas (LPG)	Compressed Natural Gas (CNG)	Liquefied Natural Gas (LNG)	85% Methanol (M85)	Neat Methanol (M100)	85% Ethanol (E85)[2]	95% Ethanol (E95)	Electric[3]	Hybrid Electric[4]	Hydrogen	Total[5]
1992	NA	23,191	90	4,850	404	172	38	1,607		0	NA
1993	NA	32,714	299	10,263	414	441	27	1,690		0	NA
1994	NA	41,227	484	15,484	415	605	33	2,224		0	NA
1995	172,806	50,218	603	18,319	386	1,527	136	2,860		0	246,855
1996	175,585	60,144	663	20,265	172	4,536	361	3,280		0	265,006
1997	175,679	68,571	813	21,040	172	9,130	347	4,453		0	280,205
1998	177,183	78,782	1,172	19,648	200	12,788	14	5,243		0	295,030
1999	178,610	91,267	1,681	18,964	198	24,604	14	6,964	17	0	322,319
2000	181,994	100,750	2,090	10,426	0	87,570	4	11,830	9,350	0	404,014
2001	185,053	111,851	2,576	7,827	0	100,303	0	17,847	20,282	0	445,739
2002	187,680	120,839	2,708	5,873	0	120,951	0	33,047	36,035	0	507,133
2003	190,369	114,406	2,640	0	0	179,090	0	47,485	65,667	9	599,666
2004	182,864	118,532	2,717	0	0	211,800	0	49,536	86,203	43	651,695
2005	173,795	117,699	2,748	0	0	246,363	0	51,398	211,716	119	803,838
2006	164,846	116,131	2,798	0	0	297,099	0	53,526	297,919	159	932,478
2007	158,254	114,391	2,781	0	0	364,384	0	55,730	354,281	223	1,050,044
2008	151,049	113,973	3,101	0	0	450,327	0	56,901	314,394	313	1,090,058
2009	147,030	114,270	3,176	0	0	504,297	0	57,185	290,271	357	1,116,586
2010	143,037	115,863	3,354	0	0	618,506	0	57,462	274,555	421	1,213,198

Data Source: EIA's Alternatives to Traditional Transportation Fuels, Table V1. Available at eia.gov/renewable/

[1] Vehicles in Use represent accumulated acquisitions, less retirements, as of the end of each calendar year. They do not include concept and demonstration vehicles.
[2] Includes only those E85 vehicles believed to be using E85. Primarily fleet-operated vehicles; excludes other vehicles with E85-fueling capability. For total number of E85 vehicles on the road, see "E85 FFVs in Use."
[3] Includes low-speed electric vehicles. Does not include hybrid electric vehicles.
[4] Includes plug-in hybrids.
[5] Total does not include EIA's "other" category.

Worksheet available at afdc.energy.gov/afdc/data/. Updated on 05/17/12

CHAPTER 14

Natural Burial

The Ultimate Back-to-the-land Movement

NATURAL BURIAL IS A CONCEPT that we introduced to our readers in the previous edition of the *Solar Living Sourcebook*. The chapter was written by Cynthia Beal of The Natural Burial Company, a pioneer in this endeavor in North America. We've asked Cynthia to provide an update for this new edition. You can find natural funeral planning tools, products, and other guidance on the Natural Burial Company's website at naturalburialcompany.com, and a list of cemeteries and funeral service providers who've signed the Natural End Pledge can be found on the Natural End Map, at naturalendmap.com.

If you're unfamiliar with the idea of natural burial, you might think the subject is macabre or depressing. To the contrary, we think you'll find this information about home funerals and natural burial to be inspiring and uplifting. After all, you're probably "dying to do the right thing" during your time here on Earth, whether it's taking good care of your family, seeking right livelihood, reducing your carbon footprint, working for social justice, or doing what you can to live by the precepts of sustainability. But you may not have realized that you can also act on these values when and after you die, for the greater good of your friends and family and the planet around you. Each year, new and improved products, policies, and practices make sustainable inroads that map to the acts of our daily lives and express our values in the process. It's not too soon to start planning ahead, and we invite you to join the "ultimate back-to-the-land movement"! Check the Real Goods website for links to even more product information (realgoods.com).

Dying to Do the Right Thing

First begun by pioneers in the United Kingdom, and now with almost three decades under its belt, a compelling consumer-driven natural funeral movement that lets you "put your stuff back" continues to gain momentum, and North America is getting on board. Natural burial areas—sections of cemetery property where people are buried in biodegradable containers, without preservatives (embalming) or synthetics, and returned to the Earth to compost into soil nutrients—have appeared in hundreds of cemeteries throughout the UK, and the US, Canada, Australia, and others are following suit. Want to "be a tree"? A woodland, orchard, or wilderness burial might be perfect for you. Would you rather "push up daisies" and feed butterflies and bees? Try a meadow burial and decompose under a field of wildflow-

ers. Still want to be in the "Family Plot," next to great-great-granddad? Then ask for a vault-free natural burial in your favorite historic, church, or county-run cemetery and see if the management is now ready to agree. (You might be surprised at how many say "yes" these days.)

Burying ourselves naturally, directly into the soil, wrapped (or not) in biodegradable packaging, without embalming preservatives, is not rocket science. In fact, it's likely that "Any Cemetery Can" improve habitat, reduce resource use, and minimize potential contamination of soil and groundwater by utilizing techniques from sustainability practitioners in the landscaping, groundwater management, and horticultural disciplines. Luckily for the future, the value of our precious natural resources, and of land

> Ask for a vault-free natural burial in your favorite historic, church, or county-run cemetery and see if the cemetery management is now ready to agree.

In a natural cemetery, people and Nature can coexist, and be celebrated, side by side.

Home funeral facilitators focus on returning control over the death and dying process to individuals and families, encouraging and teaching them to take charge of their own end-of-life affairs in a proactive manner that engages family and friends, returning dignity and meaning to what has become, for many, a sterile and uncomfortable commercial process.

C.A. Beal

stewardship in general, is becoming both quantifiable and doable, thanks to public research into ecosystem services and their importance to the quality of life on Earth. Advances in understanding suggest that it's cheaper to conserve our environment than it is to consume it, and cemeteries around the world offer a unique opportunity for conservation by establishing perpetual reserves of habitat and repositories of cultural history. Increasingly, taxpayer-funded counties, municipalities, and organizations that run cemeteries without profit (and even those that do) recognize that environmentally friendly options for human disposition must be found if we're to have sustainable processes at the ends of our lives as well as during them. Similar to the organic farm movement, sustainable cemetery management—with its focus on creating and diversifying habitat, supporting soil health, and reducing resource use and contamination potential while preserving our cultural history—offers a way forward, and natural burial is the key.

Driving the Change:
Home Funeral Services and
Biodegradable Grave Goods

Concerns about pollution; appropriate use of resources, land, and energy; and the depersonalization of the dying process, as well as a fat Baby Boom demographic (with a death rate that puts 80 million Americans "over the edge" in the next couple of decades), are driving the natural burial trend. Two distinct groups stand at the forefront

of this sea change: the natural burial product and cemetery proponents themselves, making the grave goods and operating the cemeteries, and a vocal citizen counterpart found in the provocative yet practical DIY home funeral movement. Home funeral advocacy is spurred on by educational and nonprofit consumer organizations educating the public on natural end-of-life options, championed by groups like the Natural Death Centre in London,[1] the USA's Funeral Consumers Alliance,[2] and the National Home Funeral Alliance[3] based in Boulder, Colorado. The adoption of natural products by the funeral industry is also spreading, as consumer demand helps operators see—and seize—the opportunity to connect with their communities around burial again, while simultaneously waking up to their role as perpetual land stewards in the cemeteries they're required to tend—forever.

The Home Funeral Movement:
Genesis of Natural Burial

For those who really want to do it yourself, a DIY home funeral may be the ideal "way to go." Home funeral facilitators focus on returning control over the death and dying process to individuals and families, encouraging and teaching them to take charge of their own end-of-life affairs in a proactive manner that engages family and friends, returning dignity and meaning to what has become, for many, a sterile and uncomfortable commercial process. And for increasing numbers of people, in addition to hands-on participation, that means a natural burial, too.

In the early 1990s, more than 90% of people in the UK died in a hospital rather than the home, providing some of the original impetus behind the founding of the Natural Death Centre in London. It began as the project of three psychotherapists, spearheaded by Nicholas Albery, with the mission of enabling a person to die a more natural death in personal surroundings, tended by loved ones, receiving treatments that they—not the hospital system—desired. The Natural Death Centre quickly became, and remains, an indispensible source of inspiration and information for self-reliance in death, primarily through its coordinating website, naturaldeathcentre.org.uk, and the new edition of its popular guide, *The Natural Death Handbook*. The work has been picked up and expanded over the last decade by home funeral advocates such as Beth Knox, founder of the Crossings network, and Jerri Lyons, who began the Natural Death Care Project. (Both women now train home funeral guides in the

US.) Considerable stimulus for the US movement was fueled by Lisa Carlson's book, *Caring for the Dead: Your Final Act of Love*, an excellent reference still sold through the FCA site, funerals.org

Until recently, most funeral directors were reluctant to let the family get involved. However, as home funeral activists teach DIY techniques, and as competing funeral service providers make themselves available to serve more individualized and nontraditional needs, the "dismal trade" is finally getting on board. Alternative services are offered by progressive funeral directors and clergy; celebrants advertise a new profession; and memorials emphasizing the individual's secular values are becoming common. Much of the inspiration for this shift in funeral practices has come through the UK, where funeral directing does not require a license, and where the hospice movement—as a result of the conscious decision to die at home—has actively reconnected families with issues of disposition and death.

This home funeral renaissance, with its desire to return the funeral back to the purview of the family and reinstate affordable simplicity, led smoothly into public calls for natural burial: no embalming, the "plain pine box," the shroud, memorialization with a tree, or even anonymity—products, services, and rituals that express a respect for both the person and the planet. That call, in turn, has helped engender the modern natural burial movement, the "gateway drug" to cemetery sustainability; and once a cemetery starts on the path of improving habitat, reducing resource use, and minimizing future pollution, it's very unlikely to go back!

Natural Grave Goods: Filling Real Needs with Style

Side by side with the home funeral front are an equally dynamic group of sustainability-oriented business entrepreneurs and artisan manufacturers with a focus on planet and people, as well as profit. Some are producing unique biodegradable burial vessels made from natural and recycled materials, while other forward-thinking land stewards are pioneering back-to-the-earth burials in existing cemeteries or starting new conservation burial grounds as the demographics permit. These companies and individuals are doing for the industrialized funeral sector what organic farmers and food producers have done for the agricultural one: anticipating and then serving an unmet but very real consumer demand and, in the process, changing the practices of a multibillion dollar end-of-life industry that will

C.A Beal

have a detrimental environmental impact on future generations unless it's turned around. These green grave goods are stimulating a renaissance in the once-thriving burial arts. Handcrafted woven items are making a comeback in the form of willow, bamboo, and seagrass burial boxes, while fabric artists fashion imaginative shrouds of organic cotton, wool, and hemp. Unique new burial containers like the Ecopod recycled paper coffin (ecopod.co.uk), traditionally woven bamboo and willow caskets, and the ARKA Acorn ash burial urn appeal to environmentally minded folks who want to depart from life as naturally as they've lived it.

Products aren't the only things that are changing—the way cemeteries are run is changing, too. And not surprisingly, much of what a cemetery has to do to improve its sustainability mirrors what we choose to do in our daily lives to improve our own environmental footprints. Like all other businesses that create impacts and consume resources, the cemetery needs to manage its own ecological economics and bring its maintenance needs into balance. There are a number of tools, many of them mentioned throughout this book, that can help cemeteries tackle the various sectors calling for their attention. Energy and resource conservation; making power from sunlight; preserving clean water and habitat for future generations—the techniques and tools for cemetery transition are plentiful, and the Natural Burial Company and Real Goods are pleased to be at the forefront of this education and distribution network, bringing our customers the information and products you need to make even your final act a positive and self-reliant one.

Cynthia Beal at a trade show.

When It's Time, Will You "Leave No Trace"?

Increasingly, people concerned about the impacts of the conventional funeral process are beginning to question the wisdom of leaving toxic burial chemicals and synthetic substances in the ground and atmosphere for future generations to clean up.

Barely 100 years old, cemeteries are rife with poorly made monuments that create headaches for the future.

Leaving life is poignant. It can be frightening. But it doesn't mean you have to leave a mess. A popular outdoor ethics campaign, the Leave No Trace program (lnt.org), took backcountry garbage from hikers to heart in the 1970s and '80s, thoughtfully outlining objectives for individual waste management and behavior when visiting natural and wilderness areas: plan ahead, dispose of waste properly, minimize impact, respect wildlife. That ethos could be usefully applied to the ends of our lives, as well—many folks think that we're "just visiting" here on Earth, and that when it's your time to go, "Leave No Trace" doesn't seem like such a bad idea. If you're one of the first to blaze the trail in your community, however, be prepared for a little bit of activism in order to get what you want!

For decades, the end of a human life in American society has been managed by a cadre of professionals who can package our experience of death just as rigidly as others have packaged our living. Prior to the modern era, death was the exclusive province of the family. Burials were done according to custom and tradition. Respect was a matter of course, for strangers were not in charge, and dignity was conferred in the sincere acts of caring for and carrying our dead. Today, however, life moves rather mechanically—and for a hefty fee—out of the raft of boxes above ground and into more boxes below, buried on high-priced real estate that commands upwards

of 1 million dollars or more an acre for its owners. (1,000–2,000 bodies per acre at a minimum of $1,000 each for the plot is standard, sometimes stacked two or more high.) For lots of people, that double-box process looks like litter, and upon closer examination, it's not the dignified and simple close to a grateful life that most of us wish to have. Increasingly, people concerned about the impacts of the conventional funeral process are beginning to question the wisdom of leaving toxic burial chemicals and synthetic substances in the ground and atmosphere for future generations to clean up. If pressed, many of these same folks would prefer their bodies to "Leave No Trace" as well.

Preserve, Disappear, or Return to the Soil? Choose Your Disposition

Once you're done with your body, only one thing happens to it next: It goes away. Well, it never "goes away"; in the words of anthropologist and garbage guru William Rathje, "there is no away." So you do go somewhere, and something is done with you first. What's done with you immediately after the funeral is called the "Final Disposition"—that moment society agrees you're definitely finished being you—and how this is done is still largely up to you. So ask yourself what you want for a final disposition. Are you going to be buried or burned? Will you be embalmed or not? Will you manage your physical remains for preservation, disappearance, or return? In other words, will you be hanging around for as long as the chemistry lets you (preserving); using machinery to get rid of you rapidly by burning or dissolving (disappearing); or will you be buried, returning to soil and becoming earth?

Modern science now recognizes that our body's living system depends upon a complex network that coordinates independent cells so they function together as skin, organs, blood, bones, nerves, and other parts all working as a team—i.e., You—to repel the invasion of external bacteria and fungi that would otherwise colonize and consume weaker individual elements trying to make it on their own, without You. "Life" is a constant struggle to resist turning into something else's dinner, and as long as you're breathing, your side is still winning.

As soon as you check out, nature's disposition begins: The system of You collapses, and your cells, that formerly clever and fun-loving collective transforming food made from soil and sunlight into ATP (biochemical energy) and

C.A. Beal

then turning that ATP into gardens, solar arrays, and microbrew festivals, bid each other a fond farewell and take their turn as food. In the natural world, a whole host of creatures—animals, insects, fungi, and microbes—then get their spot in the sun, so to speak, and take on the very necessary work of breaking you down into smaller component parts, putting you back into the system that you built yourself from in the first place. Our food comes from soil, and we can return to the soil as food. It's an amazing cycle—or it can be, if we'd just leave it alone.

But no, WE have IDEAS.

Preservation: It's Not All It's Cracked Up to Be

The ancient Egyptians were masters of the Slow, learning to pickle and preserve human remains for reasons that are still somewhat obscure. Modern embalming came back into fashion in the mid-1800s in the United States, with arsenic, mercury, and lead-based formulas marketed by the emerging military and chemical supply industry to Union and Confederate armies, and used to preserve the bodies of dead soldiers for positive identification and burial during and after the Civil War. Although arsenic was banned in 1910, perhaps as many as half the bodies from those decades were treated with as much as several pounds each, posing potential groundwater and soil contamination challenges for the future; the contamination potential remains unexplored even today. The reasons given for embalming today remain the same as those used 150 years ago: restoring the body visually after a disfiguring death; delaying the disposition until family can arrive from long distances; or permitting the body to be transported or stored above ground in a mausoleum crypt, rather than being cremated or interred.

While formalin-free solutions are now available, most modern embalming fluid still contains toxic chemicals, including methanol, ethanol, and formalin (from formaldehyde), the latter a suspected carcinogen. Used by a large number of funeral homes to slow the body's decomposition by eradicating natural decomposers, formalin arrests the breakdown processes by "fixing" cellular proteins. It stiffens the body's tissues and, with the help of added colorants, is used to make a corpse more attractive and lifelike.

The European Union began the process of banning embalming fluid in 2006, and in 2012, its use as an approved biocide was discontinued. In 2011, the US Centers for Disease Control officially declared formaldehyde a carcinogen that's

C.A. Beal

harmful to mortuary workers.[4] It's not approved to kill dangerous human pathogens, however, and one of its biggest dangers, outside of the toxicity of its primary ingredients, is the myth—often promoted by embalmers—that it does. Its core danger, however, is as a workplace toxin, since exposure to formaldehyde poses significant health risks to funeral industry workers. Nasal and lung cancers have been indicated in scientific studies, and while some industrial research still disputes these claims, the Occupational Safety and Health Administration (OSHA) has joined the CDC in tightening up controls related to its occupational use. (For the CDC evidence of carcinogenicity in formaldehyde, see cdc.gov/niosh/docs/81-111/.)

The impact of formaldehyde on the environment when buried is still unknown, and little research has been done. However, it's said to break down rapidly in soil instead of bio-accumulating, and the primary concerns remain the effect on worker health, the interference with the body's own decomposition process post-burial, and the expense of a procedure that's rarely required for the protection of human health. Another related issue is that, because embalming fluid is used to replace the blood and organs of a body, that blood has to go somewhere. You guessed it; those 1.5 million embalming procedures performed in the US each year produce 2½–3 gallons of blood and excess embalming fluid per body. That fluid, along with the organs and internal parts suctioned out of the corpse during the process, goes down the drain and into the water supply. Not a pretty picture, and with little science to verify that public water systems are up to the task of processing the bacterial and viral load, people who question the public health impacts of natural

Tarn Moor is a pioneering meadow burial ground in the United Kingdom.

NATURAL BURIAL

While formalin-free solutions are now available, most modern embalming fluid still contains toxic chemicals, including methanol, ethanol, and formalin (from formaldehyde), the latter a suspected carcinogen.... No state in the US requires embalming except in special circumstances, such as death from a reportable and communicable disease.

Unlike a vaulted burial, natural burial requires a small amount of maintenance early on, but none at all down the road. Eventually the graves disappear altogether.

burial might well start questioning the conventional process instead!

Embalming is rarely required by law

Today embalming is a common practice in the US, not because most people want it, but because it has been considered so customary and beyond question that many have assumed it's required by law. No state in the US requires embalming except in special circumstances, such as death from a reportable and communicable disease. Adding to the confusion, states differ on this, with some requiring and others prohibiting the practice. The Funeral Consumers Alliance states that a number of funeral homes still regularly imply to their customers (and to legislators) that embalming is "necessary" for public health and safety even though the federal Funeral Rule explicitly prohibits implying that such a law exists.[5] Contrary to industry opinions (still prevalent in mortuary education today), embalming is not necessary to prevent decomposition in the first few days after death; chilling the body sufficiently does the job. Embalming does not prevent the spread of disease. Its use as a sanitizer is overrated (precisely because it does not disinfect, a stricter standard than sanitation that formalin itself cannot attain), and according to the CDC, embalming serves little appreciable sanitizing or public health purpose that couldn't be handled effectively with more natural techniques.

C.A. Beal

But should we be burying THIS stuff? What's in the box besides you

We think it's appropriate to ask hard questions about toxic burial chemicals or by-products and promote research and development of alternatives. Under most state and federal regulations, any company would be hard-pressed to get permits to bury almost 2–3 million gallons of embalming fluid in the soil annually, and yet that's exactly what happens with most of the 1.5 million bodies that are embalmed and buried in US cemeteries every year. Guess what else is buried along with these embalmed bodies every year? Even with a cremation rate approaching 45%, it's estimated that over 100 million pounds of steel, bronze, copper, and brass; 30 million board-feet of hardwood timber; uncounted tons of plastic, vinyl, fiberglass, adhesives, paints, finishes, and synthetic fabrics; and 1.5 million tons of concrete *annually* accompany Americans to their underground afterlife—an unattainable disposal permit indeed, unless you're burying the conventional American casket suite one grave at a time![6]

The Casket & Funeral Supply Association of America estimates that almost two-thirds of American caskets are stamped steel, with veneered chipboard and fiberglass making up most of the rest. These caskets are designed to resist—or at least appear to resist—decomposition. Holding the soil or water out of the grave for even a little while is a comforting thought for some, and product makers use that as a selling point, with the most durable caskets and vaults bringing in the highest prices. Caskets touted for their resistance to breaking down may also be sold with an optional rubber or plastic seal installed between the bottom and the lid, designed to prevent mold and rot—our friendly decomposing fungi and bacteria at work. In reality, however, this seal fosters an anaerobic environment and causes the body to putrefy rather than decompose, a possible challenge for future generations who may have to deal with this someday.

A box for the box

After the packaging is complete, the conventional casket isn't lowered directly into the ground but is instead placed inside a concrete, steel, or fiberglass vault, or "grave box." American cemeteries often require grave liners, primarily to keep the casket from deteriorating and then collapsing under the weight of heavy excavating equipment. Grave liners delay what's known in cemetery landscape maintenance parlance as "subsidence," the slight sinking of the earth after decomposition that creates bumpy ground if not manually filled in once or twice post-burial, and can detract from the golf course-smooth surfaces sought by some lawn-style cemeteries. But gaskets break down eventually; seals fail, concrete cracks, and eventually the liners, and their graves, will collapse. Practices like vaulting keep the funeral home's revenue up, and they help ensure that the collapse is down the road, in someone else's future, at some future cemetery owner's expense.

Given America's obsession with packaging, preservatives, and increasing shelf life over the last one hundred years, it's not surprising that habits of plastic and preservation would have made their way into our last products, as well. The innovations did solve problems of the time: The funeral director shared his home for funerals when people didn't have one; he sold caskets, and then vaults and markers, when people lacked the tools and skills to make their own; he preserved the body until family members could come long distances for the funeral. When these options were introduced, they were improvements. But once the well-meaning professional joined up with the marketing plans and sales pitches of the funeral industry consolidators, the chemical-intensive double-box casket-and-liner system became a highly profitable enterprise, its slick uniformities smoothing out and sanitizing the uncomfortable and (very human) experience that accompanies death, while ignoring the flaws of excess expense, wasted resources, and deferred maintenance. It's an unintended error; an originally compassionate but increasingly commercialized effort to ease our emotional pain, where things are double- and triple-wrapped, awkward sights and smells are whisked away, and the vacuum filled with smooth and shiny services that temporarily mask the reality of death. The box is pretty, the lawns are neat, and nature can't get a word in edgewise.

Disappear: To Burn or Not to Burn

Until recently, the environmentalist's response to this industry machinery and product-intensive process has been to opt for cremation. However, as Baby Boomers age and the actual experience of managing our deaths (and those of our parents and friends) comes to each of us, we've learned that cremation—now complicated by legitimate questions around energy use, mercury vapor and carbon emissions, the lack enforceable environmental standards, uneven or nonexistent filtration requirements, and aging industrial crematoria—is not necessarily the "no-muss, no-fuss" disposition option it first appeared to be, once its environmental footprint is considered.

It's not certain which tradition is older, burning or burial. Archaeological evidence exists for both scenarios, and each has a longer history than preservation. The oldest arts we know of are the burial arts, and the practice of cremation is thousands of years old. In times of disease and mass death, cremation has often been the method of choice, especially in landscapes where the soil was not suitable for rapid breakdown. But cremation takes fuel—wood, gas, or electricity today—and these days, fuel is something we don't spend quite as casually as we may have done before.

The choice to burn or not to burn may be made for spiritual reasons. Some religions teach about the impermanence of life and back that up with a ritualized display of public burning, proving to the community that the person cannot come back—once they're burned to ash, they're truly "gone." Other groups consider burning the harshest of punishments, depriving the soul of a body to either return to or use in an afterlife, and thus reserved for criminals and heretics. In either case, the goal of cremation is to make the body quickly disappear, and for various reasons, this can seem the logical choice. The illusion of "disappearance" has led many who are disenchanted with modern industrial burial to opt for cremation, almost as if the elimination of the body could somehow remove the negative human impacts that so disturb us. But it's not that easy.

Cremation impacts

Whether it's cremation through incineration (conventional cremation) or dissolution (alkaline hydrolysis), mechanically driven accelerated disposition has its own array of issues and impacts. Chief among them are the energy used for complete combustion or reduction; the emissions or waste products that result from burning or otherwise disposing of synthetic materials and body implants; and, in the case of cremation especially, the volatilization of mercury fillings. Emissions from crematoria contains a varying degree of pollutants such as particulate matter, volatile organic compounds, carbon monoxide, nitrogen oxides, sulfur dioxides, hydrogen chloride, heavy metals (cadmium, mercury, and lead), and dioxins and furans. It's been estimated that vaporized dental amalgam accounts for up to 16% of the airborne mercury pollution in the UK;[7] in the US, with over 1,900 crematoria across the country and a steeply rising cremation rate, "a mercury flow worksheet developed for EPA's Chicago office… estimated that in the United States in 2005, almost 3,000 kilograms (6,613 lbs.) of mercury were released to the environment from crematoria.

Good empirical data on the magnitude of mercury emissions from crematoria, however, are lacking. At this time, no federal or state regulations restrict mercury emissions from crematoria."[8] American cremation industry reports state that there are no significant emissions.[9]

Whether it's cremation through incineration (conventional cremation) or dissolution (alkaline hydrolysis), mechanically driven accelerated disposition has its own array of issues and impacts. Chief among them are the energy used for complete combustion or reduction; the emissions or waste products that result from burning or otherwise disposing of synthetic materials and body implants; and, in the case of cremation especially, the volatilization of mercury fillings. Emissions from crematoria contains a varying degree of pollutants.

Disagreements over standards

Agreements on emissions standards (and research results) are difficult to achieve, especially given that a large number of the older polluting crematoria facilities remain in operation. The US isn't the only slacker when it comes to crematoria, and the lack of crematoria standards worldwide, the carbon footprint that accompanies cremation and accelerated disposition, and increasing population all suggest that a cremation rethink is in order. Crematorium makers are quick to tout their progress: Increasingly efficient filtration systems capture more emissions than they once did; multiple burners combust more emissions; and energy-efficient designs are increasingly available. However, those comprehensive filters are very expensive, not yet required in the US, and the trapped pollutants must still be disposed of once the filters are full. In fact, according to sources in Europe, the contaminants found in used crematory filters are so hazardous they need to be stored like nuclear waste.[10] So, until clean air standards also apply to crematoria, be burned with care.

Shortchanging the funeral process

One reason many funeral directors don't care for cremation is that people tend to forego the funeral, and professionals in the field believe that the lack of a funeral, a ritual of closure with the body present, has an impact on the psyche, so that cremation makes it possible to put off or perhaps never experience emotional closure around a death. This may be true; cremation *doesn't* force one to complete the letting-go process around death the way a body burial does, and many people never get around to scattering, or even claiming, the remains. In fact, estimates suggest that perhaps as much as a quarter to a third of all cremated remains are still on the shelf somewhere, perhaps still in the original "temporary" box or bag. Whether this is a psychological requirement or not is up to the experts, but circumstantial evidence suggests that *something* is being avoided!

Unlike cremation, when a person chooses burial for the final disposition, the ritual of closure is tangible and complete; the body is literally "laid to rest," and funeral directors report that family and friends have a deeper emotional connection with the event. Additionally, buried bodies in cemeteries tend to be memorialized, or at least buried in an accessible place designed for remembrance, a physical place that can be visited and provides a link between generations of a family. For cremated remains that end up on a shelf in the pantry, or up in the attic, the future—and the link with future generations they could speak to—is much less certain.

You only die once: Making cremation a gentler alternative

Even so, cremation is still a viable alternative. You can offset the carbon and remove your mercury (along with other ways to gentle the environmental impact of your death) before burning, so don't let anyone talk you out of it if that's what you prefer. The trend worldwide is toward cremation, and there's little ground for argument if inputs and emissions are managed properly and the only other available method is the resource-intensive conventional industrial model of burial. Creative options that support underwater habitat (Eternal Reefs) or forest preserves (Eco-Eternity Forests) are opening up; universities and public institutions are adding alumni columbaria, providing niches for cremated remains while funding the creation of campus parks and greenspace; cemeteries are offering memorial habitat hedgerows and wildflower meadow gardens. Today, those who wish to be cremated can support the same sustainability principles as those who opt for burial. And for many people—especially those whose deaths involve complex organ donation, serious infectious disease, limited funds that preclude supporting a forestland, or dying far away from your chosen place of burial without the funds to fly there—cremation may be the best option. Clean cremation wins over an embalmed body and nondegradable casket system any day.

Scattering or burying cremated remains

When a modern crematorium (busy, filtered, and energy-efficient) can be located and cremation is still your method of choice, cremated remains offer a chance for multiple survivors to honor a loved one after they're gone, with the remains divvied up among family and friends. Many people are surprised at the amount of ash, and even bone, that remains after a cremation, and scattering it around can feel awkward for some people. (The rule of thumb is about 1 cubic inch of remains for every pound of lean body weight, since fat burns.) Cremated remains are pulverized bone, converted by high temperatures to a form of calcium phosphate that's more difficult to assimilate, with a nutritional value to the soil decomposer ecosystem much less than that of a full body burial. (SOILWEB TIP: *instead of scattering the bulk of the remains, consider burying them, mixed in with about 4 or 5 parts healthy soil, to make the calcium more bio-available and less like a pile of rock; the plant roots will love you for it.*)

With the rise in cremation rates, scattering ashes in wilderness venues has become so popular in some national parks that special use permits to perform the scattering are usually required, and visitors have to be reminded to spread the alkaline ash out of sight, and disperse it widely to avoid harming the plants or disturbing other park goers with visible remains.[11] While a little bit of calcium phosphate "ash" is fine for any landscape, too much dust can eventually clog the pores of plant leaves or over-alkalinize the soil—so scatter with care. To address this issue, as well as create a place for family visitation and permanent memorialization, many cemeteries now offer burial areas for cremated remains, and the more progressive ones are using the burials to sponsor areas of habitat within the cemetery that would otherwise go unfunded. Biodegradable burial urns, or ocean release urns, make personalized forest or sea burials of the ash an earth-centered and ceremonial option, so the loss of the ritual isn't a given. In fact, if you still have an aunt or two "on the shelf," consider taking them down to your favorite local cemetery on some special day, and propose using the ash burials to create habitat in honor of your family members.

Accelerated Disposition: Raising the Environmental Footprint Questions

Spurred on by arguments against conventional cremation, new disposition technologies that claim to be more environmentally benign and make the body disappear are coming into view. One proposed method that's caught the public's imagination is called "Promession," envisioned by Swedish soil scientist (and former organic gardener) Susanne Wiigh-Mäsak. Now independently marketed by Promessa Organic AB company (promessa.se/en/), this method proposes a cryogenic process that freeze-dries the body immediately after death. Frozen solid, the intent is to vibrate the frozen body apart using ultrasound, reducing it to a moist powder. Theoretically, the moisture—70% of a body's mass—is evaporated off, and the various metals and nondegradables sifted out. What remains afterward, according to the inventor, will be a dry, silt-like, and nutrient-dense substance suitable for burial and use as a fertilizer.

In another process that's gradually becoming legal state by state, machines that combine high-temperature water and chemical treatments to dissolve bodies with potassium hydroxide (lye) are being developed to break the body down through a process called alkaline hydrolysis (AH). This chemical action relies on water, heat, and the alkaline lye to remove flesh from bone. Other techniques then remove the liquid to create a dry biological residue that's returned to the family like other remains, much as the Promession group proposes.

Contrary to claims that the AH process was developed by this or that company or inventor, the technology has been in use for centuries, with much of its origin in soap-making with animal fat. An early patent spelling out the use of an alkaline solution under pressure to remove gelatin from bones and create a fertilizer out of the remainder was issued to Amos Hobson in England in 1888. Later, when intensive factory farming and meat-processing facilities generated massive quantities of carcasses that required disposition, alkaline hydrolysis was used to render them down into disposable forms that included usable by-products. Now extended to humans, patented variations of this process are marketed under trade names like "Resomation," "Eco-Green Cremation System," "Bio-Response Alkaline Hydrolysis system," "Bio-Liquidator," and "Bio-Cremation."

The jury is still out on the environmental friendliness of either process, cryogenic or alkali-based. So far, the most prominent groups claim

Planted in a forest and becoming dinner for the regenerating planetary system, we can remain fully present, albeit transformed, nourish the soil, enlarge the habitat, and rekindle the life of meadows and forests, feeding and becoming plants, animals, and trees.

the AH process is environmentally superior to cremation by incineration, with leading cremation companies buying up patent rights and lobbying for legal exceptions and rulings. The numbers should still be taken with a grain of salt: The full emissions and embodied energy footprints may not be calculated properly yet, and emerging technology claims are usually subject to change, so it will take time for independent third-party analysts to get truly objective assessments. That said, alkaline hydrolysis is a process with a solid technological basis and few variables, and it can be evaluated on its own merits as it evolves. The cryogenics technologies likely have their place, too. Dense urban areas without suitable cemetery soils and medically difficult dispositions that don't lend themselves to burial come to mind. Expect more scientists to get involved as the technology advances, clarifying language in the process. And although their earliest claims may be a bit off base, it's likely that at least some of the new systems will be able to address environmental concerns related to current disposition methods.

So, yes, a case can be made for cremation, and other forms of accelerated disposition. The environmental footprints can be addressed, and even offset. But the least talked about and perhaps most compelling argument against cremation may be that, in disappearing completely—in using machines to rapidly evaporate, oxidize, or dissolve our earthly forms away—we deprive the

landscape of our bodies, including the wide range of decomposers who take their turns at our table, and a rapid dissolution closes off our last chance to continue participating in this physical life in such a useful way, as food. As cremated ash, it's true that we can be scattered to the winds or on the waters, or remain cherished and elemental in an art piece on the mantle, a comforting tangible presence in our descendants' lives. But planted in a forest and becoming dinner for the regenerating planetary system, we can still do one last thing with our bodies that may be much more significant than a disappearing act: We can remain fully present, albeit transformed, nourish the soil, enlarge the habitat, and rekindle the life of meadows and forests, feeding and becoming plants, animals, and trees. For many, it seems right to someday "be a tree."

"In the meantime, I dream of the cemetery of the future, full of fruit and nut trees and ornamental plantings, some of which yield food, too, or holiday decorations like pinecones and bittersweet. At the entrance there would be a farmers' market to sell the surplus of food from the cemetery grounds. There might be wood for fuel or for carpentry from the trees that in time grow old and need to be replaced. I imagine a family picking up hickory nuts from Grandmother's gravesite, remembering the pies she made from them."

—Gene Logsdon[12]

Natural Burial:
The Traditional Alternative

A number of methods are available to us for reintegration with Earth's biological systems in natural ways. Some of them, such as the Tibetan "Sky Burial," the Beaker People's "barrow burial," or the more familiar "burial at sea" (as long as the body is in a weighted shroud and not a nondegradable casket!), are older than our recorded histories. Others, like the accelerated dispositions now being developed, are attempts to address problems created by the old ways that often generate new impacts in their wake. Burial is the oldest known form of intentional disposition, and still one of the most common. Burial in soil breaks the body down via biological, geological, and chemical processes in the environment, producing the elemental reactions, the weathering, and the natural succession of large and small creatures that eventually consume the body. Full skeletonization is the goal; average soil can achieve that in 5–10 years, and active soil can do the job in as little as 18 months, once the soil has direct contact with the body. It's a natural.

Tree burial in Yorkshire.

C.A. Beal

Soil Disposition: Making the Case for a Biological Return

On the continuum of processes, a direct earth burial that makes one's body available as a full-spectrum nutrient source for the soil web does more for the planet's biological system than cremation. According to Dorian Sagan, author of *Into the Cool* and student/teacher of the thermodynamics of living systems, the longer our biological web can keep life forms "in play," transferring energy from one creature to another in the Great Chain of Being, the more resilient our planetary system can remain. The complex and self-organizing, self-regulating biological and geophysical systems that help to balance temperature, moisture, and atmospheric gases and support life as we know it on Earth are created and maintained by the continuous recycling of the organic and inorganic matter that are the elemental building blocks of all animate beings. Sterilization (from embalming) and the combustion of cremation destroy the integrity of fundamental molecules, enzymes, and microbes present in your body, and the former may even affect the soil it's buried in, depending upon the chemicals present in the embalming solution. In contrast to the chemical-intensive practice of preservation or the energy-intensive process of combustion, returning bodies to the Earth's natural system makes a strategic use of our parts for the greatest number of beings, over the longest period of time.

Soil Quality: Building the Living Web

Burial in a biodegradable container presumes and encourages decomposition. Decomposition requires active (that is, "alive") soil, and according to soil scientists, the same conditions that are necessary for proper decomposition—nutrients cycling at the right rates for complete breakdown to occur—are required for healthy plant systems, too. Organic carbon is the key to these processes, as it is constantly recycled from organism to or-

ganism, including trees and other plants that absorb it out of the air. And so it's not enough to just plant the tree. The soil web has to be healthy enough to grow the tree well.

The US Department of Agriculture's Natural Resources Conservation Service provides a tremendous amount of free online information related to building and maintaining soil quality, with in-depth sections on soil health, soil assessments, and maintaining the "soil food web," a term coined by soil scientist Dr. Elaine Ingham, now at the Rodale Institute (nrcs.usda.gov/wps /portal/nrcs/main/soils/health/). The NRCS defines soil health as "*the continued capacity of soil to function as a vital living ecosystem that sustains plants, animals, and humans. This definition speaks to the importance of managing soils so they are sustainable for future generations.... Only 'living' things can have health, so viewing soil as a living ecosystem reflects a fundamental shift in the way we care for our nation's soils.*"[13]

Ideally, cemetery disposition should support and sustain the cycle of life, not compromise it. Modern biology is only now beginning to deeply connect with other scientific disciplines—geology, climatology, physiology, and thermodynamics—to quantify the energy transfer that interdependent living systems generate and manage in the complex soup of life. Is it really such a big leap to imagine that your own death can be a doorway back into that natural and elemental world? For those of us who've been frustrated by the difficulty of living an integral life in this forest of synthetic industrial marvels, a natural death may be the easiest lifestyle choice we'll ever make. In the end, all we leave is energy. Good, useful energy still available in the form of complex molecules—fat, bone, and blood—there to be wrestled apart and turned into good little worms and beetles (who eventually also take their turn in feeding the small). Or as one organic gardener insisted he wanted on his headstone—"WORM PARTY!"

> Is it really such a big leap to imagine that your own death can be a doorway back into that natural and elemental world?

Cemetery Stewardship's Triple Bottom Line

Experts from all quarters have said that the key to getting through the next several decades—when population, energy requirements, and the level of resource consumption to meet our needs must become sustainable, or drastic changes to our lifestyles and cultures will take place—lies in achieving sustainability throughout all levels of human life. Sustainability is the 21st century's watchword, and it needs to be a part of cemetery management, too. As a forward-thinking version of the

Golden Rule says, "Do unto future generations as you would have them do unto you."

Sustainability has three primary components: social, environmental, and fiscal. The collective evaluation of costs and benefits based on financial, social, and environmental factors is known as the "triple bottom line." This conceptualization helps the operators of businesses and organizations address all three areas of activity simultaneously, identifying critical elements in each with the goal of balancing all three in the course of operation. Each of these categories affects the other two when a "full-cost accounting" is done, and the overall sustainability of an endeavor—i.e., its likelihood of success—is best served when all three are in balance and no one aspect damages the other two. For example, selling products and services below cost may create short-term social benefit (popularity and service to the disadvantaged) but financial calamity in the long run, ending the social benefit altogether (and perhaps the company). Polluting the environment may help the immediate financial picture, but costs in environmental fines and social "badwill" can exceed the gain or jeopardize an entire industry, exposing it to nationalization, regulation, or excessive consolidation. In sustainability parlance, "stewardship" is paying attention to all three aspects of the triple bottom line (TBL).

Finding the Cemetery's Triple Bottom Line

The Financial TBL

Like all businesses, a cemetery must make a profit, or receive donations and subsidies in excess of its costs, to survive. The financial TBL is probably the most familiar, and the easiest to calculate; it's what's left—the "bottom line" at the end of the balance sheet—after all the costs are subtracted from revenues. No social or environmental benefit is worth very much if the lack of profit kills the operation, and so the social and environmental factors, while important, can't be so excessive that the operation dies financially. Because many social and environmental actions have a financial cost, and because a cemetery's financial obligations are ongoing forever, the balance between the three is always carefully managed, requiring that many of the social and environmental elements return at least some income for their support. Calculating the financial value of ecosystem services may help that prospect immensely, and could help a cemetery qualify for grants and other assistance. However, nothing is more important financially to a cemetery than to have a properly sized endowment care fund.

The Social TBL

The social aspects of the TBL are sometimes the most difficult to see. They include issues of ethnic diversity, worker fairness, responsiveness to the local demographic (cultural, ethnic, age), and cultural or historic stewardship. Answering questions like the ones that follow can help you understand where the cemetery you're interested

in stands with respect to social TBL criteria, and also show how a cemetery can gain financially (i.e., attract socially responsible customers) by supporting its social TBL:

Does the cemetery meet the needs of the community without discrimination? For example, the Muslim community has burial practices that are out of step with conventional Christian ones. They require a direct earth burial; they use a shroud; they perform the handling of their dead themselves; they bury North/South instead of East/West.

- *Fiscal benefit: more customers; good customer relations; less/different kind of work to do a Muslim burial.*
- *Environmental benefit: maintaining natural burials are easier on the future and the cemetery's soil.*

Are its workers fairly treated? The US and the UK have minimum wage and worker protection laws that also apply to cemetery workers. Workers may, however, be denied union opportunities; they may be part-time employees without benefits; and long-term employees may not receive pensions. In cases where the cemetery ownership is large and wealthy while its employees are many and poor, this aspect of social justice may be important to a customer, and advertising employee treatment can be useful to serve this preference.

- *Fiscal benefit: Fair treatment = employee retention = more income; happy workers = lower medical bills.*
- *Environmental benefit: When people are healthy and stable, they impact shared resources less; well-treated employees make fewer mistakes and waste less.*

Does the cemetery fulfill its obligation as the historic custodian? Cemeteries are eventually historic sites. Most are sitting on treasure troves of community culture. Is the cemetery connected to its historic society? Does it provide information to the public about who's buried there? Does it care for any historic documents appropriately, including proper storage and emergency plans for maps and archived information?

- *Fiscal benefit: Reconstructing lost historic records is expensive; cemeteries that connect with the public on the basis of local history have higher sales and donations.*
- *Environmental benefit: When people value the history of a cemetery, they're more likely to take care of its environment as well.*

The Environmental TBL

Difficult to measure financially, the environmental aspect of the TBL is represented by those activities that help to renew, regenerate, rebuild, and conserve ecosystem services that are of value to living things. Sometimes leaving an area alone is of tremendous value to the local ecosystem, and the act of NOT impacting an area should be counted when accounting for the cemetery's environmental TBL. Common elements of sustainability programs that can be implemented in the cemetery, clearly connecting to recognized ecosystem services (and thus counting as assets and positives in the financial TBL) include:

- Soil and water conservation
- Fish and wildlife habitat
- Public health and environmental safety
- Animal health and welfare
- Energy intensity, frequency of use, and renewability

People, Planet, and Profit

When viewed through sustainability's TBL lens, a cemetery needs to consider all three categories of Planet (Environmental), People (Social), and Profit (Financial) in order to make sound and well-balanced decisions that don't seriously compromise one stewardship role in favor of another. Eventually, cemeteries of the future will need to have good answers for most, if not all, of these questions:

Financial Stewardship

- Does the cemetery have an Endowment Care Fund?
- Can the cemetery pay its bills, conduct maintenance, and fulfill its contracts?
- Will the cemetery become a future burden on taxpayers?
- Does the cemetery have a multi-generational financial plan?
- Is the cemetery facing future liability or risk from degradation?

Social Stewardship

- Does the cemetery meet the needs of the community without discrimination?
- Are its workers fairly treated?
- Does the cemetery fulfill its obligation as the historic custodian?

Environmental Stewardship

- Does the cemetery pose a future pollution or public nuisance threat to its community or neighbors?
- Does the cemetery manage its landscape to rebuild soil and support habitat and wildlife?
- Does the cemetery reduce its resource use whenever possible?

No matter what their ownership, mission, business organization, or marketing budget, cemeteries that meet the above benchmarks are making a difference, and are in the process of transitioning to sustainability. Encouraging cemeteries to take these steps—by purchasing grave or cremated remains space there—will go a long way to supporting this shift.

Grave Reuse: A Practical Solution for Urbanizing Areas

One significant environmental cost that's almost never calculated is the cost of perpetually occupying the grave space. On top of the issues created by thousands of containers-in-containers holding non-decomposed bodies, the caskets and headstones are placed in cemeteries or churchyards "in perpetuity" and require ongoing maintenance, ostensibly forever. Grave reuse, common in Europe, has yet to take hold in North America—but it's probably on the way. Ken West, telling the story of the UK's natural burial movement in *A Guide to Natural Burial*, cites a technique for reusing abandoned grave space now being tested in the UK called "Lift and Deepen." The technique involves opening the grave, reburying any skeletal remains below the floor of the grave (or returning

Reclaiming graves in Queen's Road Cemetery, south of London.

C.A. Beal

them to the family), and then performing the new burial in the vacated soil cell. (It's important to note that this is a *reuse* of the space—meaning a burial has already taken place there—and not simply a reselling of rights that have been abandoned without originally using the grave.)

In 2012, the ICCM (The UK's Institute for Cemetery and Crematory Management) issued a letter encouraging the minister of justice to look into the possibilities of changing UK law to allow for grave reuse. This is significant coming from the leading cemetery industry trade association that represents municipal as well as for-profit cemetery companies. Their position is based on the association's firm grasp of cemetery economics and its understanding that the easy availability of perpetual grave space has come to an end. We would do well to heed these considerations:

"If these practices were instituted—and especially…[grave reuse]—the need for cemeteries to expand onto new land would be dramatically curtailed. Since there's no proven health and safety reason why this practice can't be engaged in, and since it's questionable whether or not arable land will continue to find market as a cemetery, operators are wise to keep this possibility open as a 'game changer' with respect to the cemetery of the future."[14]

At this time, long-term maintenance costs, the resources consumed, and the true environmental and taxpayer costs of aging cemeteries have not yet been factored into many cost-benefit equations, even though they can be easily calculated with budget planning software available today. City planners, corporate cemetery stockholders, and their insurers are only now beginning to appreciate the expense accruing as they run out of space and are faced with tighter regulatory controls on the burial and discharge of potential pollutants and nondegradables into the

The Natural End Play in Three Acts

The Natural End Play offers a discussion framework that separates the various EOL (end of life) activities into meaningful segments by focusing on who does what, and where and when they do it, to help us think about and discuss them more clearly. Grouping tasks in this way helps professionals, policymakers, friends, and families talk about the tasks one by one and plan them in sequence, a step at a time.

The Natural End Play
Act I: The Body, the Family, and the Funeral
Act II: The Final Disposition
Act III: Everything Else and After…

- **Act I is about the Body, the Family, and the Funeral**—managing the deceased's body naturally, gathering friends, religious community, and family, and the other activities that take place before the final disposition.
- **Act II centers on the Final Disposition** itself, a legally defined method of body disposal, usually spelled out in government statute, with the place and type of disposition entered into public record.
- **Act III is everything that happens post-disposition**, after the burial or cremation takes place. Act III involves the burial of either the body or its cremated remains, and the perpetual care of the landscape that the burials take place in.

Using this framework, it is possible to identify who is responsible for (and who is in control of) the various parts of the process when someone dies, especially when there's a need to purchase unfamiliar items and services like funerals, coffins, dispositions, and burial plots. Dividing the tasks into Act I, Act II, and Act III elements makes it easier to think about and shop for products and services at fair prices and, in general, provide what most people seem to want: meaningful, affordable, and, in an increasing number of cases, environmentally responsible funerals and celebrations of life that reflect the values and personality of the deceased as well as the family.

This three-act framework also makes it easier to focus on the most important things to each individual. Some folks care a lot about the funeral, but not so much about what happens after. For others, the only thing that matters is returning to earth and "being a tree."

Creating Accountability
For those who do want to "do death" differently than the current industrial paradigm has dictated, once we separate these activities from one another, it's simpler to decide upon the changes that are desired, what things could easily be changed, and what things must stay the same. Knowing who is legally responsible helps. For example, to change how we manage the body at death—perhaps no embalming, perhaps a different sort of funeral, or using a homemade coffin—it doesn't do much good to talk to the cemetery manager, since that

environment. While the UK ministry has not yet issued a final decision as of this writing, Australia is in the process of approving the practice. Given the clear TBL advantages of the practice—with social, ecosystem, and financial benefits that happen immediately—any cemetery-using society with urbanizing areas and doubling populations can't be that far behind.

Crafting the Fond Farewell: It Takes a Plan

When a loved one dies, multiple issues—the body, the family, the casket, the disposition, the cemetery, the money—need to be managed properly and quickly, and that takes planning. Funeral businesses tell us that all the time, and they're not kidding. Most of us don't have a lot of experience with death; the language is unfamiliar, and we work to avoid it for as long as we can. But, as experience eventually teaches us, while planning for death may be uncomfortable, *not* planning can be miserable, especially for the family and friends that have to sort it out when we neglect to do so in advance. This is where funeral directors and home funeral guides can come in handy—especially since it costs nothing to plan. (The Natural End Play breaks these activities out into sections: Act I, Act II and Act III. See the accompanying sidebar.)

Funeral consumer activism over the last several decades has led to the formation of a number of organizations and services that can assist with the planning job, making it much less difficult than it once was. In addition to the home funeral advocates mentioned earlier, the Funeral

NATURAL BURIAL

person handles the body *after* it's been buried. Likewise, if we're concerned about how the cemetery is going to be cared for in the long term or what sort of tree we'd like on our grave, the funeral director or the crematory operator is not the resource to consult. *Sequencing the End-of-Life activities in this way lets us know who to talk to and helps to keep these last things straight.*

Natural Packaging
Body packaging and preservation choices made during Act I can exert their greatest environmental impacts during Act II, when whatever items were used during that first act are consumed in the burial, cremation, or other disposition method chosen. It's at this point that natural coffin and casket materials show their true value, with their qualities of renewability, biodegradability, and minimal impact on the environment, whether buried or burned, lightening death's last footprint. The conventional funeral industry has little experience in natural materials, and it's up to natural products consumers to request and insist upon products and services that meet their needs. Companies like the Natural Burial Company sell natural grave goods both retail and wholesale, ensuring that anyone who wants a biodegradable coffin for a more natural disposition can have one.

natural coffins, urns and shrouds

NaturalBurialCompany.com

serving a more natural end

"...if it's the last thing you do..."

Act III: The Sustainably Managed Cemetery
In Act III, everything that comes after the final disposition—from the decomposition of the corpse and the coffin, to the growing of trees and the placing of stones—can now take place, and does so over decades and centuries, from this point on. Cemeteries are generally considered permanent sites for disposition and memorialization, and for many, the cemetery provides a focal point of remembrance—a physical place to go, to memorialize for a time the life of someone they've loved.

Placement of the body in the cemetery marks the beginning of a much slower process than the first two acts, and includes bereavement and grieving on the one hand, and the functional storage of the remains and the memorial, coupled with the long-term maintenance of the cemetery site, on the other. Whether or not grave reuse becomes as common in the US and the UK as it currently is in Europe, a cemetery is clearly an important community space, and most are likely to endure for a couple of centuries, at least. Even if the practice of burial is abandoned, most of the cemeteries that exist now won't be dug up and moved or destroyed. Act III—ongoing and "forever" as far as the cemetery is concerned—is here to stay. Consequently, to the extent that a cemetery's practices are redirected so that it minimizes its resource use and future maintenance costs, refrains from contributing to pollution, and turns its landscapes into habitat-worthy micro-ecologies that benefit the area it's located in—and to the extent that it connects with the history of the community it's a part of—it will likely sustain and pay its way.

To the extent that a cemetery's practices are redirected so that it minimizes its resource use and future maintenance costs, refrains from contributing to pollution, and turns its landscapes into habitat-worthy micro-ecologies that benefit the area it's located in—and to the extent that it connects with the history of the community it's a part of —it will likely sustain and pay its way.

Writing out what you want and leaving it somewhere that it can be found easily will make someone praise your name when the time comes.

Consumers Alliance, an organization with chapters all over the country, produces planning pamphlets that can be downloaded and shared with friends and family (funerals.org).[15] Funeral planning websites for those Boomer-age and younger (Get Your Shit Together, getyourshittogether.org), periodicals on death and dying (*Natural Transitions* magazine), discussion group gatherings like the Death Cafes, and popular mortician celebrities like Caitlin Doughty ("Ask a Mortician" on YouTube) make doing and learning about death both interesting and fun. And everywhere that people are talking about a more natural funeral, they're talking about what they want: something that's lighter on the Earth, with minimal impact and preferably some benefit to the environment. And that talk, and the actions that follow, is the key to change.

Death has Paperwork!

Death has paperwork. There's no getting around it. But since other people have to do it (because you can't), make it easy on them. While the specifics may vary slightly from state to state—and some places still have onerous laws prohibiting personal involvement with a loved one's body, so check this out with your state first—some general principles apply. Before the body can be moved, the doctor or medical examiner certifies the cause of death and signs a death certificate, a form that requires a lot of information (that should already be noted in your funeral plan). The paperwork, and the person in charge of it, often dictates the remaining hours of the body's management before disposition. If the death was natural and expected, the management of the body is usually still up to you—or rather, it's up to the person you've designated as your "Personal Funeral Director," the "person in charge of interment," who manages the "disposition of the body," officially, in advance, on a notarized piece of paper. Really.

Whether or not a funeral director is hired to manage your body post-death, if you haven't legally named a personal funeral director to manage your disposition (often a close and trusted friend, and still permitted in most states, and different from the executor of your will or the person with a general power of attorney), the decisions and arrangement tasks fall to your "official next of kin." Absent the next of kin, the only others who are legally able to transport your body around and do things with it are certified licensed professionals. Anyone else caught dead with you—sorry, you dead with them—and without the necessary permits could have a problem. Writing out what you want and leaving it somewhere that it can

be found easily will make someone praise your name when the time comes. Finally, don't forget the other aspects of bureaucratic closure: a living will, a personal will, and an advance directive, at minimum, along with a comprehensive listing of all the bits someone needs to know if they're going to have to dig through your files and piece together what you were supposed to be paying for next week but couldn't. Yes, it's a big job, but someone has to do it, and it ought to be you.[16]

Tell your Family and Friends

Aside from the paperwork, the most demanding part of your death (provided you've arranged for everything else in advance) is the preparation of your body, since you're no longer very good at it. In addition to your instructions on embalming, services, containers and disposition choice, you'll also want to think about what's currently inside you. Modern bodies tend to go out with more than they came in with. Teeth are often filled with mercury amalgam—stable when cool and in the ground, but not so good if you've chosen cremation. Silicone and artificial joint implants are increasingly common, and bodies may have pacemakers (they explode in crematoriums, and silicone pools in the kiln). Unless you leave instructions that you know are workable, it's unlikely these items will be handled responsibly after your death. This is generally one of the least pleasant tasks left to be managed and should, if at all possible, be arranged in advance, by you.

If you're planning on a home funeral, put together a group of committed friends and loved ones who are willing to handle you properly when the time comes and support your wishes, and make this group known to your biological family. Church groups and extended family units are great for helping out here. This is also where the help of an experienced consultant can come in handy—someone trained as a home funeral guide, or a sympathetic funeral director—since your personal group will need to understand how to bathe, chill, and dress you, how to carry you and when to move you, where and how to place you, and, in general, to be there to help others feel okay about being there with you when the time comes. People have been doing it for millennia, but the cultural chains have been broken, and it helps to have the guidance of those who've been through it before.

Neutral groups like hospice can be helpful, but they tend to shy away from advocacy of businesses or services used after a death, especially when those needs deviate from traditional death management practices and utilize alternative providers like celebrants, biodegradable coffin companies, and home funeral guides! Generally, the hospice role is to assist until just before you die, and then turn the final steps over to professionals, unless the family and friends are clearly participating post-death. And, like a growing number of people today, if you're facing a terminal illness and are choosing to be cognizant of (or even determining the time of) your end, discussing this process openly with your family, friends, and/or group will be a relief for all concerned, especially those who may not be able to cope easily with your passing. Your wishes will be known, your group will become as comfortable as possible under the circumstances, and in the process, they will become guides for the rest of your friends and family, turning a typically disengaged experience into a fully empowered one. Here's one revealing testimonial: *"I know the discussions of funerals may sound a bit morbid to many out there. However, you cannot believe the change in my father's attitude once my mom, dad, and I sat down and discussed some of this stuff. Suddenly, he was able to discuss everything regarding his cancer more easily, which eventually led him to realizing that his chances for survival are very good."* Catie Jay Bee, 2002, Online Organic Gardening Forum.[17]

Tools to help the planning:

- Download a natural funeral planner for free at (naturalburialcompany.com)
- Play "My Gift of Grace," a conversation game for living and dying well (mygiftofgrace.com)
- Read *Final Rights: Reclaiming the American Way of Death* (upperaccess.com)
- Ponder a bit on how to "Be a Tree"

> By planning ahead, by choosing your process and your container in advance, and spelling out your wishes clearly, you'll go a long way toward improving what might otherwise be quite the opposite of what you'd wish, if someone could ask you after the fact.

What You Need for a Natural Burial

You don't have to be buried in a dedicated natural burial ground to make your last moments more natural. By planning ahead, by choosing your process and your container in advance, and spelling out your wishes clearly, you'll go a long way toward improving what might otherwise be quite the opposite of what you'd wish, if someone could ask you after the fact. The key elements of a natural burial are:

- A preservative-free body
- A biodegradable container, or none at all
- A cemetery that accepts a vault-free burial
- People to put you there
- Laws to support your right to be there
- A community to tend the habitat as you're decomposing

The Last Stuff

Whether you're buried in a coffin or wrapped in a shroud, the main thing to insist on is the use of biodegradable materials in everything that accompanies you "out the door" or into the earth, no matter where you end up. Just by using a natural container, you'll minimize your impact on the environment because of all the conventional casket materials you won't be buying or burying, and you avoid the polluting or energy-inefficient processes used to make them. Your container is a great place to start. Even if you (or your parents) are buried in a conventional cemetery, in a vault, or in a mausoleum, you'll still lessen the ecological footprint of burial boxes simply by choosing the natural ones—and it just gets cleaner from there.

As of 2014, we're just beginning to accept natural coffins in the USA, and natural coffins don't yet register with American casket company trade associations. But if successful competitors are any guarantee of markets shifting and options emerging, help is on the way. The templates for the "cleanest" natural coffins we've seen this century—rapidly biodegradable, cleanly combustible, made from natural and renewable materials, produced by local makers—found their first fertile soil in the UK. (A densely packed island with predictable death rates and a lot of bird lovers is a great place to birth a natural burial movement.) For the past several decades, UK-based coffin makers, designers, and weavers have produced high-quality willow, wool, and recycled

How much is that coffin in the window?

C. A. Beal

The Ecopod is made out of recycled paper.

newspaper coffins, their success due in no small part to alternative funeral home operators like the Green Funeral Company in Devon and ARKA Original Funerals in Brighton. Firms like these promote environmental friendly techniques and products, write passionate blogs, put coffins in their shop windows, and encourage natural home funerals for the family. Today, with over 50,000 "green" funerals conducted in the UK annually,[18] consumers are well past the tipping point, and the trend shows no sign of abating.

With hopes of recreating the UK success, the Natural Burial Company in 2006 hosted some of the first successful UK makers into the US, introducing woven coffins of willow, cane, bamboo, and seagrass, handmade paper urns and pet coffins, and recycled paper Ecopods. Shortly thereafter, Passages International, a seasoned US supplier, and E-Coffins, a UK-based company, started supplying low-priced wicker coffins from Asia. Dozens more producers have come along since, and today's range of biodegradable and low-impact burial containers and wrappings offers more variety in design, production techniques, and materials than the funeral sector has seen in some time. Shrouds of silk, linen, hemp, and cotton; urns of earth, paper, and sand; coffins of wool and weave; artful wooden and traditional plain pine boxes. Even the Kraft-wrap alternative container threatens to become trendy, thanks to the fact that everyone knows that cardboard biodegrades.

Since almost anything can be ordered online today, once someone knows about the natural

Reclaimed pine casket kit from Northwoods Casket.

possibilities it's hard to imagine *not* getting what you want. And while biodegradability is important, it's not everything. Although the debate continues to evolve as to what will and won't biodegrade in the presence of healthy soil microbes, hungry trees, or introduced fungi, guidelines from the natural products world like those above readily appeal to our common sense, and they work for burial goods, too. Consumer advocates and natural product companies collaborate on public educational events like the Green Festivals (greenfestivals.org) and other product shows. These trade shows provide a lot of information about the environmental and social impacts and benefits of various materials, products, and processes, and much of what they know translates over to the world of natural funerals. At the rate these ideas and products are spreading, don't be surprised if your local natural foods co-op, garden center, or favorite online eco-retailer begins to offer a selection of "final furnishings" in your own not-too-distant future. Supply is no longer an impediment to change.

Saving *Your* Money, Supporting *Your* Values

The "freedom to shop"—to choose from a range of products and services that best reflect one's values and don't waste money—is one that Americans have a tendency to take for granted. Even so, getting access to alternatives isn't always easy, especially in an industry with a lot of regulations, more than a few of which seem to protect the businesses rather than the public the rules are designed to defend. Fortunately, the right of US consumers to supply their own burial containers rather than those purchased through the funeral home is protected by the Federal Trade Commission's Funeral Rule[19] and can't be countermanded by states. Even so, the price of a funeral continues to climb, suggesting to economist David Harrington[20] that the casket manufacturer isn't the culprit. In 1959, according to *Time* magazine, $1.5 billion was spent on burial annually, at an average of about $900 per death. As of 2013, the average funeral in America, including embalming and a metal casket, priced out at around $6,600 (not counting cemetery costs), with cremation at about half of that amount. Conventional cemetery services are, on average, an additional $2,000–$10,000 or more, depending upon whether or not the grave is vault-free, the coffin is biodegradable, the type of monument, and how much the cemetery charges for the rights to use burial and memorial space. Harrington and others contend that improper regulation of the end-of-life industry by narrowing consumer choice and limiting business activity while minimizing public oversight results in artificially high prices. He suggests that once the public has access to a greater variety of funeral products, competition in other services will also emerge, and he claims that the Internet is the key to this expanded competition.

And while natural funerals don't *have* to be inexpensive (what's the price of a great party these days?), they *can* be. Home funerals make the costs of gathering more controllable, whereas renting the services of a funeral home for body management and gatherings, with rush services and non-essential but attractive extras, ups the price accordingly. Burial in a cemetery always involves the cost of the grave space and basic fees, but maintenance of a natural burial plot is much less intensive than a conventional one over the long run, so expect direct costs to eventually be lower, especially once the municipal and public cemeteries get involved. Some cemeteries (and even some states) still require the graveside presence of a paid and certified professional during burial, but if the requirement is a law, expect it to be challenged in the future by the growing funeral consumer movement as an unjustifiable cost, and if the requirement is a business practice, expect the free market and competition to change it.

Natural funeral products don't have to be expensive either. The most popular coffin style in the UK, for natural burial or cremation, is the biodegradable cardboard box (now available with custom photo finishing!), usually costing several hundred pounds, and a real bargain when contrasted with the metal and hardwood veneered caskets typically sold in the US that can retail for multiple thousands of dollars. But you get what you pay for, and don't let that cardboard coffin get too wet in the rain! Woven wicker has a price range that depends upon a lot of variables: Did the coffin come from Asia (made with lower-wage skilled labor), or was it woven in the UK or Europe, made by equally skilled people where the costs of production and the wages are much higher? Is the quality just passing, or is the workmanship to higher standards? Woven coffins, no matter where they're made, are more expensive than cardboard or plywood. However, unlike cardboard, willow and other wicker fibers are produced without the use of industrial papermaking facilities. Woven coffins biodegrade much faster than solid wood, and the materials are nontoxic to produce and renewable. Their production keeps an important suite of artisan skills alive—production-quality journeyman weaving is a skill

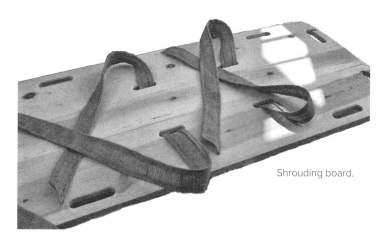

Shrouding board.

that's been lost in the US; perhaps the spread of these coffins will bring it back again.

Considerations like these affect the price, just as they would anything else, but they're also discretionary, and the market has room for them all. Supporting artists, traditional handicrafts, and a natural cemetery environment takes money, and many people think those are causes worth spending money on, dead or alive. The basics of a funeral should be as affordable as possible, however, with services or products required by law only if they preserve public health, safety, or critical areas of the environment. With freedom to choose, you can put your money where your values are, and with the money you save, someone might be able to throw a darned good party in your honor!

Do You Have to Have a Box?

Shroud burial, where the body is wrapped in fabric of some kind, is a perfectly acceptable form of natural burial that is still common in much of the world. The Jewish community traditionally buries their dead ritually wrapped in an unhemmed shroud, in the classic plain pine box that is, per orthodox rule, "unadorned." Many Muslims do the same, and may or may not use a box. Buddhists, Baha'i, Pagans, and plain ole Grandmas may all prefer to go box-free. Hemmed shrouds in creative designs, made of organic hemp, cotton, wool, and other natural fibers are available. But there *are* practical considerations with shrouding, especially when it comes to handling and moving a body, so plan ahead. Rigor mortis fades after 24 hours, the body softens again, and lowering a shrouded person gracefully into even a shallow grave takes some skill and forethought—or a Shrouding Board(TM).

Whether you're required by law to be buried in a container varies from state to state in the US, although this decision is usually left up to the

Fortunately, the right of US consumers to supply their own burial containers rather than those purchased through the funeral home is protected by the Federal Trade Commission's Funeral Rule[21] and can't be countermanded by states.

cemetery, and a cemetery is free to have a policy that requires one. Because sustainable landscape management has yet to catch on, many cemeteries still have long-standing rules favoring the double-box casket-and-liner package and haven't even thought about changing. If in doubt about when caskets are required by law, consult the Funeral Consumers Alliance website, funerals.org/; they have chapters in every state and can point you to sources of local information. Or call the state's cemetery board to get the statute or administrative rule.

Beyond Decomposition

Decomposition is important, but for many it's only the start. Keeping in mind the TBL of the products your death and funeral will consume, we suggest you begin with synthetic-free items, focusing first on products made of natural materials so that decomposition is assured. From there, go "up the ladder" of what's important to you, choosing from qualities like enhanced bio-

degradability, recycled and non-virgin materials, and sustainable production characterized by local handicrafting, family businesses, fair trade, and economic justice. Some possibilities include:

- Avoid synthetic and non-natural materials in your container and clothing
- Choose products designed to break down in the soil web
- Favor items from recycled and waste material instead of virgin resources
- Support sustainably produced burial goods with organic, fair trade, and eco-certifications as they begin to appear in the marketplace if you're not making your own

Any additional requirements can be spelled out in your final instructions and should include asking the family to leave your favorite gadget at home (or better, give it away!) and not burying you in synthetic clothing. The natural products section at the end of this chapter lists a number of natural grave goods and information on how to use them.

A Place to Go

Citizen-driven movements in support of natural burial can now be found in Europe, China, Japan, Germany, and Africa. Some groups have started new burial grounds to fill the gap left by conventional cemeteries slow to recognize this new demand.

Once you've decided upon your method of disposition and your container of choice, finding the right place to plant you, and folks who will do it, is next on the list. Since 2005, when we first began documenting this trend, hundreds of sites offering some form of natural burial—vault-free at minimum—have emerged in the UK, Australia, New Zealand, the US, and Canada, with other countries coming on fast. In the UK, and within just two decades of the first municipal cemetery advertising woodland burial, over two hundred dedicated natural burial sites are listed by the ANBG;[22] the ICCM lists hundreds of cemetery members offering natural burial, both municipal and private; and most of the cemeteries throughout the UK serve its 65 million citizens with vault-free, no-embalming funeral and burial services. A large number of these grounds are owned and run by city councils with public funds, and the natural funeral movement has provided the perfect impetus to bring rapid change to taxpayer-owned cemeteries across the country. The US, Canada, Australia, and others seem poised to follow.

Citizen-driven movements in support of natural burial can now be found in Europe, China, Japan, Germany, and Africa. Some groups have started new burial grounds to fill the gap left by conventional cemeteries slow to recognize this new demand. In the UK, where the process is more mature, exemplars of the shift to sustain-

ability include Tarn Moor Memorial Woodland (a city-run natural burial ground that discounts graves to residents); the Meadow at Usk Castle Chase (winner of the UK's Cemetery of the Year award); and the city-run Brighton & Hove woodland burial offerings, with its first groves now sold out and new sites established in environmentally strategic areas important to the community. While some grounds are new, many of the UK's natural burial opportunities are run by managers of existing cemeteries who recognize that, by simply returning to the way they used to bury and manage their graves, they can offer the natural fundamentals many people are now seeking, rescue their struggling cemetery operations, and ensure grave space far into the future. Sounds good, yes? So how did they do it? And how can we do it over here?

The Living Churchyard Project and the UK Pioneers

When Ken West, M.B.E., established the first "official" woodland burial site in the UK in 1993, it was designed to be an environmentally sound alternative to conventional burial that would be less expensive for the taxpayer to maintain and, ultimately, financially, environmentally, and culturally sustainable. West, then Bereavement Services Manager for the City of Carlisle Cemetery, proposed to city management that the most cost-

NATURAL BURIAL

366   REAL GOODS

efficient solution to issues stacking up around the cemetery—monuments toppling, mausoleums crumbling, lack of space, and vandalism—was a natural burial program. He claimed that burying citizens simply, in grasslands and under trees, would restore habitat, reduce resource use, rekindle community support for burial, and enhance the performance of the cemetery. He was right, as his fast-selling natural sections soon proved that natural burial was less expensive to maintain and more sustainable for the taxpayer. A number of cemeteries, public and private, followed suit, and the UK's natural burial movement was born. West's success didn't arise out of a vacuum. National and international programs from the 1970s and 1980s calling for environmental and cultural responsibility helped set the stage for his successful appeal to his city council. The timing was good; by 1989, the cumulative disrepair in old Victorian churchyard cemeteries had communities in a quandary. The responsible governments, churches, nonprofit organizations, and other owners were under pressure to clean them up, but without funds to perform the maintenance there was little to be done. Threats to a closed cemetery don't have a lot of weight, and the taxpayer eventually owns the abandoned cemetery, whether it cares to or not.

The national government knew it needed to do something, but it wasn't quite sure what. In 1987, about 15 years prior to West's first natural burial ground, the Arthur Rank Centre and leading conservation organizations launched the Church & Conservation Project, with an educational program called the "Living Churchyard" developed primarily to guide volunteer cemetery friends groups to arouse an "interest in the value of churchyards, chapel yards and cemeteries for nature conservation"[23] in the general public. In the subsequent 25 years, multiple Living Churchyard programs and natural burial programs in both new and established cemeteries across the UK have become valuable outdoor classrooms and museums, educating neighbors and the community about the wildlife and the local human history. To date, more than 6,000 cemeteries have participated in some fashion, and thousands of additional cemeteries worldwide have access to the same information via workshops and the Internet. Their materials are accessible online and available for cemeteries to use, limited only by their manpower, funds, and creativity.

Seeking to replicate the success of the Living Churchyard project, West's original public cemetery efforts, and the natural cemetery operators leading the trend, a growing number of facilities

A Living Churchyard conservation area in Carlisle's City Cemetery, begun by Ken West.

C.A. Beal

around the world are adding natural features that appeal to the environmentally conscious burial buyer. The UK still provides the largest number of models, however, including formal community sections (for consecration and religious burials with natural traditions); community, family, individual, and pet-human plots; orchard burials, meadow sections, remembrance and scattering gardens; and ashes burial. Rules vary and flow with the market. Some cemeteries advocate choice while others restrict burial to non-embalmed bodies and biodegradable caskets. Most encourage families to be actively involved in the organization of the funeral, and follow the Charter for the Bereaved,[24] and many of the cemeteries operate detailed websites that help others duplicate their work (and positive impact!).

"Any Cemetery Can"

Unlike their US counterparts, UK cemeteries (and many European ones, as well) comply with stringent environmental, cemetery-specific operational, and public health standards. They're more directly influenced by taxpayers, and often subject to restrictive environmental regulations that US cemeteries have yet to be constrained by. To open or expand a cemetery in the UK and many other countries today often requires hydrogeological and habitat impact assessments, neighbor approval, and a population-based proof of need (although that's likely to change in the US as cemetery impacts become better understood). Even so, the clear success of 6,000 largely volunteer-run Living Churchyard projects in *closed* cemeteries, promoting their reinvigoration as places of ecology, history, and final disposition, suggest that a workable model is in hand, and it's not hard to believe that "Any Cemetery Can."

Although the means to create more sustainability—or at least lighten environmental impact—is easily within reach of most cemeteries, the bulk of them still remain archaic and out of touch with current consumer trends, afraid to alter the status quo even though a conventional burial, with its nondegradable caskets, concrete vaults, and lawned landscapes, is clearly losing its consumer appeal. This bureaucratic short-sightedness dooms these cemeteries to fewer and fewer customers and eventual closure, bankruptcy, and abandonment (or taxpayer bailout) unless they develop ecological appeal. Fortunately, as many champions in the natural end-of-life movement point out, consumer demand for environmentally friendly disposition is real, and once the cemeteries and other related businesses begin to serve the public's desire by supplying the natural settings, products, and services that are wanted, many of their financial pressures will be relieved. By coupling Living Churchyard practices with sustainable cemetery management techniques that improve resource use, mitigate potential pollution issues, and minimize damage to soil health, cemeteries have a roadmap to improvement that can be pursued incrementally, with little risk, and with a strong likelihood of success. With hundreds of thousands of cemeteries in the US alone, and given the simplicity of

the first transition steps, we should use the cemetery to create valuable living habitat, greenspace, and cultural connection for the community. We call this collection of characteristics the "Living Cemetery Style."

Look for the "Living Cemetery Style"

The "Living Cemetery Style" can be identified by its practices, its techniques and tools, and its participants. Its practices will be focused on reducing environmental impacts, diversifying and enlarging habitat, and emphasizing comfort from nature, remembrance, culture, and community history where humans are concerned. Its tools and techniques will be the ones that make the improvements possible, many easy to spot just by walking through the cemetery (are there flowers? is the natural world respected? do you see a person with a shovel, or pushing a reel mower?). And its participants will be the humans and the wildlife that come to interact there over time, improving every decade as the landscape matures. Like a museum, the cemetery will take a multigenerational view, and evaluate its actions within a context that covers centuries, not just fiscal quarters or annual reports. It will have passionate staff and volunteers, and connect with the community in unique and engaging ways. That's how you'll know if you're "home."

As you can see, almost any cemetery probably already has or does—or is thinking about doing—at least some of the elements mentioned above. They're the same techniques and practices we're all becoming familiar with in our homes and city parks and schoolyards; the same techniques increasingly used on farms and golf courses; the same products used in our daily lives translated to the environmental requirements of the 21st century. Supporting these cemeteries—noticing them, observing what they do, and buying plots from them—will prove the concepts, build the models, and produce the resources needed for making the shift. And when Living Cemetery elements are included in the business plan and offered to the community, as Ken West and his colleagues in the UK have been doing for over 25 years, many cemetery operators will find that the cemetery can take on a whole new "persona," reinventing itself as a place of environmental and cultural significance and able to thrive over time. (The taxpayers and consumers will be happy, too.)

In-depth discussions of many of these topics—renewable energy, green building, permaculture, rainwater harvesting and greywater reuse, the importance of relocalization—can be found in other chapters of this *Sourcebook*. The relevant

Carlisle's reduced-mowing regime has the "Living Cemetery Style."

C.A. Beal

products sold by Real Goods are available at realgoods.com.

Tracking Down a More Natural End

As of 2014, several dedicated natural burial grounds have been announced in North America, and many more are in the planning stages. The number of cemeteries that operate on an exclusively natural basis is increasing slowly, but new start-up cemeteries aren't your only option. Cities and counties own or manage a lot of cemeteries, and thanks to hundreds of outreach programs in conservation and natural resource management, cities around the country are embracing sustainable landscape management, lowering their pesticide use, and enhancing greenspace. And they're starting to pay attention to concepts like ecosystem services and sustainability's triple bottom line.

The smart cemetery shopper looks for cemeteries that: 1) offer natural burial alongside their conventional offerings (adding new and needed income), and 2) simultaneously steward plants and build habitat (improving customer appeal), while 3) conserving resources (saving money and the environment). If you're in the market for a natural burial, you may even find yourself becoming a cemetery activist, lobbying for a natural burial option in your own local cemetery and volunteering to help with the transition. Any municipal cemetery can incorporate many Living Cemetery elements into their practices, and it's a rare mayor or councilor who would disagree.

Jewish and Muslim burial customs are natural by tradition, and cemeteries that serve these populations will be familiar with Living Cemetery concepts. Most municipalities have at least one cemetery with a dedicated Jewish section. Historic pioneer cemeteries often have more lenient regulations than conventional ones. Many of these older cemeteries are managed by fraternal orders like the Oddfellows or small volunteer boards of directors, with policies that can be easily changed. However, while these may provide hopeful opportunities, most cemetery operators may not realize how far along the prospects of natural burial have come. Visiting them with a copy of this article might be just the impetus they need for considering a change.

Don't assume that you know who's offering natural burial and who isn't; you need to do your homework. Just because you haven't heard about it doesn't mean they're not doing it. The cemetery business is highly competitive—not everyone advertises what they're going to do before they do it, nor do they brag about how well they're doing

Elements of the Living Cemetery Style (in no particular order)

- vault-free burial in biodegradable containers
- decreased mowed areas; cutting with reel mowers and scythes
- wildflower plantings
- living memorials on and off graves (plants and trees)
- low-input multi-species turfs and groundcovers
- proper water management, including conservation and contaminant mitigation
- native vegetation and xeriscaping
- rainwater harvesting and greywater reuse
- care for the soil web
- toxics use reduction (pesticides, herbicides, synthetic fertilizers)
- habitat support for flagship species (snags, thickets, year-round environments)
- interpretive signage describing the habitat
- advertised nature and bird walks
- composting of tributes and clippings
- Permaculture-aware master planning
- support of DIY funerals and family participation at graveside
- use of biodegradable containers for burial
- use of renewable energy in cemetery operations (solar, biofuel, hydro, wind)
- locally sourced monument stone, sculptors, and masons
- locally made burial goods
- cemetery hedgerow buffers
- employment of disadvantaged persons
- low-income/indigent burial services
- support of multicultural practices
- creating and caring for historic trees
- historic buildings that provide community history
- use of alternative building materials and techniques
- monuments of interest and/or unobtrusively marked graves
- use of volunteers for improving habitat health
- brochures and websites to explain the native flora and fauna of the cemetery
- conservation lands as part of the cemetery's burial reserve
- secondary income sources from land-based products
- collaborative relationships with local churches, extension offices, government services, and conservation organizations

with it, and historically cemeteries get their business by word of mouth. Fortunately, the cemetery business is also inherently local; as with your CSAs and your local natural food stores, the local cemetery may offer just what you're looking for, and all you've got to do is ask.

Online resources for cemetery and funeral service providers are growing. Those who provide basic minimum services like a vault-free burial, a funeral service without embalming, and

"There are compelling reasons for us to return to the web of life in a literal, as well as a figurative, sense. Becoming a tree, if for no other reason than to offset our own lifetime CO_2 emissions and kickstart some habitat in a likely-to-be-forgotten cemetery, might just be the best "last thing you do.""

The Orchard at Memorial Woodlands.

assistance finding natural burial goods can sign the Natural End Pledge and get into directories like the Natural End Map (naturalendmap.com). The ICCM in the UK has guidelines for both the operation of natural burial grounds and the treatment of customers looking for natural funerals, and its website lists hundreds of cemeteries that have signed the Charter for the Bereaved or are adopting their guidelines for natural burial. The Green Burial Council, made up largely of US funeral directors advertising natural services, lists cemetery members that engage in a range of sustainable practices. Finding a vault-free burial option or a funeral director that conducts no-embalming services in natural coffins grows simpler every year.

Information Sources:

Association of Natural Burial Grounds: naturaldeath.org.uk

The Natural End Map: naturalendmap.com

Funeral Consumers Alliance: funerals.org

Green Burial Council: greenburialcouncil.org

National Funeral Directors Association: nfda .org/green-funeral-practices-certificate.html

Institute for Cemetery and Crematory Management-UK: iccm-uk.com/naturalburial.php ?type=nat

So You Want to Be a Tree

Once you find a cemetery that will accommodate your wishes, what happens to you and your

tree (if you choose to plant yourself under one) should be governed by a contract signed between you and the burial ground proprietors when you purchase your plot—*meaning that you both need to think this through.* For example, the contract could allow for a tree to be planted at the grave to serve as a marker—but who cares for it, what kind is it, and what happens when it dies? In smaller grounds with tighter budgets, or large ones with different planting philosophies, one tree per grave isn't always practical, but any diverse woodland consists of shrubs and meadow areas as well as trees, and some people really would enjoy just pushing up daisies!

Some burial grounds may have provisions for cutting the timber after a certain number of years, harvesting any produce, or rotating the plots as a means of paying for the land and services and maintaining the site. Others put the land into permanent trust and let your tree or sod grow undisturbed, in perpetuity. Small sites with no room for expansion, and conventional cemeteries with high-density plot schemes, may find it problematic to plant a tree for each individual; larger grounds with the goals of reforestation and habitat creation may be happy with a low-density plan and more likely to support your wish to be a tree. For some, being part of an orchard or a garden and turning into dinner makes perfect sense. For others, the thought of being harvested is an abomination. Only you can know what will work for you and your friends and relations, and it should all be spelled out in the contract when you purchase your plot. Alternatively, you can let those terms remain vague and be released to the needs of future generations.

Because everyone is fairly new at natural burial in the US, the details of these eco-cemetery contracts will probably vary widely, and many questions will arise. Some contract terms will be governed by federal, state, and local regulations. In certain cases, though—if the burial grounds are run by a recognized religious organization— the grounds may be exempt from such rules. To get an idea of what you should negotiate for and expect when choosing your site, consulting the websites and terms of UK cemeteries that offer natural burials provides a good overview of what's successfully been done. Your desires should be spelled out in writing with the people you pay to manage your interment for the long haul. No matter how you word it—even if you plant yourself and then simply say "I don't care; let the cemetery decide"—the fact remains that something like this won't happen unless you take a direct hand in making it so. It could be well worth it. With the human contributions to climate change looming large, our continually increasing understanding of the science of Gaia continues to generate compelling reasons for us to return to the web of life in a literal, as well as a figurative, sense. Becoming a tree, if for no other reason than to offset our own lifetime CO_2 emissions and kickstart some habitat in a likely-to-be-forgotten cemetery, might just be the best "last thing you do."

The Ultimate
Back-to-the-Land Movement

Probably the most compelling model for greenspace and habitat advocates is the "conservation burial area," a piece of land simultaneously dedicated to natural burial and legally committed to the act of environmental conservation. This cemetery is either part of an existing cemetery that already holds the land in legal dedication to cemetery purposes via a government's cemetery statues, or it is newly zoned cemetery land with formal conservation easements attached.

The latter adds the impacts of burial and graveside services to a landscape in exchange for the income stream from cemetery revenue that will be used to deliver cemetery services and conserve the land. Favored by the mission-oriented ideals of natural burial advocates like Memorial Ecosystems' founder Dr. Billy Campbell and Greensprings Natural Cemetery founder Mary Woodsen, the cemetery may include the partnership of a private (or public) landowner who holds title and puts the land into trust, perhaps even contracting with a conservation group for ecological management and/or oversight. This type of project is generally run by a board of directors and has a land management plan that includes written guidelines about who or what can be buried, when, where, and how. When the demographics support a new start-up, and when existing cemetery businesses are not harmed by the loss of business, positive uses for this scheme include urban brownfields rehabilitation, logged lands restoration, and the preservation of sensitive ecological areas that won't be harmed by the increased human impact.

The former—the existing cemetery with a landscape to be transitioned to sustainability—is also in need of the income to survive, and conservation sections are an important part of the Living Cemetery style. Many existing cemeteries are in a position to create near-instant habitat, and they've already got the cemetery business and facility infrastructure to support it. Operating cemeteries often either have excess land (and once grave reuse is legal, they'll have a lot more!), or

they may have plans to expand into neighboring land. An existing cemetery can create a conservation burial area just like any other landowner, but with less legal hassle and with current contracts, personnel, and customer support infrastructure already in place. In either case, however, conserved land is conserved land—and the environment is often better off for it.

Making the Business Case—Why Not?

The commercial potential for operating a natural burial ground is becoming interesting to entrepreneurs who see a market opening, and this includes environmental groups looking for income streams to support their activities. Cemeteries, whether for public good or private gain, are still businesses with significant financial requirements, and starting from scratch may not be as sustainable as converting an existing cemetery, especially once all the other cemeteries nearby get into the act. The main difference between normal business operations and the end-of-life business is that in one of them you get repeat customers; in the other, you don't. Nobody is going to like natural burial so much that they come back and do it again. Consequently, demographics—how many people live nearby, and how many of those are dying—are everything in the cemetery business, and newbies should proceed with caution.

A brand-new conservation start-up that's surrounded on all sides by too many cemeteries is going to have a hard row to hoe; within 10–20 years, the competition will be doing "natural" too, and the newcomer will need a better business plan, unless the population growth and type supports it, or the marketing is very well done. Starting up a new cemetery will also bring new

A Watershed Sciences class uses one of Beal's cemeteries as an outdoor classroom.

C.A. Beal

The urgent creativity of the 21st century is rising to the challenges we face from overpopulation, peak oil, and climate change, and solutions for our bodies' final end that support the values of greenspace preservation, carbon sequestration, habitat creation, nutrient cycling, and resource use reduction are becoming apparent.

human impacts (bodies, traffic, visitors), and therefore starting from scratch—rather than repurposing an existing cemetery and using the infrastructure already there—may actually create more environmental impact than it reduces. Existing cemeteries are cemeteries forever, and if there are too many cemeteries in an area, and one of them is suitable for conversion, it may make sense to focus on one of *them* before adding new human impacts to land.

Staying Informed

New territory requires both research and education, and sustainable cemetery operators will need to connect with educational resources that can foster the changes the new market requires. Trade associations like those run by the ICCM and the ICCFA (US) offer continuing education courses, workshops, and professional training. The Natural End Map currently serves as a por-

tal to multiple sources of information for transitioning cemeteries. In addition to the resources listed in this article, the map links to the Sustainable Cemetery Studies Lab of Oregon State University[25] (founded by Cynthia Beal and Dr. Jay Noller in 2013); the Sustainable Cemetery Management Regional Resource List; and relevant online course work for academic degrees and professional continuing education. Begun in the fall of 2013, OSU's Introduction to Sustainable Cemetery Management course (created and taught by myself) is online and open to a worldwide public.[26] Thanks to publications like the Real Goods *Solar Living Sourcebook* and the spread of information on the Internet, this ultimate back-to-the-land movement is moving into a new and more vigorous phase of development—so look for an increasing number of products, and the voices that promote them, to make concepts, guidance, and new tools available for a more natural end.

Last Acts That Make a Difference: How Your Support of Environmentally Friendly Products Can Turn an Industry Around

Typically, about 2 million people die annually in the US. The post–World War II Baby Boom generation began to turn 60 in 2006, creating a bulge in the upcoming death demographic that will put over 20% of Americans over age 65 by 2030. That means that more of us will be hitting the end of the line for the next 15–30 years, which will cause our society to focus more intently on how we die, and what we do as we do it, than ever before.

The Everybody Coffin Kit, designed by Netherlander Gijs Zilstra.

Over $20 billion dollars a year is spent on death management in the US, much of it for industrial burial packaging products (caskets and liners), cemetery land purchase, and maintenance. Industry estimates place funeral sales at $11 billion annually, but this does not include cemetery fees and burial plot sales. The true environmental costs of aging cemeteries have not yet been factored into many equations, and city planners, corporate cemetery stockholders, and their insurers are only now beginning to appreciate the expense accruing as they run out of space and are faced with tighter regulatory controls on the burial and discharge of pollutants and nondegradables into the environment.

When we change our purchasing behavior, we send a signal to the industries we want to change. Asking that our caskets be free of toxins and pollutants, that our cemeteries get creative and end the use of liners and nonsustainable land management practices, or that our communities follow the lead of the UK and provide low-cost burial options as the public utility that the service rightfully could be, are not unreasonable requests.

The urgent creativity of the 21st century is rising to the challenges we face from overpopulation, peak oil, and climate change, and solutions for our bodies' final end that support the values of greenspace preservation, carbon sequestration, habitat creation, nutrient cycling, and resource use reduction are becoming apparent. Understanding of the importance of forests and the usefulness of trees in the form of ecosystem services, along with the power of the soil to transform natural elements and return them to utility for the web of life itself, is becoming more widespread every year. The demand to "Leave No Trace" is increasing. On the heels of these developments, and the emergence of a consumer who is looking for "a clean death" to accompany a conscious, low-environmental-impact life, natural burial is an exciting possibility. It was once thought impossible, but now is considered by many in the industry as just a matter of time, bringing win-win scenarios for individuals, communities, and our wildlife friends at every turn.

Taking the Natural Step

This article has shown that natural burial is not a dream of the future but is happening here and now; natural burial grounds are possible today. The community of people who look ahead is in-

What's Buried Along with Our Loved Ones in US Cemeteries Every Year

- 827,060 gallons of embalming fluid
- 90,272 tons of steel (caskets)
- 2,700 tons of copper and bronze (caskets)
- 1,636,000 tons of reinforced concrete (vaults)
- 14,000 tons of steel (vaults)
- 30+ million board-feet of hardwoods, much tropical (caskets)

Source: Greensprings Natural Cemetery FAQ, May 2014; naturalburial.org /greensprings-faq/

creasingly putting its money where its mouth is, and now we're putting our bodies there, too. It's hard to do the daily things right, every day, all the time. No one can. But dying is a once-in-a-lifetime experience, and each of us can take the time to plan it out and do it right.

Planting forests is a lot of work, and every community needs at least one natural burial sanctuary, in our opinion. Check out your local arboretum; are they strapped for cash? Does your city

Number and Percentage of Cremations

This chart shows for the United States:
1. The percentage of cremations from 2000 projected to 2017 (circles).
2. Total number of deaths each year, 2000 projected to 2017.
3. The increase in cremation trends in both

percentages and numbers from 2002 to 2017 (middle section).
4. Cremation rate if it were to remain constant. 2000–2017 in number of cremations and percentages.

Source: funeralbusinessadvisor.com/cremation-tends -and-opportunities/funeral-business-advisor

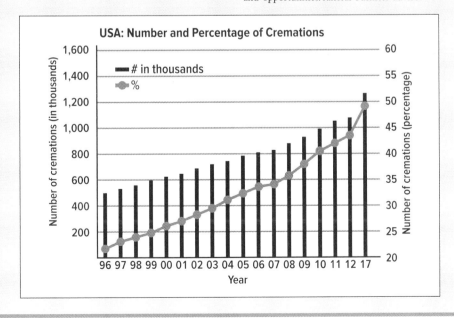

USA: Number and Percentage of Cremations

have an urban growth boundary or brownfield areas that could use some healthy greenspace? Is there a pioneer cemetery nearby with room for a Bioneer or two? Is it possible that taking responsibility for our own deaths may make us more aware of the unintended deaths we bring to others throughout the webs of life? And could we, in managing our own ends properly and in advance, plan an exit that reduces, or even reverses, the toll our lives have taken on natural resource systems up to this point?

It's an exciting thought. The emerging natural burial movement offers unexpected and overlooked opportunities to make choices that just might nudge our culture in another direction, if we make the time to do what no one else can do for us—plan ahead to exit stage left, and do it right.

All that's left to make the leap is you.

CHAPTER 15

The Real Goods Sustainable Living Library

Reading for a Renewable Future

"KNOWLEDGE IS POWER" the old saying goes, but at Real Goods, knowledge is even more: It's our most important product.

The *Solar Living Sourcebook* provides the technical and practical information about renewable energy, right livelihood, and the best ways to live in a post-carbon world while mitigating climate change. Our webstore, realgoods.com, allows you to take all that knowledge and put it into practice with deeply researched products that represent the best that our staff can find. The real value of this book is that it is a hub that connects you to the entire world of solar living, ranging from renewable energy to green building to the pond to the permaculture garden to sustainable transportation to natural burial and all points in between. Because our basic mission is to move the world toward the paradigm shift of low-impact, small-footprint, carbon-neutral sustainable living, ideas are just as important as equipment.

Even in the midst of an ongoing information revolution, the traditional book still reigns as the heavyweight champ. The book is amazingly efficient, portable, durable, and even sensuous in its ability to catalog, organize, and present data. Many of us have tried to read ebooks, only to discover that we like the feel and heft of a real book! Data accumulates into information, information transforms into knowledge, and knowledge matures into wisdom. No offense intended to all those crazy new technologies—they all shine in one way or another—but none of them has come close to challenging the book's grip on the championship belt. That is why we have put significant energy into making this *Sourcebook* a "book of books." The library we have assembled represents the end result of countless hours of research and selection. Each title, in its turn, leads you into a new journey into the world of knowledge.

In this edition of the *Sourcebook*, as we celebrate 36 years of progress and optimism, we have greatly expanded our book offerings, in service to you and to the goal we share of building a sustainable future. You won't find such a comprehensive library as we have cataloged anywhere, not on Amazon, not in all the megabookstores combined. If you come to our Solar Living Center in Hopland, however, please drop by and browse through the shelves, where our library comes alive and where you'll find thousands more book titles. But if you can't, then do your browsing here and on realgoods.com. You will be giving yourself the gift of power through the product of knowledge, which will ultimately result in wisdom.

Garden, Farm, Food, and Community

GARDEN AND FARM

Grow Your Own Food Made Easy: Nutritious Organic Produce from Your Own Garden, A Step-by-Step Guide

C. Forrest McDowell & Tricia Clark-McDowell

Learn how to save as much as $200 a month in food costs by growing highly nutritious produce in as little as 100 square feet! This color-illustrated guide gives accurate, concise, and easy step-by-step instructions on more than 25 Earth-friendly gardening subjects, including soil preparation, raised beds, natural fertilizers, companion planting, cover crops, mulching, natural pest control, and more! Included are sample food garden designs and 16 pages of health, nutrition, harvesting, storage, and eating tips for over 40 vegetables and fruits. The Nutrition Connection offers invaluable strategies to reduce cancer risk, aid in weight control, and increase ingestion of protein, calcium, iron, and vitamins.

The Backyard Homestead: Produce All the Food You Need on Just a Quarter Acre!

Carleen Madigan

Put your backyard to work! Enjoy fresher, organic, better-tasting food all the time. The solution is as close as your own backyard. Grow the vegetables and fruits your family loves; keep bees; raise chickens, goats, or even a cow. *The Backyard Homestead* shows you how it's done. And when the harvest is in, you'll learn how to cook, preserve, cure, brew, or pickle the fruits of your labor. From a quarter of an acre, you can harvest 1,400 eggs, 50 pounds of wheat, 60 pounds of fruit, 2,000 pounds of vegetables, 280 pounds of pork, and 75 pounds of nuts.

How to Grow More Vegetables

Decades before the terms "eco-friendly" and "sustainable growing" entered the vernacular, *How to Grow More Vegetables* demonstrated that small-scale, high-yield, all-organic gardening methods could yield bountiful crops over multiple growing cycles using minimal resources in a suburban environment. The concept that John Jeavons and the team at Ecology Action launched more than 40 years ago has been embraced by the mainstream and continues to gather momentum. Today, *How to Grow More Vegetables*, now in its fully revised

and updated 8th edition, is the go-to reference for food growers at every level: from home gardeners dedicated to nurturing their backyard edibles in maximum harmony with nature's cycles, to small-scale commercial producers interested in optimizing soil fertility and increasing plant productivity. Whether you hope to harvest your first tomatoes next summer or are planning to grow enough to feed your whole family in years to come, *How to Grow More Vegetables* is your indispensable sustainable garden guide.

The Vegetable Gardener's Bible

Edward C. Smith

Ed Smith's W-O-R-D system has helped countless gardeners grow an abundance of vegetables and herbs. This 10th Anniversary Edition of *The Vegetable Gardener's Bible* is essential reading for locavores in every corner of North America. Everything you loved about the first edition is still here: accessible language, full-color photography, comprehensive vegetable-specific information in the A-to-Z section, ahead-of-its-time commitment to organic methods, and much more. With every difficult question we face about the financial and environmental costs of conventional agriculture, more people think, "Maybe I should grow a few vegetables of my own." This book will continue to answer all their vegetable gardening questions.

The Year-Round Vegetable Gardener: How to Grow Your Own Food 365 Days a Year, No Matter Where You Live

Niki Jabbour

The first frost used to be the end of the vegetable gardening season—but not anymore! In *The Year-Round Vegetable Gardener*, Nova Scotia-based gardener and writer Niki Jabbour shares her secrets for growing food during every month of the year. Her season-defying techniques, developed in her own home garden where short summers and low levels of winter sunlight create the ultimate challenge, are doable, affordable, and rewarding for gardeners in any location where frost has traditionally ended the growing season. Jabbour explains how to make every month a vegetable-gardening month. She provides in-depth instruction for all of her time-tested tech-

All books featured here are available at competitive prices at
WWW.REALGOODS.COM

niques, including selecting the best varieties for each season, mastering the art of succession planting, and maximizing the use of space throughout the year to increase production. She also offers complete instructions for making affordable protective structures that keep vegetables viable and delicious throughout the colder months. What could be more amazing than harvesting fresh greens in February? Jabbour's proven, accessible methods make this dream possible for food gardeners everywhere.

Successful Small-Scale Farming: An Organic Approach

Karl Schwenke

"My advice is as old as the plow." So says author, Karl Schwenke of his guide to making a full- or part-time living on the land, a book for anyone who plans to own a small farm. With sections on soil management, farm practices, cash crop selections, machinery, and many other topics, as well as comprehensive series of appendices, the author touches upon the basics of getting started with one's own small-scale farm. Schwenke, himself a small farm owner, has provided a great practical resource for the beginning cash crop grower. Get started on acquiring "the hodgepodge of knowledge blended with a plethora of skills" necessary to becoming a successful organic farmer.

All New Square Foot Gardening: The Revolutionary Way to Grow More In Less Space, Second Edition

Mel Bartholomew

Square foot gardening is the most practical, foolproof way to grow a home garden. That explains why author and gardening innovator Mel Bartholomew has sold more than two million books describing how to become a successful DIY square foot gardener. Now, with the publication of *All New Square Foot Gardening*, the essential guide to his unique step-by-step method has become even better. Bartholomew furthers his discussion of one of the most popular gardening trends today: vertical gardening. He also explains how you can make gardening fun for kids by teaching them the square foot method. Finally, an expanded section on pest control helps you protect your precious produce. Rich with new full-color images and updated tips for selecting materials, this new edition is perfect for brand-new gardeners as well as the millions of square

foot gardeners who are already dedicated to Mel's industry-changing insights.

The Quarter-Acre Farm: How I Kept the Patio, Lost the Lawn, and Fed My Family for a Year

Spring Warren and Jesse Pruet

When Spring Warren told her husband and two teenage boys that she wanted to grow 75% of all the food they consumed for one year—and that she wanted to do it in their yard—they told her she was crazy. She did it anyway. *The Quarter-Acre Farm* is Warren's account of deciding, despite all resistance, to take control of her family's food choices, get her hands dirty, and create a garden in her suburban yard. It's a story of bugs, worms, rot, and failure; of learning, replanting, harvesting, and eating. The road is long and riddled with mistakes, but by the end of her yearlong experiment, Warren's sons and husband have become her biggest fans. Full of tips and recipes to help anyone interested in growing and preparing at least a small part of their diet at home, *The Quarter-Acre Farm* is a warm, witty tale about family, food, and the incredible gratification that accompanies self-sufficiency.

Straw Bale Gardens: The Breakthrough Method for Growing Vegetables Anywhere, Earlier and with No Weeding

Joel Karsten

You'll find a bumper crop of vegetable gardening books on the shelves today, but it is a very rare title that actually contains new information. Straw Bale Gardens teaches gardening in a way that isn't only new but is thoroughly innovative and revolutionary to home gardening. It solves every impediment today's home gardeners face: bad soil, weeds, a short growing season, watering problems, limited garden space, and even physical difficulty working at ground level. Developed and pioneered by garden expert Joel Karsten, straw bale gardens create their own growing medium and heat source so you can get an earlier start. It couldn't be simpler or more effective: all you need is a few bales of straw, some fertilizer, and some seeds or plants, and you can create a weedless vegetable garden anywhere—even in your driveway. Karsten's step-by-step guide offers all the information you need to make your own straw bale garden today.

Mini Farming:
Self-Sufficiency on ¼ Acre

Brett L. Markham

Mini Farming describes a holistic approach to small-area farming that will show you how to produce 85% of an average family's food on just a quarter acre—and earn $10,000 in cash annually while spending less than half the time that an ordinary job would require. Even if you have never been a farmer or a gardener, this book covers everything you need to know to get started: buying and saving seeds, starting seedlings, establishing raised beds, soil fertility practices, composting, dealing with pest and disease problems, crop rotation, farm planning, and much more. Because self-sufficiency is the objective, subjects such as raising backyard chickens and home canning are also covered along with numerous methods for keeping costs down and production high. Materials, tools, and techniques are detailed with photographs, tables, diagrams, and illustrations.

Don't Throw It, Grow It!:
68 Windowsill Plants from Kitchen Scraps

Deborah Peterson

Magic and wonder hide in unexpected places—a leftover piece of ginger, a wrinkled potato left too long in its bag, a humdrum kitchen spice rack. In *Don't Throw It, Grow It!* Deborah Peterson reveals the hidden possibilities in everyday foods. She shows how common kitchen staples—pits, nuts, beans, seeds, and tubers—can be coaxed into lush, vibrant houseplants that are as attractive as they are fascinating. The book offers growing instructions for 68 plants in four broad categories: vegetables, fruits and nuts, herbs and spices, and more exotic plants from ethnic markets. The book is enhanced with beautiful illustrations, and its at-a-glance format makes it a quick and easy reference. Best of all, every featured plant can be grown in a kitchen, making this handy guide a must-have for avid gardeners and apartment-dwellers.

Building Chicken Coops—
Storey's Country Wisdom Bulletin

Since 1973, Storey's Country Wisdom Bulletins have offered practical, hands-on instructions designed to help readers master dozens of country living skills quickly and easily. There are now more than 170 titles in this series, and their re-markable popularity reflects the common desire of country and city dwellers alike to cultivate personal independence in everyday life.

Chicken Coops: 45 Building
Ideas for Housing Your Flock

Judy Pangman

Bring your chickens home to roost in comfort and style! Whether you're keeping one hen in a small backyard or 1,000 birds in a large free-range pasture, this delightful collection of hen hideaways will spark your imagination and inspire you to begin building. The coops range from fashionable backyard structures featured in the annual Seattle Tilth City Chickens Tour and the Mad City Chickens Tour in Madison, Wisconsin, to the large-scale, moveable shelters Joel Salatin has fashioned for Polyface Farm in Virginia. You'll also find ideas for converting trailer frames, greenhouses, and backyard sheds; low-budget alternatives for working with found and recycled materials; and simple ways to make waterers, feeders, and nestboxes. With basic building skills, a little elbow grease, and this book of conceptual plans and how-to drawings, you've got all you need to shelter your flock.

Reinventing the Chicken Coop:
14 Original Designs with Step-by-Step
Building Instructions

Kevin McElroy

Backyard chickens meet contemporary design! Matthew Wolpe and Kevin McElroy give you 14 complete building plans for chicken coops that range from the purely functional to the outrageously fabulous. One has a water-capturing roof; one is a great homage to mid-Modern architecture; and another has a built-in composting system. Some designs are suitable for beginning builders, and some are challenging enough for experts. Step-by-step building plans are accompanied by full-color photographs and detailed construction illustrations.

Eggs and Chickens—
Storey's Country Wisdom Bulletin

John Vivian

Since 1973, Storey's Country Wisdom Bulletins have offered practical, hands-on instructions designed to help readers master dozens of country living skills quickly and easily. There are now more than 170 titles in this series, and their remarkable popularity reflects the common desire of country and city dwellers alike to cultivate personal independence in everyday life.

Free-Range Chicken Gardens: How to Create a Beautiful, Chicken-Friendly Yard

Jessi Bloom

Many gardeners fear chickens will peck away at their landscape, and chicken lovers often shy away from gardening for the same reason. But you can keep chickens and have a beautiful garden, too! Fresh eggs aren't the only benefit—chickens can actually help your garden grow and thrive, even as your garden does the same for your chickens. In this essential handbook, award-winning garden designer Jessi Bloom covers everything a gardener needs to know, including chicken-keeping basics, simple garden plans to get you started, tips on attractive fencing options, the best plants and plants to avoid, and step-by-step instructions for getting your chicken garden up and running.

Storey's Guide to Raising Chickens, 3rd Edition

Gail Damerow

Here is all the information you need to successfully raise chickens—from choosing breeds and hatching chicks to building coops, keeping the birds healthy, and protecting them from predators. This revised third edition contains a new chapter on training chickens and understanding their intelligence, expanded coverage of hobby farming, and up-to-date information on chicken health issues, including avian influenza and fowl first aid.

The Small-Scale Poultry Flock: An All-Natural Approach to Raising Chickens and Other Fowl for Home and Market Growers

Harvey Ussery

The most comprehensive guide to date on raising all-natural poultry for the small-scale farmer, homesteader, and professional grower. *The Small-Scale Poultry Flock* offers a practical and integrative model for working with chickens and other domestic fowl, based entirely on natural systems. Readers will find information on growing (and sourcing) feed on a small scale, brooding (and breeding) at home, and using poultry as insect and weed managers in the garden and orchard. Ussery's model presents an entirely sustainable system that can be adapted and utilized in a variety of scales, and will prove invaluable for beginner homesteaders, growers looking to incorporate poultry into their farm, or poultry farmers seeking to close their loop. He even includes one of the best step-by-step poultry butchering guides available, complete with extensive illustrative photos. No other book on raising poultry takes an entirely whole-systems approach, or discusses producing homegrown feed and breeding in such detail. This is a truly invaluable guide that will lead farmers and homesteaders into a new world of self-reliance and enjoyment.

Aquaponic Gardening: A Step-By- Step Guide to Raising Vegetables and Fish Together

Sylvia Bernstein

Aquaponics is a revolutionary system for growing plants by fertilizing them with the wastewater from fish in a sustainable closed system. A combination of aquaculture and hydroponics, aquaponic gardening is an amazingly productive way to grow organic vegetables, greens, herbs, and fruits, while providing the added benefits of fresh fish as a safe, healthy source of protein. On a larger scale, it is a key solution to mitigating food insecurity, climate change, groundwater pollution, and the impacts of overfishing on our oceans. *Aquaponic Gardening* is the definitive do-it-yourself home manual, focused on giving you all the tools you need to create your own aquaponic system and enjoy healthy, safe, fresh, and delicious food all year-round.

Composting: An Easy Household Guide

Nicky Scott

Did you know that up to two-thirds of most household trash can be composted? That composting reduces the need for more landfills? Composting is fun and easy! And you can make compost even if you live in an apartment and don't have access to a garden. This book provides all the information you need for successful composting—a satisfying way to live lightly on Earth.

How to Make and Use Compost: The Ultimate Guide

Nicky Scott

Composting is easy, fun, saves you money, and helps you to grow lovely plants. Whether you live in an apartment with no garden or have a family and garden that generate large amounts of food and garden waste, this book shows you how to compost everything that can be composted at home, work, or school, and in spaces big and small. *How to Make and Use Compost* features an A-Z guide that includes a comprehensive list of what you can and can't compost, concepts and techniques, compost systems, and common problems and solutions.

Compost, Vermicompost, and Compost Tea: Feeding the Soil on the Organic Farm

Grace Gershuny

Part of the NOFA Guides series. Information on composting techniques, including: Principles and biology of composting; Temperature, aeration, and moisture control; Composting methods; Materials (additives and inoculants, biodynamic preparations); About costs (site preparation, equipment, labor and time); What do you do with it? And when is it finished?; Compost tea and other brewed microbial cultures; Compost and the law. With extended appendices including a recipe calculator, potting mix recipes, and a sample compost production budget sheet.

Home Composting Made Easy

C. Forrest McDowell and Tricia Clark-McDowell

Home Composting Made Easy is considered to be the best educational guide on home composting for the public. It is used throughout North America by hundreds of municipalities, counties, waste management districts, states, organizations, and businesses. Fully illustrated with step-by-step, no-nonsense instructions and state-of-the-art advice by gardening and compost experts.

Let It Rot!: The Gardener's Guide to Composting

Stu Campbell

In 1975, *Let It Rot* helped start the composting movement and taught gardeners everywhere how

to recycle waste to create soil-nourishing compost. Contains advice for starting and maintaining a composting system, building bins, and using compost. Third Edition. 267,000 copies in print.

Easy Composters You Can Build— Storey's Country Wisdom Bulletin

Since 1973, Storey's Country Wisdom Bulletins have offered practical, hands-on instructions designed to help readers master dozens of country living skills quickly and easily. There are now more than 170 titles in this series, and their remarkable popularity reflects the common desire of country and city dwellers alike to cultivate personal independence in everyday life.

The Complete Compost Gardening Guide

Barbara Pleasant and Deborah L. Martin

Pleasant and Martin turn the compost bin upside down with their liberating system of keeping compost heaps right in the garden. The compost and the plants live together from the beginning in a nourishing, organic environment. Pleasant and Martin bring readers on a thorough, informative tour of materials and innovative techniques, leading the way to an efficient and rewarding home gardening system. Their methods are sure to help gardeners turn average vegetable plots into rich incubators of healthy produce, bursting with fresh flavor, and flowerbeds into rich tapestries of bountiful blooms all season long.

What Every Gardener Should Know About Earthworms— Storey's Country Wisdom Bulletin

Since 1973, Storey's Country Wisdom Bulletins have offered practical, hands-on instructions designed to help readers master dozens of country living skills quickly and easily. There are now more than 170 titles in this series, and their remarkable popularity reflects the common desire of country and city dwellers alike to cultivate personal independence in everyday life.

The Worm Book: The Complete Guide to Gardening and Composting with Worms

Loren Nancarrow and Janet Hogan Taylor

Worms are the latest (as well as, of course, perhaps the oldest!) trend in earth-friendly gardening, and in this handy guide, the authors of *Dead Snails Leave No Trails* demystify the world of worm wrangling, with everything you need to know to build your own worm bin, make your garden worm-friendly, pamper your soil, and much much more.

Worms Eat My Garbage:
How to Set Up and Maintain a Worm Composting System, 2nd Edition

Mary Appelhof

A new edition of the definitive guide to vermicomposting—a process using redworms to recycle human food waste into nutrient-rich fertilizer for plants. Mary Appelhof provides complete illustrated instructions on setting up and maintaining small-scale worm composting systems. Internationally recognized as an authority on vermicomposting, Appelhof has worked with worms for over three decades. Topics include: bin types, worm species, reproduction, care and feeding of worms, harvesting, and how to make the finished product of potting soil.

Growing Food in a Hotter, Drier Land:
Lessons from Desert Farmers on Adapting to Climate Uncertainty

Gary Paul Nabhan

Because climatic uncertainty has now become "the new normal," many farmers, gardeners, and orchard keepers in North America are seeking ways to adapt their food production to become more resilient in the face of such "global weirding." This book draws upon the wisdom and technical knowledge from desert farming traditions around the world to offer time-tried strategies for: building greater moisture-holding capacity and nutrients in soils; protecting fields from damaging winds, drought, and floods; harvesting water from uplands to use in rain gardens and terraces filled with perennial crops; selecting fruits, nuts, succulents, and herbaceous perennials that are best suited to warmer, drier climates. Gary Paul Nabhan is one of the world's experts on the agricultural traditions of arid lands. For this book, he has visited indigenous and traditional farmers in the Gobi Desert, the Arabian Peninsula, the Sahara Desert, and Andalusia, as well as the Sonoran, Chihuahuan, and Painted deserts of North America, to learn firsthand their techniques and designs aimed at reducing heat and drought stress on orchards, fields, and dooryard gardens. This

practical book is replete with detailed descriptions and diagrams of how to implement these desert-adapted practices in your own backyard, orchard, or farm.

The Biochar Solution:
Carbon Farming and Climate Change

Albert Bates

Civilization as we know it is at a crossroads. For the past 10,000 years, we have turned a growing understanding of physics, chemistry, and biology to our advantage in producing more energy and more food and as a consequence have produced exponential population surges, resource depletion, ocean acidification, desertification, and climate change. The path we are following began with long-ago discoveries in agriculture, but it divided into two branches, about 8,000 years ago. The branch we have been following for the most part is conventional farming—irrigation, tilling the soil, and removing weeds and pests. That branch has degraded soil carbon levels by as much as 80% in most of the world's breadbaskets, sending all that carbon skyward with each pass of the plow. The other branch disappeared from our view some 500 years ago, although archaeologists are starting to pick up its trail now. At one time, it achieved success as great as the agriculture that we know, producing exponential population surges and great cities, but all that was lost in a fluke historical event borne of a single genetic quirk. It vanished when European and Asian diseases arrived in the Americas. While conventional agriculture leads to deserts, this other, older style, brings fertile soils, plant and animal diversity, and birdsong. The agriculture that was nearly lost moves carbon from sky to soil and crops. The needed shift can be profound and immediate. We could once more become a garden planet. We can heal our atmosphere and oceans. Come along on this journey of rediscovery with *The Biochar Solution*.

Growing & Using Lavender—
Storey's Country Wisdom Bulletin

Since 1973, Storey's Country Wisdom Bulletins have offered practical, hands-on instructions designed to help readers master dozens of country living skills quickly and easily. There are now more than 170 titles in this series, and their remarkable popularity reflects the common desire of country and city dwellers alike to cultivate personal independence in everyday life.

How to Build Your Own Greenhouse: Designs and Plans to Meet Your Growing Needs

Rodger Marshall

From the simplest cold frame to the most elaborate tropical paradise, this collection has a greenhouse plan to suit your needs. Master gardener and builder Roger Marshall guides you through the various options, the details of choosing materials and a site, and every step of the building process, from foundation to glazing. He even covers plumbing, heating, lighting, misting, and automatic venting systems, and he includes detailed plans for nine different greenhouses.

Outdoor Water Features: 16 Easy-to Build Projects for Your Yard and Garden

Alan Bridgewater

No one can deny the power and attraction of water features. And it's never been easier for gardeners to add the sparkle and serenity of water to their own landscapes. Garden centers and flower shows throughout North America provide both the tools and the inspiration; this book will give readers the know-how. Projects range from simple to more advanced. Readers can choose among a classic wall-mounted fountain spout, a traditional or contemporary cascade and garden pond, an authentic Japanese-style bamboo water pump, witty water sculptures, and more. This book has something to offer all tastes, budgets, and abilities. Each project is accompanied by clear, step-by-step photographs and easy-to-follow instructions, plus helpful tips and techniques.

Natural Beekeeping: Organic Approaches to Modern Apiculture

Ross Conrad

The various chemicals used in beekeeping have, for the past decades, held Varroa Destructor, a mite, and other major pests at bay, but chemical resistance is building and evolution threatens to overtake the best that laboratory chemists have to offer. In fact, there is evidence that chemical treatments are making the problem worse. *Natural Beekeeping* flips the script on traditional approaches by proposing a program of selective breeding and natural hive management. Conrad brings together the best organic and natural approaches to keeping honeybees healthy and

productive here in one book. Readers will learn about nontoxic methods of controlling mites, eliminating American foulbrood disease (without the use of antibiotics), breeding strategies, and many other tips and techniques for maintaining healthy hives. *Natural Beekeeping* describes opportunities for the seasoned professional to modify existing operations to improve the quality of hive products, increase profits, and eliminate the use of chemical treatments. Beginners will need no other book to guide them.

Storey's Guide to Keeping Honey Bees: Honey Production, Pollination, Bee Health

Malcolm Sanford

Urban beekeeping is on the rise as swarms of people do their part to help nurture local food systems, make gardens more productive, connect with nature, and rescue honeybee populations from colony collapse disorder. Here Sanford presents a thorough overview of these industrious and critically important insects. With this book as their guide, beekeepers will understand how to plan a hive, acquire bees, install a colony, keep bees healthy, maintain a healthy hive, understand and prevent new diseases, and harvest honey crops. The book also provides insights into honeybee life and behavior, an exploration of apiary equipment and tools, season-by-season beekeeper responsibilities, instructions for harvesting honey, and up-to-date information about diseases and other risks to bees. This comprehensive reference will appeal to both the experienced beekeeper who seeks help with specific issues and the novice eager to get started.

The Holistic Orchard: Tree Fruits and Berries the Biological Way

Michael Phillips

Many people want to grow fruit on a small scale but lack the insight to be successful orchardists. *The Holistic Orchard* demystifies the basic skills everybody should know about the inner workings of the orchard ecosystem, as well as orchard design, soil biology, and organic health management. Detailed insights on grafting, planting, pruning, and choosing the right varieties for your climate are also included, along with a step-by-step instructional calendar to guide growers through the entire orchard year. The extensive profiles of pome fruits (apples, pears, asian pears, quinces), stone fruits (cherries, peaches, nectarines, apricots, plums), and berries (raspberries, blackberries, blueberries, gooseberries, currants, and elderberries) will quickly have you savoring the prospects. Phillips draws connections between home orcharding and permaculture; explains the importance of native pollinators; describes the world of understory plantings with shade-tolerant berry bushes and other insectary plants; and presents detailed information on cover crops, biodiversity, and the newest research on safe, homegrown solutions to pest and disease challenges.

Pruning Trees, Shrubs & Vines—Storey's Country Wisdom Bulletin

Do you have an old grapevine that needs training? A forsythia that has become too leggy? Evergreen shrubs that threaten to push through your first-floor windows? Have you just purchased a new home and feel mystified by the pruning chores ahead? If so, help is here in *Pruning Trees, Shrubs & Vines*. Pruning is one of the best things you can do for your plants and trees, if it's done the right way. When pruned regularly and properly, your plants stay vibrantly healthy. Flowering shrubs and vines produce more blossoms, fruit trees and berry patches yield more fruit, and hedges and evergreens remain full and well shaped. This book offers all the advice you need to start pruning fruit and shade trees, evergreens, grapevines, berry bushes, ornamental shrubs and vines, hedges, and roses like a professional. For quick reference, you'll find a pruning timetable, a condensed plant-by-plant pruning guide, and an A-to-Z glossary of pruning terms.

The Complete Guide to Saving Seeds: 322 Vegetables, Herbs, Fruits, Flowers, Trees, and Shrubs

Robert E. Gough and Cheryl Moore-Gough

Learn how to collect, save, and cultivate the seeds from more than 300 vegetables, herbs, fruits, flowers, trees, and shrubs. It's easy, and it's fun! Authors Robert Gough and Cheryl Moore-Gough thoroughly explain every step in the seed-saving process. Descriptions of seed biology; tips on how to select plants for the best seeds; and advice on harvesting and cleaning, proper storage and care, and propagating and caring for new seedlings are all presented with clear, easy-to-follow instructions. Chapters dedicated to individual plants contain species-specific directions and detailed information. Gardeners of any experience level will find all the information they need to extend the life of their favorite plants to the next generation and beyond.

Homegrown and Handmade: A Practical Guide to More Self-Reliant Living

Deborah Niemann

Our food system is dominated by industrial agriculture and has become economically and environmentally unsustainable. The incidence of diet-related diseases, including obesity, diabetes, hypertension, cancer, and heart disease, has skyrocketed to unprecedented levels. Whether you have forty acres and a mule or a condo with a balcony, you can do more than you think to safeguard your health, your money, and the planet. *Homegrown and Handmade* shows how making things from scratch and growing at least some of your own food can help you eliminate artificial ingredients from your diet, reduce your carbon footprint, and create a more authentic life. Written from the perspective of a successful self-taught modern homesteader, this well-illustrated, practical, and accessible manual will appeal to anyone who dreams of a simpler life.

BIODYNAMICS

What Is Biodynamics?: A Way to Heal and Revitalize the Earth

Rudolf Steiner

Created from indications by Rudolf Steiner around 1924, farming and gardening with biodynamic methods are spiritual, artistic, and sophisticated forms of organic horticulture that nurture and enhance the earth. They emphasize the interdependence and unity of all the elements of an ecosystem or landscape—including soil, plants, animals, and weather. Biodynamic methods use special herbal "preparations" to increase the energetic quality of the soil, stimulating plant growth and health. Typical of the biodynamic approach are companion planting, crop rotation, cover crops, green manures, liquid manures, compost, the integration of crops and livestock, and planting and harvesting in harmony with the lunar and planetary cycles. This introduction to biodynamic methods contains five lectures by Rudolf Steiner and an extensive introduction by Hugh Courtney, who unravels the practice of biodynamics and its spiritual and esoteric background.

Culture and Horticulture: The Classic Guide to Biodynamic and Organic Gardening

Wolf D. Storl

Studies have shown that small organic farms and home gardens are capable of producing more food per acre with less fossil energy than large-scale commercial agricultural installations dependent on machines and toxic chemical fertilizers and pesticides. This classic book by Wolf Storl details how food is grown holistically and beautifully by traditional communities around the world, and shows how to apply their ancient wisdom to our own gardens. The book explains how to build the soil to maintain fertility; how to produce compost; how to plant, sow, and tend the various fruit and vegetable plants; how to rotate crops and practice companion planting; how to set up a favorable microclimate; how to deal with so-called weeds and pests; how to harvest at the right time; and finally how to store vegetables and herbs. The reader is introduced to the wider aspects of horticulture, to its historical, philosophical, and cosmological contexts and social relevance.

The Biodynamic Year: Increasing Yield, Quality and Flavour: 100 Helpful Tips for the Gardener or Smallholder

Maria Thun

Maria Thun, a pre-eminent expert in biodynamic cultivation, has here compiled over 100 of her best gardening tips based on 50 years' research. Find out: How to produce abundant and tasty crops; how special preparations can transform your soil and produce; how the moon affects planting and growth; the difference between root, leaf, blossom and fruit plants; what the best storage methods are; and much more. Accompany the author on a journey through the seasons and discover lots of new tips and suggestions. There is a wealth of advice here for gardeners seeking to manage nature responsibly and successfully.

A Biodynamic Farm: For Growing Wholesome Food

Hugh Lovel

The inventory of knowledge that is generally warehoused under the classification of *biodynamic* is rich and timeless, and yet very few farmers have even a nodding acquaintance with the subject. This book performs a rescue operation. A practical, how-to guide to making all the biodynamic preparations, this book will provide what you need for putting these proven techniques to work in your fields. Further explains how to achieve success through CSA-style market garden marketing.

PERMACULTURE

The Resilient Farm and Homestead: An Innovative Permaculture and Whole Systems Design Approach

Ben Falk

The Resilient Farm and Homestead is a manual for developing durable, beautiful, and highly functional human habitat systems fit to handle an age of rapid transition. Ben Falk is a land designer and site developer whose permaculture research farm has drawn national attention. The site is a terraced paradise on a hillside in Vermont that would otherwise be overlooked by conventional farmers as unworthy farmland. Falk's wide array of fruit trees, rice paddies (relatively unheard of in the Northeast), ducks, nuts, and earth-inspired buildings is a hopeful image for the future of regenerative agriculture and modern homesteading. The book covers nearly every strategy Falk and his team have been testing over the past decade. It includes detailed information on earthworks; gravity-fed water systems; species composition; the site design process; site management; fuel wood hedge production and processing; human health and nutrient-dense production strategies; rapid topsoil formation and remineralization; agroforestry/silvopasture/grazing; ecosystem services, especially regarding flood mitigation; fertility management; human labor and social systems aspects; tools/equipment/appropriate technology; and much more, complete with photographs and detailed design drawings. It presents a viable home-scale model for an intentional food-producing ecosystem in cold climates, and beyond.

The Basics of Permaculture Design

Ross Mars

The Basics of Permaculture Design, first published in Australia in 1996, is an excellent introduction to the principles of permaculture, design processes, and the tools needed for designing sustainable gardens, farms, and larger communities. Packed with useful tips, clear illustrations, and a wealth of experience, it guides you through designs for gardens, urban and rural properties, water harvesting systems, animal systems, permaculture in small spaces like balconies and patios, and on farms, schools, and ecovillages. This is both a do-it-yourself guide for the enthusiast and a useful reference for permaculture designers.

The Permaculture Handbook:
Garden Farming for Town and Country

Peter Bane

Imagine how much more self-reliant our communities would be if thirty million acres of lawns were made productive again. *The Permaculture Handbook* is a step-by-step, beautifully illustrated guide to creating resilient and prosperous households and neighborhoods, complemented by extensive case studies of three successful farmsteads and market gardens. It shows how, by mimicking the intelligence of nature and applying appropriate technologies such as solar and environmental design, permaculture can offer numerous benefits. A must-read for anyone concerned about creating food security, resilience, and a legacy of abundance rather than depletion.

The Vegetable Gardener's Guide to Permaculture: Creating an Edible Ecosystem

Christopher Shein

Once a fringe topic, permaculture is moving to the mainstream as gardeners who are ready to take their organic gardening to the next level are discovering the wisdom of a simple system that emphasizes the idea that by taking care of the earth, the earth takes care of you. *The Vegetable Gardener's Guide to Permaculture* teaches gardeners of every skill, with any size space, how to live in harmony with both nature and neighbors to produce and share an abundant food supply with minimal effort. Permaculture teacher Christopher Shein highlights everything you need to know to start living off the land lightly, including how to create rich, healthy, and low-cost soil, blend a functional food garden and decorative landscape, share the bounty with others, and much more.

Sepp Holzer's Permaculture:
A Practical Guide to Small-Scale, Integrative Farming and Gardening

Sepp Holzer

Sepp Holzer farms steep mountainsides in Austria 1,500 meters above sea level. His farm is an intricate network of terraces, raised beds, ponds, waterways and tracks, well covered with productive fruit trees and other vegetation, with the farmhouse neatly nestling among them. In this book, Holzer shares the skill and knowledge acquired over his lifetime. He covers every aspect of his farming methods, not just how to create a holistic system on the farm itself, but how to make a living from it. He offers a wealth of information for the gardener, smallholder, or alternative farmer, yet the book's greatest value is the attitudes it teaches. He reveals the thinking processes based on principles found in nature that create his productive systems. These can be applied anywhere.

FOOD AND FOOD SECURITY

Rebuilding the Foodshed:
How to Create Local, Sustainable, and Secure Food Systems

Philip Ackerman-Leist

Droves of people have turned to local food as a way to retreat from our broken industrial food system. From rural outposts to city streets, they are sowing, growing, selling, and eating food produced close to home—and crying out for agricultural reform. All this has made "local food" into everything from a movement buzzword to the newest darling of food trendsters. But now it's time to take the conversation to the next level. That's exactly what Ackerman-Leist does in *Rebuilding the Foodshed*, where he refocuses the local food lens on the broad issue of rebuilding regional food systems that can replace the destructive aspects of industrial agriculture, meet food demands affordably and sustainably, and be resilient enough to endure potentially rough times ahead. Showcasing some of the most promising, replicable models for growing, processing, and distributing sustainably grown food, this book points the reader toward the next stages of the food revolution.

Emergency Food Storage & Survival Handbook: Everything You Need to Know to Keep Your Family Safe in a Crisis

Peggy Layton

What if your life was disrupted by a natural disaster, food or water supply contamination, or any other type of emergency? Do you have the essentials for you and your family? Do you have a plan in the event that your power, telephone, water and food supply are cut off for an extended amount of time? With this guide by your side, you and your family will learn how to plan, purchase, and store a three-month supply of all the necessities—food, water, fuel, first-aid supplies, clothing, bedding, and more—simply and economically. In other words, this book may be a lifesaver.

SOLAR CUISINE AND SUSTAINABLE DIET

Cooking with Sunshine: The Complete Guide to Solar Cuisine with 150 Easy Sun-Cooked Recipes

Lorraine Anderson and Rick Palkovic

What could be more entertaining and magical than putting food into a cardboard box outdoors on a sunny day and taking it out fully cooked a few hours later? Solar cooking—a safe, simple cooking method using the sun's rays as the sole heat source—has been known for centuries and can be done at least during the summer in just about any place where there's sun. *Cooking with Sunshine* provides everything you need to know to cook great sun-fueled meals. The book describes how to build your own inexpensive solar cooker, explains how solar cooking works and its benefits over traditional methods, offers more than 100 tasty recipes emphasizing healthy ingredients, and suggests a month's worth of menu ideas.

Good Meat: The Complete Guide to Sourcing and Cooking Sustainable Meat

Deborah Krasner

Good Meat is a comprehensive guide to sourcing and enjoying sustainable meat. With the rising popularity of the locavore and organic food movements—and the terms "grass fed" and "free range" commonly seen on menus and in grocery stores—people across the country are turning their attention to where their meat comes from. Whether for environmental reasons, health ben-

efits, or the astounding difference in taste, consumers want to know that their meat was raised well.

How to Roast a Pig: From Oven-Roasted Tenderloin to Slow-Roasted Pulled Pork Shoulder to the Spit-Roasted Whole Hog

Tom Rea

How to Roast a Pig teaches you the five main methods for cooking the perfect pork, and how to choose what to cook with each method. Whether you're looking for whole hog roast or a pulled pork sandwich, author Tom Rea has you covered.

The Mexican Slow Cooker: Recipes for Mole, Enchiladas, Carnitas, Chile Verde Pork, and More Favorites

Deborah Schneider

When acclaimed chef and cookbook author Deborah Schneider discovered that using her trusty slow cooker to make authentic Mexican recipes actually enhanced their flavor while dramatically reducing active cooking time, it was a revelation. Packed with Schneider's favorite south-of-the-border recipes such as Tortilla Soup, Zesty Shredded Beef (Barbacoa), famed Mole Negro, the best tamales she has ever made, and more, *The Mexican Slow Cooker* delivers sophisticated meals and complex flavors, all with the ease and convenience that have made slow cookers enormously popular. This is a Real Goods in-store favorite!

Ball Complete Book of Home Preserving

Edited by Judi Kingry and Lauren Devine

From the experts, Ball Home Canning Products, the new bible of home preserving. Home canning and preserving has increased in popularity for the benefits it offers: Cooks gain control of the ingredients, including organic fruits and vegetables; preserving foods at their freshest point locks in nutrition; the final product is free of chemical additives and preservatives. These 400 innovative and enticing recipes include everything from salsas and savory sauces to pickling, chutneys, relishes, and of course, jams, jellies, and fruit spreads. The book includes comprehensive directions on safe canning and preserving methods plus lists of required equipment and utensils. Specific instructions for first-timers and handy tips for the experienced make the *Ball Complete Book of Home Preserving* a valuable addition to any kitchen library.

Food in Jars:
Preserving in Small Batches Year-Round

Marisa McClellan

Popular food blogger Marisa McClellan takes you through all manner of food in jars, storing away the tastes of all seasons for later. Basics like jams and jellies are accompanied by pickles, chutneys, conserves, whole fruit, tomato sauces, salsas, marmalades, nut butters, seasonings, and more. Small batches make them easy projects for a canning novice to tackle, and the flavors of vanilla bean, sage, and pepper will keep more experienced jammers coming back for more. Stories of wild blackberry jam and California Meyer lemon marmalade from McClellan's childhood make for a read as pleasurable as it is delicious; her home-canned food learned from generations of the original "foodies" feeds the soul as well as the body in more than 100 recipes.

The Big Book of Preserving the Harvest:
150 Recipes for Freezing, Canning, Drying and Pickling Fruits and Vegetables

Carol W. Costenbader

Learn how to preserve the season's bounty in this classic primer on drying, freezing, canning, and pickling techniques. You'll learn everything you need to know to stock your pantry with fruits, vegetables, herbs, meats, vinegars, pickles, chutneys, and seasonings. Carol Costenbader presents more than 150 simple, step-by-step recipes for delicious creations such as Green Chile Salsa, Tomato Leather, Spiced Pear Butter, Peach Pie Filling, Eggplant Caviar, Blueberry Marmalade, Yellow Tomato Jam, Cranberry-Lime Curd, Preserved Lemons, and much more.

Food Drying Techniques—
Storey's Country Wisdom Bulletin

Carol W. Costenbader

Since 1973, Storey's Country Wisdom Bulletins have offered practical, hands-on instructions designed to help readers master dozens of country living skills quickly and easily. There are now more than 170 titles in this series, and their remarkable popularity reflects the common desire

of country and city dwellers alike to cultivate personal independence in everyday life.

Making Cheese, Butter & Yogurt—
Storey's Country Wisdom Bulletin

Ricki Carroll

Ten thousand years of cheese making (as well as butter and yogurt making) wisdom is distilled in this detailed bulletin. Expert cheese maker Ricki Carroll teaches you the latest, most healthful methods for whipping up delicious dairy products in your own kitchen. Here is everything you need to know—from what equipment and ingredients are required to how to pasteurize your own milk. And, of course, you get recipes for making cheeses and other dairy products that surpass anything you could buy in the store.

Wild Fermentation: The Flavor, Nutrition, and Craft of Live-Culture Foods

Sandor Ellix Katz

Bread. Cheese. Wine. Beer. Coffee. Chocolate. Most people consume fermented foods and drinks every day. For thousands of years, humans have enjoyed the distinctive flavors and nutrition resulting from the transformative power of microscopic bacteria and fungi. *Wild Fermentation* is the first cookbook to widely explore the culinary magic of fermentation. This book takes readers on a whirlwind trip through the wide world of fermentation, providing readers with basic and delicious recipes—some familiar, others exotic—that are easy to make at home.

The Art of Fermentation:
An In-Depth Exploration of Essential Concepts and Processes from Around the World

Sandor Ellix Katz

The Art of Fermentation is the most comprehensive guide to do-it-yourself home fermentation ever published. Sandor Katz presents the concepts and processes behind fermentation in ways that are simple enough to guide readers through their first experience making sauerkraut or yogurt, and in-depth enough to provide greater understanding and insight for experienced practitioners. While Katz expertly contextualizes fermentation in terms of biological and cultural evolution, health and nutrition, and even economics, this is primarily a compendium of practical information—how the processes work; parameters for safety; techniques for effective preservation; troubleshooting; and more. With illustrations and extended resources, this book provides essential wisdom for cooks, homesteaders, farmers, gleaners, foragers, and food lovers of any kind who want to develop a deeper understanding and appreciation for arguably the oldest form of food preservation, and part of the roots of culture itself. Readers will find detailed information on fermenting vegetables; sugars into alcohol (meads, wines, and ciders); sour tonic beverages; milk; grains and starchy tubers; beers (and other grain-based alcoholic beverages); beans; seeds; nuts; fish; meat; and eggs, as well as growing mold cultures; using fermentation in agriculture, art, and energy production; and considerations for commercial enterprises.

Mastering Fermentation: Recipes for Making and Cooking with Fermented Foods

Mary Karlin

A beautifully illustrated and authoritative guide to the art and science of fermented foods, featuring 70+ recipes that progress from simple condiments like vinegars and mustards to more advanced techniques for using wild yeast, fermenting meats, and curing fish. Fermented foods are currently experiencing a renaissance: kombucha, kefir, sauerkraut, and other potent fermentables appeal not only for their health benefits, but also because they are fun, adventurous DIY projects for home cooks of every level. Cooking instructor and author Mary Karlin begins with a solid introduction to the wide world of fermentation, explaining essential equipment, ingredients, processes, and techniques. The diverse chapters cover everything from fermented dairy to grains and breads; legumes, nuts, and aromatics; and fermented beverages.

Making Sauerkraut and Pickled Vegetables at Home: Creative Recipes for Lactic Fermented Food to Improve Your Health (Natural Health Guide)

Klaus Kaufmann

Homemade sauerkraut, pickles, and other lactic-acid-fermented foods are superior to their store-bought equivalents, both in flavor and healing properties. Step-by-step recipes guide the mod-ern reader through centuries-old methods. Includes full-color photos.

Preserving Food Without Freezing or Canning: Traditional Techniques Using Salt, Oil, Sugar, Alcohol, Vinegar, Drying, Cold Storage, and Lactic Fermentation

The Gardeners and Farmers at Centre Terre Vivant, with Deborah Madison and Eliot Coleman

Typical books about preserving garden produce nearly always assume that modern "kitchen gardeners" will boil or freeze their vegetables and fruits. Yet here is a book that goes back to the future, celebrating traditional but little-known French techniques for storing and preserving edibles in ways that maximize flavor and nutrition. With a new foreword by Deborah Madison, this book deliberately ignores freezing and high-temperature canning in favor of methods that are superior because they are less costly and more energy-efficient. *Preserving Food Without Freezing or Canning* offers more than 250 easy and enjoyable recipes featuring locally grown and minimally refined ingredients. It is an essential guide for those who seek healthy food for a healthy world.

Real Cidermaking on a Small Scale: An Introduction to Producing Cider at Home

Michael Pooley and John Lomax

Cidermaking is part of a rich tradition that dates back hundreds of years. Now you can make yourself a part of this tradition by becoming an at-home cider maker. *Real Cidermaking on a Small Scale* will teach you everything you need to know about the process of making hard cider from any kind of apple. You will learn how to build your own cider press, how to ferment the cider, and how to store it for enjoyment year-round. You will also discover delicious recipes, tips for preserving apple juice that you don't want to ferment, and instructions for making perry, or pear cider.

Strong Waters: A Simple Guide to Making Beer, Wine, Cider and Other Spirited Beverages at Home

Scott Mansfield

Today's renewed interest in making wine and beer at home amounts to nothing less than a renaissance. No matter why you want to join the new generation of homebrewers, *Strong Waters* will tell you how. This do-it-yourself guide makes a grand tradition accessible for today's enthusiasts. Beginners will welcome his tips for getting started inexpensively with everyday materials, and experienced hobbyists will be inspired by recipes for longtime favorites and forgotten delights. Worried that making your own spirits is complicated? Don't be! *Strong Waters* covers everything from the basics of bottling to the science of sweetening. It's surprisingly easy, and as eight pages of color photos illustrate, the results are tantalizing. Cheers!

The Kings County Distillery Guide to Urban Moonshining: How to Make and Drink Whiskey

Colin Spoelman and David Haskell

A new generation of urban bootleggers is distilling whiskey at home, and cocktail enthusiasts have embraced the nuances of brown liquors. Written by the founders of Kings County Distillery, New York City's first distillery since Prohibition, this spirited illustrated book explores America's age-old love affair with whiskey. It begins with chapters on whiskey's history and culture. For those thirsty for practical information, the book next provides a detailed, easy-to-follow guide to safe home distilling, complete with a list of supplies, step-by-step instructions, and helpful pictures, anecdotes, and tips. The final section focuses on the contemporary whiskey scene, featuring a list of microdistillers, cocktail and food recipes from the country's hottest mixologists and chefs, and an opinionated guide to building your own whiskey collection.

Build Your Own Barrel Oven: A Guide for Making a Versatile, Efficient, and Easy to Use Wood-Fired Oven

Eva Edleson, Max Edleson

Max and Eva Edleson offer a comprehensive guide for planning and building a practical, efficient, and affordable wood-fired oven. The Barrel Oven offers surprising convenience because it is hot and ready to bake in within 15–20 minutes and is easy to maintain at a constant temperature. It can be the seed for a small-scale baking enterprise or the heart of a community's wood-fired cuisine. All kinds of food can be baked in the Barrel Oven including bread, roasts, pizza, cookies cakes, pies, casseroles, and stews. Follow this step-by-step guide to transform local, low-cost materials and the sun's energy into good food.

Build Your Own Earth Oven

Kiko Denzer and Hannah Field

Denzer and Field, maker and baker, invite you into the artisan tradition. First, build a masonry oven out of mud. Then mix flour and water for real bread "better than anything you can buy." Total cost? Hardly more than a baking stone—and it can cook everything else, from 2-minute pizza to holiday fowl, or a week's meals. Clear, abundant drawings and photos clarify every step of the process. Informative text puts it all into context with artisan traditions of many ages and cultures. Beautifully sculpted ovens will inspire the artist in anyone. And the simple 4-step recipe (based on professional and homestead experience) promises authentic hearth loaves for anyone, on any schedule.

All That the Rain Promises and More: A Hip Pocket Guide to Western Mushrooms

David Arora

Full-color illustrated guide to identifying 200 Western mushrooms by their key features.

The Forager's Harvest: A Guide to Identifying, Harvesting, and Preparing Edible Wild Plants

Samuel Thayer

A practical guide to all aspects of edible wild plants: finding and identifying them, their seasons of harvest, and their methods of collection and preparation. Each plant is discussed in great detail and accompanied by excellent color photographs. Includes an index, illustrated glossary, bibliography, and harvest calendar. The perfect guide for all experience levels.

The Joy of Foraging: Gary Lincoff's Illustrated Guide to Finding, Harvesting, and Enjoying a World of Wild Food

Gary Lincoff

Discover the edible riches in your backyard, local parks, woods, and even roadside! In *The Joy of Foraging*, Lincoff shows you how to find fiddlehead ferns, rose hips, beach plums, bee balm, and more, whether you are foraging in the urban jungle or the wild, wild woods. You will also learn about fellow foragers—experts, folk healers, hobbyists, or novices like you—who collect wild things and are learning new things to do with them every day. Along with a world of edible wild plants—wherever you live, any season, any climate—you'll find essential tips on where to look for native plants, and how to know without a doubt the difference between edibles and toxic look-alikes. There are even ideas and recipes for preparing and preserving the wild harvest year-round, all with full-color photography.

Land and Shelter

Buying Country Land— Storey's Country Wisdom Bulletin

Peggy Tonseth

Since 1973, Storey's Country Wisdom Bulletins have offered practical, hands-on instructions designed to help readers master dozens of country living skills quickly and easily. There are now more than 170 titles in this series, and their remarkable popularity reflects the common desire of country and city dwellers alike to cultivate personal independence in everyday life.

Shelter

Lloyd Kahn

Shelter is many things—a visually dynamic, over-sized compendium of organic architecture past and present; a how-to book that includes over 1,250 illustrations; and a *Whole Earth Catalog*-type sourcebook for living in harmony with the Earth by using every conceivable material. First published in 1973, *Shelter* remains a source of inspiration and invention. Including the nuts-and-bolts aspects of building, the book covers such topics as dwellings from Iron Age huts to Bedouin tents to Togo's tin-and-thatch houses; nomadic shelters from tipis to "housecars"; and domes, dome cities, sod igloos, and even treehouses. The authors recount personal stories about alternative dwellings that illustrate sensible solutions to problems associated with using materials found in the environment—with fascinating, often surprising results.

The Barefoot Architect

Johan Van Lengen

A former UN worker and prominent architect, Johan van Lengen has seen firsthand the desperate need for a greener approach to housing in impoverished tropical climates. This comprehensive book clearly explains every aspect of this endeavor, including design (siting, orientation, climate consideration), materials (sisal, cactus, bamboo, earth), and implementation. The author emphasizes throughout the book what is inexpensive and sustainable. Included are sections discussing urban planning, small-scale energy production, cleaning and storing drinking water, and dealing with septic waste, and all information is applied to three distinct tropical regions: humid areas, temperate areas, and desert climates. Hundreds of explanatory drawings by van Lengen allow even novice builders to get started.

A Solar Buyer's Guide for the Home and Office: Navigating the Maze of Solar Options, Incentives, and Installers

Stephen Hren, Rebekah Hren

Many home and business owners are curious about solar electric and solar thermal systems, and wonder how to go about getting a clean energy generation system of their own. The vast majority will hire a professional installer to do the job. But what should they be asking of these installers? What system makes the most sense for their home or office: solar electric, solar hot water, solar heating, or some combination of these? This book explains the options so that property owners can make the right choices both for their energy needs and their financial security.

The Solar House: Passive Heating and Cooling

Daniel D. Chiras

Passive solar heating and passive cooling provide comfort throughout the year by reducing or eliminating the need for fossil fuel. Yet while heat from sunlight and ventilation from breezes is free for the taking, few modern architects or builders really understand the principles involved. Dan Chiras brings those principles up to date for a new generation of solar enthusiasts and explains in methodical detail how today's home builders can succeed with solar designs.

Home Sweet Zero Energy Home: What It Takes to Develop Great Homes that Won't Cost Anything to Heat, Cool or Light Up, Without Going Broke or Crazy

Barry Rehfeld

Zero energy homes produce at least as much energy as they consume through a combination of energy efficiencies, passive design, and renewable energy production. California has adopted zero net energy as the new residential standard for 2020; many other governments are considering similar policies. Developing zero energy homes is the first step toward making all buildings zero energy—a critical step in mitigating climate change, since buildings account for 40% of material and energy use worldwide. *Home Sweet Zero Energy Home* is the first practical guidebook that clearly identifies all the pieces of the zero energy puzzle, and explains how homeowners and buyers can also take smaller steps toward sharply reducing the energy use of existing buildings. Focusing on real costs and savings, this book takes an in-depth look at all the relevant topics, from materials to systems to financing.

Tiny Homes: Simple Shelter

Lloyd Kahn

There's a grassroots movement in tiny homes these days. The real estate collapse, the economic downturn, burning out on 12-hour workdays—many people are rethinking their ideas about shelter, seeking an alternative to high rents or a lifelong mortgage debt to a bank on an overpriced home. This book profiles 150 people who have created tiny homes (under 500 sq. ft.). Homes on land, homes on wheels, homes on the road, homes on water, even homes in the trees. There are also studios, saunas, garden sheds, and greenhouses. If you're thinking of scaling back, you'll find plenty of inspiration here, from builders, designers, architects, dreamers, artists, road gypsies, and water dwellers who've achieved a measure of freedom and independence by taking shelter into their own hands.

Tiny Homes on the Move: Wheels and Water

Lloyd Kahn

Tiny Homes on the Move chronicles 21st-century nomads—people who inhabit homes that are compact and mobile, either on wheels or in the water. In photos and stories, this fascinating book explores modern travelers who live in vans, pickup trucks, buses, trailers, sailboats, and houseboats that combine the comforts of home with the convenience of being able to pick up and go at any time. With over 1,000 color photos accompanying the stories and descriptions of these moveable sanctuaries, this is a valuable and inspirational book for anyone thinking outside the box about shelter.

Compact Cabins: Simple Living in 1000 Square Feet or Less

Gerald Rowan

The setting might be a sparkling lakefront, a cool clearing in the woods, a breathtaking mountaintop, or an expansive beach, but the dream of a modest retreat from everyday life often includes a simple little cabin. *Compact Cabins* presents 62 design interpretations of the getaway dream, with something to please every taste, complete with floor plans. Best of all, these small footprint designs are affordable and energy-efficient without skimping on comfort and style; all include sleeping accommodations, kitchen and bath facilities, and a heat source. Complete chapters on low-maintenance building materials, utilities and appliances, and alternative energy sources supply readers with the options for living efficiently in small spaces.

Rustic Retreats: A Build-It-Yourself Guide

Jeanie Stiles, David Stiles

Campers, anglers, hunters, and nature lovers—the woodland hideaway you've dreamed of is in these pages! Whether you long for an elegant water gazebo, a luxurious sauna hut, or a traditional tipi, you'll find illustrated, step-by-step instructions to help you build it yourself. *Rustic Retreats* includes more than 20 low-cost, sturdy, beautiful outdoor projects, including a garden pavilion, a grape arbor, a triangular tree hut, a log cabin, a wigwam, a river raft, and an Adirondack lean-to.

Projects to Get You Off the Grid: Rain Barrels, Chicken Coops, and Solar Panels

Instructables.com

Instructables is back with this compact book focused on a series of projects designed to get you thinking creatively about thinking green. Twenty Instructables illustrate just how simple it can be to make your own backyard chicken coop, or turn a wine barrel into a rainwater collector. Illustrated with dozens of full-color photographs per project accompanying easy-to-follow instructions, this Instructables collection utilizes the best that the online community has to offer, turning a far-reaching group of people into a mammoth database churning out ideas to make life better, easier, and in this case, greener, as this volume exemplifies.

The Septic System Owner's Manual: Subterranean Mysteries Revealed (Revised)

Lloyd Kahn

The gravity-powered septic system is so quiet, so natural, and so energy-free that we tend to forget the vital function it serves. This detailed book will show you how to understand, maintain, troubleshoot, and fix the septic system buried in your yard. Contains glossary, index, bibliography, sources, and hundreds of diagrams.

Treehouses and Playhouses You Can Build

David & Jeanie Stiles

This book shows how average do-it-yourself families can easily and affordably bring to life a "Hobbit's Treehouse," a "Pirate's Playhouse," or a "Crow's Nest" in their own backyards! The authors have created a straightforward how-to-build book filled with beautiful hand-drawn step-by-step illustrations that are easy to follow and describe in detail how to create each project. They include tips on budgeting, using basic tools, buying materials, and kid- and adult-friendly instructions. Even for DIY novice types, this book simplifies the building process and inspires families of all types to work together and build cool stuff.

Energy-Wise Landscape Design: A New Approach for Your Home and Garden

Sue Reed

Residential consumption represents nearly one-quarter of North America's total energy use, and the average homeowner spends thousands of dollars a year on power bills. To help alleviate this problem, *Energy-Wise Landscape Design* presents hundreds of practical ways everyone can save money, time, and effort while making their landscapes more environmentally healthy, ecologically rich, and energy-efficient. Readers will learn how to: Lower a home's heating and cooling costs; minimize fuel used in landscape construction, maintenance, and everyday use; choose landscape products and materials with lower embedded energy costs. Intended for homeowners, gardeners, landscape professionals, and students, the design ideas in this book will work in every type of setting. Written in non-scientific language with clear explanations and an easy conversational style, *Energy-Wise Landscape Design* is an essential resource for everyone who wants to shrink their energy footprint while enhancing their property and adding value to their home.

The Backyard Homestead Book of Building Projects

Spike Carlsen

Homesteaders, gardeners, small farmers, and outdoor living enthusiasts will love these 76 DIY projects for practical outdoor items designed to help you live more sustainably and independently. Expert woodworker Spike Carlsen offers clear, simple, fully illustrated instructions for everything from plant supports and a clothesline to a potting bench, a chicken coop, a hoop greenhouse, a cold frame, a beehive, a root cellar with storage bins, and an outdoor shower. Most of the projects are suitable for complete novices, and all use just basic tools and standard building materials.

Making Home: Adapting Our Homes and Our Lives to Settle in Place

Sharon Astyk

Other books tell us how to live the good life—but you might have to win the lottery to do it. *Making Home* is about improving life with the real people around us and the resources we already have. While encouraging us to be more resilient in the face of hard times, Sharon Astyk also points out the beauty, grace, and elegance that result, because getting the most out of everything we use is a way of transforming our lives into something more fulfilling. Written from the perspective of a family who has already made this transition, *Making Home* shows readers how to turn the challenge of living with less into settling for more: more happiness, more security, and more peace of mind. Learn simple but effective strategies to: Save money on everything from heating and cooling to refrigeration, laundry, water, sanitation, cooking, and cleaning; create a stronger, more resilient family; and preserve more for future generations.

Making Home takes the fear out of the prospect of change, and invites us to embrace a simpler, more abundant reality.

Renewable Energy in Theory and Practice

Photovoltaics: Design and Installation Manual

Solar Energy International (SEI)

Produced by a world-class solar energy training and education provider, this book has made available the critical information to successfully design, install, and maintain PV systems. It contains an overview of photovoltaic electricity and a detailed description of PV system components, including modules, batteries, controllers, and inverters. It also includes chapters on sizing photovoltaic systems, analyzing sites, and installing PV systems, as well as detailed appendices on PV system maintenance, troubleshooting, and solar insolation data for over 300 sites around the world.

Battery Book for Your PV Home

New England Solar Electric, Inc.

This is a good investment for anyone with a lead-acid battery bank. After reading this book, you will understand the internal workings of a battery bank. You will learn how to maintain your batteries, equalize your batteries' cells, and rejuvenate batteries in the case that they have been misused.

Masonry Heaters: Designing, Building, and Living with a Piece of the Sun

Ken Matesz

A complete guide to designing and living with one of the oldest, and yet one of the newest, heating devices. A masonry heater's design, placement in the home, and luxurious radiant heat redefine the hearth for the modern era. The value of a masonry heater lies in its durability, quality, service-ability, dependability, and health-supporting features. And it is an investment in self-sufficiency and freedom from fossil fuels. The book discusses different masonry heater designs, including variations extant in Europe. Those who are looking to build, add onto, or remodel a house will find comprehensive and practical advice for designing and installing a masonry heater, including detailed discussion of materials, code considerations, and many photos and illustrations. While this is not a do-it-yourself guide for building a masonry heater, it provides facts every heater builder should know. Professional contractors will find this a useful tool to consult, and homeowners considering a new method of home heating will find all they need to know about masonry heaters within these pages.

Serious Microhydro: Water Power Solutions from the Experts

Scott Davis

Waterpower is the largest source of renewable energy in the world today, and microhydro is a mature, proven technology that can provide clean, inexpensive, renewable energy with little or no impact on the environment. *Serious Microhydro* brings you dozens of firsthand stories of energy independence covering a complete range of systems, from household pressure sites to higher-pressure installations capable of powering a farm, business, or small neighborhood.

These case studies represent the most comprehensive collection of knowledge and experience available for tailoring an installation to meet the needs of a site and its owner or operators. If you are considering building a system, you are bound to find a wealth of creative solutions appropriate to your own circumstances.

Solar Home Heating Basics: A Green Energy Guide

Dan Chiras

As fossil fuel supplies dwindle, home heating will be one of the major challenges in temperate and cold climates in upcoming years. The reserves of natural gas used to heat the majority of North American buildings are rapidly being depleted. This latest *Green Energy Guide* helps readers who want to slash their energy bills and reduce their dependence on scarce resources to navigate the sometimes confusing maze of clean, reliable, and affordable options.

The Homeowner's Guide to Renewable Energy: Achieving Energy Independence Through Solar, Wind, Biomass, and Hydropower (Second Edition)

Dan Chiras

Energy bills have skyrocketed, and traditional energy sources can be as damaging to the environment as they are to your pocketbook. *The Homeowner's Guide to Renewable Energy* will show you how to slash your home energy costs while dramatically reducing your carbon footprint. Completely revised and updated, this new edition describes the most practical and affordable methods for making significant improvements in home energy efficiency and tapping into clean, affordable, renewable energy resources. If implemented, these measures will save the average homeowner tens of thousands of dollars over the coming decades. Focusing on the latest technological advances in residential renewable energy, this guide examines each alternative energy option available including: Solar hot water and solar hot air systems; passive and active solar retrofits for heating and cooling; electricity from solar, wind, and microhydro; hydrogen, fuel cells, methane digesters, and biodiesel.

Building Types and Materials

STRAW BALE, EARTH, STONE, AND CONCRETE

The Straw Bale House

Athena Swentzell Steen, Bill Steen, David Bainbridge

Imagine building a house with superior seismic stability, fire resistance, and thermal insulation, using an annually renewable resource, for half the cost of a comparable conventional home. Welcome to the straw bale house! Whether you build an entire house or something more modest—a home office or studio, a retreat cabin or guest cottage—plastered straw bale construction is an exceptionally durable and inexpensive option. What's more, it's fun, because the technique is easy to learn and easy to do yourself. And the resulting living spaces are unusually quiet and comfortable. *The Straw Bale House* describes the many benefits of building with straw bales: super insulation, with R-values as high as R-50; good indoor air quality and noise reduction; a speedy construction process; construction costs as low as $10-per-square-foot; use of abundant renewable resources; and a better solution than burning agricultural waste straw, which creates tons of air pollutants.

The Cob Builders Handbook: You Can Hand-Sculpt Your Own Home, 3rd Edition

Becky Bee

Cob (an old English word for lump) is old-fashioned concrete, made out of a mixture of clay, sand, and straw. Becky Bee's manual is a friendly guide to making your own earth structure, with chapters on design, foundations, floors, windows and doors, finishes, and of course, making glorious cob. "I believe that building with cob is a way to recreate community and experience the joy of working together while taking back the right to build our own homes and look after our Mother Earth." She loves doing something that makes sense in a world where lots of things don't.

The Hand-Sculpted House: A Practical and Philosophical Guide to Building a Cob Cottage

Ianto Evans, Michael G. Smith, Linda Smiley, and Deanne Bednar

Are you ready for the Cob Cottage? This is a building method so old and so simple that it has been all but forgotten in the rush to synthetics. A cob cottage, however, might be the ultimate expression of ecological design, a structure so attuned to its surroundings that its creators refer to it as "an ecstatic house." The authors build a house the way others create a natural garden. They use the oldest, most available materials imaginable—earth, clay, sand, straw, and water—and blend them to redefine the future of building. Building with cob requires no forms, no cement, and no machinery of any kind: Builders actually sculpt their structures by hand. Cob houses (or cottages, since they are always efficiently small by American construction standards) are not only compatible with their surroundings, they ARE their surroundings, literally rising up from the earth. They are full of light, energy-efficient, and cozy, with curved walls and built-in whimsical touches. They are delightful. They are ecstatic.

Adobe Homes for All Climates:
Simple, Affordable, and Earthquake-Resistant Natural Building Techniques

Lisa Schroder and Vince Ogletree

The lay-up of adobe bricks is an easy, forgiving way to achieve a solid masonry wall system. Contrary to stereotypes, adobe is perfectly adaptable for use in cold, wet climates as well as hot and dry ones, and for areas prone to earthquakes. With its efficient use of energy, natural resources for construction, and minimal effort for long-term maintenance, it's clear that the humble adobe brick is an ideal option for constructing eco-friendly structures throughout the world. All aspects of adobe construction are covered, including making and laying adobe bricks, installing lintels and arches, conduits and pipes, doors and windows, top plates and bondbeams, ideal wall dimensions, adobe finishes, and other adobe construction components.

Building Stone Walls—
Storey's Country Wisdom Bulletin

Charles McRaven

Since 1973, Storey's Country Wisdom Bulletins have offered practical, hands-on instructions designed to help readers master dozens of country living skills quickly and easily. There are now more than 170 titles in this series, and their remarkable popularity reflects the common desire of country and city dwellers alike to cultivate personal independence in everyday life.

Stonework: Techniques and Projects

Charles McRaven

Discover the lasting satisfaction of working with stone and learn the tricks of the trade from a master craftsman with this collection of 22 fully illustrated, step-by-step instructions for garden paths and walls, porches, pools, seats, waterfalls, a bridge, and more.

Concrete Garden Projects:
Easy & Inexpensive Containers,
Furniture, Water Features & More

Douglas Kent

For gardeners and backyard do-it-yourselfers, concrete is a revelation. It's durable, weatherproof, impossible to steal, and it provides much-needed insulation for outdoor plants. Concrete weathers beautifully, softening around the edges, developing moss, and becoming more picturesque with age. *Concrete Garden Projects* contains step-by-step instructions for dozens of easy, do-it-yourself ideas including containers of all shapes and sizes, elegant benches and stools, miniature ponds and birdbaths, stepping stones, a barbecue, and a fire pit. The authors use a variety of molds easily found or made, household items like bowls and baking pans, and simple wooden frames and boxes. At pennies per pound, and so simple to use—just mix with water and pour—concrete is the key to handcrafted backyard.

GENERAL BUILDING/OTHER

A Place in the Sun:
The Evolution of the Real Goods
Solar Living Center

John Schaeffer and the collaborative design/construction team

The Real Goods Solar Living Center, in Hopland, California (two hours north of San Francisco), has hosted hundreds of thousands of visitors since opening in June 1996. They come to see: Windmills and solar modules that produce enough electricity to allow Real Goods to sell excess power to the local utility; abundant solar-pumped water flowing through a lush oasis that was formerly a toxic patch of abused ground; the world's largest straw bale structure, beautifully illuminated by daylight, and heated and cooled without using fossil fuels; a central courtyard that is really a giant solar calendar. This story, told through the eyes of the participants, is not a worshipful tale, but rather a fascinating behind-the-scenes look at what it takes to bring a vision to reality.

Yurts: Living in the Round

Becky Kemery

This book journeys from Central Asia to modern America and reveals the history, evolution, and contemporary benefits of yurt living. One of the oldest forms of indigenous shelter still in use today, yurts have exploded into the 21st century as a multi-faceted, thoroughly modern, utterly versatile, and immensely popular modern structure whose possibilities are still being explored. Kemery introduces the innovators who redesigned the yurt and took it from backcountry trekking and campground uses to modern permanent homes and offices. *Yurts* shows how to build, insulate, ventilate, and transport a yurt, plus shares invaluable information on everything from foundations and heating to building codes and floor plans. Inspiring and imaginative photographs plus an extensive resource section offer all the information needed to take the next step.

Backyard Sheds & Tiny Houses: Build Your Own Guest Cottage, Writing Studio, Home Office, Craft Workshop, or Personal Retreat

Jay Shafer

Good things do come in small packages. Jay Shafer's small buildings have appeared on CNN, *Oprah*, *Fine Homebuilding*, and *This Old House*. Ranging in size from 100 to 120 square feet, these tiny backyard buildings can be used as guest cottages, art or writing studios, home offices, craft workshops, vacation retreats, or a full-time residence. Filled with photos, elevation drawings, and door/window schedules for six box bungalows, the book also includes an extensive how-to set of instructions that can be applied to any backyard building project. Though conventionally built, these handsome little buildings have real doors, windows, and skylights with interesting and practical details throughout. Extra attention is given to energy and space efficiency in their design.

The Handbuilt Home: 34 Simple Stylish and Budget-Friendly Woodworking Projects for Every Room

Ana White

Create a beautiful, modern home with one-of-a-kind DIY furniture. It's easy to build inexpensive, quality furnishings with this indispensible collection of woodworking projects from Ana White, the popular blogger who has inspired millions of homemakers with her stylish furniture plans and DIY spirit. In this reference for woodworkers of all skill levels, you'll find plans for 34 versatile furniture projects for every room in your house—from beginner-friendly home accessories to sturdy tables, a media center, kids' items, and storage solutions. Easy-to-follow instructions, costs, and time estimates to guide even the most amateur of carpenters through any project. All you need is the determination to create a better home for yourself or your family and the confidence to say, "I can build that."

The Homebuilt Winery: 43 Projects for Building and Using Winemaking Equipment

Steve Hughes

All the information you need to set up a home winery and build all of the basic equipment—for just a fraction of the cost of store-bought. Steve Hughes includes building plans and step-by-step instructions for more than 30 winemaking essentials, including a crusher, a de-stemmer, presses, pumps, and a bottle filler. He even offers a range of options for cellar racking. Along the way, Hughes leads readers through the entire process of winemaking—how to use the equipment, how to set up a winery, the best ways to store and analyze wine, and the best ways to filter, bottle, cork, and label. With this guide, you'll have everything you need to affordably enjoy delicious, high-quality, homemade wine.

The Green Roof Manual: A Professional Guide to Design, Installation, and Maintenance

Edmund C. Snodgrass, Linda McIntyre

Green roofs continue to generate enormous interest and enthusiasm among architects, landscape designers, and urban planners. Until now, no book has taken a comprehensive look at how to effectively adapt green roof technology to the variable and extreme North American climate, and how to design projects that will function and endure as successfully as those in Germany, Switzerland, and other European countries. This book fills the gap by providing an overview of practices and techniques that have been effective in North America. The authors offer options regarding structure, function, horticulture, and logistics, as well as surveys of actual projects and analyses of why they have or haven't succeeded. Photographs highlight the range of design possibilities and show green roofs both during construction and at various stages of maturity.

Peak Oil, Relocalization, and Politics

Limits to Growth: The 30-Year Update

Donella H. Meadows, Dennis Meadows, Jurgen Randers

In 1972, three scientists from MIT created a computer model that analyzed global resource consumption and production. Their results shocked the world and created stirring conversation about global "overshoot," or resource use beyond the carrying capacity of the planet. Now, preeminent environmental scientists Donella Meadows, Jorgen Randers, and Dennis Meadows have teamed up again to update and expand their original findings. Over the past three decades, population growth and global warming have forged on with a striking semblance to the scenarios laid out by the model in the original *Limits to Growth*. In many ways, the message contained in the 30-year update is a warning. Overshoot cannot be sustained without collapse. But, as the authors are careful to point out, there is reason to believe that humanity can still reverse some of its damage to Earth if it takes appropriate measures to reduce inefficiency and waste.

The Party's Over: Oil, War, and the Fate of Industrial Societies

Richard Heinberg

The world is about to run out of cheap oil and change dramatically. Thereafter, even if industrial societies begin to switch to alternative energy sources, they will have less net energy each year to do all the work essential to the survival of complex societies. We are entering a new era, as different from the industrial era as the latter was from medieval times. In *The Party's Over*, Richard Heinberg places this momentous transition in historical context, showing how industrialism arose from the harnessing of fossil fuels, how competition to control access to oil shaped the geopolitics of the 20th century, and how conten-tion for dwindling energy resources in the 21st century will lead to resource wars around the world. He describes the likely impacts of oil depletion and all of the energy alternatives. Predicting chaos unless the United States—the world's foremost oil consumer—is willing to join with other countries to implement a global program of resource conservation and sharing, he also recommends a "managed collapse" that might make way for a slower-paced, low-energy, sustainable society in the future.

The Ecology of Commerce: A Declaration of Sustainability, Revised Edition

Paul Hawken

The Ecology of Commerce is the provocative national bestseller that addresses the necessity of merging good business practices with common sense environmental concerns. Nearly two decades after its initial publication, this controversial work by Paul Hawken has been revised and updated, arguing why business success and sustainable environmental practices need not—and, for the sake of our planet, must not—be mutually exclusive any longer. An essential work, *The Ecology of Commerce* belongs on the bookshelf of every concerned citizen, alongside *Capitalism at the Crossroads* by Stuart Hart and Al Gore's *Earth in the Balance* and *An Inconvenient Truth.*

Reinventing Fire: Bold Business Solutions for the New Energy Era

Amory Lovins, Rocky Mountain Institute

Reinventing Fire is a comprehensive introduction to the issues and challenges tied to our nation's energy use. Lovins—a noted authority on energy, especially its efficient use and sustainable supply—

outlines the current state of energy use, including the nation's addiction to fossil fuels, and propose an array of transformational solutions. HIs long-term view emphasizes smart business strategy over public policy as the route to the "new energy era." The book sets the stage with two contrasting scenarios for energy consumption in 2050, one that is "business as usual" and one that "reinvents fire." The optimal scenario would reduce overall energy consumption through innovation and efficiency, while increasing use of renewable sources and bringing a multitude of benefits to the economy and the environment, as well as to our health and national security.

The Upcycle: Beyond Sustainability

William McDonough and Michael Braungart

The Upcycle is the eagerly awaited follow-up to *Cradle to Cradle*, one of the most consequential ecological manifestos of our time. Drawing on the lessons gained from 10 years of putting the Cradle to Cradle concept into practice, McDonough and Braungart envision the next step in the solution to our ecological crisis: We don't just use or reuse resources with greater effectiveness, we actually improve the world as we live, create, and build. They are practical-minded visionaries: They envision beneficial designs of products, buildings, and business practices—and they show us these ideas being put to use around the world as everyday objects like chairs, cars, and factories are being reimagined not just to sustain life on the planet but to grow it. Instead of protecting the planet from human impact, why not redesign our activity to improve the environment? We can have a beneficial footprint. Abundance for all. The goal is within our reach.

Tales From the Sustainable Underground: A Wild Journey with People Who Care More About the Planet than the Law

Stephen Hren

Activists striving for any type of social change often find themselves operating on the fringes of legal and social norms. Many experience difficulties when their innovative ideas run afoul of antiquated laws and regulations that favor a big business, energy- and material-intensive approach. *Tales From the Sustainable Underground* is packed with the stories of some of these pioneers who care more for the planet than the rules, whether

they're engaged in natural building, permaculture, community development, or ecologically based art. The profiles in this highly original book provide a unique lens through which to view deeper questions about the societal structures that are preventing us from attaining a more sustainable world. By examining such issues as the nature of property rights and the function of art in society, the author raises profound questions about how our social attitudes and mores have contributed to our current destructive paradigm.

The Homework Myth: Why Our Kids Get Too Much of a Bad Thing

Alfie Kohn

So why do we continue to administer this modern cod liver oil—or even demand a larger dose? Kohn's incisive analysis reveals how a set of misconceptions about learning and a misguided focus on competitiveness has left our kids with less free time, and our families with more conflict. Pointing to stories of parents who have fought back—and schools that have proved educational excellence is possible without homework—Kohn demonstrates how we can rethink what happens during and after school in order to rescue our families and our children's love of learning.

SUSTAINABLE TRANSPORTATION

Plug-in Hybrids: The Cars That Will Recharge America

Sherry Boschert

A politically polarized America is coming together over a new kind of car—the plug-in hybrid that will save drivers money, reduce pollution, and increase US security by reducing dependence on imported oil. *Plug-in Hybrids* points out that, where hydrogen fuel cell cars won't be ready for decades, the technology for plug-in hybrids exists today. Unlike conventional hybrid cars that can't run without gasoline, plug-in hybrids use gasoline or cheaper, cleaner, domestic electricity—or both. Cautioning that the oil and auto companies know how to undermine the success of plug-in car programs to protect their interests, the book gives readers tools to ensure that plug-in hybrids will thrive in the market.

Alcohol Can Be a Gas!: Fueling an Ethanol Revolution for the 21st Century

David Blume, Foreword by R. Buckminster Fuller

Alcohol, the first automotive fuel, is poised to replace gasoline and diesel in the rapidly approaching peak oil world. It's 98% pollution free and reverses global warming, and you can make it for less than a buck a gallon. This is the definitive book about on alcohol fuel production and use for home and farm. It explores the history of alcohol as fuel, conversion of gasoline and diesel vehicles, technical details, and more.

Biodiesel America: How to Achieve Energy Security, Free America from Middle-east Oil Dependence and Make Money Growing Fuel

Josh Tickell

With a new slant on biodiesel, here's a readable overview of the biodiesel world. Whether you're a seasoned pro or a newbie, you'll find straightforward language in an industry that abounds with complicated concepts. It's a necessary read for those interested in renewable fuels.

Water Development

WATER SYSTEMS, PUMPING, AND STORAGE

The Home Water Supply: How to Find, Filter, Store, and Conserve It

Stu Campbell

If you live in the country or suburbs, you've had, are having, or will have water problems. What's yours? Not enough water? Tastes terrible? The pump quits? The water's contaminated? No matter what it is, Stu Campbell addresses it in this book, offering down-to-earth solutions in clear, understandable language. He provides concrete and money-saving answers to questions that range from how to locate water to how to dig a pond to how to hook up the plumbing in your house. You'll learn how to find water, how to move it, how to purify it, and how to store and distribute it in your home.

Water Storage: Tanks, Cisterns, Aquifers, and Ponds for Domestic Supply, Fire and Emergency Use

Art Ludwig

Water Storage shows how to make your storage—and your entire water system—perform better, especially if there are water quality issues, severe supply constraints, or highly variable demand. You'll learn what kind of storage will serve best—tanks, ponds, groundwater; how much storage you need; where to install it; how to properly plumb it; which accessories would benefit your home, farm, or community; and how to sustainably manage groundwater—in short, how to make

your access to clean water more secure. *Water Storage* includes original design innovations, real-life examples, and complete instructions for constructing tanks from ferrocement. It is aimed at small communities, farms, ranches, back-to-the-landers, and anyone interested in water security.

Build a Pond for Food & Fun— Storey's Country Wisdom Bulletin

Since 1973, Storey's Country Wisdom Bulletins have offered practical, hands-on instructions designed to help readers master dozens of country living skills quickly and easily. There are now more than 170 titles in this series, and their remarkable popularity reflects the common desire of country and city dwellers alike to cultivate personal independence in everyday life.

WATER HEATING

Solar Water Heating, Revised & Expanded Edition: A Comprehensive Guide to Solar Water and Space Heating Systems

Bob Ramlow and Benjamin Nusz

Heating water with the sun is a practice almost as old as humankind itself. *Solar Water Heating*, now completely revised and expanded, is the definitive guide to this clean and cost-effective technology. This book presents an introduction to modern solar energy systems, energy conservation, and energy economics. Drawing on the authors' experiences as designers and installers of these systems, the book covers: Types of solar collectors, solar water, and space heating systems and solar pool heating systems, including their advantages and disadvantages; system components, their installation, operation, and maintenance; system sizing and siting; choosing the appropriate system. It also focuses on the financial aspects of solar water or space heating systems, clearly showing that such systems generate significant savings in the long run. With many diagrams and illustrations to complement the clearly written text, this book is designed for a wide readership ranging from the curious homeowner to the serious student or professional.

The Outdoor Shower: Creative Design Ideas for Backyard Living, from the Functional to the Fantastic

Ethan Fierro

Designer/builder Ethan Fierro celebrates summer with this collection of original outdoor shower designs that he's discovered across North America. He offers dozens of designs to choose from, all pictured in stunning color photographs, ranging from a simple showerhead on the side of your cottage to semi-detached and freestanding structures. Fierro also includes an overview of basic plumbing and structural requirements, along with suggestions for using a wide range of building materials, such as stone, wood, metal, frosted glass blocks, earth, concrete, tile, stucco, and landscape vegetation. His plans are environmentally sound and incorporate cost-saving techniques, with ideas for repurposing found and recycled materials.

DRINKING WATER

Rainwater Harvesting for Drylands and Beyond, Volume 1: Guiding Principles to Welcome Rain into Your Life and Landscape, 2nd Edition

Brad Lancaster

The award-winning *Rainwater Harvesting for Drylands and Beyond* is the first book in a three-volume guide that teaches you how to conceptualize, design, and implement sustainable water harvesting systems for your home, landscape, and community. The lessons in this volume will enable you to assess your onsite resources, give you a diverse array of strategies to maximize their potential, and empower you with guiding principles to create an integrated, multi-functional water

harvesting plan specific to your site and needs. This revised and expanded edition increases potential for onsite harvests with more integrated tools and strategies for solar design, a primer on your water/energy/carbon connections, descriptions of water/erosion flow patterns, and updated illustrations to show you how to do it all.

Design for Water: Rainwater Harvesting, Stormwater Catchment, and Alternate Water Reuse

Heather Kinkade-Levario

In an era of dwindling resources, water is poised to become the new oil. The entire world now faces the reality of a decreasing supply of clean water. To avert a devastating shortage, we must not only look at alternate water sources for existing structures but must plan our new developments differently. *Design for Water* is an accessible and clearly written guide to alternate water collection, with a focus on rainwater harvesting in the urban environment. The book outlines the process of water collection from multiple sources; provides numerous case studies; details the assembly and actual application of equipment; and includes specific details, schematics, and references. All aspects of rainwater harvesting are outlined, including passive and active system setup, storage, storm water reuse, distribution, purification, analysis, and filtration. Heather Kinkade-Levario is a land-use planner in Arizona and the author of the award-winning *Forgotten Rain*.

COMPOSTING TOILETS AND GREYWATER

The Toilet Papers: Recycling Waste and Conserving Water

Sim Van der Ryn, Foreword by Wendell Berry

A classic is back in print! One of the favorite books of 1970s back-to-the-landers, *The Toilet Papers* is an informative, inspiring, and irreverent look at how people have dealt with their wastes through the centuries. In a historical survey, Van der Ryn provides the basic facts concerning human wastes, and describes safe designs for toilets that reduce water consumption and avert the necessity for expensive and unreliable treatment systems. *The Toilet Papers* provides do-it-yourself plans for a basic compost privy and a variety of greywater systems.

The New Create an Oasis with Greywater: Choosing, Building and Using Greywater Systems

Art Ludwig

Create an Oasis describes how to quickly and easily choose, build, and use a simple greywater system. Some can be completed in an afternoon for under $30. It also provides complete instructions for more complex installations, how to deal with freezing, flooding, drought, failing septics, low perk soil, non-industrialized world conditions, coordinating a team of professionals to get optimum results on high-end projects, and "radical plumbing" that uses 90% less resources.

Builder's Greywater Guide: Installation of Greywater Systems in New Construction and Remodeling

Art Ludwig

The Builder's Greywater Guide (a supplement to *Create an Oasis with Greywater*) will help building professionals or homeowners work within or around building codes to successfully include greywater systems in new construction or remodeling, even if they have little prior greywater experience. It is also a great resource for regulators interested in improving oversight of greywater systems. Topics include: Special reasons for builders to install or not install a greywater system, flow chart for choosing a system, suggestions for dealing with inspectors, legal requirements checklist, detailed review of system options with respect to new laws, latest construction details and design tips, maintenance suggestions, equations for estimating irrigation demand, and the complete text of new US greywater law with suggested improvements. Note: The book *Create an Oasis with Greywater* is required in order to effectively use the *Builder's Greywater Guide*.

The Humanure Handbook: A Guide to Composting Human Manure, Third Edition

Joseph Jenkins

The 10th Anniversary Edition of the most comprehensive, up-to-date, and thoroughly researched book on the topic of composting human manure available anywhere. It includes a review of the historical, cultural, and environmental issues pertaining to "human waste," as well as an in-depth look at the potential health risks related to humanure recycling, with clear instructions on how to eliminate those dangers in order to safely convert humanure into garden soil. Written by a humanure composter with over thirty years experience, this classic work now includes illustrated, step-by-step instructions on how to build a "$25 humanure toilet," a chapter on alternative greywater systems, photos of owner-built humanure toilets from around the world, and an overview of commercial composting toilets and systems.

How to Shit in the Woods: An Environmentally Sound Approach to a Lost Art, Third Edition

Kathleen Meyer

It's the feisty third edition of *How to Shit in the Woods*, jam-packed with new information for outdoor enthusiasts of every stripe. Hailed in its first edition as the most important environmental book of the decade by Books of the Southwest, and in its second as the real shit by the late, great, outdoor photographer Galen Rowell, this best-selling guide is often called the backpacker's bible and has sold more than 2.5 million copies in eight languages. Kathleen Meyer continues pioneering the way with her inimitable voice at once humorous, irreverent, and direct, examining the latest techniques for graceful backcountry elimination, and answering a desperate cry from nature concerning environmental precautions in our ever-shrinking wilds. To assist with responsible human waste disposal, Meyer's new edition features the latest in product innovations, from classy high-tech to inexpensive do-it-yourself. She covers the most current solutions to the health risks of drinking straight from wilderness waterways; presents a raft of natural substitutes for the purist swearing off toilet tissue; and offers a wealth of new recommendations for ladies who must make do without a loo.

References and Resources

Consumer Guide to Home Energy Savings: Save Money, Save the Earth 10th Edition

Jennifer Thorne Amann, Katie Ackerly, Alex Wilson

Increasing the energy efficiency of your home can save you money, help the environment, and enhance your comfort, but how do you decide which improvements are the most beneficial and cost-effective? Completely revised to incorporate the latest developments in green technology, *Consumer Guide to Home Energy Savings* is the definitive resource for consumers who want to better their home's performance while reducing their energy bills. The book begins with an overview of the relationships between energy use, economics, and the environment. Updated and expanded chapters focus on specific aspects of any home, such as heating and cooling, ventilation, electronics, lighting, cooking, and laundry, and provide helpful explanations for each.

Country Wisdom & Know-How

From the Editors of Storey Books

Reminiscent in both spirit and design of the beloved Whole Earth Catalog, *Country Wisdom & Know-How* is an unprecedented collection of information on nearly 200 individual topics of country and self-sustained living. Compiled from the information in Storey Publishing's landmark series of "Country Wisdom Bulletins," this book is the most thorough and reliable volume of its kind. Organized by general topic including animals, cooking, crafts, gardening, health and well-being, and home, it is further broken down to cover dozens of specifics from "Building Chicken Coops" to "Making Cheese, Butter, and Yogurt" to "Improving Your Soil" to "Restoring Hardwood Floors." Nearly 1,000 black-and-white illustrations and photographs run throughout, and fascinating projects and trusted advice crowd every page.

Cool Tools: A Catalog of Possibilities

Kevin Kelly

Cool Tools is a highly curated selection of the best tools available for individuals and small groups. Tools include hand tools, maps, how-to books, vehicles, software, specialized devices, gizmos, websites—and anything useful. Tools are selected and presented in the book if they are the best of kind, the cheapest, or the only thing available that will do the job. This is an oversized book that reviews over 1,500 different tools, explaining why each one is great, and what its benefits are. Indirectly the book illuminates the possibilities contained in such tools, and the whole catalog serves an education outside the classroom. The content in this book was derived from ten years of user reviews published at the Cool Tools website, cool-tools.org.

Made by Dad: 67 Blueprints for Making Cool Stuff

Scott Bedford

The Snail Soup Can Decoy to keep the candy stash safe. The Customizable "Keep Out" Sign to deter meddlesome siblings and parents. A Bunk Bed Communicator made from cardboard tubes. Clever, whimsical, and kind of genius, these unique projects will turn any dad (or mom) with DIY leanings into a mad scientist hero that his kid(s) will adore. No screens, no high-tech gadgetry. *Made by Dad* combines the rough-edged, handmade ethos of a Boy Scout manual or *The Dangerous Book for Boys* with a sly sense of humor that kids love. Scott Bedford wields an X-ACTO knife, magic marker, and prodigious imagination to create endlessly delightful projects. He knows that kids like contraptions and gadgets, and things that are surprising—a chair that appears to be balanced on eggshells. Things that are complex—a multilevel city, with buildings, tunnels, and roads, built from old boxes around the legs of a table. And especially things with humor—the Snappy Toast Rack, made to resemble a crocodile's gaping mouth. The projects are shown in full-color photographs, and the instructions are illustrated in detailed line drawings that exude personality. Some are quick and simple enough to be done in a coffee shop; others are more of an afternoon project—yielding hours and hours of rich, imaginative playtime.

Nontoxic Housecleaning

Amy Kolb Noyes

When it comes to cleaning products, society often values convenience over personal and planetary

SUSTAINABLE LIVING LIBRARY

health, thanks to decades of advertising propaganda from the chemical companies that market overpriced and dangerous concoctions. But awareness is changing: Not only are homemade and nontoxic cleaners strong enough for the toughest grunge, they are often as convenient as their commercial counterparts. *Nontoxic Housecleaning* provides a way for people to improve their immediate environment every day. Pregnant women, parents of young children, pet owners, people with health concerns, and those who simply care about a healthy environment—and a sensible budget—can all benefit from the recipes and tips in this guide.

Tan Your Hide!: Home Tanning Leathers & Furs

Phyllis Hobson

A step-by-step guide to making vests, belts, and wallets by home tanning and hand-working furs and leathers. 138,000 copies in print.

Mycelium Running: How Mushrooms Can Help Save the World

Paul Stamets

Mycelium Running is a manual for the mycological rescue of the planet. Based upon the premise that habitats and humans (animals) have immune systems, and that mushrooms are the beneficial bridges for both, mycologist extraordinaire Paul Stamets explains the mycorestoration revolution that can help strengthen the sustainability of habitats. Linking mushroom cultivation, permaculture, ecoforestry, bioremediation, and soil enhancement, *Mycelium Running* marks the dawn of a new era: the use of mycelial membranes for ecological health.

The Mushroom Cultivator: A Practical Guide to Growing Mushrooms at Home

Paul Stamets and J. S. Chilton

Extensively used as a mycological textbook and known throughout the world as "The Grower's Bible," this book details the cultivation of 16 edible (including the Button/Portobello mushrooms) and psychoactive species and control measures for 40 genera of contaminants. Illustrated with 250 black and white photographs, diagrams, and scanning electron micrographs.

Growing Gourmet and Medicinal Mushrooms

Paul Stamets

If *The Mushroom Cultivator* is considered "the grower's Bible," this book is commonly referred to as "The New Testament" by amateur and professional mycologists alike. It covers in detail state-of-the-art commercial cultivation techniques, liquid culture inoculation methods, mycological landscaping, growing room and lab designs, troubleshooting, and more. Contains more than 500 photographs and diagrams.

SUSTAINABLE LIVING LIBRARY

Appendix

Real Goods' Mission Statement

Through our products, publications, and educational demonstrations, Real Goods promotes and inspires an environmentally healthy and sustainable future.

Part I: Who We Are and How We Got Here

Who Put the "Real" in Real Goods

As did many of his contemporaries in the 1960s and early 1970s, John Schaeffer, founder of Real Goods, experimented with an alternative lifestyle. After protracted exposure to nearly every strand of the lunatic fringe, he graduated in Anthropology from U.C. Berkeley in 1971 and moved to an archetypical hippie commune called "Rainbow" in Mendocino County just outside of Boonville, California. There, in an isolated 290-acre mountain-top community, John pursued a picturesque life of enlightened self-sufficiency.

John Schaeffer, founder of Real Goods.

Despite the idyllic surroundings of undeveloped gorgeous wilderness, John soon found that certain key elements of life were missing. After several years of reading bedtime stories to his children by the flickering light of a kerosene lamp, John began to squint. He grew tired of melted ice cream and lukewarm beer. He began to miss some of the creature comforts his family was lacking due to their "off-the-grid" lifestyle. He yearned for just a tiny amount of energy to strike a balance between the lifestyle he had grown up with and the relative deprivation of the commune. In other words, John came to the realization that self-sufficiency was more appealing as a concept than a reality.

Then one day, he discovered 12-volt power. John hooked up an extra battery to his car that he charged while commuting to work in his Volkswagen bug with a redwood tree stump for a driver's seat (pre-seat belt era!), with just enough juice to power lights, a radio, and the occasional television broadcast. Despite his departure from a pure ascetic lifestyle, each and every time that *Saturday Night Live* aired, John's home became the most popular place on the commune. John took a job as a computer operator in Ukiah, some 35 twisty miles from Boonville, but ample charging time for his off-grid energy system, powered by a Volkswagen alternator in the days before photovoltaics were on the market.

Once the word got out that John would be making the trek over the mountain to the "big city" daily, he became a one-man pick-up and delivery service, procuring the wood stoves, fertilizer, chicken wire, bone meal, gardening seeds, tools, and supplies needed for the commune. As a conscientious and naturally frugal person, John spent hours scrutinizing the hardware stores and home centers of Ukiah, searching for the best deals on the real goods needed for the communards' close-to-the-earth lifestyle. One day, while driving his VW bug back to the commune after a particularly vexing shopping trip, a thought occurred to John. "Wouldn't it be great," he mused, "if there was one store that sold all the products needed for independent, off-the-grid living, and sold them at fair prices?" That day in 1977, the idea of Real Goods was born, and it remains today, 37 years hence. The company thrived opening up retail stores and many minds, and eventually morphed into a mail-order company through many iterations. Today Real Goods remains a robust and innovative store in Hopland at the Solar Living Center as well as a strong eCommerce business (realgoods.com). The residential and commercial solar installation business has morphed into "RGS Energy" and is a larger than $100 million national business with over 15 offices around the country, doing thousands of solar installations. John continues to sit on the RGS Energy board of directors.

Real Goods in the New Millennium

From its humble beginnings in 1978, Real Goods became a Real Business, with Real Employees serving Real Customers. In the 1990s, Real Goods pioneered the "direct public offering" process, whereby it raised investment capital from its customers without the need for investment bankers or other financial middlemen. Before the Internet and cell phones had really caught on, Real Goods was selling stock electronically and allowing its customers to print virtual stock certificates in the privacy of their own homes. It was revolutionary. The company now can lay claim to the title of the Oldest and Largest catalog/eCommerce firm devoted to the sale and service of renewable energy products in the world. Real Goods (Nasdaq: RGSE), and its installation wing, RGS Energy, is still devoted to the same principles that guided its founding—quality, innovative, well-made products for fair prices, education and deep knowledge about the environment, and unsurpassed customer service with courtesy and dignity.

Early on, John managed to turn his personal commitment to right livelihood into company policy, pioneering the concept of a socially conscious and environmentally responsible business. The company consistently has been honored and awarded for its ethical and environmental business standards. Plaudits include Corporate

APPENDIX

Conscience Awards (from the Council on Economic Priorities); inclusion in *Inc.* magazine's list of America's 500 Fastest-Growing Companies; three consecutive Robert Rodale Awards for Environmental Education; Northern California Small Business of the Year Winner; finalist for Entrepreneur of the year two years running; John's induction into the International Green Industry Hall of Fame, recent congressional awards for sustainability, and countless other awards; news coverage in *Time, Fortune, The Wall Street Journal*, and *Mother Earth News*; numerous TV appearances; countless Japanese magazines; and many thick scrapbooks full of press clippings.

Five Principles to Live By

Real Goods is considered newsworthy not because its methods reflect the latest trends in corporate or business school thinking, but because it unwittingly has helped to birth an astonishingly healthy "baby"—an ethical corporate culture based on environmental and social responsibility. Led by a certain naiveté and affection for simplicity, Real Goods has discovered some simple principles that, by comparison to the "straight" business world, are wildly innovative. This has not been the work of commercial gurus or public relations mavens, but rather the result of realizing that business need not be so complicated that the average person cannot understand its workings. Real Goods is built around five simple core principles.

Principle #1: We Are a Business and We Must Be Profitable to Be Sustainable

A business is, first and foremost, a financial institution. You can have the most noble social mission on the planet, but if you can't maintain financial viability, you cease to exist. And so does your mission. The survival instinct is very strong at Real Goods, and that reality governs many decisions. To say it another way, you can't be truly sustainable if your business isn't economically sustainable (read profitable). To ensure the continued flourishing of our mission, we pursue profitability through our retail store, webstore (real goods.com), and our residential and commercial solar sales and installation business. We learned long ago that nonprofits aren't driven by profit. That's why we spun off the Solar Living Institute in 1998 to become a 501(c)(3). The SLI furthers the original educational mission of Real Goods without the constraints of the profit motive. Both organizations have a symbiotic relationship and help each other out immensely, but there are no financial or legal ties between them.

John Schaeffer accepts a Sustainability Award for the Solar Living Institute from Congressman Jared Huffman.

Principle #2: Know Your Stuff: Knowledge Is Our Most Important Product

Knowledge seldom turns a profit, yet our social and environmental missions cannot be achieved without it. The independent lifestyle we advocate relies largely on technologies that often require a high degree of understanding, and a level of interaction that has been largely forgotten during our nation's nearly full-century binge on cheap power. We aren't interested in selling things to people that they aren't well-informed enough about to live comfortably and happily with. We want people to understand not only what we are selling and how it is used, but how a particular piece of hardware contributes to the larger goal of a sustainable lifestyle.

It goes against the grain of mainstream business to give anything away. Even a "loss leader" is designed to suck you into the store to buy other, higher-profit items. At Real Goods, knowledge is our most important product, yet we give it away daily, through our webstore (there's a lot in there to learn, even if you never buy a thing); through our Solar Living Center (free self-guided

APPENDIX

and group tours), now run by the nonprofit Solar Living Institute; through free workshops at our Hopland store, and through workshops we support through the SLI's annual SolFest renewable energy celebration. Our webstore acts as a launching pad for renewable energy research, leading you to fascinating information on sustainability topics of all kinds. We believe that as our collective knowledge of sustainability principles and renewable energy technology increases, the chance of achieving the Real Goods mission increases, too.

Principle #3: Give Folks a Way to Get Involved

Just about all of our employees are also our customers, and a huge proportion of them live with solar—some, like our founder John Schaeffer, live completely off the grid. Our Solar Living Center parking lot looks like an advertisement for biofuels and electric vehicles with all the VW diesel cars and plug-in Priuses driven by our employees. Real Goods has become a real community, acting in concert toward the common goal of a sustainable future. With the knowledge that the age of oil is likely soon coming to an end, it's comforting to know that we are all in this together.

Principle #4: Walk the Walk

At Real Goods, we conduct our business in a way that is consistent with our social and environmental mission. We use the renewable energy systems we sell, and we sell what works. Our merchandising team makes absolutely sure the merchandise we sell performs as expected, is safe and nontoxic when used as directed, and is made from the highest-quality sustainable materials. In 1990, we challenged our customers to help us rid the atmosphere of one billion pounds of CO_2 by the year 2000, and we achieved our goal three years ahead of schedule. Again in 2007, we set an ambitious goal to offset the production of another one billion pounds of CO_2—this time even more quickly (in only two years!). Real Goods was recognized by being awarded the Rodale Award, as the business making the most positive contribution to the environment in America, for three years running. We don't just talk the talk at Real Goods, we walk the walk. Come visit us at the Solar Living Center in Hopland, California, and see some of our innovative practices. Fill up your biodiesel vehicle onsite, see water pumped from the sun, ride our bicycle generators and see how much human power it takes to power the average American home, and see 150 kW of PV solar powering the site and much more. It's one of the top tourist attractions in Northern California, and is now found in tour books and on maps and has had over 3 million visitors since opening in 1996.

Principle #5: Have Fun and Enjoy Balance in Your Life

We strongly support the best party of the year for thousands of our closest friends every summer in August or September on the grounds of our Solar Living Center in Hopland. We look forward to the Solar Living Institute's annual SolFest renewable energy celebration all year. If we've learned anything since 1978, it's that all work and no inspiration makes Jack and Jill a couple of burnt-out zombies. SolFest is our little reminder to take care of ourselves with some good clean fun, so we'll be rejuvenated and reinvested in the hard work of creating a sustainable future. And we believe in taking time for our families and personal lives rather than fostering a workforce of 80-hour-per-week workaholics.

Part II: Living the Dream: The Real Goods' Solar Living Center

Imagine a destination where ethical business is conducted daily amidst a diverse and bountiful landscape, where the gurgle of water flowing through its naturally revitalizing cycle heightens your perception of these ponds, these gardens, these living sculptures. You follow the sensuous curve of the hill and lazy meanders of the watercourse to a structure of sweeping beauty, where floor-to-ceiling windows and soaring architecture clearly proclaim this building's purpose—to take every advantage of the power of the sun throughout its seasonal phases. A few more steps and the spidery legs of a water-pumping windmill come into view, and the top of a tree that looks as though it might be planted in the rusted shell of a vintage Cadillac in the "Grow through Trees" grove. An awesome sense of place begins to reveal itself to you. Inside the building, sunlight and rainbows play across the walls and floors of a 5,000-square-foot showroom built of over 600 rice straw bales, and you begin to understand that all of this, even the offices and cash registers, are powered by the energy of the sun. Welcome to the Solar Living Center in Hopland, California, the crowning achievement of the Real Goods mission.

Our Solar Living Center began as the vision of Real Goods founder and board member John Schaeffer. His dream was to create an oasis of biodiversity, where the company could demonstrate the culture and technology of solar living, where the grounds and structures were designed to embody the sustainable living philosophy of Real Goods' business. With the opening of the Solar Living Center in April 1996, John's vision is now a reality. As of mid-2014, over three million people have visited the center, and have left this place with an overwhelming sense of inspiration and possibility. As the sign on the gate upon leaving the Solar Living Center states, "Turn Inspiration into Action" and take your newfound knowledge back to your communities.

In 1998, the Real Goods Solar Living Institute split off from its parent Real Goods Trading Corporation and became a legal 501(c)(3) nonprofit called the Solar Living Institute (SLI). Since then the SLI has nurtured and developed the 12-acre permaculture site that has flourished with fecundity.

Form and Function United: Designing for the Here and Now

If the "weird restrooms" sign doesn't grab them first, the 40,000 daily passersby on busy Highway 101 are bound to notice the striking appearance of the company showroom, which also serves as one of the most innovative retail stores we know of in the world. This does not look like business as usual! The building design and the construction materials were selected with an eye toward merging efficiency of function, educational value, and stunning beauty.

The architect chosen to design the building was Sim Van der Ryn of the Ecological Design Institute of Sausalito, California (once the State Architect of California under Jerry Brown). His associate, David Arkin, served as project architect, and Jeff Oldham of Real Goods managed the building of the project. Their creation is a tall and gracefully curving single-story building that is so adept in its capture of the varying hourly and seasonal angles of the sun that additional heat and light are virtually unnecessary. A wood-burning stove provides backup heating for the coldest winter mornings, and solar-powered fluorescent lighting is available, but is rarely used. Through a combination of overhangs and manually controlled hemp awnings, excess insolation during

The Solar Living Center from the air during SolFest 2005. Note the shadow of the airplane on the solar array at the bottom.

The central oasis and "stonehenge of the future" at the Hopland Solar Living Center (picture taken with a camera on a kite!).

the hot-weather months has been avoided. Energy-efficient fans provide a low-energy alternative to air conditioning, and are also used to flush the building with cool night air, storing "coolth" in the 600 tons of thermal mass of the building's walls, columns, and floor. Grape arbors and a central fountain with a "drip ring" for evaporative cooling are positioned along the southern exposure of the building to serve as a first line of defense against the many over-one-hundred-degree days that occur during the summer in this part of California.

Many of the materials used in the construction of the building were donated by companies and providers with a commitment similar to Real Goods'. As an example, the walls of the SLC were built with more than 600 rice straw bales donated by the California Rice Industries Association. Previously, rice straw has routinely been disposed of by open burning, a practice that contributes to the production of carbon dioxide, the greenhouse gas that is the leading cause of global climate change. By using this agricultural by-product as a building material, everyone benefits. The farmers receive income for their straw bales, carbon is sequestered in the walls rather than produced by burning, and the builder benefits from a low-cost, highly efficient building material that minimizes energy consumption.

At the SLC, visitors experience the practicality of applied solar power technology, including the generation of electricity and solar water pumping. The electrical system for the facility comprises nearly 150 kilowatts of photovoltaic power. Through an intertie with the Pacific Gas and Electric Company, the SLI sells the excess power it generates to the electric company, mak-

ing the SLC 100% independent from the grid. Once again, like-minded companies have shared in the costs of developing the Solar Living Center as a demonstration site. Siemens Solar (now Solar World) donated more than 10 kilowatts of the latest state-of-the-art photovoltaic modules to the center, and has periodically used the SLC as a test site for new modules. Trace Engineering (now Schneider Electric) contributed four intertie inverters, which are on display behind the glass window of the SLC's "engine room" so that visitors can see the inner workings of the electrical system.

In November 1999, a partnership between GPU, Astropower, Real Goods, and the Solar Living Institute installed a 132-kilowatt PV array on campus, at the time one of the largest in power-hungry Northern California. This direct-intertie array delivers 163,000 kWh of power annually, enough to power fifty average California homes. On tours, either self-guided or with Solar Living Institute intern tour guides, visitors learn about the guiding principles of sustainable living, and are offered a chance to appreciate the beauty that lies in the details of the project. The site also provides a wonderful space for presentations by guest speakers and special events, and serves as the main campus and classroom for the workshop series staged by the Solar Living Institute, a nonprofit dedicated to education and inspiration toward sustainable living.

The Natural World Reclaimed: The Grounds and Gardens

Learning potential is intrinsic to the award-winning landscape, designed by Chris and Stephanie Tebbutt of Land and Place in Boonville, California. For this project, the design of the grounds, gardens, and waterworks was the first phase of construction and contributed much to establishing the character of the site. This is a radically different approach than most commercial building projects, where the landscaping appears to be a cosmetic afterthought. At the SLC, the gardens are a synthesis of the practical and the profound. Most of the plantings produce edible and/or useful crops, and the vegetation is utilized to maximize the site's energy efficiency while portraying the dramatic aspects of the solar year. Plantings and natural stone markers follow the lines of sunrise and sunset for each equinox and solstice, emanating from a sundial at the exact center of the oasis. More sundials and unique solar calendars scattered throughout the site encourage visitors to establish a feeling for the relationship between this specific location and the sun. Those of us who

work and play here daily have discovered an organic connection with the seasonal shifts of the solar year and the natural rhythms of the Earth.

The gardens themselves follow the sun's journey through the seasons, with zones planted to represent the ecosystems of different latitudes. Woodland, Wetland, Grassland, and Dryland zones are manifested through plantings moving from north to south, with the availability of water being the definitive element. Trees are planted to indicate the four cardinal directions. The fruit garden, perennial beds, herbs, and grasses reflect the abundance and fertility of a home-based garden economy and are utilized to feed the 8–12 interns who inhabit, learn from, and nurture the site. Visitors discover aesthetic statements in design and landscape tucked into nooks and crannies all over the grounds, and unexpected simple pleasures, too, like shallow water channels for cooling aching feet, and perfect hidden spots for picnics and conversation.

Unique to these gardens are the "Living Structures," which reveal their architectural nature according to the turn of the seasons. Through annual pruning, plants are coaxed into various dynamic forms, such as a willow dome, a hops tipi, and a pyramid of timber bamboo. These living structures grow, quite literally, out of the garden itself. Visitors unaccustomed to the heat of a Hopland summer find relief inside the "agave cooling tower," where the turn of the dial releases a gentle mist into the welcome shade of vines and agave plants. By the time visitors leave the center, they've begun to understand the subtle humor of the "memorial car grove," where the rusting hulks

of '50s and '60s "gas hog muscle" cars have been turned into planter boxes for trees. These "grow-through" cars make a fascinating juxtaposition to the famous Northern California "drive-through" redwood trees! Truly, nature bats last here as one can witness these trees, planted at ½" caliper in 1994, now "eating" the cars with their ample 2' tree trunks!

A Place to Play

In case this all sounds awfully serious, it should be pointed out that the SLC is a wonderful place to play! Upon entering the showroom, one is greeted by a delightful rainbow spectrum created by a large prism mounted in the roof of the building. Visitors need not understand on a conscious level that this rainbow functions throughout the year as a "solar calendar," or that the prism's bright hues mark the daily "solar noon"—this is a deeper learning that ignites the place where inspiration

Bicycle generators at the Solar Living Center demonstrate that the average human can generate 150 sustained watts per hour. Therefore it would take six people peddling for 24 hours to produce 21.6 kWh per day, which is 2⁄3 of the energy that the average American home uses. This is the most popular interactive display onsite.

At an Earth Day for kids event, school children learn about solar hot water technology.

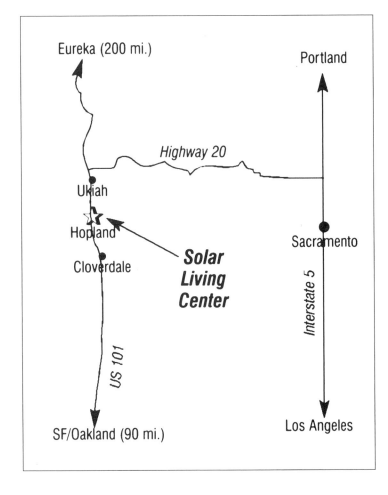

happens, not a raw scientific dissertation. In actuality, one can tell time from this amazingly engineered solar calendar as well as the actual season as the sun's projection moves daily along the north-south path highlighting the equinoxes, solstices, and even "cross-quarter" points. Outside, interactive games and play areas tempt the young at heart to forget about the theory and enjoy the pleasure of pure exploration. A six-station bicycle "generator" allows riders to see how much energy humans are capable of producing compared to the average American home's energy needs. It doesn't take long for riders to really feel how much energy is required to create the tiniest bit of energy; almost everyone takes the time to compare the aching results of muscle power to the ease with which the same amount of energy is harvested from the sun with a solar panel.

The hands-down favorite for kids is the sand and water area. A solar-powered pump provides a water source that can then be channeled, diverted, dammed, and flooded through whatever sandy topography emerges from the maker's imagination. A shadow across the solar panel stops the flow of water, and it doesn't take long for kids to become immersed in starting and stopping the flow at will. Without even realizing it, these little scientists are learning about engineering, hydrology, erosion, and renewable energy theory!

These are only a very few of the dozens of hands-on interactive displays available to the public at the Solar Living Center.

Where to Find the Solar Living Center

The Real Goods Solar Living Center is located 94 miles north of San Francisco on Highway 101 and is open every day except Thanksgiving and Christmas. There is no admission charge (although donations to the Solar Living Institute are strongly encouraged!) for regularly scheduled or self-guided tours, and picnicking alongside the gorgeous ponds is strongly encouraged.

Customized group tours for students of all ages, architects, gardeners, or others with special interests are available on a fee basis by advanced reservation, through the nonprofit Solar Living Institute. The Institute also offers a variety of structured learning opportunities, including intensive, hands-on, one-day to one-week seminars on a variety of renewable energy, sustainable living, and permaculture gardening topics. Please call the Institute at 707-472-2450 for more information, or visit its website at solarliving.org for a complete syllabus of current workshop and class offerings.

REAL GOODS

Part III: Spreading the Word: The Solar Living Institute

In April 1998, the Solar Living Institute separated from Real Goods Trading Corporation to become a legal nonprofit 501(c)(3). By severing financial ties to a for-profit corporation, the Institute became free to develop its educational mission without the constraint of required profitability, and to focus its efforts solely on environmental education.

The Solar Living Institute (SLI) has received several generous donations from Real Goods and RGS Energy which have enabled it to expand its interactive displays and programs, to beautify its landscape, and heighten its impact on environmental education. This generosity is a part of the SLI's annual fundraising program, with the bulk coming from donations and partnerships from individuals like you.

The SLI's Four-fold Mission

The Institute's mission is to teach people of all ages how to live more sustainably on the Earth, to teach interdependence between people and the environment, to replace fossil fuels with renewable energy, and to honor biodiversity in all its forms. To further its mission, the Institute is focusing its resources on three endeavors: Sustainable Living Workshops; the Solar Living Center's Interactive Displays, Exhibits, and Educational Tours for children and adults; and the annual SolFest Energy Festival, Educational Celebrations, and Earth Day events. In short, the Institute's mission is to promote sustainable living through inspirational environmental education.

#1: Sustainable Living Workshops

The Solar Living Institute is a leader in educating our society to prepare for a fossil fuel-free, sustainable future. The Institute's workshops approach the themes of shelter, energy, transportation, urban homesteading, and food from an ecologically sustainable perspective, teaching students from around the world how to rethink these basic necessities in a manner that is in harmony with our planet. Workshop topics include green and natural building; biofuels and alternative transportation; solar and hydroelectric energy; and permaculture, organic farming, and urban homesteading. Now in its 22nd season with a few thousand students annually, Institute workshops provide an ideal opportunity to learn hands-on skills, meet people of similar interests, work be-

side them, and learn the most cutting-edge techniques in sustainability.

As the issues of peak oil, the fracking mess, and climate change become more accepted by the mainstream, entrepreneurial and career opportunities in various sectors of the green economy abound, creating a need for a workforce trained in regenerative and sustainable techniques. One of the main focuses of the Institute's workshop program is to prepare its students to be forward-thinking individuals who take the skills they've learned and utilize them to find sustainable solutions to the numerous challenges our society faces. After graduating from the Solar Living Institute's classes, many of the Institute's workshop students go on to make career changes into a green job, or found their own environmentally friendly businesses.

Workshops are offered mostly at the main campus at the Solar Living Center (SLC) in Hopland, California. The main campus is not only an immensely inspirational setting, but a living, breathing model for sustainable development, restorative permaculture landscaping, and renewable technologies. Many of the hands-on workshops at the SLC enable students to participate in actual projects like the building of a straw bale structure in the intern village, the erecting of a windmill to pump water, or the installation of a PV array to power the volunteer kitchen. Students often camp at the SLC, and are not only offered the opportunity to practice what they've learned,

A Solar Living Institute workshop entitled "Women in Solar" teaches solar energy hands-on in a supportive environment by women and for women.

APPENDIX

Build a straw bale house from scratch, including mixing the natural plaster.

but also live temporarily in a sustainable setting. One graduate expressed it best: "I can't decide if I've just had the best short vacation of my life, or the best learning experience! Could it be both?" If students are unable to make the pilgrimage to the Solar Living Center, the SLI also offers classes online at solarliving.org.

Product demonstrations, cost-benefit analyses, and a tour of applied technologies at the SLC provide students with a practical grasp on the tools and techniques available for taking their energy lives in hand. Those who arrive with some technical skill leave qualified to put together an independent solar energy system, to begin the design of an energy-efficient home made of alternative and sustainable building materials, to know what to look for in a piece of property, or to create balance and harmony in their own garden. The mission of each workshop is to share excitement, sense of purpose, and knowledge with others interested in living sensibly and lightly on the Earth.

The Solar Living Institute's expert faculty has decades of experience and a strong passion for passing along its expertise to information-hungry students. The practical skills that they teach inspire students to become informed consumers and practitioners, and able to ask the right ques-

tions and to understand the answers. They report that the experience inspires them to go further forward along the path toward sustainable living.

Typical Classes Offered at the Solar Living Institute

Classes range from one-day introductions to weeklong intensives, and cost between $50 and $135 per day. Workshops are scheduled year-round. Here is a representative list of workshops that were scheduled for last year's workshop season. Because the Institute is strongly committed to expanding its workshop series, this list will always continue to develop. Potential students are encouraged to visit the Solar Living Institute's beautifully redesigned website at solarliving.org and to let them know which additional topics you might be interested in pursuing.

Solar Training

The SLI offers IREC and NABCEP approved solar training courses from beginning to advanced levels. Whether you are just curious about solar, preparing for a career transition, desiring quality training for your employees, looking to start or expand a business, or advance within the solar industry, there is a course to meet your needs.

PV100 Solar Training Intro to Photovoltaics

If you are looking for training to find employment in the ever-expanding solar PV market, this workshop is the first step to familiarize yourself with the concepts of solar electricity. Topics include: Basics of electricity; The nuts and bolts of solar components; Simple ways to save money by improving energy efficiency; Determine ideal system size based on equipment costs, incentives/rebates, and grid utility rates.

PV 200 Solar Training Design and Installation Intensive

PV 200 is a 5-day intensive solar training course that offers 40 hours of critical knowledge and hands-on practice in the essentials of photovoltaic technology (PV). The course content has been honed through 15 years of experience and is taught by experienced industry professionals. Designed to meet the learning objectives in preparation for the NABCEP entry level exam.

PV 300 Solar Training Advanced PV Systems

Looking to become a Certified PV Installer? This advanced course has been accredited by the Institute for Sustainable Power Quality (ISP) and covers the NABCEP PV Installer Certification Task Analysis. Knowledge and experience in PV is assumed, and a minimum of basic algebra proficiency is expected.

Online Solar Training
The SLI's PV Design and Installation course continues to provide all you need as an education prerequisite for the NABCEP entry-level exam. Whether your career track is management, installation, finance, sales, marketing, customer service, or administrative support, this eCourse series provides a solid foundation to build or jump-start your solar industry career.

Solar PV Design and Installation Training
This online eCourse is a highly interactive and innovative approach to learning based on the curriculum that offered in the popular in-person courses.

Solar PV Sales Fundamentals
This interactive, self-paced online course provides the essential knowledge and skills needed to enter the fast-growing field of solar sales.

Solar PV Sales Economics and Financing
Develop greater understanding of and expertise in the factors that affect the cost and profitability of a solar PV System. Explore the deeper details on how to calculate and provide data in regard to system sizing, cost, production, and payback to present a compelling value proposition to a potential solar customer.

Advanced Solar PV Marketing and Sales
To establish and grow a successful solar sales business, you need to go beyond the design of a PV installation and master the crafting of a successful sales process, the intricacies of ROI calculations, and navigating the complex world of solar finance. These skills will differentiate you from your competition.

Sustainable Living
The Solar Living Institute teaches practices, skills, and philosophies that allow individuals and communities to live more sustainably. Producing and preparing organic food, raising animals and using their by-products, building and maintaining a home, using natural and earthen materials for building, reusing and conserving water, cultivating herbal medicine, and using primitive technologies are just a few of the many ways to live sustainably.

Sustainable Beekeeping
This informative hands-on workshop is for any-

one interested in bees. Students will learn what is needed to prepare for acquiring bees and the skills to successfully manage your own bee colony sustainably at home.

Build a Straw Bale House
This course will give you both the knowledge and practice you need to tackle your own straw bale construction project. The instructor, a leading expert in straw bale construction and a building inspector, will teach you the latest innovations in this constantly evolving building technique. You will help build a small load-bearing structure, including foundation and roofing. In addition, you will have the opportunity to mix and apply various plasters.

NB200 Natural Building Intensive
This course offers an immersion in the tools and techniques of natural building. Useful both for

A hands-on wiring lesson during a solar "boot camp" at the Solar Living Center. One-week boot camps teach the complete course in solar energy for those embarking on solar careers.

Kids learn about our hybrid house on a tour.

building professionals and for first-time builders. This class focuses on the natural materials most commonly used in construction: clay soil, sand, stone, fibers, and bamboo. As a complement to the hands-on portion of the course, included are discussions of the theoretical aspects of natural building.

Camp Solar—Kids Summer Camp

Harness the Sun this Summer! The SLI's summer camp will be putting the sun to work baking, making and racing… Over the 3 days, you will get a chance to make a sculpture from natural building materials, race solar cars, and run your own farmers market! Designed for kids between the ages of 7 and 11.

Introduction to Aquaponics
Using Applied Permaculture

What if there was a way to produce an abundance of organic food using up to 98% less water, while producing up to ten times more food in the same amount of area and time? What if you could achieve this using ¼ the amount of energy, in a system that is totally scalable, did not require fertile soil, saved you time and labor, made you a good living, and was incredibly sustainable? The exciting news is that such a system of farming exists! It's called Aquaponics and is a combination of Aquaculture and Hydroponics.

Install a Greywater System

Residential greywater, which is wastewater generated from domestic processes such as washing dishes, laundry, and bathing, is a valuable water resource that can be recycled to irrigate plants. Decentralized, low-maintenance, and low-energy methods can be used to move, treat, and use water.

Over 2,000 K-12 children visit us every year.

Mushroom Cultivation

Join the SLI to learn how to grow your own mushrooms. This class covers information on the many roles of fungi and its importance in today's world. It discusses basic propagation techniques using resources found at home, outdoor/indoor cultivation, and possibilities of myco-remediation of polluted soil.

Feedback from Solar Living Institute Graduates

"I am so excited to get home and get started that I can't sit still. One night I was sizing my system all night. Last night I dreamt I had five different PV systems laid out around the top of a green meadow. All night I was choosing the best one— a combination—for my needs. Jeff and Ross and Nancy were there talking to other people then answering all my questions!"

"An honest possibility even for a conservative Republican that voted twice for Reagan. It is not as difficult as others want you to believe."

"Good spiritual company, great food, wonderful setting, excellent information."

"You guys are great people, and your knowledge and experience is invaluable to the rest of us— that is why we came to you. Please concentrate on conveying what is unique to you and your experience. The rest we can get from the books you sell."

"Very informative introductory course…I feel rejuvenated and motivated to take action!"

"Super course—lots of practical information. Instructor is excellent, his enthusiasm is catching!"

To register for classes, or for more information or to request a workshop catalog, call 707-472-2450, write the Solar Living Institute at P.O. Box 836, Hopland, CA 95449, or check out our website at solarliving.org.

#2: Interactive Displays and Curriculum)

The Solar Living Institute is continually expanding and enriching the Solar Living Center's interactive and educational displays. The staff won't rest until the SLC is recognized globally as a major learning campus and educational tour destination. As a nonprofit organization, the SLI seeks sources of funding in keeping with its educational goals. For example, the Institute has applied for grant monies to fund additional interactive exhibits on hydro power, solar water heating, photovoltaics, wind power, an observation beehive, and hydrogen fuel cells at the SLC. The SLI has engaged professional designers to create more engaging, effective exhibits and interactive displays on the site. The SLI is also in the process of further promoting the SLC as a destination for school groups. The SLI is developing partnerships with educators to bring sustainable living education into local schools, with the vision to expand the programs across the country. In addition to improving the onsite educational demonstrations, the SLI is working toward bringing its educational programs to the world, recognizing that the future limits of readily available cheap fossil fuels will make it more difficult for people to come to the SLC.

#3: SolFest—Annual Late Summer Energy Festival and Educational Fair

The Solar Living Institute's largest annual event is the SolFest Energy Festival, scheduled each year toward the end of August or early September. The first ever SolFest took place in June 1996, celebrating the Grand Opening of the Real Goods Solar Living Center in Hopland, California. That gala three-day event was attended by 10,000 people, inaugurating the Solar Living Center as the premier destination for those interested in learning about renewable energy and other sustainable living technologies. The educational theme of the event, in concert with the uniquely beautiful setting of the Solar Living Center, has inspired thousands of visitors. World-class speakers, unique entertainment, educational workshops, exhibitor booths with a renewable energy orientation, along with a parade and display of electric vehicles, all helped to create a lively and very successful event.

Over the years, SolFest has featured incredible speakers, including Amory Lovins of the Rocky Mountain Institute, Paul Hawken, Wes Jackson of the Land Institute, Ralph Nader, Jim Hightower, Ben Cohen of Ben and Jerry's Homemade, Julia Butterfly Hill, Alice Walker, Helen Caldicott, Amy Goodman ("Democracy Now"), Ed Begley Jr., Darryl Hannah, and many others. World-class entertainers who have graced the SolFest stage include Bruce Cockburn, Michelle Shocked, Spearhead, Michael Franti, David Grisman, Mickey Hart, and more. SolFest also features dozens of ongoing educational workshops exploring topics like solar energy, straw bale construction, building with bamboo, biodiesel, electric vehicles, eco-design, socially responsible investing, unconventional financing, climate change, industrial hemp, creative reuse, preparing for peak oil, solar cooking, and much more. The family stage and educational area provides fun and educational games, performances, and interactive children's workshops. SolFest provides a unique opportunity for renewable energy and sustainable living aficionados to gather and swap stories, and to have a great time for two days in the sun while learning and playing. Call the SLI at 707-472-2450 or visit the website (solarliving.org) to find the exact date and bill of entertainment.

The SLI Internship and Volunteer Program

The Solar Living Institute has a well-developed and well-established internship program for students wanting to continue their studies in organic and biodynamic gardening, permaculture, renewable energy, green and natural building, alternative fuels, urban homesteading, and

The annual SolFest has drawn about 200 vendors with solar-powered booths.

Bruce Cockburn brought his unique blend of politics and amazing music to two SolFests, in 2005 and 2007.

Our tiny house can be rented on AirBnB!

Solar Living Institute interns manage the organic farm and harvest and market the produce.

other sustainability education topics. SLI interns are honored and appreciated for their contributions, and come away with a lasting sense of achievement. The SLI has an extensive internship program with students of all ages coming to the Solar Living Center from all over the world to live, work, learn, and experience the joys of sustainable living. Interns come for 12-week to 24-week stints and consistently depart with the remark that their internships have been one of their life's peak experiences. Many interns have gone on to secure lucrative employment in the solar and sustainability industries. One prominent company commented that "an internship from the Solar Living Institute to us is more valuable on a resume than a degree from Stanford." For more information on the SLI's internship program check out the website at solarliving.org.

It takes a lot of energy to manifest the SLI's vision, and is not the kind that is measured in kilowatts. The Institute extends a warm welcome to any volunteers willing to give their time and energy to promote sustainable living through inspirational environmental education. You do not need to live in California to help; volunteering could take the form of consultation, publicity, physical labor, graphic art work, day-to-day support, or wherever your talents lie and however you'd like to help.

The Institute's Membership Program

The other kind of energy needed to sustain the SLI is the green kind. The Solar Living Institute (SLI) depends upon, appreciates, and thrives on the support of individuals like you! The programs of SLI, a 501(c)(3) nonprofit organization, are funded by its constituents—people who believe in our mission and who want to help us make a difference for the planet. You can support the SLI by making an annual contribution, volunteering, or joining us by becoming a member. Membership in the SLI for costs as little as $35/year, and includes benefits like magazine subscriptions and discounts on workshops and books.

Your annual membership support helps us to realize our goals of providing renewable energy and sustainable living education to school groups, young adults in the internship program, mature adults seeking a career change, and the general public when visiting our 12-acre demonstration site, the Solar Living Center. Be part of the solution and join SLI by becoming a member. Please contact the SLI at 707-472-2450, check out our website at solarliving.org, or email sli@solarliving.org for complete details.

Part IV: Getting Down to Business

Remember our first principle? We never forget it. When the profitability principle is met, we are free to pursue our educational mission effectively. We always seek to achieve profitability through our eCommerce site and retail store (realgoods.com). We support the principles of knowledge, involvement, credibility, and even fun, through a variety of innovative programs, from educational opportunities to networking. When our business goals and principles work in tandem with our educational goals and principles, we sleep easier at night. These are the elements and programs which, taken together, make our business real.

Webstore (realgoods.com)

The Real Goods webstore includes products and information geared to folks who have been thinking a lot about bringing elements of sustainable living and renewable energy into their lives. Our focus is on high-impact, user-friendly merchandise like LED and compact fluorescent lighting, air and water purification systems, energy-saving household implements, and mainstream products that introduce the concepts behind renewable energy and energy efficiency without being dauntingly technical. We also include lots of information on sustainable living, environmental responsibility, gardening and permaculture, and books, books, books. The Real Goods webstore also has evolved into our technical source, for individuals who have made a serious commitment to reducing their impact on the planet, and need the tools to get on with it. Along with renewable energy system components (like PV modules, wind generators, hydroelectric pumps, cables, inverters, and controllers), our catalog offers off-the-grid and DC appliances, energy-saving climate control products, solar ovens, air and water purification systems, composting toilets, energy efficiency, and even more books.

Retail Store

As an Internet business, Real Goods can reach almost every nook and cranny in America. Even so, there will always be people who want to "kick the tires" before making a purchase. Our 5,000-square-foot built-of-straw-bales Hopland Solar Living Center flagship retail store is second to none. Not only do customers have an opportunity to visit 12 gorgeous acres of permaculture gardens and enjoy dozens of interactive displays, but they get to see all the products referred to in this *Sourcebook* and our catalogs come to life. Our retail store offers an opportunity to snag curious passers-by and show them how sustainable living can enhance their lives.

Real Goods Technical Staff

The Real Goods technical staff are experts at designing, procuring, and installing residential renewable energy systems—both off-the-grid and grid-intertie systems to your utility. You don't want your project to be a test bed for unproved

The Real Goods store staff is experienced, friendly, and knows how to have a good time.

technology or experimental system design. You want information, products, and service that are of the highest quality. You want to leave the details to a company with the capability to do the job right the first time. We welcome the opportunity to design and plan entire systems. Here's how the Real Goods technical services work:

1. To assess your needs, capabilities, limitations, and working budget, we ask you to complete a specially created worksheet (see page 421). The information required includes a list of your energy needs, an inventory of desired appliances, site information, and potential for hydroelectric and wind development. Note that we only need this detailed energy use information for off-the-grid systems.

2. A member of our technical staff will determine your wattage requirements, and design an appropriate system with you. Or, for intertie systems we will refer you to the "energy broker" at RGS Energy who services your location from one of our offices around the US. The sales representative will determine either how much you want to spend or how much utility power you wish to offset, and then design an appropriate system.

Real Goods has been providing clean, reliable, renewable energy to people all over the planet for over 36 years and has provided solar energy systems for over 22,500 homes and businesses in that time, amounting to more than 235 megawatts. With the creation of our separate residential and commercial solar installation divisions, our renewable energy technicians now have a tighter focus on off-the-grid residential and farm systems; this specialization means we are better prepared than ever to refer you to an "energy bro-

ker" who will design a utility intertie system that meets your unique needs.

You will leave your Real Goods technical consultation with the information you need to make informed decisions on which technologies are best suited for your particular application. You'll learn how to utilize proven technologies and techniques to reap the best environmental advantages for the lowest possible cost. And we won't leave you hanging once your system is installed; our technicians will help you with troubleshooting, ongoing maintenance, and future upgrades. The Real Goods commitment to custom design, proven technology, and ongoing support has resulted in more than two decades of exceptionally high customer satisfaction. Real Goods specializes in residential and commercial systems from 100 watts to several megawatts, including:

- Utility Intertie with Renewable Energy
- Off-the-Grid Renewable Energy Systems Design (solar, wind, hydro)
- Large-Scale Uninterruptible Backup Power
- High-Efficiency Appliances
- Biological Waste Treatment Systems (composting toilets)
- Power Quality Enhancement
- Whole Systems Integration
- Water Quality and Management

Real Goods technical services and sales for residential and commercial renewable energy systems are available by phone at 800-919-2400, from 8:00 AM to 5:00 PM, Pacific Time, Monday through Friday, and 10 AM to 6 PM on Saturday and Sunday through our retail store. We endeavor to answer technical email within 48 hours, and snail mail (USPS) within one week.

Log on to solarliving. org for the latest on events like courses, workshops, Earth Day for Kids programs, and lots more happenings at the Solar Living Institute.

SYSTEM SIZING WORKSHEET

AC device	Device watts	×	Hours of daily use	×	Days of use per week	÷	7	=	Average watt-hours per day
		×		×		÷	7	=	
		×		×		÷	7	=	
		×		×		÷	7	=	
		×		×		÷	7	=	
		×		×		÷	7	=	
		×		×		÷	7	=	
		×		×		÷	7	=	
		×		×		÷	7	=	
		×		×		÷	7	=	
		×		×		÷	7	=	
		×		×		÷	7	=	
		×		×		÷	7	=	
		×		×		÷	7	=	
		×		×		÷	7	=	
		×		×		÷	7	=	
		×		×		÷	7	=	
		×		×		÷	7	=	
		×		×		÷	7	=	
		×		×		÷	7	=	
		×		×		÷	7	=	
		×		×		÷	7	=	

1 Total AC watt-hours/day

2 × 1.1 = Total corrected DC watt-hours/day

DC device	Device watts	×	Hours of daily use	×	Days of use per week	÷	7	=	Average watt-hours per day
		×		×		÷	7	=	
		×		×		÷	7	=	
		×		×		÷	7	=	
		×		×		÷	7	=	
		×		×		÷	7	=	
		×		×		÷	7	=	
		×		×		÷	7	=	
		×		×		÷	7	=	
		×		×		÷	7	=	

3 Total DC watt-hours/day

3 (from previous page)	Total DC watt-hours/day	
4	Total corrected DC watt-hours/day from line 2 +	
5	Total household DC watt-hours/day =	
6	System nominal voltage (usually 12 or 24) ÷	
7	Total DC amp-hours/day =	
8	Battery losses, wiring losses, safety factor × 1.2	
9	Total daily amp-hour requirement =	
10	Estimated design insolation (hours per day of sun, see map on page 424) ÷	
11	Total PV array current in amps =	
12	Select a photovoltaic module for your system	
13	Module rated power amps ÷	
14	Number of modules required in parallel =	
15	System nominal voltage (from line 6 above)	
16	Module nominal voltage (usually 12) ÷	
17	Number of modules required in series =	
18	Number of modules required in parallel (from line 14 above) ×	
19	Total modules required =	

BATTERY SIZING

20	Total daily amp-hour requirement (from line 9)	
21	Reserve time in days ×	
22	Percent of useable battery capacity ÷	
23	Minimum battery capacity in amp-hours =	
24	Select a battery for your system, enter amp-hour capacity ÷	
25	Number of batteries in parallel =	
26	System nominal voltage (from line 6)	
27	Voltage of your chosen battery (usually 6 or 12) ÷	
28	Number of batteries in series =	
29	Number of batteries in parallel (from line 25 above) ×	
30	Total number of batteries required	

Power Consumption Table

Appliance	Watts	Appliance	Watts	Appliance	Watts
Coffeepot	200	Electric blanket	2,000	Compact fluorescent	
Coffee maker	800	Blow dryer	1,000–1,500	Incandescent equivalents	
Toaster	800–1,500	Shaver	15	40 watt equiv.	11
Popcorn popper	250	WaterPik	100	60 watt equiv.	16
Blender	300	Computer		75 watt equiv.	20
Microwave	600–1,700	Laptop	50–75	100 watt equiv.	30
Waffle iron	1,200	PC	200–600	Ceiling fan	10–50
Hot plate	1,200	Printer	100–500	Table fan	10–25
Frying pan	1,200	System (CPU, monitor, laser printer)	up to 1,500	Electric mower	1,500
Dishwasher	1,200–1,500	Fax	35	Hedge trimmer	450
Sink disposal	450	Typewriter	80–200	Weed eater	450
Washing machine		DVD Player	25	¼" drill	250
Automatic	500	TV 25" color	150+	½" drill	750
Manual	300	19" color	70	1" drill	1,000
Vacuum cleaner		12" b&w	20	9" disc sander	1,200
Upright	200–700	VCR	40–100	3" belt sander	1,000
Hand	100	CD player	35–100	12" chain saw	1,100
Sewing machine	100	Stereo	10–100	14" band saw	1,100
Iron	1,000	Clock radio	1	7¼" circular saw	900
Clothes dryer		AM/FM car tape	8	8¼" circular saw	1,400
Electric	4,000	Satellite dish/Internet	30–65	Refrigerator/freezer— Conventional ENERGY STAR	
Gas heated	300–400				
		CB radio	5	23 cu. ft.	540 kWh/yr.
Heater		Electric clock	3	20 cu. ft.	390 kWh/yr.
Engine block	150–1,000	Radiotelephone		16 cu. ft.	370 kWh/yr.
Portable	1,500	Receive	5	Sun Frost	
Waterbed	400	Transmit	40–150	16 cu. ft. DC	112
Stock tank	100	Lights		12 cu. ft. DC	70
Furnace blower	300–1,000	100 W incandescent	100	Freezer—Conventional	
Air conditioner		25 W compact fluorescent	28	14 cu. ft.	440
Room	1,500	50 W DC incandescent	50	14 cu. ft.	350
Central	2,000–5,000	40 W DC halogen	40	Sun Frost freezer	
Garage door opener	350	20 W DC compact fluorescent	22	19 cu. ft.	112

Solar Insolation Maps

The maps below show the sun-hours per day for the U.S. Charts courtesy of the D.O.E.

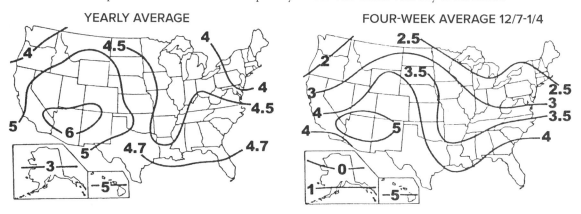

YEARLY AVERAGE

FOUR-WEEK AVERAGE 12/7-1/4

Solar Insolation by U.S. City

This chart shows solar insolation in kilowatt-hours per square meter per day in many U.S. locations. For simplicity, we call this figure "Sun Hours/Day." To find average Sun hours per day (last column) in your area, check local weather data, look at the maps above, or find a city in the table that has similar weather to your location. If you want year-round autonomy, use the figure in the Low column. If you want 100% autonomy only in summer, use the figure in the High column.

State	City	High	Low	Avg.	State	City	High	Low	Avg.	State	City	High	Low	Avg.
AK	Fairbanks	5.87	2.12	3.99	KS	Manhattan	5.08	3.62	4.57	NY	Schenectady	3.92	2.53	3.55
AK	Matanuska	5.24	1.74	3.55	KS	Dodge City	6.5	4.2	5.6	NY	Rochester	4.22	1.58	3.31
AL	Montgomery	4.69	3.37	4.23	KY	Lexington	5.97	3.6	4.94	NY	New York City	4.97	3.03	4.08
AR	Bethel	6.29	2.37	3.81	LA	Lake Charles	5.73	4.29	4.93	OH	Columbus	5.26	2.66	4.15
AR	Little Rock	5.29	3.88	4.69	LA	New Orleans	5.71	3.63	4.92	OH	Cleveland	4.79	2.69	3.94
AZ	Tucson	7.42	6.01	6.57	LA	Shreveport	4.99	3.87	4.63	OK	Stillwater	5.52	4.22	4.99
AZ	Page	7.3	5.65	6.36	MA	East Wareham	4.48	3.06	3.99	OK	Oklahoma City	6.26	4.98	5.59
AZ	Phoenix	7.13	5.78	6.58	MA	Boston	4.27	2.99	3.84	OR	Astoria	4.76	1.99	3.72
CA	Santa Maria	6.52	5.42	5.94	MA	Blue Hill	4.38	3.33	4.05	OR	Corvallis	5.71	1.9	4.03
CA	Riverside	6.35	5.35	5.87	MA	Natick	4.62	3.09	4.1	OR	Medford	5.84	2.02	4.51
CA	Davis	6.09	3.31	5.1	MA	Lynn	4.6	2.33	3.79	PA	Pittsburgh	4.19	1.45	3.28
CA	Fresno	6.19	3.42	5.38	MD	Silver Hill	4.71	3.84	4.47	PA	State College	4.44	2.79	3.91
CA	Los Angeles	6.14	5.03	5.62	ME	Caribou	5.62	2.57	4.19	RI	Newport	4.69	3.58	4.23
CA	Soda Springs	6.47	4.4	5.6	ME	Portland	5.23	3.56	4.51	SC	Charleston	5.72	4.23	5.06
CA	La Jolla	5.24	4.29	4.77	MI	Sault Ste. Marie	4.83	2.33	4.2	SD	Rapid City	5.91	4.56	5.23
CA	Inyokern	8.7	6.87	7.66	MI	East Lansing	4.71	2.7	4.0	TN	Nashville	5.2	3.14	4.45
CO	Granby	7.47	5.15	5.69	MN	St. Cloud	5.43	3.53	4.53	TN	Oak Ridge	5.06	3.22	4.37
CO	Grand Lake	5.86	3.56	5.08	MO	Columbia	5.5	3.97	4.73	TX	San Antonio	5.88	4.65	5.3
CO	Grand Junction	6.34	5.23	5.85	MO	St. Louis	4.87	3.24	4.38	TX	Brownsville	5.49	4.42	4.92
CO	Boulder	5.72	4.44	4.87	MS	Meridian	4.86	3.64	4.43	TX	El Paso	7.42	5.87	6.72
DC	Washington	4.69	3.37	4.23	MT	Glasgow	5.97	4.09	5.15	TX	Midland	6.33	5.23	5.83
FL	Apalachicola	5.98	4.92	5.49	MT	Great Falls	5.7	3.66	4.93	TX	Fort Worth	6	4.8	5.43
FL	Belle Isle	5.31	4.58	4.99	MT	Summit	5.17	2.36	3.99	UT	Salt Lake City	6.09	3.78	5.26
FL	Miami	6.26	5.05	5.62	NM	Albuquerque	7.16	6.21	6.77	UT	Flaming Gorge	6.63	5.48	5.83
FL	Gainesville	5.81	4.71	5.27	NB	Lincoln	5.4	4.38	4.79	VA	Richmond	4.5	3.37	4.13
FL	Tampa	6.16	5.26	5.67	NB	North Omaha	5.28	4.26	4.9	WA	Seattle	4.83	1.6	3.57
GA	Atlanta	5.16	4.09	4.74	NC	Cape Hatteras	5.81	4.69	5.31	WA	Richland	6.13	2.01	4.44
GA	Griffin	5.41	4.26	4.99	NC	Greensboro	5.05	4	4.71	WA	Pullman	6.07	2.9	4.73
HI	Honolulu	6.71	5.59	6.02	ND	Bismarck	5.48	3.97	5.01	WA	Spokane	5.53	1.16	4.48
IA	Ames	4.8	3.73	4.4	NJ	Seabrook	4.76	3.2	4.21	WA	Prosser	6.21	3.06	5.03
ID	Boise	5.83	3.33	4.92	NV	Las Vegas	7.13	5.84	6.41	WI	Madison	4.85	3.28	4.29
ID	Twin Falls	5.42	3.42	4.7	NV	Ely	6.48	5.49	5.98	WV	Charleston	4.12	2.47	3.65
IL	Chicago	4.08	1.47	3.14	NY	Binghamton	3.93	1.62	3.16	WY	Lander	6.81	5.5	6.06
IN	Indianapolis	5.02	2.55	4.21	NY	Ithaca	4.57	2.29	3.79					

Magnetic Declinations in the United States

Figure indicates correction of compass reading to find true north. For example, in Wasington State when your compass reads 22°E, it is pointing due north.

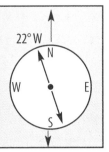

TRUE SOUTH IF STANDING
IN NORTH WASHINGTON

TRUE SOUTH IF STANDING
IN NEW BRUNSWICK

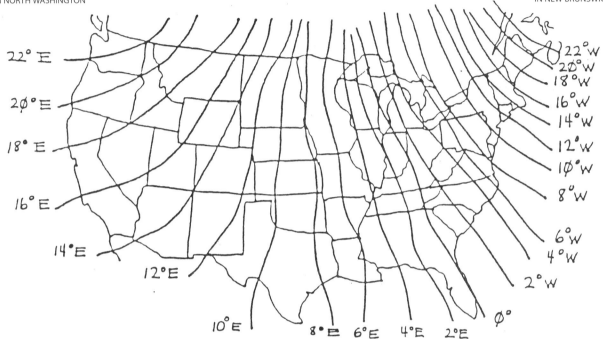

MAXIMUM NUMBER OF CONDUCTORS FOR A GIVEN CONDUIT SIZE

Conduit size		½"	¾"	1"	1¼"	1½"	2"
Conductor size	#12	10	18	29	51	70	114
	#10	6	11	18	32	44	73
	#8	3	5	9	16	22	36
	#6	1	4	6	11	15	26
	#4	1	2	4	7	9	16
	#2	1	1	3	5	7	11
	#1		1	1	3	5	8
	#1/0		1	1	3	4	7
	#2/0		1	1	2	3	6
	#3/0		1	1	1	3	5
	#4/0		1	1	1	2	4

Battery Wiring Diagrams

The following diagrams show how 2-, 6-, and 12-volt batteries are connected for 12-, 24-, and 48-volt operation.

─⌁⌐ Fuse symbol (always use appropriate fusing at your battery)

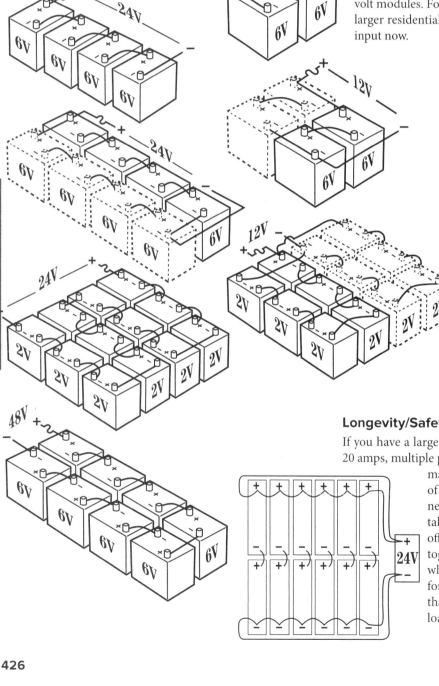

Wiring Basics

We answer a lot of basic wiring questions over the phone, which we're always happy to do, but there's nothing like a picture or two to make things apparent.

Battery Wiring

The batteries for your energy system may be supplied as 2-volt, 6-volt, or 12-volt cells. Your system voltage is probably 12, 24, or 48 volts. You'll need to series wire enough batteries to reach your system voltage, then parallel wire to another series group as needed to boost amperage capacity. See our drawings for correct series wiring. Paralleled groups are shown in dotted outline.

PV Module Wiring

PV modules are almost universally produced as nominal 12-volt modules. For smaller 12-volt systems, this is fine. Most larger residential systems are configured for 24- or 48-volt input now.

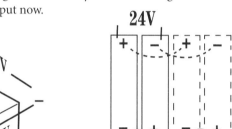

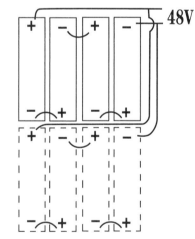

Longevity/Safety Tip for Wiring Larger PV Arrays

If you have a large PV array that produces close to, or over, 20 amps, multiple power take-off leads are a good idea. They may prevent toasted terminal boxes. Instead of taking only a single pair of positive and negative leads off some point on the array, take off one pair at one end, and another pair off the opposite end. Then join them back together at the array-mounted junction box where you're going to the larger wire needed for transmission. This divides up the routes that outgoing power can take, and eases the load on any single PV junction box.

APPENDIX

Wire Sizing Chart/Formula

This chart is useful for finding the correct wire size for any voltage, length, or amperage flow in any AC or DC circuit. For most DC circuits, particularly between the PV modules and the batteries, we try to keep the voltage drop to 3% or less. There's no sense using your expensive PV wattage to heat wires. You want that power in your batteries!

Note that this formula doesn't directly yield a wire gauge size, but rather a "VDI" number, which is then compared to the nearest number in the VDI column, and then read across to the wire gauge size column.

1. Calculate the Voltage Drop Index (VDI) using the following formula:

VDI = AMPS x FEET ÷ (% VOLT DROP x VOLTAGE)

Amps = Watts divided by volts

Feet = One-way wire distance

% Volt Drop = Percentage of voltage drop acceptable for this circuit (typically 2% to 5%)

2. Determine the appropriate wire size from the chart below.
   A. Take the VDI number you just calculated and find the nearest number in the VDI column, then read to the left for AWG wire gauge size.
   B. Be sure that your circuit amperage does not exceed the figure in the Ampacity column for that wire size. (This is not usually a problem in low-voltage circuits.)

Example: Your PV array consisting of four Sharp 80-watt modules is 60 feet from your 12-volt battery. This is actual wiring distance, up pole mounts, around obstacles, etc. These modules are rated at 4.63 amps × 4 modules = 18.5 amps maximum. We'll shoot for a 3% voltage drop. So our formula looks like:

VDI = (18.5 A x 60 ft) ÷ (3% x 12 V) = 30.8

Looking at our chart, a VDI of 31 means we'd better use #2 wire in copper, or #0 wire in aluminum. Hmmm. Pretty big wire.

What if this system was 24-volt? The modules would be wired in series, so each pair of modules would produce 4.4 amps. Two pairs × 4.63 amps = 9.3 amps max.

VDI = (9.3 A x 60 ft) ÷ (3% x 24 V) = 7.8

Wow! What a difference! At 24-volt input you could wire your array with little ol' #8 copper wire.

Wire Size	Copper Wire		Aluminum Wire	
AWG	VDI	Ampacity	VDI	Ampacity
0000	99	260	62	205
000	78	225	49	175
00	62	195	39	150
0	49	170	31	135
2	31	130	20a	100
4	20	95	12	75
6	12	75	•	•
8	8	55	•	•
10	5	30	•	•
12	3	20	•	•
14	2	15	•	•
16	1	•	•	•

Chart developed by John Davey and Windy Dankoff. Used with permission.

Why There's No National Electrical Code in This *Sourcebook*

Previous editions of this *Sourcebook* have printed in full the Suggested Practices of the National Electrical Code with respect to Photovoltaic Power Systems. This time around, to save paper, trees, and unnecessary costs and because the NEC is so easily accessible (and free!) on the Internet, we recommend you download it at: nmsu.edu/~tdi/pdf-resources/NEC.pdf.

Friction Loss Charts for Water Pumping

How to Use Plumbing Friction Charts

If you try to push too much water through too small a pipe, you're going to get pipe friction. Don't worry, your pipes won't catch fire. But it will make your pump work harder than it needs to, and it will reduce your available pressure at the outlets, so sprinklers and showers won't work very well. These charts can tell you if friction is going to be a problem. Here's how to use them:

PVC or black poly pipe? The rates vary, so first be sure you're looking at the chart for your type of supply pipe. Next, figure out how many gallons per minute you might need to move. For a normal house, 10–15 gpm is probably plenty. But gardens and hoses really add up. Give yourself about 5 gpm for each sprinkler or hose that might be running. Find your total (or something close to it) in the "Flow GPM" column. Read across to the column for your pipe diameter. This is how much pressure loss you'll suffer for every 100 feet of pipe. Smaller numbers are better.

Example: You need to pump or move 20 gpm through 500 feet of PVC between your storage tank and your house. Reading across, 1-inch pipe is obviously a problem. How about 1¼ inches? 9.7 psi times 5 (for your 500 feet) = 48.5 psi loss. Well, that won't work! With 1½-inch pipe, you'd lose 20 psi…still pretty bad. But with 2-inch pipe, you'd lose only 4 psi…ah! Happy garden sprinklers! Generally, you want to keep pressure losses under about 10 psi.

Friction Loss in PSI per 100 Feet of Scheduled 40 PVC Pipe

Flow GPM	Nominal Pipe Diameter in Inches							
	½"	¾"	1"	1¼"	1½"	2"	3"	4"
1	3.3	0.5	0.1					
2	11.9	1.7	0.4	0.1				
3	25.3	3.5	0.9	0.3	0.1			
4	43.0	6.0	1.5	0.5	0.2	0.1		
5	65.0	9.0	2.2	0.7	0.3	0.1		
10		32.5	8.0	2.7	1.1	0.3		
15		68.9	17.0	5.7	2.4	0.6	0.1	
20			28.9	9.7	4.0	1.0	0.1	
30			61.2	20.6	8.5	2.1	0.3	0.1
40				35.1	14.5	3.6	0.5	0.1
50				53.1	21.8	5.4	0.7	0.2
60				74.4	30.6	7.5	1.0	.03
70					40.7	10.0	1.4	0.3
80					52.1	12.8	1.8	0.4
90					64.8	16.0	2.2	0.5
100					78.7	19.4	2.7	0.7
150						41.1	5.7	1.4
200						69.9	9.7	2.4
250							14.7	3.6
300							20.6	5.1
400							35.0	8.6

Friction Loss in PSI per 100 Feet of Polyethylene (PE) SDR-Pressure Rated Pipe

Flow GPM	Nominal Pipe Diameter in Inches							
	½"	¾"	1"	1¼"	1½"	2"	2½"	3
1	0.49	0.12	0.04	0.01				
2	1.76	0.45	0.14	0.04	0.02			
3	3.73	0.95	0.29	0.08	0.04	0.01		
4	**6.35**	1.62	0.50	0.13	0.06	0.02		
5	9.60	2.44	0.76	0.20	0.09	0.03		
6	13.46	3.43	1.06	0.28	0.13	0.04	0.02	
7	17.91	4.56	1.41	0.37	0.18	0.05	0.02	
8	22.93	**5.84**	1.80	0.47	0.22	0.07	0.03	
9		7.26	2.24	0.59	0.28	0.08	0.03	
10		8.82	2.73	0.72	0.34	0.10	0.04	0.01
12		12.37	**3.82**	1.01	0.48	0.14	0.06	0.02
14		16.46	5.08	1.34	0.63	0.19	0.08	0.03
16			6.51	1.71	0.81	0.24	0.10	0.04
18			8.10	2.13	1.01	0.30	0.13	0.04
20			9.84	2.59	1.22	0.36	0.15	0.05
22			11.74	**3.09**	1.46	0.43	0.18	0.06
24			13.79	3.63	1.72	0.51	0.21	0.07
26			16.00	4.21	1.99	0.59	0.25	0.09
28				4.83	2.28	0.68	0.29	0.10
30				5.49	**2.59**	0.77	0.32	0.11
35				7.31	3.45	1.02	0.43	0.15
40				9.36	4.42	1.31	0.55	0.19
45				11.64	5.50	1.63	0.69	0.24
50				14.14	6.68	**1.98**	0.83	0.29
55					7.97	2.36	0.85	0.35
60					9.36	2.78	1.17	0.41
65					10.36	3.22	1.36	0.47
70					12.46	3.69	**1.56**	0.54
75					14.16	4.20	1.77	0.61
80						4.73	1.99	0.69
85						5.29	2.23	0.77
90						5.88	2.48	0.86
95						6.50	2.74	0.95
100						7.15	3.01	**1.05**
150						15.15	6.38	2.22
200							10.87	3.78
300								8.01

Temperature Conversions

°C = Degrees Celsius. 1 degree is 1/100 of the difference between the temperature of melting ice and boiling water.
°F = Degrees Fahrenheit. 1 degree is 1/180 of difference between the temperature of melting ice and boiling water.

°C	°F	°C	°F	°C	°F	°C	°F
200	392	140	284	80	176	15	59
195	383	135	275	75	167	10	50
190	374	130	266	70	158	5	41
185	365	125	257	65	149	0	32
180	356	120	248	60	140	−5	23
175	347	115	239	50	122	−10	14
170	338	110	230	45	113	−15	5
165	329	105	221	40	104	−20	−4
160	320	100	212	35	95	−25	−13
155	311	95	203	30	86	−30	−22
150	302	90	194	25	77	−35	−31
145	293	85	185	20	68	−40	−40

Temperature Conversion Chart

Nominal Pipe Size versus Actual Outside Diameter for Steel and Plastic Pipe

Nominal Size	Actual Size	Nominal Size	Actual Size
½"	0.840"	2½"	2.875"
¾"	1.050"	3"	3.500"
1"	1.315"	3½"	4.000"
1¼"	1.660"	4"	4.500"
1½"	1.900"	5"	5.563"
2"	2.375"	6"	6.625"

Nominal Pipe Size vs. Actual Outside Diameter for Steel and Plastic Pipe

The Real Goods' Resource List

Although this *Sourcebook* is a great source for renewable energy and environmental products and information, we can't be everything to everyone. Here is our current list of other trusted resources for information and products. We've selected these organizations carefully, and since we have worked directly with many of them, we are giving you the benefit of our experience. Still, we'll offer the standard disclaimer that Real Goods does not necessarily endorse all the actions of each group listed, nor are we responsible for what they say or do.

This list can never be complete. One good place to look for additional resources is wiserearth.org, an amazing compendium of tens of thousands of nonprofits compiled by Paul Hawken over many years We apologize for resources we may have overlooked and for contact information that may have changed since this 14th edition of the *Sourcebook* went to press. We welcome your suggestions. Please email a brief description of the organization and access info to: john@realgoods.com.

100Fires Books
Website: 100fires.com
Contact: via webform: 100fires.com/cgi-bin/contact.cgi?cart=1400631919
Anti-corporate personhood advocate Paul Cienfuegos is a tireless speaker, activist, and bookseller. His selection of progressive titles is extensive, and his focus on non-corporate publishers offers a rare glimpse into a range of media available nowhere else.

Advanced Buildings
Website: advancedbuildings.org/
Email: advancedbuildings@enermodal.com
Tremendous cache of resources and case studies, geared toward building professionals interested in technologies and practices for environmentally appropriate and energy-efficient construction.

American Council for an Energy-Efficient Economy (ACEEE)
Website: aceee.org
Email: info@aceee.org
Publishes books, papers, yearly guides, and comparisons of appliances and vehicles based on energy efficiency. Their website is an excellent source of efficient appliance info.

The American Hydrogen Association
Website: clean-air.org/
Email: contact@clean-air.org, question@clean-air.org
Phone: 602-328-4238
A nonprofit organization that promotes the use of hydrogen for fuel and energy storage. Publishes Hydrogen Today, a bimonthly newsletter.

American Society of Landscape Architects (ASLA)
Website: asla.org
Phone: 202-898-2444
This professional organization advocates on public policy issues such as livable communities, surface transportation, the environment, historic preservation, and small business affairs.

American Solar Energy Society (ASES)
Website: ases.org/
Email: ases@ases.org
Phone: 303-443-3130
ASES is the United States section of the International Solar Energy Society, a national organization dedicated to advancing the use of solar energy for the benefit of US citizens and the global environment. Publishers of the bimonthly Solar Today magazine for members, and sponsors of the yearly National Tour of Solar Homes.

The American Wind Energy Association
Website: awea.org/
Email: windmail@awea.org
Phone: 202-383-2500
A national trade association that represents wind power plant developers, wind turbine manufacturers, utilities, consultants, insurers, financiers, researchers, and others involved in the wind industry. These folks primarily work with utility-level wind systems. Not a good source for residential info.

Architects/Designers/Planners for Social Responsibility (ADPSR)
Website: adpsr.org
Email: forum@adpsr.org
Phone: 510-845-1000
ADPSR works for peace, environmental protection, ecological building, social justice, and the development of healthy communities. It has chapters around the country and publishes books on community arts and development through New Village Press (newvillage.net).

Artha Sustainable Living Center
Website: arthaonline.com
Email: info@arthaonline.com
Phone: 715-824-3463
Offers solar thermal (water and/or space heating) site assessments for large applications worldwide. Offers hands-on workshops and training at their site or yours on solar thermal and sustainable living.

Battery Recycling (for NiCads)
Rechargeable Battery Recycling Corp.
Website: rbrc.org
Email: recycling@rbrc.com
Phone: 678-419-9990
A nonprofit public service organization to promote the recycling of NiCad batteries. Just type your zip code in the website and get a list of local stores that will accept your old NiCads for recycling.

Bioneers
Website: bioneers.org
Bioneers, whose slogan is Revolution from the Heart of Nature, is "a forum for connecting the environment, health, social justice, and spirit within a broad progressive framework." Their annual conference, focused on practical and visionary solutions for restoring imperiled ecosystems and healing human communities, is extraordinary.

The Borrower's Guide to Financing Solar Energy Systems
Website: nrel.gov/docs/fy99osti/26242.pdf
A downloadable publication of the National Renewable Energy Lab.

BuildingGreen, Inc.
Website: buildinggreen.com
Email: info@buildinggreen.com
Phone: 802-257-7300
A leading source of objective information on green building and renewable energy. Publishes the journal Environmental Building News and the GreenSpec Directory of green building products.

Build It Green
Website: builditgreen.org
Email: contact@builditgreen.org
Oakland Office: Phone: 510-590-3360; Fax: 510-590-3361
Los Angeles Office: Phone: 213-688-0070; Fax: 213-402-2002
Build It Green is a membership-supported Bay Area nonprofit established in 2005. The organization works with building and real estate professionals, local and state governments, and homeowners to increase awareness and adoption of green building practices.

Building Education Center
Website: bldgeductr.org/
Email: syd@bldgeductr.org
Phone: 510-525-7610
BEC offers lectures, classes, and workshops of all varieties in building and remodeling in Berkeley, California.

California Energy Commission
Media and Public Communications Office
Website: energy.ca.gov
Email: mediaoffice@energy.state.ca.us
Phone: 800-555-7794 (inside CA); 916-654-4058 (outside CA)
The state of California's primary energy policy and planning agency. Strongly supports energy efficiency and small-scale utility intertie projects. A good source of honest grid-intertie information.

California Straw Building Association (CASBA)
Website: strawbuilding.org
Email: via webform: strawbuilding.org/contact
Phone: 209-785-7077
CASBA is a nonprofit project of the Tides Center. Its primary objective is to further the practice of straw bale construction by exchanging current information and practical experience, conducting research and testing, and making that body of knowledge available to working professionals and the public at large.

CalStart
Northern California Office:
Phone: 510-307-8700
Southern California Office:
Phone: 626-744-5600
Colorado Office:
Phone: 303-825-7550
Website: calstart.org
Email: calstart@calstart.org
The latest information on electric, natural gas, and hybrid electric vehicles, and other intelligent transportation technologies.

Center for Energy and Climate Solutions/Cool Companies
Website: cool-companies.org/homepage.cfm
CECS promotes clean and efficient energy technologies as a money-saving tool for reducing greenhouse gas emissions and other pollutants. Its website contains good information, including skeptical views about a hydrogen-based economy.

Center for Environmental Health
Website: ceh.org/
Mission: We envision a world where everyone lives, works, learns, and plays in a healthy environment; we protect people from toxic chemicals by working with communities, businesses, and the government to demand and support business practices that are safe for human health and the environment.

Center for Renewable Energy and Sustainable Technology (CREST)
See Renewable Energy Policy Project
Chelsea Green Publishing
Website: chelseagreen.com
Email: staff directory: chelseagreen.com /company/staff/
Phone: 802-295-6300.
One of the world's premier publishers of books on renewable energy, natural building, and sustainable living.

DCAT—Development Center for Appropriate Technology
Website: dcat.net
Email: office@dcat.net
Phone: 520-624-6628
The pioneering organization putting straw bale construction on the map, addressing building codes and standards for natural building, and now assisting tribal peoples to create their own building codes appropriate to their cultural needs.

Ecological Building Network
Website: ecobuildnetwork.org
Email: bruce@ecobuildnetwork.org
Phone: 415-491-4802
In the straw bale world, Bruce King's name is synonymous with the most complete understanding of structural engineering available. The consulting engineer on the construction of the Real Goods Solar Living Center has put together an online resource of immense breadth and value. From the most current codes and testing on bale building, to an online BuildWell library and annual conference.

Electric Auto Association
Website: eaaev.org
Email: contact@eaaev.org
The national electric vehicle association, with local chapters in most states and Canada. Dues are $39/year with an excellent monthly newsletter. The website has links to practically everything in the EV biz.

Energy Efficiency and Renewable Energy Clearinghouse (EREC)
Website: eere.energy.gov/
Email: EEREMailbox@EE.DOE.Gov
Phone: 877-337-3463
The best source of information about renewable energy technologies and energy efficiency. These folks seem to be able to find good information about anything energy related. One of the best overall websites about energy.

The Energy & Environmental Building Association (EEBA)
Website: eeba.org
Email: inquiry@eeba.org
Phone: 952-881-1098
Provides education and resources to transform the residential design and construction industry to profitably deliver energy-efficient and environmentally responsible buildings and communities.

Energy Information Administration
Website: eia.doe.gov
Email: infoctr@eia.doe.gov
Phone: 202-586-8800
Official energy statistics from the US government. Gas, oil, electricity, you name it, there's more supply, pricing, and use info here than you can shake a stick at.

ENERGY STAR®
Environmental Protection Agency (EPA)
ENERGY STAR Hotline
Website: energystar.gov/
Email: info@energystar.gov
Phone: 888-STAR-YES (888-782-7937)
Good, up-to-date listings of the most efficient appliances, lights, windows, home and office electronics, and more.

Environment California
Website: environmentcalifornia.org
Email: info@environmentcalifornia.org
Phone: 213-251-3688
Maintains a website committed to providing California consumers with up-to-date information on solar electric and solar water heating systems for new and existing homes.

The Environmental and Energy Study Institute
Website: eesi.org
Email: eesi@eesi.org
Phone: 202-628-1400
This nonprofit dedicated to sustainability has programs on climate and energy, agriculture and energy, transportation and energy, and energy and smart growth. The website contains many publications, press releases, and updates and offers a great way to keep current with relevant public policy deliberations.

Environmental Building News (online and print)
Website: www2.buildinggreen.com/
Email: webform at www2.buildinggreen.com /contact
The most reputable source of in-depth analysis of green building materials, techniques, and systems. Geared toward the professional but accessible to anyone, their free and subscription-based resources are the best and most universally acknowledged, including their comprehensive resource, GreenSpec.

Environmental News Network
Website: enn.com
ENN rounds up the most important and compelling environmental news stories of the week.

Environmental Protection Agency (EPA)
Website: epa.gov
Email: public-access@epamail.epa.gov
Phone: 202-272-0167
Findsolar.com
Website: Findsolar.com

This is a free public service to help home and building owners estimate the cost and benefits of a solar energy system, and to help them select a qualified installer, designer, or other solar professional.

Fungi Perfecti
Website: fungi.com
Host Defense Organic Mushrooms
Website: hostdefense.com
These websites of mycologist extraordinaire Paul Stamets offer information, resources, and products for growing mushrooms, immune system support, mycoremedation, and more.

Florida Solar Energy Center (FSEC)
Website: fsec.ucf.edu/en
Email: info@fsec.ucf.edu
Phone: 321-638-1000
Provides highly regarded independent third-party testing and certification of solar hot water systems and other solar or energy-efficiency goods.

Gas Appliance Manufacturers Association
Website: gamanet.org
Email: info@gamanet.org
Phone: 703-525-7060

Geothermal Resources Council
Website: geothermal.org
Email: grc@geothermal.org
Phone: 530-758-2360
A nonprofit organization dedicated to geothermal research and development.

Go Solar California
California Public Utilities Commission
Website: gosolarcalifornia.ca.gov
Phone: 415-703-2782
Joint website of the California Public Utilities Commission and the California Energy Commission, providing a portal into California's Million Solar Roofs Initiative, with resources for homeowners, businesses, schools, and public buildings.

Green Home Guide
Website: greenhomeguide.com
Email: infor@greenhomeguide.com
The Green Home Guide is a great resource for those looking for green building and remodeling information. Produced by the US Green Building Council, the site offers a blog that covers many topics with informative content; also lists applicable rating systems.

Green Key Real Estate
Website: greenkeyrealestate.com/
Email: info@greenkeyrealestate.com
Phone: 415-750-1120
Serves those interested in buying or selling green properties. "We have a vision that real estate can come back down to earth and begin participating in the rebuilding of local communities, and the recreating of homes that are in balance with the natural environment."

Green Living Journal: A Practical Journal for Friends of the Environment
Website: greenlivingjournal.com
Phone: 802-234-9101
Green Living is a quarterly, grassroots publication that is published in local editions (currently Vermont, New Hampshire, Massachusetts, southern Oregon) to serve "friends of the environment," businesses, and individuals who value the protection of natural resources.

GreenMoneyJournal.com
greenmoneyjournal.com/
A leading periodical about socially responsible investing and green business, "From the stock market to the supermarket." They've been doing it for 20 years.

The Green Power Network
Net Metering and Green Power information
Website: eere.energy.gov/greenpower/
Information on the electric power industry's green power efforts. Includes up-to-date info on green power availability and pricing. Also the best state-by-state net metering details on the Net.

Green Science Policy Institute
Website: greensciencepolicy.org/
Email: info@greensciencepolicy.org/
Phone: 510-898-1704
A powerhouse organization bringing the best science to bear on chemical threats in consumer products. Recent work and victories have focused on eliminating toxic flame retardants in furniture and rigid foam building insulation. More at saferinsulation.org.

Grist.org
Website: grist.org/
An excellent online periodical dedicated to environmental news and commentary.

Healthy Building Network
Website: healthybuilding.net/
Email: info@healthybuilding.net
Phone: 202-741-5717 or 877-974-2767
Fax: 202-898-1612
A vital and necessary resource for those concerned about chemical sensitivity issues in building materials. HBN is a leader in research and advocacy.

The Heartwood School for the Homebuilding Crafts
Website: heartwoodschool.com
Email: request@heartwoodschool.com
Phone: 413-623-6677
Offers a variety of hands-on building workshops. Specializes primarily in timber frames.

Home Energy Magazine
Website: homeenergy.org
Email: contact@ homeenergy.org
Phone: 510-524-5405
Fax: 510-981-1406
The magazine's mission is to disseminate objective and practical information on residential energy efficiency, performance, comfort, and affordability. Published bimonthly. US subscription is $55/yr a year and up for combination of print and electronic versions.

Home Power Magazine
Website: homepower.com
Email: hp@homepower.com
Phone: 800-707-6585
The journal of the renewable energy industry. Published bimonthly. US subscription is $24.95/yr.

Inside EVs
Website: insideevs.com
Email: insideevs@gmail.com

International Institute for Ecological Agriculture (IIEA)
Website: permaculture.com
Email: info@permaculture.com
Phone: 888-Permaculture (737-6228)
Provides courses, consulting, and books on permaculture design and alcohol fuel. Publishers of Alcohol Can Be a Gas, the only comprehensive manual on farm-scale alcohol production and its use. Also provides services for design and planting of high-value timber retirement or endowment forests that mature in 15–20 years.

International Living Future Institute
Website: living-future.org/
Phone: 206-223-2028
Known for its creation of the Living Building Challenge, ILFI is recognized as the organization pushing the cutting edge of building approaches to surpass all other existing rating systems. Living Building Challenge structures must provide net positive benefits to the environment and the building occupants.

The Intelligent Optimist Magazine
Website: theoptimist.com
This inspirational magazine focuses on positive new people and ideas that are changing the world for the better.

The Last Straw Journal Online
Website: thelaststraw.org
Contact via webform: thelaststraw.org/contact-us/
Phone: 970-704-5828
The first and most definitive journal of straw bale construction and related info. Newly revitalized by Jeff Ruppert in Paonia, CO, TLS is now both digital and print, and offers the most comprehensive cross-section of in-depth natural building articles from around the world. Subscriptions vary depending upon print/digital. Full archives available back to 1992!

Midwest Renewable Energy Association
Website: the-mrea.org
Email: info@the-mrea.org
Phone: 715-592-6595
A regional organization that promotes renewable energy, energy efficiency, and sustainable living through education and demonstration.

Mother Jones
Website: motherjones.com
A leading progressive magazine of politics and analysis: "Smart, fearless journalism."

National Association of Home Builders (NAHB)
Website: nahb.org
Email: info@nahb.com
Phone: 800-368-5242
Now has an active green building section.

National Association of State Energy Officials (NASEO)
Website: naseo.org
Email: mnew@naseo.org
Phone: 703-299-8800
An excellent portal site with access info for all state energy offices.

National Biodiesel Board
Website: www@biodiesel.org
Email: info@biodiesel.org
Phone: 800-841-5849
This advocacy organization has an excellent website brimming with all things biodiesel.

National Center for Appropriate Technology
Website: ncat.org
Email: info@ncat.org
Phone: 800-275-6228
A nonprofit organization and information clearinghouse for sustainable energy, low-income energy, resource-efficient housing, and sustainable agriculture.

National Renewable Energies Laboratory (NREL)
Website: nrel.gov
Email: webmaster@nrel.gov
Phone: 303-275-3000
The nation's leading center for renewable energy research. Tests products, conducts experiments, and provides information on renewable energy.

Natural Building Network
Website: nbnetwork.org
Contact: via webform: nbnetwork.org/contact
A collaborative project among many natural building advocates, instructors, and practitioners. Includes member directories, event listings, white papers, and many other resources. A great way to network for new and seasoned natural builders!

New Society Publishers
Website: newsociety.com
Email: info@newsociety.com
Phone: 250-247-9737
Leading publisher of books for renewable energy, alternative fuels, and sustainable living, including this Solar Living Sourcebook. New Society has recently gone carbon-neutral!

The North American Board of Certified Energy Practitioners (NABCEP)
Website: nabcep.org
Phone: 518-899-8186
Offers certification for renewable energy installers.

Oikos Green Building Source
Website: oikos.com/index.php
Oikos is a website geared to professionals whose work promotes sustainable design and construction. Lots of good books, resources, and links here.

Plenty Magazine
Website: plentymagazine.com
This green living site is more than a magazine, including podcasts, information on "green gear," blogs, and more.

Plug In America
Website: pluginamerica.org
Phone: (415) 323-3329
The Public Press
Website: thepublicpress.com
Phone: 802-234-9101
The Public Press is a publisher of books that are too special, too controversial, or too experimental for conventional publishers. The company goal is to protect freedom of speech "word by word."

The Rahus Institute
Website: rahus.org
A nonprofit research and educational organization that focuses on resource efficiency. Website has news, products, and books, and offers consulting.

The Regenerative Design Institute
Website: regenerativedesign.org
Email: info@regenerativedesign.org
RDI offers courses in permaculture and regenerative design, leadership, and deep communion with nature. One of the most comprehensive learning environments for permaculture in this hemisphere.

RenewableEnergyAccess.com
Website: renewableenergyaccess.com/rea/home
News, podcasts, products, job opportunities, events, and all things renewable energy.

Rocky Mountain Institute
Website: rmi.org
Email: outreach@rmi.org
Phone: 970-927-3851
A terrific source of books, papers, and research on renewable energy, energy-efficient building design and components, and sustainable transportation.

Sandia National Laboratory's Photovoltaic Systems Program
Website: sandia.gov/pv/
Check out the Design Assistance Center.

Small-Scale Sustainable Infrastructure Development Fund
Website: s3idf.org
Email: info@s3idf.org
Phone: 91-80-56902558
Known affectionately as S3IDF, this is a "social merchant bank" located in India that helps small enterprises provide modern energy and other infrastructural services to poor people in developing countries in ways that are financially sustainable and environmentally responsible.

Solar Cookers International
Website: solarcookers.org/
Email: info@solarcookers.org
Phone: 916-455-4499
A nonprofit organization that promotes and distributes simple solar cookers in developing countries to relieve the strain of firewood collection. Check out an amazing online archive: solarcooking.org; newsletter and small products catalog available.

Solar Electric Light Fund
Website: self.org/
Email: info@self.org
Phone: 202-234-7265
A nonprofit organization promoting PV rural electrification in developing countries.

Solar Energy Industries Association (SEIA)
Website: seia.org
Email: info@seia.org
Phone: 202-628-0556

Solar Energy Info for Planet Earth
Website: eosweb.larc.nasa.gov/sse/
Complete solar energy data for anyplace on the planet! Thanks to NASA's Earth Science Enterprise program, there's more solar data here than even the Real Goods techies can find a use for. Just point to your location on a world map.

Solar Energy International
Website: solarenergy.org
Email: sei@solarenergy.org
Phone: 970-963-8855
Offers hands-on and online classes on renewable technology.

Solar Living Institute
Website: solarliving.org
Email: sli@solarliving.org
Phone: 707-472-2450
A nonprofit educational organization that offers world-class workshops on renewable energy, sustainable living, green building, and permaculture. Produces the annual SolFest Energy Festival. Sign up for electronic newsletter: solarliving.org.

Solar Today
Website: solartoday.com
A leading magazine in the solar field: "Delivering today's news on solar energy technology."

SustainableABC.com, Sustainable Architecture, Building, and Culture
Website: SustainableABC.com
Email: royprince@sustainableabc.com
A unique compendium of links and content oriented to the global community of ecological and natural building proponents. Has a free newsletter.

SustainableBusiness.com
Website: sustainablebusiness.com/
On online newsletter and news service dedicated to green business, including sections about progressive investing, green dream jobs, and a great resource directory.

Sustainable By Design
Website: susdesign.com/tools.php
Email: christopher@susdesign.com
Phone: 206-925-9290
Provides shareware tools for sustainable design.

The Union of Concerned Scientists
Website: ucsusa.org
Email: ucs@ucsusa.org
Phone: 617-547-5552
Organization of scientists and citizens concerned with the impact of advanced technology on society. Programs focus on energy policy, climate change, and national security.

Voice of the Environment (VOTE)
Website: voiceoftheenvironment.org
Email: vote@pacific.net
Phone: 707-467-0329; 415-250-4115
Voice of the Environment educates the public regarding environmental, political, and social justice issues.

Wind Energy Maps for US
Website: rredc.nrel.gov/wind/pubs/atlas
The complete Wind Energy Resource Atlas for the United States in downloadable format. Offers averages, regions, states, seasonal; more slicing and dicing than you can imagine. Includes Alaska, Hawaii, Puerto Rico, and Virgin Islands.

WiserEarth
Website: wiserearth.org
Paul Hawken's website is a community directory and networking forum for people working on the critical issues of the day and all things sustainable. The network is growing rapidly, and there are many different ways to plug in.

Worldwatch Institute
Website: worldwatch.org
Email: worldwatch@worldwatch.org
Phone: 202-452-1999
A nonprofit public policy research organization dedicated to informing policymakers and the public about emerging global problems and trends. Offers a magazine and publishes a variety of books.

State Energy Offices Access Information

Here is all the straight, up-to-date information on utility intertie and on any state programs that might help pay for a renewable energy system. All states have provided web, email, phone, and fax access.

Also check dsireusa.org for the most up-to-date information on rebates, net metering, and other state programs.

Alabama Energy Office
Department of Economic and Community Affairs
Website: adeca.alabama.gov
Phone: 334-242-5100
Fax: 334-242-5099

Alaska Energy Authority
Website: akenergyauthority.org
Phone: 907-771-3000
Fax: 907-269-3044

American Samoa Energy Office
Territorial Energy Office
American Samoa Government
Website: asgteo.com
Phone: 684-699-1101
Fax: 684-699-2835

Arizona Energy Office
Arizona Department of Commerce
Website: azenergy.gov
Email: oep@az.gov
Phone: 602-771-1137
Fax: 602-771-1203

Arkansas Energy Office
Arkansas Industrial Development Commission
Website: arkansasenergy.org
Email: energyinfo@arkansasedc.com
Phone: 800-558-2633

California Energy Office
California Energy Commission
Website: energy.ca.gov
Email: mediaoffice@energy.ca.us; renewable@energy.ca.gov
Phone: 916-654-4287

Colorado Energy Office
Website: colorado.gov/energy
Email: Coloradoenergyoffice@state.co.us
Phone: 303-866-2100
Fax: 303-866-2930

Connecticut Energy Office
Policy Development and Planning—Energy Management
Connecticut Office of Policy and Management
Website: opm.state.ct.us/pdpd2/energy/enserv.htm
Email: deep.webmaster@ct.gov
Phone: 860-424-3000

Delaware Energy Office
Website: delaware-energy.com
Email: staff directory at dnrec.delaware.gov/energy/Pages/Contact-Us.aspx
Phone: 302-735-3480
Fax: 302-739-1840

District of Columbia Energy Office
District Department of the Environment Energy Office
Website: ddoe.dc.gov/energy
Email: dceo@dc.gov
Phone: 202-535-2600
Fax: 202-535-2881

DSIRE—Database of State Incentives for Renewables and Efficiency
Website: dsireusa.org/
A project of the US Department of Energy. A one-stop shop for the most current information on rebates and other incentives in your location.

Florida Energy Office
Florida Department of Agriculture and Consumer Services
Website: freshfromflorida.com/Divisions-Offices/Energy
Email: energy@freshfromflorida.com
Phone: 850-617-7470
Fax: 850-617-7471

Georgia Energy Office
Division of Energy Resources
Georgia Environmental Finance Authority
Website: gefa.georgia.gov/renewable-energy-and-alternative-fuels
Phone: 404-584-1000
Fax: 404-584-1069

Guam Energy Office
Website: guamenergy.com/
Email: energy@ns.gov.gu
Phone: 671-646-4361
Fax: 671-649-1215

Hawaii Energy Office
Department of Business, Economic Development, and Tourism
Website: energy.hawaii.gov/
Email: energyoffice@dbedt.hawaii.gov
Phone: 808-587-3807
Fax: 808-586-2536

Idaho Energy Office
Governor's Office of Energy Resources
Idaho Department of Water Resources
Website: energy.idaho.gov/
Email: staff directory at energy.idaho.gov/contact.htm
Phone: 208-332-1660
Fax: 208-332-1661

Illinois Energy Office
Bureau of Energy & Recycling
Illinois Department of Commerce and Economic Opportunity
Website: illinois.gov/dceo/whyillinois/Key Industries/Energy/Pages/default.aspx
Phone: 217-785-3420

Indiana Energy Office
Office of Energy Development
Website: in.gov/oed/
Email: staff directory at in.gov/oed/2390.htm
Phone: 317-232-8939

Iowa Energy Center
Iowa Economic Development Agency
Website: iowaeconomicdevelopment.com/Programs/Energy
Email: iec@energy.iastate.edu
Phone: 515-725-3000

Kansas Energy Office
Kansas Corporation Commission
Website: kcc.state.ks.us/energy/index.htm
Email: public.affairs@kcc.ks.us
Phone: 785-271-3100

Kentucky Energy Office
Department for Energy Development and Independence
Division on Renewable Energy
Website: energy.ky.gov/renewable/Pages/default.aspx
Email: kenya.stump@ky.gov
Phone: 502-564-7192

Louisiana Energy Office
Technology Assessment Division
Department of Natural Resources
Website: dnr.louisiana.gov/index.cfm?md=pagebuilder&tmp=home&pid=35&ngid=2PO Box 44156
Email: techasmt@la.gov
Phone: 225-342-1399
Fax: 225-342-1397

Maine Energy Office
Efficiency Maine
Website: efficiencymaine.com
Email: efficiencymaine@efficiencymaine.com
Phone: 866-376-2463

Maryland Energy Office
Maryland Energy Administration
Website: energy.maryland.gov
Email: dlinfo_MEA@maryland.gov
Phone: 410-260-7655

Massachusetts Energy Office
Division of Energy Resources
Executive Office of Energy and Environmental Affairs
Website: mass.gov/eea/energy-utilities-clean-tech/
Email: env.internet@state.ma.us
Phone: 617-626-1000

Michigan Energy Office
Michigan Public Service Commission
Department of Licensing and Regulatory Affairs
Website: michigan.gov/mpsc/0,4639,7-159-16393---,00.htmlLansing, MI 48909
Email: mpsc_commissioners@michigan.gov
Phone: 517-241-6180
Fax: 517-241-6181

Minnesota Energy Office
Minnesota Department of Commerce
Website: mn.gov/commerce/energy/topics/clean-energy/Solar/
Email: general.commerce@state.mn.us
Phone: 651-539-1886

Mississippi Energy Office
Mississippi Development Authority
Website: mississippi.org/energy/
Email: staff directory at mississippi.org/energy
/energy-and-natural-resources-team.html
Phone: 601-359-3449

Missouri Energy Office
Environmental Improvement and Energy Re-
sources Authority
Department of Natural Resources
Website: eiera.mo.gov/
Email: eiera@dnr.mo.gov
Phone: 573-751-4919

Montana Energy Office
Energize Montana
Department of Environmental Quality
Website: deq.mt.gov/Energy/Renewable/Solar
Geothermal.mcpx
Email: staff directory at svc.mt.gov/deq/staffdir.asp
Phone: 406-445-2544

Nebraska Energy Office
Website: neo.ne.gov/
Email: energy@nebraska.gov
Phone: 402-471-2867
Fax: Phone: 402-471-3064

Nevada Energy Office
Nevada Governor's Office of Energy
Website: energy.nv.gov
Phone: 775-687-1850
Fax: 775-687-1869
Email: staff directory at energy.nv.gov/About/Staff/

New Hampshire Energy Office
NH Office of Energy and Planning
Website: nh.gov/oep/index.htm
Email: oepinfo@nh.gov
Phone: 603-271-2155

New Jersey Energy Office
Office of Clean Energy
New Jersey Board of Public Utilities
Website: bpu.state.nj.us
Phone: 8008-624-0241

New Mexico Energy Office
Energy Conservation and Management Division
Energy, Minerals, and Natural Resources De-
partment
Website: emnrd.state.nm.us/ECMD/
Email: staff directory at emnrd.state.nm.us
/ECMD/STAFF/contact.html
Phone: 505-476-3318

New York Energy Office
New York State Energy Research and Develop-
ment Authority
Website: nyserda.ny.gov/Energy-Efficiency-and
-Renewable-Programs/Renewables.aspx
Email: staff directory at nyserda.ny.gov/About
/Contacts.aspx
Phone: 518-862-1090

North Carolina Energy Office
State Energy Office
North Carolina Department of Environment
and Natural Resources
Website: energync.net/about-us/state-energy
-office
Phone: 919-707-9238

North Dakota Energy Office
Division of Community Services
North Dakota Department of Commerce
Website: communityservices.nd.gov/renewable
energyprograms/RenewableEnergyProgram/
PO Box 2057
Email: ndicinfo@nd.gov
Phone: 701-328-3722

Northern Mariana Islands Energy Office
Commonwealth of the Northern Mariana
Islands
Website: cnmienergy.com
Phone: 670-664-4480

Ohio Energy Office
Energy Assistance Programs
Ohio Development Services Agency
Website: development.ohio.gov/is/is_energy
assist.htm
Email: Inquiries at development.ohio.gov
/contactus/ContactInfo.aspx
Phone: 800-848-1300

Oklahoma Energy Office
Division of Community Development
Oklahoma Department of Commerce
Website: okcommerce.gov/state-energy-office/
Email: kylah_mcnabb@okcommerce.gov
Phone: 405-815-5249

Oregon Energy Office
Oregon Department of Energy
Website: oregon.gov/ENERGY/RENEW/pages
/index.aspx
Email: energy.in.internet@odoe.state.or.us
Phone: 503-378-4040

Pennsylvania Energy Office
Office of Pollution Prevention and Energy
Assistance
Department of Environmental Protection
Website: depweb.state.pa.us/energy/cwp/view
.asp?a=3&q=482723
Email: ra-epenergy@pa.gov
Phone: 717-783-2300

Puerto Rico Energy Office
Energy Affairs Administration
Website: prgef.com
Email: infoprgef@prlohacemejor.com
Phone: 787-999-2000
Fax: Phone: 787-999-2246

Rhode Island Energy Office
Office of Energy Resources
Website: riseo.state.ri.us/
Email: energy.resources@energy.ri.gov
Phone: 401-574-9100
Fax: 401-574-9125

South Carolina Energy Office
South Carolina Budget and Control Board
Website: energy.sc.gov
Email: staff directory at energy.sc.gov/contact
Phone: 803-737-8030

South Dakota Energy Office
Energy Management Office
South Dakota Bureau of Administration
Website: boa.sd.gov/divisions/energy/
Email: boageneralinformation@state.sd.us
Phone: 605-773-3899
Fax: 605-773-5980

Tennessee Energy Office
Department of Environment & Conservation
Office of Energy Programs
Website: tn.gov/environment/energy.shtml
Email: ask.tdec@tn.gov
Phone: 615-891-8332

Texas Energy Office
State Energy Conservation Office
Texas Comptroller of Public Accounts
Website: seco.cpa.state.tx.us/
Email: seco@cpa.state.tx.us
Phone: 512-463-1931
Fax: (512) 475-2569

Utah Energy Office
Office of Energy Development
Website: energy.utah.gov
Phone: 801-538-8732
Fax: 855-271-4373

Vermont Energy Office
Efficiency Vermont
Website: efficiencyvermont.com
Email: webform at efficiencyvermont.com/
About-Us/Contact-Us
Phone: 888-921-5990
Fax: 802-658-1643

Virginia Energy Office
Division of Energy
Department of Mines, Minerals & Energy
Website: dmme.virginia.gov/DE/DELanding
Page.shtml
Email: dmmeinfo@dmme.virginia.gov
Phone: 804-692-3200
Fax: 804-692-3237

Virgin Islands Energy Office
Website: vienergy.org/
Email: don.buchanan@eo.vi.gov
Phone: 340-713-8436

Washington Energy Office
Washington State University Extension Energy
Program
Website: energy.wsu.edu/RenewableEnergy.aspx
Email: info@energy.wsu.edu (Olympia);
hacklanderm@energy.wsu.edu
Phone: 360-956-2000 (Olympia); 509-443-4355
(Spokane)
Fax: 360-956-2217 (Olympia); 509-474-1954
(Spokane)

West Virginia Energy Office
Energy Efficiency Program
West Virginia Development Office
Website: wvcommerce.org/energy/default.aspx
Phone: 304-558-2234

Wisconsin State Energy Office
Website: energyindependence.wi.gov/
Email: seo@wisconsin.gov
Phone: 608-261-6609

Wyoming Energy Office
Wyoming Business Council
Website: wyomingbusiness.org/energy
Email: info.wbc@wyo.gov
Phone: 307-777-2800
Fax: 307-777-2837

Glossary

A

AC: alternating current, electricity that changes voltage periodically, typically 60 times a second (or 50 in Europe). This kind of electricity is easier to move.

activated stand life: the period of time, at a specified temperature, that a battery can be left stored in the charged condition before its capacity fails.

active solar: any solar scheme employing pumps and controls that use power while harvesting solar energy.

A-frame: a building that looks like the capital letter A in cross-section.

air lock: two doors with space between, like a mud room, to keep the weather outside.

alternating current: AC electricity that changes voltage periodically, typically 60 times a second.

alternative energy: "voodoo" energy not purchased from a power company, usually coming from photovoltaic, micro-hydro, or wind.

ambient: the prevailing temperature, usually outdoors.

amorphous silicon: a type of PV cell manufactured without a crystalline structure. Compare with single-crystal and multi- (or poly-) crystalline silicon.

ampere: an instantaneous measure of the flow of electric current; abbreviated and more commonly spoken of as an "amp."

amp-hour: a one-ampere flow of electrical current for one hour; a measure of electrical quantity; two 60-watt 120-volt bulbs burning for one hour consume one amp-hour.

angle of incidence: the angle at which a ray of light (usually sunlight) strikes a planar surface (usually of a PV module). Angles of incidence close to perpendicularity (90°) are desirable.

anode: the positive electrode in an electro-chemical cell (battery) toward which current flows; the earth ground in a cathodic protection system.

antifreeze: a chemical, usually liquid and often toxic, that keeps things from freezing.

array: an orderly collection, usually of photovoltaic modules connected electrically and mechanically secure; array current: the amperage produced by an array in full sun.

avoided cost: the amount utilities must pay for independently produced power; in theory, this was to be the whole cost, including capital share to produce peak demand power, but over the years supply-side weaseling redefined it to be something more like the cost of the fuel the utility avoided burning.

azimuth: horizontal angle measured clockwise from true north; the equator is at 90°.

B

backup: a secondary source of energy to pick up the slack when the primary source is inadequate. In alternatively powered homes, fossil fuel generators are often used as "backups" when extra power is required to run power tools or when the primary sources—sun, wind, water—are not providing sufficient energy.

balance of system: (BOS) equipment that controls the flow of electricity during generation and storage baseline: a statistical term for a starting point; the "before" in a before-and-after energy conservation analysis.

baseload: the smallest amount of electricity required to keep utility customers operating at the time of lowest demand; a utility's minimum load.

battery: a collection of cells that store electrical energy; each cell converts chemical energy into electricity or vice versa, and is interconnected with other cells to form a battery for storing useful quantities of electricity.

battery capacity: the total number of ampere-hours that can be withdrawn from a fully charged battery, usually over a standard period.

battery cycle life: the number of cycles that a battery can sustain before failing.

berm: earth mounded in an artificial hill.

bioregion: an area, usually fairly large, with generally homogeneous flora and fauna.

biosphere: the thin layer of water, soil, and air that supports all known life on Earth.

blackwater: what gets flushed down the toilet.

blocking diode: a diode that prevents loss of energy to an inactive PV array (rarely used with modern charge controllers).

boneyard: a peculiar location at Real Goods where "experienced" products may be had for ridiculously low prices.

Btu: British thermal unit, the amount of heat required to raise the temperature of 1 pound of water 1 degree Fahrenheit. 3,411 Btus equals one kilowatt-hour.

bus bar: the point where all energy sources and loads connect to each other; often a metal bar with connections on it.

bus bar cost: the average cost of electricity delivered to the customer's distribution point.

buy-back agreement or contract: an agreement between the utility and a customer that any excess electricity generated by the customer will be bought back for an agreed-upon price.

C

cathode: the negative electrode in an electro-chemical cell.

cell: a unit for storing or harvesting energy. In a battery, a cell is a single chemical storage unit consisting of electrodes and electrolyte, typically producing 1.5 volts; several cells are usually arranged inside a single container called a battery. Flashlight batteries are really flashlight cells. A photovoltaic cell is a single assembly of doped silicon and electrical contacts that allow it to take advantage of the photovoltaic effect, typically producing 0.5 volts; several PV cells are usually connected together and packaged as a module.

CF: compact fluorescent, a modern form of lightbulb using an integral ballast.

CFCs: chlorinated fluorocarbons, an industrial solvent and material widely used until implicated as a cause of ozone depletion in the atmosphere.

charge controller: device for managing the charging rate and state of charge of a battery bank.

charge rate: the rate at which a battery is recharged, expressed as a ratio of battery capacity to charging current flow.

clear-cutting: a forestry practice, cutting all trees in a relatively large plot.

cloud enhancement: the increase in sunlight due to direct rays plus refracted or reflected sunlight from partial cloud cover.

compact fluorescent: a modern energy-efficient form of light bulb using an integral ballast.

compost: the process by which organic materials break down, or the materials in the process of being broken down.

concentrator: mirror or lens-like additions to a PV array that focus sunlight on smaller cells; a very promising way to improve PV yield.

conductance: a material's ability to allow electricity to flow through it; gold has very high conductance.

controller terminology:

adjustable set point: allows adjustment of voltage disconnect levels.

high-voltage disconnect: the battery voltage at which the charge controller disconnects the batteries from the array to prevent over-charging.

low-voltage disconnect: the voltage at which the controller disconnects the batteries to prevent overdischarging.

low-voltage warning: a buzzer or light that indicates low battery voltage.

maximum power tracking: a circuit that maintains array voltage for maximal current.

multistage controller: a unit that allows multi-level control of battery charging or loading.

reverse current protection: prevents current flow from battery to array.

single-stage controller: a unit with only one level of control for charging or load control.

temperature compensation: a circuit that adjusts setpoints to ambient temperature in order to optimize charge.

conversion efficiency: the ratio of energy input to energy output across the conversion boundary. For example, batteries typically are able to store and provide 90% of the charging energy applied and are said to have a 90% energy efficiency.

cookie-cutter houses: houses all alike and all-in-a-row. Daly City, south of San Francisco, is a particularly depressing example. Runs in direct contradiction to our third guiding principle, "encourage diversity."

core/coil-ballasted: the materials-rich device required to drive some fluorescent lights; usually contains radioactive Americium (see electronic ballasts).

cost-effectiveness: an economic measure of the worthiness of an investment; if an innovative solution costs less than a conventional alternative, it is more cost effective.

cross-section: a "view" or drawing of a slice through a structure.

cross-ventilation: an arrangement of openings allowing air to pass through a structure.

crystalline silicon: the material from which most photovoltaic cells are made; in a single crystal cell, the entire cell is a slice of a single crystal of silicon, while a multicrystalline cell is cut from a block of smaller (centimeter-sized) crystals. The larger the crystal, the more exacting and expensive the manufacturing process.

cut-in: the condition at which a control connects its device.

cutoff voltage: in a charge controller, the voltage at which the array is disconnected from the battery to prevent overcharging.

cut-out: the condition at which a control interrupts the action.

cycle: in a battery, from a state of complete charge through discharge and recharge back to a fully charged state.

D

days of autonomy: the length of time (in days) that a system's storage can supply normal requirements without replenishment from a source; also called days of storage.

DC: direct current, the complement of AC, or alternating current, presents one unvarying voltage to a load.

deep cycle: a battery type manufactured to sustain many cycles of deep discharge that are in excess of 50% of total capacity.

degree-days: a term used to calculate heating and cooling loads; the sum, taken over an average year, of the lowest (for heating) or highest (for cooling) ambient daily temperatures. Example: if the target is 68° and the ambient on a given day is 58°, this would account for 10 degree-days.

depth of discharge: the percent of rated capacity that has been withdrawn from the battery; also called DOD.

design month: the month in which the combination of insolation and loading require the maximum array output.

diode: an electrical component that permits current to pass in only one direction.

direct current: the complement of AC, or alternating current, presents one unvarying voltage to a load.

discharge: electrical term for withdrawing energy from a storage system.

discharge rate: the rate at which current is withdrawn from a battery expressed as a ratio to the battery's capacity; also known as C rate.

disconnect: a switch or other control used to complete or interrupt an electrical flow between components.

doping, dopant: small, minutely controlled amounts of specific chemicals introduced into the semiconductor matrix to control the density and probabilistic movement of free electrons.

downhole: a piece of equipment, usually a pump, that is lowered down the hole (the well or shaft) to do its work.

drip irrigation: a technique that precisely delivers measured amounts of water through small tubes; an exceedingly efficient way to water plants.

dry cell: a cell with captive electrolyte.

duty cycle: the ratio between active time and total time; used to describe the operating regime of an appliance in an electrical system.

duty rating: the amount of time that an appliance can be run at its full rated output before failure can be expected.

E

earthship: a rammed-earth structure based on tires filled with tamped earth; the term was coined by Michael Reynolds.

Eco Desk: an ecology information resource maintained by many ecologically minded companies.

edison base: a bulb base designed by (a) Enrico Fermi, (b) John Schaeffer, (c) Thomas Edison, (d) none of the above. The familiar standard residential light bulb base.

efficiency: a mathematical measure of actual as a percentage of the theoretical best. See conversion efficiency.

electrolyte: the chemical medium, usually liquid, in a battery that conveys charge between the positive and negative electrodes; in a lead-acid battery, the electrodes are lead and the electrolyte is acid.

electromagnetic radiation: EMR, the invisible field around an electric device. Not much is known about the effects of EMR, but it makes many of us nervous.

electronic ballasts: an improvement over core/coil ballasts, used to drive compact fluorescent lamps; contains no radioactivity.

embodied: of energy, meaning literally the amount of energy required to produce an object in its present form. Example: an inflated balloon's embodied energy includes the energy required to blow it up.

EMR: electromagnetic radiation, the invisible field around an electric device. Not much is known about the effects of EMR, but it makes many of us nervous.

energy density: the ratio of stored energy to storage volume or weight.

energy-efficient: one of the best ways to use energy to accomplish a task; for example, heating with electricity is never energy-efficient, while lighting with compact fluorescents is.

equalizing: periodic overcharging of batteries to make sure that all cells are reaching a good state of charge.

externalities: considerations, often subtle or remote, that should be accounted for when evaluating a process or product, but usually are not. For example, externalities for a power plant may include downwind particulate fallout and acid rain, damage to life forms in the cooling water intake and effluent streams, and many other factors.

F

fail-safe: a system designed in such a way that it will always fail into a safe condition.

feng shui: an Asian system of placement that pays special attention to wind, water, and the cardinal directions.

ferro-cement: a construction technique; an armature of iron contained in a cement body, often a wall, slab, or tank.

fill factor: of a photovoltaic module's I-V (current/voltage) curve, this number expresses the product of the open circuit voltage and the short-circuit current, and is therefore a measure of the "squareness" of the I-V curve's shape.

firebox: the structure within which combustion takes place.

fixed-tilt array: a PV array set in a fixed position.

flat-plate array: a PV array consisting of non-concentrating modules.

float charge: a charge applied to a battery equal to or slightly larger than the battery's natural tendency to self-discharge.

FNC: fiber-nickel-cadmium, a new battery technology.

frequency: of a wave, the number of peaks in a period. For example, alternating current presents 60 peaks per second, so its frequency is 60 hertz. Hertz is the standard unit for frequency when the period in question is one second.

G

gassifier: a heating device which burns so hotly that the fuel sublimes directly from its solid to its gaseous state and burns very cleanly.

gassing: when a battery is charged, gasses are often given out; also called outgassing.

golf cart batteries: industrial batteries tolerant of deep cycling, often used in mobile vehicles.

gotcha!s: an unexpected outcome or effect, or the points at which, no matter how hard you wriggle, you can't escape.

gravity-fed: water storage far enough above the point of use (usually 50 feet) so that the weight of the water provides sufficient pressure.

greywater: all other household effluents besides blackwater (toilet water); greywater may be reused with much less processing than blackwater.

grid: a utility term for the network of transmission lines that distribute electricity from a variety of sources across a large area.

grid-connected system: a house, office, or other electrical system that can draw its energy from the grid; although usually grid-power-consumers, grid-connected systems can provide power to the grid.

groundwater: as distinct from water pumped up from the depths, groundwater is run off from precipitation, agriculture, or other sources.

H

heat exchanger: device that passes heat from one substance to another; in a solar hot water heater, for example, the heat exchanger takes heat harvested by a fluid circulating through the solar panel and transfers it to domestic hot water.

high-tech glass: window constructions made of two sheets of glass, sometimes treated with a metallic deposition, sealed together hermetically, with the cavity filled by an inert gas and, often, a further plastic membrane. High-tech glass can have an R-value as high as 10.

homeschooling: educating children at home instead of entrusting them to public or private schools; a growing trend quite often linked to off-the-grid-powered homes.

homestead: the house and surrounding lands.

homesteaders: people who consciously and intentionally develop their homestead.

hot tub: a quasi-religious object in California; a large bathtub for several people at once; an energy hog of serious proportions.

house current: in the United States, 117 volts root mean square of alternating current, plus or minus 7 volts; nominally 110-volt power; what comes out of most wall outlets.

HVAC: heating, ventilation, and air conditioning; space conditioning.

hydro turbine: a device that converts a stream of water into rotational energy.

hydrometer: tool used to measure the specific gravity of a liquid.

hydronic: contraction of hydro and electronic, usually applied to radiant in-floor heating systems and their associated sensors and pumps.

hysteresis: the lag between cause and effect, between stimulus and response.

incandescent bulb: a light source that produces light by heating a filament until it emits photons—quite an energy-intensive task.

incident solar radiation: or insolation, the amount of sunlight falling on a place.

indigenous plantings: gardening with plants native to the bioregion.

inductive transformer/rectifier: the little transformer device that powers many household appliances; an "energy criminal" that takes an unreasonably large amount of alternating electricity (house current) and converts it into a much smaller amount of current with different properties; for example, much lower-voltage direct current.

infiltration: air, at ambient temperature, blowing through cracks and holes in a house wall and spoiling the space conditioning.

infrared: light just outside the visible spectrum, usually associated with heat radiation.

infrastructure: a buzz word for the underpinnings of civilization; roads, water mains, power and phone lines, fire suppression, ambulance, education, and governmental services are all infrastructure. *Infra* is Latin for "beneath." In a more technical sense, the repair infrastructure is local existence of repair personnel and parts for a given technology.

insolation: a word coined from incident solar radiation, the amount of sunlight falling on a place.

insulation: a material that keeps energy from crossing from one place to another. On electrical wire, it is the plastic or rubber that covers the conductor. In a building, insulation makes the walls, floor, and roof more resistant to the outside (ambient) temperature.

Integrated Resource Planning: an effort by the utility industry to consider all resources and requirements in order to produce electricity as efficiently as possible.

interconnect: to connect two systems, often an independent power producer and the grid; see also intertie.

interface: the point where two different flows or energies interact; for example, a power system's interface with the human world is manifested as meters, which show system status, and controls, with which that status can be manipulated.

internal combustion engines: gasoline engines, typically in automobiles, small stand-alone devices like chain saws, lawn mowers, and generators.

intertie: the electrical connection between an independent power producer—for example, a PV-powered household—and the utility's distribution lines, in such a way that each can supply or draw from the other.

inverter: the electrical device that changes direct current into alternating current.

irradiance: the instantaneous solar radiation incident on a surface; usually expressed in Langleys (a small amount) or in kilowatts per square meter. The definition of "one sun" of irradiance is one kilowatt per square meter.

irreverence: the measure, difficult to quantify, of the seriousness of Real Goods techs when talking with utility suits.

IRP: See Integrated Resource Planning.

I-V curve: a plot of current against voltage to show the operating characteristics of a photovoltaic cell, module, or array.

K

kilowatt: 1,000 watts, a measure of instantaneous work. Ten 100-watt bulbs require a kilowatt of energy to light up.

kilowatt-hour: the standard measure of household electrical energy use. If the 10 bulbs left unfrugally burning in the preceding example are on for an hour, they consume 1 kilowatt-hour of electricity.

L

landfill: another word for dump.

Langley: the unit of solar irradiance; 1 gram-calorie per square centimeter.

lead-acid: the standard type of battery for use in home energy systems and automobiles.

LED: light-emitting diode. A very efficient source of electrical lighting, typically lasting 50,000 to 100,000 hours.

life: You expect an answer here? Well, when speaking of electrical systems, this term is used to quantify the time the system can be expected to function at or above a specified performance level.

life-cycle cost: the estimated cost of owning and operating a system over its useful life.

line extensions: what the power company does to bring their power lines to the consumer.

line-tied system: an electrical system connected to the power lines, usually having domestic power generating capacity and the ability to draw power from the grid or return power to the grid, depending on load and generator status.

load: an electrical device, or the amount of energy consumed by such a device.

load circuit: the wiring that provides the path for the current that powers the load.

load current: expressed in amps, the current required by the device to operate.

low-emissivity: applied to high-tech windows, meaning that infrared or heat energy will not pass back out through the glass.

low-flush: a toilet using a smaller amount (usually about 6 quarts) of water to accomplish its function.

low pressure: usually of water, meaning that the head, or pressurization, is relatively small.

low-voltage: usually another term for 12- or 24-volt direct current.

M

maintenance-free battery: a battery to which water cannot be added to maintain electrolyte volume. All batteries require routine inspection and maintenance.

maximum power point: the point at which a power conditioner continuously controls PV source voltage in order to hold it at its maximum output current.

ME: mechanical engineer; the engineers who usually work with heating and cooling, elevators, and the other mechanical devices in a large building.

meteorological: pertaining to weather; meteorology is the study of weather.

microclimate: the climate in a small area, sometimes as small as a garden or the interior of a house. Climate is distinct from weather in that it speaks for trends taken over a period of at least a year, while weather describes immediate conditions.

micro-hydro: small hydro (falling water) generation.

millennia: 1,000 years.

milliamp: one thousandth (1/1000) of an ampere.

module: a manufactured panel of photovoltaic cells. A module typically houses 36 cells in an aluminum frame covered with glass or acrylic, organizes their wiring, and provides a junction box for connection between itself, other modules in the array, and the system.

N

naturopathic: a form of medicine devoted to natural remedies and procedures.

net metering: a desirable form of buy-back agreement in which the line-tied house's electric meter turns in the utility's favor when grid power is being drawn, and in the system owner's favor when the house generation exceeds its needs and electricity is flowing into the grid. At the end of the payment period, when the meter is read, the system owner pays (or is paid by) the utility depending on the net metering.

NEC: the National Electrical Code, guidelines for all types of electrical installations including (since 1984) PV systems.

NiCad: slang for nickel-cadmium, a form of chemical storage often used in rechargeable batteries.

nominal voltage: the terminal voltage of a cell or battery discharging at a specified rate and at a specified temperature; in normal systems, this is usually a multiple of 12.

nontoxic: having no known poisonous qualities.

normal operating cell temperature: defined as the standard operating temperature of a PV module at 800 W/m°, 20° C ambient, 1 meter per second wind speed; used to estimate the nominal operating temperature of a module in its working environment.

N-type silicon: silicon doped (containing impurities), which gives the lattice a net negative charge, repelling electrons.

O

off-peak energy: electricity during the baseload period, which is usually cheaper. Utilities often must keep generators turning, and are eager to find users during these periods, and so sell off-peak energy for less.

off-peak kilowatt: a kilowatt hour of off-peak energy.

off-the-grid: not connected to the power lines, energy self-sufficient.

ohm: the basic unit of electrical resistance; I=RV, or Current (amperes) equals Resistance (ohms) times Voltage (volts).

on-line: connected to the system, ready for work.

on-the-grid: where most of America lives and works, connected to a continent-spanning web of electrical distribution lines.

open-circuit voltage: the maximum voltage measurable at the terminals of a photovoltaic cell, module, or array with no load applied.

operating point: the current and voltage that a module or array produces under load, as determined by the I-V curve.

order of magnitude: multiplied or divided by 10; 100 is an order of magnitude smaller than 1,000, and an order of magnitude larger than 10.

orientation: placement with respect to the cardinal directions north, east, south, and west; azimuth is the measure of orientation.

outgas: of any material, the production of gasses; batteries outgas during charging; new synthetic rugs outgas when struck by sunlight, or when warm, or whenever they feel like it.

overcharge: forcing current into a fully charged battery; a bad idea except during equalization.

overcurrent: too much current for the wiring; overcurrent protection, in the form of fuses and circuit breakers, guards against this.

owner-builder: one of the few printable things building inspectors call people who build their own homes.

P

panel: any flat modular structure; solar panels may collect solar energy by many means; a number of photovoltaic modules may be assembled into a panel using a mechanical frame, but this should more properly be called an array or subarray.

parallel: connecting the like poles to like in an electrical circuit, so plus connects to plus and minus to minus; this arrangement increases current without affecting voltage.

particulates: particles that are so small that they persist in suspension in air or water.

passive solar: a shelter that maintains a comfortable inside temperature simply by accepting, storing, and preserving the heat from sunlight.

passively heated: a shelter that has its space heated by the sun without using any other energy.

patch-cutting: clear-cutting (cutting all trees) on a small scale usually less than an acre.

pathetic fallacy: attributing human motivations to inanimate objects or lower animals.

payback: the time it takes to recoup the cost of improved technology as compared to the conventional solution. Payback on a compact fluorescent bulb (as compared to an incandescent bulb) may take a year or two, but over the whole life of the CF, the savings probably will exceed the original cost of the bulb, and payback will take place several times over.

peak demand: the largest amount of electricity demanded by a utility's customers; typically, peak demand happens in early afternoon on the hottest weekday of the year.

peak kilowatt: a kilowatt-hour of electricity taken during peak demand, usually the most expensive electricity money can buy.

peak load: the same as peak demand but on a smaller scale, the maximum load demanded of a single system.

peak power current: the amperage produced by a photovoltaic module operating at the "knee" of its I-V curve.

peak sun hours: the equivalent number of hours per day when solar irradiance averages one sun (1 kW/m°). "Six peak sun hours" means that the energy received during total daylight hours equals the energy that would have been received if the sun had shone for six hours at a rate of 1,000 watts per square meter.

peak watt: the manufacturer's measure of the best possible output of a module under ideal laboratory conditions.

Pelton wheel: a special turbine, designed by someone named Pelton, for converting flowing water into rotational energy.

periodic table of elements: a chart showing the chemical elements organized by the number of protons in their nuclei and the number of electrons in their outer, or valence, band.

PG&E: Pacific Gas and Electric, the local and sometimes beloved utility for much of northern California.

phantom loads: "energy criminals" that are on even when you turn them off: instant-on TVs, microwaves with clocks; symptomatic of impatience and our sloppy preference for immediacy over efficiency.

photon: the theoretical particle used to explain light.

photophobic: fear of light (or preference for darkness), usually used of insects and animals. The opposite, phototropic, means light-seeking.

photovoltaic cell: the proper name for a device manufactured to pump electricity when light falls on it.

photovoltaics: PVs or modules that utilize the photovoltaic effect to generate useable amounts of electricity.

photovoltaic system: the modules, controls, storage, and other components that constitute a stand-alone solar energy system.

plates: the thin pieces of metal or other material used to collect electrical energy in a battery.

plug-loads: the appliances and other devices plugged into a power system.

plutonium: a particularly nasty radioactive material used in nuclear generation of electricity. One atom is enough to kill you.

pn-junction: the plane within a photovoltaic cell where the positively and negatively doped silicon layers meet.

pocket plate: a plate for a battery in which active materials are held in a perforated metal pocket on a support strip.

pollution: any dumping of toxic or unpleasant materials into air or water.

polyurethane: a long-chain carbon molecule, a good basis for sealants, paints, and plastics.

power: kinetic, or moving energy, actually performing work; in an electrical system, power is measured in watts.

power-conditioning equipment: electrical devices that change electrical forms (an inverter is an example) or assure that the electricity is of the correct form and reliability for the equipment consuming it; a surge protector is another example.

power density: ratio of a battery's rated power available to its volume (in liters) or weight (in kilograms).

PUC: Public Utilities Commission; many states call it something else, but this is the agency responsible for regulating utility rates and practices.

PURPA: this 1978 legislation, the *Public Utility Regulatory Policy Act*, requires utilities to purchase power from anyone at the utility's avoided cost.

PVs: photovoltaic modules.

R

radioactive material: a substance which, left to itself, sheds tiny, highly energetic pieces that put anyone nearby at great risk. Plutonium is one of these. Radioactive materials remain active indefinitely, but the time over which they are active is measured in terms of half-life, the time it takes them to become half as active as they are now; plutonium's half-life is a little over 22,000 years.

ram pump: a water-pumping machine that uses a water-hammer effect (based on the inertia of flowing water) to lift water.

rated battery capacity: manufacturer's term indicating the maximum energy that can be withdrawn from a battery at a standard specified rate.

rated module current: manufacturer's indication of module current under standard laboratory test conditions.

renewable energy: an energy source that renews itself without effort; fossil fuels, once consumed, are gone forever, while solar energy is renewable in that the sun we harvest today has no effect on the sun we can harvest tomorrow.

renewables: shorthand term for renewable energy or materials sources.

resistance: the ability of a substance to resist electrical flow; in electricity, resistance is measured in ohms.

retrofit: to install new equipment in a structure that was not originally designed for it. For example, we may retrofit a lamp with a compact fluorescent bulb, but the new bulb's shape may not fit well with the lamp's design.

romex: an electrician's term for common two-conductor-with-ground wire, the kind houses are wired with.

root mean square: RMS, the effective voltage of alternating current, usually about 70% (the square root of two over two) of the peak voltage. House current typically has an RMS of 117 volts and a peak voltage of 167 volts.

RPM: rotations per minute.

R-value: Resistance value, used specifically of materials used for insulating structures. Fiberglass insulation three inches thick has an R-value of 13.

S

seasonal depth of discharge: an adjustment for long-term seasonal battery discharge resulting in a smaller array and a battery bank matched to low-insolation season needs.

secondary battery: a battery that can be repeatedly discharged and fully recharged.

self-discharge: the tendency of a battery to lose its charge through internal chemical activity.

semiconductor: the chief ingredient in a photovoltaic cell, a normal insulating substance that conducts electricity under certain circumstances.

series connection: wiring devices with alternating poles, plus to minus, plus to minus; this arrangement increases voltage (potential) without increasing current.

set-back thermostat: combines a clock and a thermostat so that a zone (like a bedroom) may be kept comfortable only when in use.

setpoint: electrical condition, usually voltage, at which controls are adjusted to change their action.

shallow-cycle battery: like an automotive battery, designed to be kept nearly fully charged; such batteries perform poorly when discharged by more than 25% of their capacity.

shelf life: the period of time a device can be expected to be stored and still perform to specifications.

short circuit: an electrical path that connects opposite sides of a source without any appreciable load, thereby allowing maximum (and possibly disastrous) current flow.

short-circuit current: current produced by a photovoltaic cell, module, or array when its output terminals are connected to each other, or "short-circuited."

showerhead: in common usage, a device for wasting energy by using too much hot water; in the Real Goods home, low-flow showerheads prevent this undesirable result.

silicon: one of the most abundant elements on the planet, commonly found as sand and used to make photovoltaic cells.

single-crystal silicon: silicon carefully melted and grown in large boules, then sliced and treated to become the most efficient photovoltaic cells.

slow-blow: a fuse that tolerates a degree of overcurrent momentarily; a good choice for motors and other devices that require initial power surges to get rolling.

slow paced: a description of the life of a Real Goods employee…not!

solar aperture: the opening to the south of a site (in the Northern Hemisphere) across which the sun passes; trees, mountains, and buildings may narrow the aperture, which also changes with the season.

solar cell: see photovoltaic cell.

solar fraction: the fraction of electricity that may be reasonable harvested from sun falling on a site. The solar fraction will be less in a foggy or cloudy site, or one with a narrower solar aperture, than an open, sunny site.

solar hot water heating: direct or indirect use of heat taken from the sun to heat domestic hot water.

solar oven: simply a box with a glass front and, optionally, reflectors and reflector coated walls, which heats up in the sun sufficiently to cook food.

solar panels: any kind of flat devices placed in the sun to harvest solar energy.

solar resource: the amount of insolation a site receives, normally expressed in kilowatt-hours per square meter per day.

specific gravity: the relative density of a substance compared to water (for liquids and solids) or air (for gases) Water is defined as 1.0; a fully charged sulfuric acid electrolyte might be as dense as 1.30, or 30% denser than water. Specific gravity is measured with a hydrometer.

stand-alone: a system, sometimes a home, that requires no imported energy.

stand-by: a device kept for times when the primary device is unable to perform; a stand-by generator is the same as a backup generator.

starved electrolyte cell: a battery cell containing little or no free fluid electrolyte.

state of charge: the real amount of energy stored in a battery as a percentage of its total rated capacity.

state-of-the-art: a term beloved by technoids to express that this is the hottest thing since sliced bread.

stratification: in a battery, when the electrolyte acid concentration varies in layers from top to bottom; seldom a problem in vehicle batteries, due to vehicular motion and vibration, this can be a problem with static batteries, and can be corrected by periodic equalization.

stepwise: a little at a time, incrementally.

subarray: part of an array, usually photovoltaic, wired to be controlled separately.

sulfating: formation of lead-sulfate crystals on the plates of a lead-acid battery, which can cause permanent damage to the battery.

superinsulated: using as much insulation as possible, usually R-50 and above.

surge capacity: the ability of a battery or inverter to sustain a short current surge in excess of rated capacity in order to start a device that requires an initial surge current.

sustainable: material or energy sources that, if managed carefully, will provide at current levels indefinitely. A theoretical example: redwood is sustainable if it is harvested sparingly (large takings and exportation to Japan not allowed) and if every tree taken is replaced with another redwood. Sustainability can be, and usually is, abused for profit by playing it like a shell game; by planting, for example, a fast-growing fir in place of the harvested redwood.

system availability: the probability or percentage of time that an energy storage system will be able to meet load demand fully.

temperature compensation: an allowance made by a charge controller to match battery charging to battery temperature.

temperature correction: applied to derive true storage capacity using battery nameplate capacity and temperature; batteries are rated at 20°C.

therm: a quantity of natural gas, 100 cubic feet; roughly 100,000 Btus of potential heat.

thermal mass: solid, usually masonry volumes inside a structure that absorb heat, then radiate it slowly when the surrounding air falls below their temperature.

thermoelectric: producing heat using electricity; a bad idea.

thermography: photography of heat loss, usually with a special video camera sensitive to the far end of the infrared spectrum.

thermosiphon: a circulation system that takes advantage of the fact that warmer substances rise. By placing the solar collector of a solar hotwater system below the tank, thermosiphoning takes care of circulating the hot water, and pumping is not required.

thin-film module: an inexpensive way of manufacturing photovoltaic modules; thin-film modules typically are less efficient than single-crystal or multi-crystal devices; also called amorphous silicon modules.

tilt angle: measures the angle of a panel from the horizontal.

tracker: a device that follows the sun and keeps the panel perpendicular to it.

transformer: a simple electrical device that changes the voltage of alternating current; most transformers are inductive, which means they set up a field around themselves, which is often a costly thing to do.

transparent energy system: a system that looks and acts like a conventional grid-connected home system, but is independent.

trickle charge: a small current intended to maintain an inactive battery in a fully charged condition.

troubleshoot: a form of recreation not unlike riding to hounds, in which the technician attempts to find, catch, and eliminate the trouble in a system.

tungsten filament: the small coil in a light bulb that glows hotly and brightly when electricity passes through it.

turbine: a vaned wheel over which a rapidly moving liquid or gas is passed, causing the wheel to spin; a device for converting flow to rotational energy.

turnkey: the jail warden; more commonly in our context, a system that is ready for the owner-occupant from the first time he or she turns the key in the front-door lock.

TV: a device for wasting time and scrambling the brain.

two-by-fours: standard building members, originally 2 by 4 inches, now 1.5 inches by 3.5 inches; often referred to as "sticks".

ultrafilter: of water, to remove all particulates and impurities down to the submicron range, about the size of giardia, and larger viruses.

uninterruptible power supply: an energy system providing ultrareliable power; essential for computers, aircraft guidance, medical, and other systems; also known as a UPS.

varistor: a voltage-dependent variable resistor, normally used to protect sensitive equipment from spikes (like lightning strike) by diverting the energy to ground.

VDTs: video display terminals, like televisions and computer screens.

vented cell: a battery cell designed with a vent mechanism for expelling gasses during charging.

volt: measure of electrical potential: 110-volt house electricity has more potential to do work than an equal flow of 12-volt electricity.

voltage drop: lost potential due to wire resistance over distance.

watt: the standard unit of electrical power; 1 ampere of current flowing with 1 volt of potential; 746 watts make 1 horsepower.

watt-hours: 1 watt for 1 hour. A 15-watt compact fluorescent consumes 15 of these in 60 minutes.

waveform: the characteristic trace of voltage over time of an alternating current when viewed on an oscilloscope; typically a smooth sine wave, although primitive inverters supply square or modified square waveforms.

wet shelf life: the time that an electrolyte-filled battery can remain unused in the charged condition before dropping below its nominal performance level.

APPENDIX

wheatgrass: a singularly delicious potation made by squeezing young wheat sprouts; said to promote purity in the digestive tract.

whole-life cost analysis: an economic procedure for evaluating all the costs of an activity from cradle to grave, that is, from extraction or culture through manufacture and use, then back to the natural state; a very difficult thing to accomplish with great accuracy, but a very instructive reckoning nonetheless.

windchill: a factor calculated based on temperature and wind speed that expresses the fact that a given ambient temperature feels colder when the wind is blowing.

wind spinners: fond name for wind machines, devices that turn wind into usable energy.

Endnotes

Chapter 1

Many of the books referenced in these notes are available at realgoods.com. Some—Meadows, et al., *Limits to Growth*; Heinberg, *The Party's Over*; and Ackerman-Leist, *Rebuilding the Foodshed*—are briefly profiled in the Real Goods Sustainable Living Library, Chapter 15.

1. Michael Shuman's 1998 book *Going Local: Creating Self-Reliant Communities in the Global Age* stands out in particular.

2. Post Carbon Institute launched the Relocalization Network (relocalize.net) soon after that think tank's founding in 2003. At its high point, the Network hosted hundreds of relocalization-oriented grassroots organizations, mostly but not exclusively in the US and Canada.

3. See centerforneweconomics.org; simplicity institute.org/ted-trainer; garretthardinsociety .org/; brtom.typepad.com/wberry/.

4. On Transition US, see transitionus.org; on Resilience Circles see localcircles.org; on Local Living Economies, see for example, bealocalist .org and neweconomyworkinggroup.org/visions /local-living-economies.

5. See Herman E. Daly and Joshua Farley, *Ecological Economics: Principles and Applications* (Island Press, 2004). Also look for popular books by Herman Daly, Brian Czech, and Richard Douthwaite.

6. See measures like the Ecological Footprint (foot printnetwork.org) and the Genuine Progress Indicator (redefiningprogress.org/projects/gpi/).

7. See storyofstuff.org/ and upstreampolicy.org/.

8. William R. Catton Jr. gives a thorough overview of ecological and social mechanisms and consequences of overshoot in his 1980 book *Overshoot: The Ecological Basis of Revolutionary Change*.

9. A great place to review standard population projections and the underlying assumptions is the United Nations Population Division website un.org/en/development/desa/population/.

10. Donella Meadows, Jorgen Randers, and Dennis Meadows, *Limits to Growth: The 30-Year Update* (Chelsea Green Publishing, 2004).

11. ecoinformatics.uvm.edu/projects/the-gumbo -model.html.

12. Of the countless books, websites, and articles about peak oil, Richard Heinberg's book *The Party's Over: Oil, War, and the Fate of Industrial Societies* (New Society Publishers, 2005, 2nd edition) is a great place to start. Online visit resilience.org.

13. An excellent review of the "peakist" and "cornucopian" predictions—and how they fared—can be found in the introduction of Richard Heinberg's book *Snake Oil: How Fracking's False Promise of Plenty Imperils Our Future* (Post Carbon Institute, 2013), also viewable at resilience.org/front -row-seat-at-the-peak-oil-games.

14. High energy prices and the technology of hydraulic fracturing (fracking) has recently produced a spike in natural gas supplies, but this is likely to be a temporary development. See Heinberg, *Snake Oil*.

15. For a slim but comprehensive book on energy and conversion factors, see John G. Howe, *The End of Fossil Energy and the Last Chance for Sustainability*, 2nd ed. (McIntire Publishing Services, 2005).

16. US Energy Information Administration, International Energy Statistics database, eia.gov/ies.

17. For a critical overview of the history of the food system's industrialization—and paths to relocalizing it—see Philip Ackerman-Leist, *Rebuilding the Foodshed* (Chelsea Green, 2013); available at realgoods.com.

18. A fundamental concept of ecology is Liebig's Law of the Minimum, which states that the growth of a population will be limited by whatever single factor of production is in short supply, not the total amount of resources. The expression "for the want of a nail" captures Liebig's Law.

19. See David Fridley, "Nine Challenges of Alternative Energy" in *The Post Carbon Reader*, Richard Heinberg and Daniel Lerch, eds. (Watershed Media, 2010), postcarbon.org/report/127153 -energy-nine-challenges-of-alternative-energy.

20. A comparison of the energy balance of different food systems is provided by David Pimentel and Marcia Pimentel, eds., *Food Energy and Society* (University Press of Colorado, 1996). An important book covering EROEI and agriculture is John Gever, Robert Kaufmann, David Skole, and Charles Vorosmarty, *Beyond Oil: The Threat to Food and Fuel in the Coming Decades* (Ballinger Publishing Company, 1986). For a view of how to transform US agriculture, see Richard Heinberg, "Fifty Million Farmers," resilience.org/stories/2006-11-17/fifty-million-farmers.

21. en.wikipedia.org/wiki/Milankovitch_cycles.

22. See Tim Flannery, *The Weather Makers: How Man Is Changing the Climate and What It Means for Life on Earth* (Atlantic Monthly Press, 2005).

23. en.wikipedia.org/wiki/Precautionary_principle.

24. John Cook et al., "Quantifying the consensus on anthropogenic global warming in the scientific literature," Environ. Res. Lett. 8 (2013). Available at skepticalscience.com/docs/Cook_2013_consensus.pdf. See also theconsensusproject.com/.

25. Lenton T. M., et al., "Tipping elements in the Earth's climate system," Proc Natl Acad Sci USA (2008) 105: 1786–1793. See also Julia Whitty, "The Thirteenth Tipping Point," *Mother Jones* (November/December 2006), motherjones.com/environment/2006/11/thirteenth-tipping-point.

26. A few books do a fine job discussing both "source" and "sink" problems with fossil fuels, including Thom Hartmann, *The Last Hours of Ancient Sunlight: Waking Up to Personal and Global Transformation*, rev. ed. (Three Rivers Press, 2004); Jeremy Leggett, *The Empty Tank: Oil, Gas, Hot Air, and the Coming Global Financial Catastrophe* (Random House, 2005); and James Howard Kunstler, *The Long Emergency: Surviving the Converging Catastrophes of the Twenty-First Century* (Grove Press, 2006).

27. Books addressing the benefits of a local economy focused on basic needs include Richard Douthwaite, *Short Circuit: Strengthening Local Economies for Security in an Unstable World* (Green Books, 1998), and Michael Shuman, *Going Local: Creating Self-Reliant Communities in a Global Age* (Routledge, 2000).

28. Sandor Ellix Katz, *The Revolution Will Not Be Microwaved: Inside America's Underground Food Movements* (Chelsea Green, 2006), Introduction.

29. Jared Diamond, *Collapse: How Societies Choose to Fail or Succeed* (Viking Penguin, 2005).

30. An example of how to sell relocalization to a community is the document called "Joint Statement toward a Sustainable, Healthy Willits" by Willits Economic Localization, old.postcarbon.org/features/willits/sustainability_statement

31. Intriguing insights into the social benefits of a more cooperative local economy can be found in Eric Brende, *Better Off: Flipping the Switch on Technology* (Harper Collins, 2004), where the author joins a traditional community in America's heartland that uses no motors, electricity, or fossil fuels.

32. See Robert D. Putnam, *Bowling Alone: The Collapse and Revival of American Community* (Simon and Schuster, 2000) for a detailed review of the loss of "social capital" in the US and what this costs us.

33. For a more in-depth discussion of what it is like to develop a community group with the goal of relocalization, see well95490.org/wp-content/uploads/archive/OutpostGuide_Bradford_0.pdf.

Chapter 2

1. usgbc.org/leed
2. builditgreen.org/greenpoint-rated
3. Seeusgbc.org/articles/infographic-leed-world, and cleantechnica.com/2012/06/01/green-home-building-booming-could-be-114-billion-market-by-2016/
4. sketchup.com/
5. apps1.eere.energy.gov/buildings/energyplus/
6. solarpathfinder.com/
7. energystar.gov/?c=home_sealing.hm_improvement_insulation_table

Chapter 11

1. Bill Mollison, *Permaculture: A Designer's Manual* (Tagari, 1988).

Chapter 12

1. Bart Anderson, "Zones and Sectors in the City," Permaculture Activist, permacultureactivist.net/articles/urbanzonesectr.htm. Much of the following disucssion of urban zones and sectors, as well as the diagrams, are inspired by Bart's smart thinking.

2. Julia Butterfly Hill, *One Makes a Difference: Inspiring Actions that Change Our World* (Harper SF, 2002).

Chapter 14

1. naturaldeathcentre.org.uk
2. funerals.org
3. homefuneralalliance.org
4. healthland.time.com/2011/06/11/why-the-federal-government-finally-acted-on-chemical-safety/
5. Lisa Carlson and Joshua Slocum, "Final Rights: Reclaiming the American Way of Death," 57–66.
6. Estimates from industry: 1) Highbeam Business Industry Report; NAICS 339995: Burial Casket Manufacturing; website: business.highbeam.com/industry-reports/food/burial-caskets; 2) Casket & Funeral Supply Association of America website: cfsaa.org/?page=CasketIndustry.
7. Memorial Society of British Columbia, *Public Health Impact of Crematoria:* memorialsocietybc.org/c/g/cremation-report.html.
8. EPA on crematoria impacts: epa.gov/earlink1/mercury/dentalamalgam-temp.html#crematoria

9. cremationassociation.org/?EmmissionsTests.

10. Ken West, author of *Guide to Natural Burial*, personal correspondence 2009.

11. Yosemite National Park website: nps.gov/yose /planyourvisit/ashes.htm.

12. *Gene Everlasting: A Contrary Farmer's Thoughts on Living Forever*, by Gene Logsdon.

13. nrcs.usda.gov/wps/portal/nrcs/main/soils /health/.

14. ICCM letter advocating grave reuse in the UK, public document (see iccm-uk.com/).

15. Funeral Consumers Alliance, Myths FAQ, funerals.org/faq/myths.htm.

16. Free natural funeral planner online at naturalburialcompany.com.

17. *Catie Jay Bee, 2002, Online Organic Gardening Forum*, forums.organicgardening.com/eve /forums.

18. Kent on Sunday, April 19, 2014. yourswale.co .uk/news/the_rise_of_the_eco_friendly_coffin _1_3565338.

19. FTC's Funeral Rule, ftc.gov/bcp/conline/pubs /services/funeral.htm.

20. David Harrington, "Breathing Life into the Funeral Market," 2004, davideharrington.com /wp-content/uploads/2012/10/Breathing-Life -into-the-Funeral-Market.pdf.

21. FTC's Funeral Rule, ftc.gov/bcp/conline/pubs /services/funeral.htm.

22. Association for Natural Burial Grounds

23. UK House of Commons; Select Committee on Environment, Transport and Regional Affairs; Year 2000, Memorandum by the Living Church-yard & Cemetery Project (CEM 60), parliament .the-stationery-office.co.uk/pa/cm200001/cm select/cmenvtra/91/91m66.htm.

24. Ken West, ICCM; "Charter for the Bereaved," iccm-uk.com/iccm/library/CharterGuiding Principles.pdf.

25. Sustainable Cemetery Studies Lab, Oregon State University; cemeterystudieslab.org.

26. Introduction to Sustainable Cemetery Manage-ment Syllabus: services.ecampus.oregonstate .edu/syllabi/downloadsyllabus.aspx?docid=5886.

Index

REAL GOODS